Air Monitoring by Spectroscopic Techniques

CHEMICAL ANALYSIS

A SERIES OF MONOGRAPHS ON
ANALYTICAL CHEMISTRY AND ITS APPLICATIONS

Editor
J. D. WINEFORDNER

VOLUME 127

A WILEY-INTERSCIENCE PUBLICATION

JOHN WILEY & SONS, INC.

New York / Chichester / Brisbane / Toronto / Singapore

Air Monitoring by Spectroscopic Techniques

Edited by

MARKUS W. SIGRIST

Infrared Physics Laboratory
Institute of Quantum Electronics
ETH Zurich
Zurich, Switzerland

A WILEY-INTERSCIENCE PUBLICATION

JOHN WILEY & SONS, INC.

New York / Chichester / Brisbane / Toronto / Singapore

This text is printed on acid-free paper.

Library of Congress Cataloging in Publication Data:

Air monitoring by spectroscopic techniques / edited by Sigrist.
p. cm. — (Chemical analysis ; v. 127)
"A Wiley-Interscience publication"
Includes bibliographical references and index.
ISBN 0-471-55875-3 (cloth : alk. paper)
1. Air—Pollution—Measurement. 2. Spectrum analysis.
I. Sigrist, Markus W., 1948– . II. Series.
TD890.A35 1993
628.5′3′0287—dc20 93-6231

Printed in the United States of America

10 9 8 7 6 5 4 3 2 1

To my family
for their love and understanding

CONTENTS

CONTRIBUTORS

J. Bechara, Unisearch Associates, Inc., 222 Snidercroft Road, Concord, Ontario L4K 1B5, Canada

David W. T. Griffith, Department of Chemistry, University of Wollongong, Northfields Avenue, Wollongong NSW 2522, Australia

Philip L. Hanst, Infrared Analysis, Inc., 1334 North Knollwood Circle, Anaheim, California 92801

Steven T. Hanst, Infrared Analysis, Inc., 1334 North Knollwood Circle, Anaheim, California 92801

G. I. Mackay, Unisearch Associates, Inc., 222 Snidercroft Road, Concord, Ontario L4K 1B5, Canada

U. Platt, Institute for Environmental Physics, University of Heidelberg, INF 366, D-69120 Heidelberg, Germany

H. I. Schiff, Unisearch Associates, Inc., 222 Snidercroft Road, Concord, Ontario L4K 1B5, Canada

Markus W. Sigrist, Infrared Physics Laboratory, Institute of Quantum Electronics, ETH Zurich, CH-8093 Zurich, Switzerland

Sune Svanberg, Department of Physics, Lund Institute of Technology, Box 118, S-22100 Lund, Sweden

PREFACE

In recent times air pollution has become an issue of worldwide concern. It has been recognized that even trace concentrations of atmospheric species can have a substantial impact in diverse areas. Well-known phenomena such as global climatic change, photochemical smog formation, acid rain, stratospheric ozone depletion, and forest decline are strongly related to increasing concentrations of trace species. Their large number and diversity pose a great challenge with regard to a thorough understanding of the complex chemical and physical processes and interactions involved.

In this respect the knowledge of the kind of species and of their spatial and temporal distribution in the atmosphere is an irrevocable prerequisite. Appropriate analytical tools have to meet several requirements, e.g., versatility, high sensitivity and selectivity, large dynamic range, good temporal resolution, easiness of operation, and field applicability. In the past, numerous techniques ranging from chemiluminescence to gas chromatography have been developed and successfully applied to trace gas monitoring. Some of them have become standard methods and are widely used on a routine basis. However, there are several requirements that cannot be fulfilled by conventional techniques. This particularly applies to the diversity of species that can be detected with a single measurement device. In more recent times, optical spectroscopic techniques have attracted considerable interest. They offer some unique features that are relevant for air monitoring.

This book is concerned with a presentation of various optical spectroscopic methods employing ultraviolet, visible, and infrared wavelengths and their applications to the detection of atmospheric trace gases in different environments. Some of the systems discussed have also been applied to measurements in the stratosphere, in the mesosphere, or in the hydrosphere. However, emphasis is put on tropospheric measurements. Although lasers have an important impact in diverse areas including environmental sensing, non-laser-based spectroscopic systems also bear a great potential and the presently most promising schemes are treated here. This book provides a comprehensive presentation of the fundamentals and the state of the art of the selected techniques illustrated by numerous examples of applications.

The individual chapters are reasonably self-contained. They are written by authors who were selected for their expertise in the field and their own original contributions. The book should thus be of value to scientists, researchers, and students concerned with environmental sensing.

Chapter 1 gives a brief introduction to some important environmental issues and their relation to atmospheric trace species. Outlined are the main characteristics of nonspectroscopic and spectroscopic detection schemes, notably sensitivity and selectivity.

Chapter 2 provides an overview of differential optical absorption spectroscopy (DOAS), a technique that is based on long-path absorption in the ultraviolet and visible wavelength region.

Chapter 3 discusses the features of light detection and ranging (lidar) and thereby demonstrates how lasers can be employed to detect trace species at a distance.

Chapter 4 represents the fundamentals and perspectives of photoacoustic spectroscopy (PAS) in combination with tunable infrared lasers and its application to in situ air monitoring.

Chapter 5 presents a comprehensive review of tunable diode laser absorption spectroscopy (TDLAS) applied to atmospheric measurements in various environments.

Chapter 6 is one of the most detailed studies on gas measurements in the fundamental infrared region. The collection of numerous molecular spectra should also be useful for further applications.

Finally, Chapter 7 discusses the features and potential of matrix isolation (MI) for air sampling combined with Fourier transform infrared spectroscopy (FTIR) or electron spin resonance (ESR) for chemical analysis.

It is hoped that this book may stimulate further research activities in the field of trace gas monitoring. Although detection techniques per se cannot solve the challenging problems associated with air pollution, they can nevertheless provide a solid basis toward a sustainable improvement of the environment.

I wish to thank the authors of the chapters for their enthusiasm and dedication as well as for their cooperation in following some guidelines designed to ensure that the book becomes a useful, comprehensive, and consistent work with little overlap between the chapters. I am also indebted to many colleagues for stimulating discussions and encouragement. Last, but not least, I thank the people at John Wiley & Sons, Inc. and at Harriet Damon Shields & Associates, New York, for their active collaboration and patience during the production of this book.

MARKUS W. SIGRIST

Zurich, Switzerland
January 1994

CHEMICAL ANALYSIS

A SERIES OF MONOGRAPHS ON ANALYTICAL CHEMISTRY AND ITS APPLICATIONS

J. D. Winefordner, *Series Editor*
I. M. Kolthoff, *Editor Emeritus*

Vol. 1. **The Analytical Chemistry of Industrial Poisons, Hazards, and Solvents.** *Second Edition.* By the late Morris B. Jacobs

Vol. 2. **Chromatographic Adsorption Analysis.** By Harold H. Strain (*out of print*)

Vol. 3. **Photometric Determination of Traces of Metals.** *Fourth Edition*
Part I: General Aspects. By E. B. Sandell and Hiroshi Onishi
Part IIA: Individual Metals, Aluminum to Lithium. By Hiroshi Onishi
Part IIB: Individual Metals, Magnesium to Zirconium. By Hiroshi Onishi

Vol. 4. **Organic Reagents Used in Gravimetric and Volumetric Analysis.** By John F. Flagg (*out of print*)

Vol. 5. **Aquametry: A Treatise on Methods for the Determination of Water.** *Second Edition (in three parts).* By John Mitchell, Jr. and Donald Milton Smith

Vol. 6. **Analysis of Insecticides and Acaricides.** By Francis A. Gunther and Roger C. Blinn (*out of print*)

Vol. 7. **Chemical Analysis of Industrial Solvents.** By the late Morris B. Jacobs and Leopold Schetlan

Vol. 8. **Colorimetric Determination of Nonmetals.** *Second Edition.* Edited by the late David F. Boltz and James A. Howell

Vol. 9. **Analytical Chemistry of Titanium Metals and Compounds.** By Maurice Codell

Vol. 10. **The Chemical Analysis of Air Pollutants.** By the late Morris B. Jacobs

Vol. 11. **X-Ray Spectrochemical Analysis.** *Second Edition.* By L. S. Birks

Vol. 12. **Systematic Analysis of Surface-Active Agents.** *Second Edition.* By Milton J. Rosen and Henry A. Goldsmith

Vol. 13. **Alternating Current Polarography and Tensammetry.** By B. Breyer and H. H. Bauer

Vol. 14. **Flame Photometry.** By R. Herrmann and J Alkemade

Vol. 15. **The Titration of Organic Compounds** (*in two parts*). By M. R. F. Ashworth

Vol. 16. **Complexation in Analytical Chemistry: A Guide for the Critical Selection of Analytical Methods Based on Complexation Reactions.** By the late Anders Ringbom

Vol. 17. **Electron Probe Microanalysis.** *Second Edition.* By L. S. Birks

Vol. 18. **Organic Complexing Reagents: Structure, Behavior, and Application to Inorganic Analysis.** By D. D. Perrin

Vol. 19. **Thermal Analysis.** *Third Edition.* By Wesley Wm. Wendlandt
Vol. 20. **Amperometric Titrations.** By John T. Stock
Vol. 21. **Reflectance Spectroscopy.** By Wesley Wm. Wendlandt and Harry G. Hecht
Vol. 22. **The Analytical Toxicology of Industrial Inorganic Poisons.** By the late Morris B. Jacobs
Vol. 23. **The Formation and Properties of Precipitates.** By Alan G. Walton
Vol. 24. **Kinetics in Analytical Chemistry.** By Harry B. Mark, Jr. and Garry A. Rechnitz
Vol. 25. **Atomic Absorption Spectroscopy.** *Second Edition.* By Morris Slavin
Vol. 26. **Characterization of Organometallic Compounds** (*in two parts*). Edited by Minoru Tsutsui
Vol. 27. **Rock and Mineral Analysis.** *Second Edition.* By Wesley M. Johnson and John A. Maxwell
Vol. 28. **The Analytical Chemistry of Nitrogen and Its Compounds** (*in two parts*). Edited by C. A. Streuli and Philip R. Averell
Vol. 29. **The Analytical Chemistry of Sulfur and Its Compounds** (*in three parts*). By J. H. Karchmer
Vol. 30. **Ultramicro Elemental Analysis.** By Günther Tölg
Vol. 31. **Photometric Organic Analysis** (*in two parts*). By Eugene Sawicki
Vol. 32. **Determination uf Organic Compounds: Methods and Procedures.** By Frederick T. Weiss
Vol. 33. **Masking and Demasking of Chemical Reactions.** By D. D. Perrin
Vol. 34. **Neutron Activation Analysis.** By D. De Soete, R. Gijbels, and J. Hoste
Vol. 35. **Laser Raman Spectroscopy.** By Marvin C. Tobin
Vol. 36. **Emission Spectrochemical Analysis.** By Morris Slavin
Vol. 37. **Analytical Chemistry of Phosphorus Compounds.** Edited by M. Halmann
Vol. 38. **Luminescence Spectrometry in Analytical Chemistry.** By J. D. Winefordner, S. G. Schulman and T. C. O'Haver
Vol. 39. **Activation Analysis with Neutron Generators.** By Sam S. Nargolwalla and Edwin P. Przybylowicz
Vol. 40. **Determination of Gaseous Elements in Metals.** Edited by Lynn L. Lewis, Laben M. Melnick, and Ben D. Holt
Vol. 41. **Analysis of Silicones.** Edited by A. Lee Smith
Vol. 42. **Foundations of Ultracentrifugal Analysis.** By H. Fujita
Vol. 43. **Chemical Infrared Fourier Transform Spectroscopy.** By Peter R. Griffiths
Vol. 44. **Microscale Manipulations in Chemistry.** By T. S. Ma and V. Horak
Vol. 45. **Thermometric Titrations.** By J. Barthel
Vol. 46. **Trace Analysis: Spectroscopic Methods for Elements.** Edited by J. D. Winefordner
Vol. 47. **Contamination Control in Trace Element Analysis.** By Morris Zief and James W. Mitchell
Vol. 48. **Analytical Applications of NMR.** By D. E. Leyden and R. H. Cox
Vol. 49. **Measurement of Dissolved Oxygen.** By Michael L. Hitchman
Vol. 50. **Analytical Laser Spectroscopy.** Edited by Nicolo Omenetto
Vol. 51. **Trace Element Analysis of Geological Materials.** By Roger D. Reeves and Robert R. Brooks

Vol. 52. **Chemical Analysis by Microwave Rotational Spectrscopy.** By Ravi Varma and Lawrence W. Hrubesh

Vol. 53. **Information Theory As Applied to Chemical Analysis.** By Karel Eckschlager and Vladimir Štěpánek

Vol. 54. **Applied Infrared Spectroscopy: Fundamentals, Techniques, and Analytical Problem-solving.** By A. Lee Smith

Vol. 55. **Archaeological Chemistry.** By Zvi Goffer

Vol. 56. **Immobilized Enzymes in Analytical and Clinical Chemistry.** By P. W. Carr and L. D. Bowers

Vol. 57. **Photoacoustics and Photoacoustic Spectroscopy.** By Allan Rosencwaig

Vol. 58. **Analysis of Pesticide Residues.** Edited by H. Anson Moye

Vol. 59. **Affinity Chromatography.** By William H. Scouten

Vol. 60. **Quality Control in Analytical Chemistry.** *Second Edition.* By G. Kateman and L. Buydens

Vol. 61. **Direct Characterization of Fineparticles.** By Brian H. Kaye

Vol. 62. **Flow Injection Analysis.** By J. Ruzicka and E. H. Hansen

Vol. 63. **Applied Electron Spectroscopy for Chemical Analysis.** Edited by Hassan Windawi and Floyd Ho

Vol. 64. **Analytical Aspects of Environmental Chemistry.** Edited by David F. S. Natusch and Philip K. Hopke

Vol. 65. **The Interpretation of Analytical Chemical Data by the Use of Cluster Analysis.** By D. Luc Massart and Leonard Kaufman

Vol. 66. **Solid Phase Biochemistry: Analytical and Synthetic Aspects.** Edited by William H. Scouten

Vol. 67. **An Introduction to Photoelectron Spectroscopy.** By Pradip K. Ghosh

Vol. 68. **Room Temperature Phosphorimetry for Chemical Analysis.** By Tuan Vo-Dinh

Vol. 69. **Potentiometry and Potentiometric Titrations.** By E. P. Serjeant

Vol. 70. **Design and Application of Process Analyzer Systems.** By Paul E. Mix

Vol. 71. **Analysis of Organic and Biological Surfaces.** Edited by Patrick Echlin

Vol. 72. **Small Bore Liquid Chromatography Columns: Their Properties and Uses.** Edited by Raymond P. W. Scott

Vol. 73. **Modern Methods of Particle Size Analysis.** Edited by Howard G. Barth

Vol. 74. **Auger Electron Spectroscopy.** By Michael Thompson, M. D. Baker, Alec Christie, and J. F. Tyson

Vol. 75. **Spot Test Analysis: Clinical, Environmental, Forensic and Geochemical Applications.** By Ervin Jungreis

Vol. 76. **Receptor Modeling in Environmental Chemistry.** By Philip K. Hopke

Vol. 77. **Molecular Luminescence Spectroscopy: Methods and Applications** (*in three parts*). Edited by Stephen G. Schulman

Vol. 78. **Inorganic Chromatographic Analysis.** Edited by John C. MacDonald

Vol. 79. **Analytical Solution Calorimetry.** Edited by J. K. Grime

Vol. 80. **Selected Methods of Trace Metal Analysis: Biological and Environmental Samples.** By Jon C. VanLoon

Vol. 81. **The Analysis of Extraterrestrial Materials.** By Isidore Adler

Vol. 82. **Chemometrics.** By Muhammad A. Sharaf, Deborah L. Illman, and Bruce R. Kowalski

Vol. 83. **Fourier Transform Infrared Spectrometry.** By Peter R. Griffiths and James A. de Haseth

Vol. 84. **Trace Analysis: Spectroscopic Methods for Molecules.** Edited by Gary Christian and James B. Callis

Vol. 85. **Ultratrace Analysis of Pharmaceuticals and Other Compounds of Interest.** Edited by S. Ahuja

Vol. 86. **Secondary Ion Mass Spectrometry: Basic Concepts, Instrumental Aspects, Applications and Trends.** By A. Benninghoven, F. G. Rüdenauer, and H. W. Werner

Vol. 87. **Analytical Applications of Lasers.** Edited by Edward H. Piepmeier

Vol. 88. **Applied Geochemical Analysis.** By C. O. Ingamells and F. F. Pitard

Vol. 89. **Detectors for Liquid Chromatography.** Edited by Edward S. Yeung

Vol. 90. **Inductively Coupled Plasma Emission Spectroscopy: Part I: Methodology, Instrumentation, and Performance; Part II: Applications and Fundamentals.** Edited by J. M. Boumans

Vol. 91. **Applications of New Mass Spectrometry Techniques in Pesticide Chemistry.** Edited by Joseph Rosen

Vol. 92. **X-Ray Absorption: Principles, Applications, Techniques of EXAFS, SEXAFS, and XANES.** Edited by D. C. Konnigsberger

Vol. 93. **Quantitative Structure–Chromatographic Retention Relationships.** By Roman Kaliszan

Vol. 94. **Laser Remote Chemical Analysis.** Edited by Raymond M. Measures

Vol. 95. **Inorganic Mass Spectrometry.** Edited by F. Adams, R. Gijbels, and R. Van Grieken

Vol. 96. **Kinetic Aspects of Analytical Chemistry.** By Horacio A. Mottola

Vol. 97. **Two-Dimensional NMR Spectroscopy.** By Jan Schraml and Jon M. Bellama

Vol. 98. **High Performance Liquid Chromatography.** Edited by Phyllis R. Brown and Richard A. Hartwick

Vol. 99. **X-Ray Fluorescence Spectrometry.** By Ron Jenkins

Vol. 100. **Analytical Aspects of Drug Testing.** Edited by Dale G. Deutsch

Vol. 101. **Chemical Analysis of Polycyclic Aromatic Compounds.** Edited by Tuan Vo-Dinh

Vol. 102. **Quadrupole Storage Mass Spectrometry.** By Raymond E. March and Richard J. Hughes

Vol. 103. **Determination of Molecular Weight.** Edited by Anthony R. Cooper

Vol. 104. **Selectivity and Detectability Optimizations in HPLC.** By Satinder Ahuja

Vol. 105. **Laser Microanalysis.** By Lieselotte Moenke-Blankenburg

Vol. 106. **Clinical Chemistry.** Edited by E. Howard Taylor

Vol. 107. **Multielement Detection Systems for Spectrochemical Analysis.** By Kenneth W. Busch and Marianna A. Busch

Vol. 108. **Planar Chromatography in the Life Sciences.** Edited by Joseph C. Touchstone

Vol. 109. **Fluorometric Analysis in Biomedical Chemistry: Trends and Techniques Including HPLC Applications.** By Norio Ichinose, George Schwedt, Frank Michael Schnepel, and Kyoko Adochi

Vol. 110. **An Introduction to Laboratory Automation.** By Victor Cerdá and Guillermo Ramis

Vol. 111. **Gas Chromatography: Biochemical, Biomedical, and Clinical Applications.** Edited by Ray E. Clement

Vol. 112. **The Analytical Chemistry of Silicones.** Edited by A. Lee Smith
Vol. 113. **Modern Methods of Polymer Characterization.** Edited by Howard G. Barth and Jimmy W. Mays
Vol. 114. **Analytical Raman Spectroscopy.** Edited by Jeannette Graselli and Bernard J. Bulkin
Vol. 115. **Trace and Ultratrace Analysis by HPLC.** By Satinder Ahuja
Vol. 116. **Radiochemistry and Nuclear Methods of Analysis.** By William D. Ehmann and Diane E. Vance
Vol. 117. **Applications of Fluorescence in Immunoassays.** By Ilkka Hemmila
Vol. 118. **Principles and Practice of Spectroscopic Calibration.** By Howard Mark
Vol. 119. **Activation Spectrometry in Chemical Analysis.** By S. J. Parry
Vol. 120. **Remote Sensing by Fourier Transform Spectrometry.** By Reinhard Beer
Vol. 121. **Detectors for Capillary Chromatography.** Edited by Herbert H. Hill and Dennis McMinn
Vol. 122. **Photochemical Vapor Deposition.** By J. G. Eden
Vol. 123. **Statistical Methods in Analytical Chemistry.** By Peter C. Meier and Richard Zund
Vol. 124. **Laser Ionization Mass Analysis.** Edited by Akos Vertes, Renaat Gijbels, and Fred Adams
Vol. 125. **Physics, Chemistry, and Technology of Solid State Gas Sensor Devices.** By Andreas Mandelis and Constantinos Christofides
Vol. 126. **Electroanalytical Stripping Methods.** By Khjena Z. Brainina and E. Neyman
Vol. 127. **Air Monitoring by Spectroscopic Techniques.** Edited By Markus W. Sigrist

Air Monitoring by Spectroscopic Techniques

CHAPTER

1

INTRODUCTION TO ENVIRONMENTAL SENSING

MARKUS W. SIGRIST

Infrared Physics Laboratory, Institute of Quantum Electronics, ETH Zurich, Switzerland

1.1. THE ROLE OF ATMOSPHERIC TRACE GASES IN THE ENVIRONMENT

Today, environmental issues are of great worldwide concern. This is manifested by a tremendous scientific effort in various fields as well as by the activities of many organizations and—last but not least—by the United Nations First Conference on Environment and Development (UNCED), held in Rio de Janeiro, Brazil, in June 1992. The hot topics to be studied range from global warming to the stratospheric ozone hole. In one or another way all these issues are related to atmospheric trace constituents, as is illustrated schematically in Fig. 1.1. Some selected trace gases, their mean current concentration and average residence time in the atmosphere, as well as their contribution to different phenomena are listed in Table 1.1 (Graedel and Crutzen, 1989). A "+" (plus) sign in the table means that the presence of the chosen trace gas enhances the corresponding effect, while a "−" (minus) sign indicates the opposite. Often, depending on the conditions, the influence varies, as shown by a "+/−" sign. As an example, the influence of CO_2, NO_x, and N_2O on the stratospheric ozone depletion depends on the altitude. Furthermore, CH_4 acts in general against the ozone depletion yet accelerates the depletion within the ozone hole.

Although measures for a reduction of the emissions are already in effect or are to be taken, the concentrations of most of the trace gases listed, except presumably NO_x and SO_2, are expected to increase further, mainly because of their long residence time in the atmosphere.

In the following subsections, the key role of trace gases is briefly outlined for

Air Monitoring by Spectroscopic Techniques, Edited by Markus W. Sigrist. Chemical Analysis Series, Vol. 127.
ISBN 0-471-55875-3

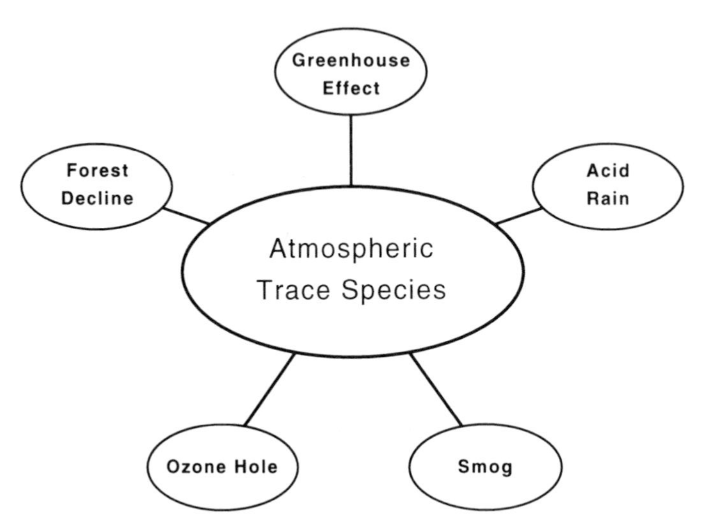

Figure 1.1. Illustration of the significance of atmospheric trace species for environmental issues.

the selected topics of global warming, stratospheric ozone depletion, and photochemical smog formation.

1.1.1. Global Warming

The greenhouse effect is well known and has been widely discussed in the past (e.g., see Bolin et al., 1986; Schneider, 1989). It is related to the radiation budget of the earth and responsible for the fact that the mean earth temperature is 33 °C above its radiation temperature as seen from space. The situation is illustrated in Fig. 1.2. The radiation budget of the earth is determined by the incident solar radiation and the thermal radiation emitted from the earth. The incident radiation essentially occurs between 0.3 and 3 μm wavelength, corresponding to the black body radiation of the sun at 5700 K, whereas the wavelength of the emitted terrestrial radiation lies between 3 and 60 μm, corresponding to the black body radiation of the earth at ca. 254 K. The radiation in the atmosphere can thus be separated into two spectral regions with short and long wavelengths. From the incident solar radiation taken as 100%, a total of approximately 30% is reflected by clouds (20%), atmospheric particles (5%), and directly by the earth's surface (5%). The remaining 70% are initially absorbed by the earth (45%), the atmosphere (20%), and the clouds (5%). The part absorbed by the earth is transported to the atmosphere by infrared (IR) radiation (104%) and atmospheric processes like convection (5%)

Table 1.1. Effects of Atmospheric Trace Gases

Gas	Mean Current Concentration	Average Residence Time	Greenhouse Effect	Stratospheric Ozone Depletion	Photochemical Smog	Acid Deposition	Decreased Visibility
Carbon dioxide (CO_2)	350 ppm	100 years	+	+/−			
Methane (CH_4)	1.7 ppm	10 years	+	+/−			
NO_x: nitric oxide (NO) and nitrogen dioxide (NO_2)	10^{-3}–10^{3} ppb	Days		+/−	+	+	+
Nitrous oxide (N_2O)	310 ppb	170 years	+	+/−			
Sulfur dioxide (SO_2)	0.03–50 ppb	Days to weeks	−			+	+
Chlorofluoro carbons (CFCs)	~3 ppb (chlorine atoms)	60–100 years	+	+			
Ozone (O_3)	20–100 ppb	Days	+		+		
Volatile organic compounds (VOCs)	1–100 ppb	Days	+		+		

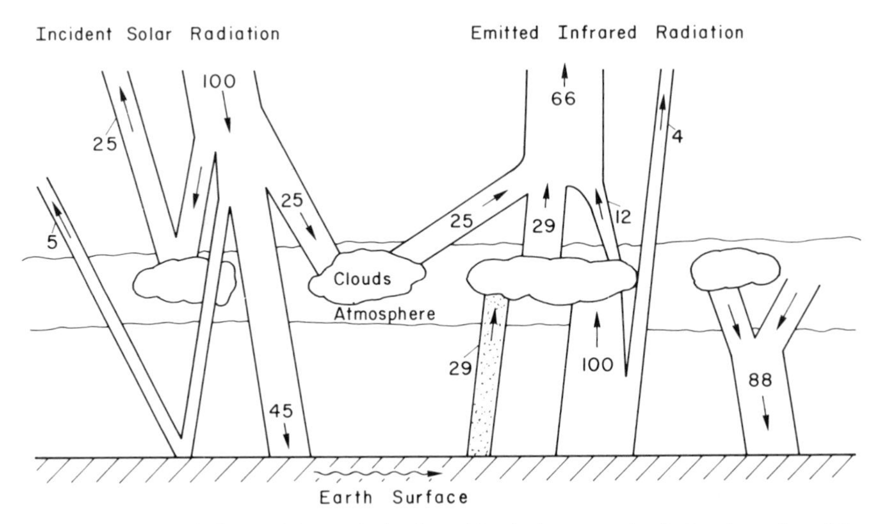

Figure 1.2. Energy balance of the earth, dominated by the heat trapping in the atmosphere. The numbers indicate the corresponding radiation fluxes, with the incident solar radiation set to 100 units at the upper edge of the atmosphere. As illustrated on the right-hand side, most of the (infrared) surface radiation from the earth is trapped by clouds and greenhouse gases and returned to the earth.

and cloud formation (24%). The predominant part (100%) of the IR radiation emitted from the earth is absorbed by the clouds and the greenhouse gases and for the most part (88%) reemitted to the earth. This trapping of IR radiation in the atmosphere is responsible for the greenhouse effect and results in an increase of the temperature of the earth surface by 33 °C compared to the case with no greenhouse effect. The remaining 70% of IR radiation is radiated into space, namely, 4% by direct radiation from the ground and 66% from the clouds and the atmosphere.

The contribution of a given gaseous compound to the greenhouse effect is determined by its IR absorptivity and its atmospheric abundance. Although it is often disregarded the major greenhouse gas is natural H_2O vapor, whereas the more widely discussed carbon dioxide (CO_2) contributes only 3% (Tomkin, 1992). However, if one considers the present *amplification* of the greenhouse effect, the situation appears as shown in Table 1.2, where the five most important *anthropogenic* trace gases and their relative contribution are listed (Dickinson and Cicerone, 1986). Obviously, CO_2 and methane (CH_4) are responsible for more than 80% of the greenhouse enhancement. Both of these trace gases have shown a rather drastic increase of their global concentration during the last 140 years, as illustrated in Fig. 1.3a,b. The CO_2 concentration has increased from 290 ppm to the present value of 350 ppm, while the CH_4

Table 1.2. The Five Most Important Anthropogenic Greenhouse Gases and Their Relative Contributions to the Present Enhancement of the Greenhouse Effect

Greenhouse Gas	Relative Contribution
Carbon dioxide (CO_2)	55%
Methane (CH_4)	26%
Tropospheric ozone (O_3)	9%
Chlorofluorocarbons (CFCs)	8%
Nitrous oxide (N_2O)	2%

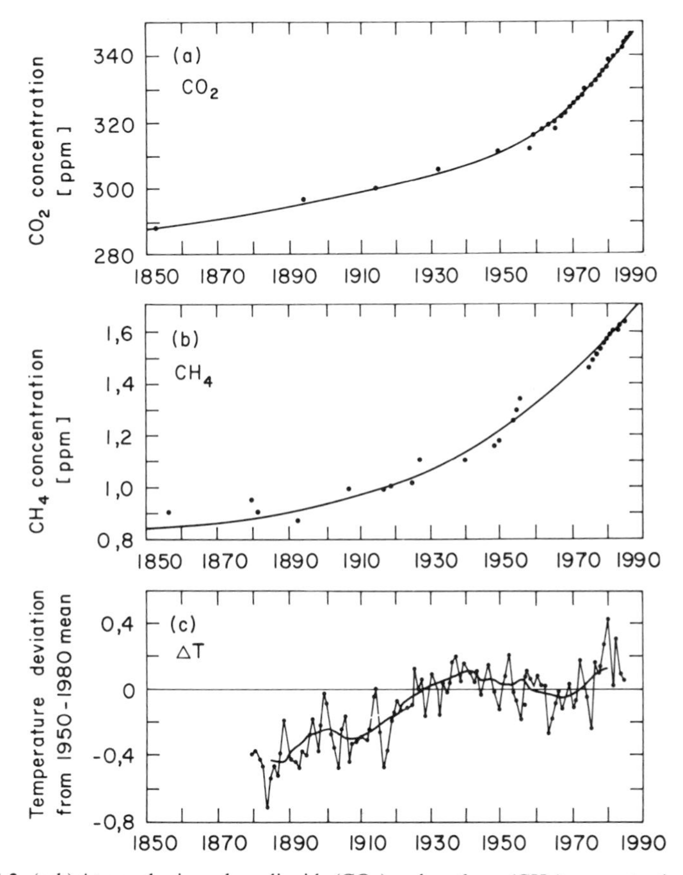

Figure 1.3. (a, b) Atmospheric carbon dioxide (CO_2) and methane (CH_4) concentrations, respectively, for the last 140 years. Pre-1958 data originate from analyses of air trapped in bubbles in glacial ice from various sites around the world. (c) Annual mean temperature (spiky curve) and the 5-year running mean (smooth curve). After Houghton and Woodwell (1989). Copyright 1989, Scientific American.

concentration has doubled from 0.85 ppm to the present 1.7 ppm. The result of an increase of the concentration of greenhouse-relevant trace gases is that initially less thermal radiation is emitted into space, i.e., more IR radiation is trapped in the atmosphere. The earth surface and the lower atmosphere heat up and emit more thermal radiation until a new radiation balance with the incident solar radiation is established. As a result the temperature of the earth has indeed increased, namely, by approximately 0.5 °C since 1850, as is evident from Fig. 1.3c. Depending on the theoretical model the predictions of the global temperature increase show some variation mainly arising from uncertainties caused by feedback mechanisms due to changes of cloud coverage (low and high clouds have contrary implications), change of the earth albedo, varying sea conditions, etc. (Cubasch, 1992; Houghton et al., 1990; Ramanathan et al., 1989; Schneider, 1987). In any case, the future concentration and distribution of atmospheric trace gases and aerosols will play a crucial role.

1.1.2. Stratospheric Ozone Depletion

Another striking phenomenon that has been of great impact in recent years is the global stratospheric ozone change. This ozone layer is of great importance for the ultraviolet (UV) radiation incident on the earth surface. Measurements of the total column amount of ozone have now been made for 13 years from the Total Ozone Mapping Spectrometer (TOMS) on the *Nimbus* 7 satellite (Stolarski, 1992). These measurements have revealed the Antarctic ozone hole, with a pronounced minimum in early October. Figure 1.4 shows a plot of the measured ozone concentration above Halley Bay (Antarctica) since 1956 in addition to TOMS data. The drastic decrease is clearly visible. Global scale studies by TOMS show no trend in total ozone near the equator yet a significant trend at northern middle and high latitudes.

A pronounced seasonal variation is observed, with the maximum in winter. The chemistry of the stratospheric ozone generation and destruction cycle is dominated by chlorofluorohydrocarbons (CFCs). These anthropogenic molecules have a medium lifetime of 60–100 years in the atmosphere and are thus transported into the stratosphere. Although their average concentration is only a few ppb, they act as reservoirs of chlorine atoms that destroy ozone molecules by catalytic processes. This hypothesis is supported by the fact that the concentration of chloromonoxide (ClO) is enhanced in spring within the ozone hole compared to the ClO concentration at middle latitudes. Other processes that involve nitric oxides can hinder the ozone destruction, which is manifested by low nitric oxide concentrations in the ozone hole in comparison to middle latitudes. A key role for the ozone depletion is played by the so-called polar stratospheric clouds. These clouds are formed at high altitudes during the winter—more often in the Antarctic than in the Arctic—at

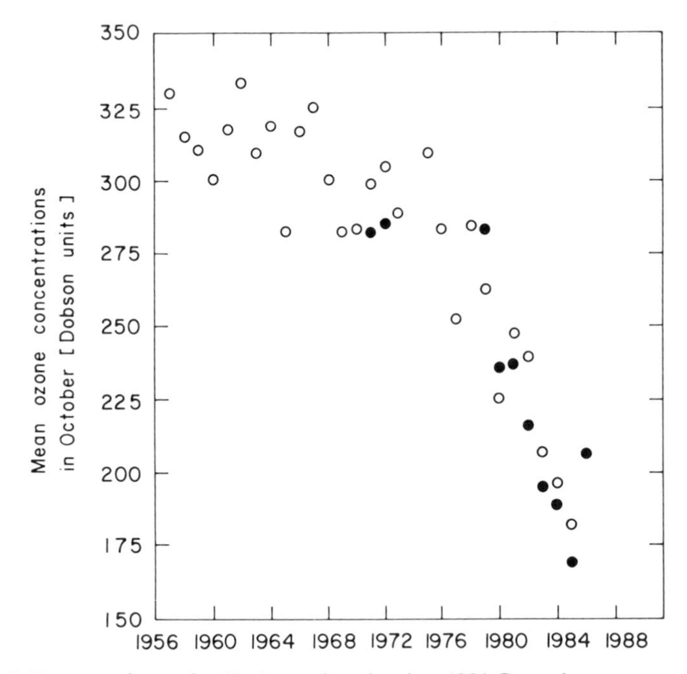

Figure 1.4. Decrease of ozone level in Antarctic spring since 1956. Ground measurements directly above Halley Bay (open circles) and NASA satellite data (solid circles). After Stolarski (1988). Copyright 1988, Scientific American.

temperatures below $-80\,^{\circ}\mathrm{C}$. The cloud particles could facilitate the change of chlorine reservoirs into active chlorine atoms since chemical reactions occur much faster at particle surfaces than in the gas phase. Further studies are needed before the stratospheric ozone depletion is fully understood. In any case, the chemistry is again controlled by trace amounts of gases and aerosols.

1.1.3. Photochemical Smog

A third example that demonstrates how small concentrations of atmospheric pollutants can have drastic effects is the process of smog formation. In earlier times, the production of smog was restricted to densely populated and industrial areas where increased concentrations of toxic compounds, mainly sulfur dioxide (SO_2), occurred near their sources. This type of smog, known as *winter smog* or *London smog*, has lost much of its previous importance owing to drastic reductions of emissions of so-called primary pollutants like sulfur compounds from heating systems and the like. In more recent times, another type of smog, known as *summer smog* or *Los Angeles smog*, has attracted a lot of interest and also became a factor in politics (BUWAL, 1989; OECD, 1982). This type of smog is produced photochemically from primary pollutants, essentially from nitric oxides (NO_x), and volatile organic compounds (VOCs) under the influence of the solar UV radiation. Numerous atmospheric chemical processes are involved in this smog formation, which is characterized by the production of tropospheric ozone (O_3) and peroxyacetyl nitrate [$CH_3C(O)O_2NO_2$; PAN]. The concentrations of these secondary compounds generally peak in the late afternoon. In contrast to winter smog, photochemical smog affects large areas and thus represents a regional problem. Since pollutants are transported over large distances, rural areas often exhibit higher ozone concentrations than urban areas.

The complex smog chemistry has been addressed by various authors (e.g., Altshuller, 1986; Crutzen, 1988; Finlayson-Pitts and Pitts, 1986; Seinfeld, 1989; van Ham, 1989). It is well accepted that the hydroxyl radicals (OH) with a tropospheric lifetime on the order of 1 s play a crucial role in photochemical smog formation (Warneck, 1988). Other important issues are the relative concentrations of the substances NO_2 and non-methane hydrocarbons (NMHC). The relative contribution of various organic compounds to the increase of the ozone production rate has been experimentally studied under controlled conditions in smog chambers (Schurath, 1988). It was found that the addition of aromatic hydrocarbons such as benzene, toluene, and xylenes results in a higher rate increase than the addition of equal amounts of olefins like ethene or else of alcohol vapors. In order to verify theoretical models and results of smog chamber experiments, comprehensive concentration measurements of numerous compounds with good spatial and temporal resolution are highly desirable. Only with a detailed knowledge of photochemical smog formation is it possible to develop and introduce reduction strategies for a sustainable ozone reduction on a large scale.

1.2. DETECTION SCHEMES FOR ATMOSPHERIC TRACE GASES

1.2.1. Requirements

The foregoing discussions of global warming, stratospheric ozone depletion, and tropospheric photochemical smog formation briefly demonstrate the great impact of comparatively low concentrations of atmospheric trace gases on the environment and on the global climate. Understanding and modeling of the complex atmospheric chemistry driven by the numerous constituents, and of the transport and exchange processes between different spheres, e.g., between the atmosphere and the hydrosphere and lithosphere, require sophisticated monitoring techniques. They should generally be of great value not only for air pollution control or control of industrial processes but also for the modeling of global pollution scenarios with the aim of taking appropriate international measures, including sanctions when necessary.

There is a great variety of gaseous atmospheric pollutants (Graedel, 1978). Their concentrations normally lie in the parts-per-billion (ppb, 10^{-9}) to parts-per-million (ppm, 10^{-6}) range yet exhibit large fluctuations both locally and temporally. Minimum and maximum concentrations extend into the parts-per-trillion (ppt, 10^{-12}) to percent (%, 10^{-2}) ranges. This situation sets a challenging standard to the detection and monitoring of air pollutants. Ideally a detection technique should fulfill the following requirements:

a. Feasibility of detecting numerous compounds with one instrument
b. High sensitivity in order to permit the detection of very low concentrations
c. High selectivity in order to differentiate between different species present in a multicomponent mixture; this permits a quantitative analysis or at least the selective monitoring of specific compounds in mixtures
d. Large dynamic range in order to monitor low and high concentrations with a single instrument
e. Good temporal resolution to enable on-line monitoring
f. Good portability for in situ measurements

Depending on the situation, the importance of these items may vary. For example, it may suffice to know the total hydrocarbon content of a sample, e.g., of car exhausts, and selectivity would thus be of minor concern. On the other hand, additional issues may become important such as the feasibility of remote rather than point monitoring.

During the past few decades various different air monitoring schemes have been developed and extensively discussed in the literature (Ewing, 1990; Finlayson-Pitts and Pitts, 1986; Fox, 1985, 1987; Frei and Albaiges, 1986; Harrison, 1984; Harrison and Perry, 1986; Keith, 1984; Thain, 1980). A continuous impetus to novel developments has arisen from progress in modern physical and chemical research, especially in surface physics and optics, as well as in electronics and computerization.

The different monitoring techniques presently available can be distinguished with respect to the sampling procedure and analytical methods that are employed for determination of the trace constituents and their concentrations. In the following discussion the main approaches and their characteristics are briefly outlined.

1.2.2. Sampling Methods

Air sampling can be performed by (a) extractive methods, (b) in situ monitoring devices, or (c) remote sensing systems.

Extractive methods are based on the collection of air samples by an appropriate container and subsequent analysis in the laboratory. These techniques usually offer high sensitivity and selectivity yet impede real-time and continuous monitoring. Furthermore, problems may arise owing to alterations of the gas composition caused by adsorption and desorption processes at the inner surface of the collecting container.

In situ monitoring methods are often less sensitive and selective than extractive methods, yet they usually offer the great advantage of real-time measurements. In practice, most techniques applied to air pollution control represent a compromise between extractive and in situ schemes. For example, the polluted air from stack emission may be pumped through a long tube to the ground-based detected instrument.

In remote sensing the analyzer is located far away from the air sample under study. Measurements result either in integrated or in three-dimensional concentration profiles of air pollution. The unique feature of three-dimensional profiling is essential for the validation of distribution models, e.g., for exhaust plumes or for vertical ozone distribution in the troposphere and stratosphere. Another advantage of remote techniques is the contactless measurement, which excludes any potential influence on the sample during monitoring. On the other hand, remote detection techniques are usually less versatile than the other methods and the measurement conditions strongly depend on the conditions of atmospheric scattering.

1.2.3. Nonspectroscopic Techniques

There exists a great variety of analytical methods for atmospheric trace gas detection that do not rely on spectroscopic features but use some chemical or physical process for the measurement. Accordingly they can be subdivided into categories according to the specific process on which they are based. Some schemes and their main characteristics are summarized in Table 1.3. Those methods with sufficient selectivity can be applied to measure a specified substance directly in the atmosphere. Such devices are called atmospheric monitors (Thain, 1980). The application of less selective methods on the other hand relies on a previous *separation* of the air components. This separation process is best performed by gas chromatography (GC). Although the principle of GC refers only to the separation of the individual components of the gas sample, their quantitative determination is generally included in the concept of GC. Theoretical aspects and practical applications of GC have

Table 1.3. Nonspectroscopic Techniques for Trace Gas Detection

Detection Scheme	Main Features	Sensitivity	Selectivity
Gas chromatography (GC), usually combined with FID	Separation/extraction; complex; low time resolution	Excellent	Excellent
Flame ionization detection (FID)	Flame ionization in combustion yields total HC concentration	Good	Poor
Photoionization detection (PID)	UV ionization; linearity depends on design	Excellent	Partial
Electron capture detection (ECD)	Ionization by β radiation; not used as direct monitoring device; limited dynamic range	Excellent (for halogenated compounds)	Poor
Mass spectrometry (MS)	Fragmentation and ionization by electron bombardment; complex; sometimes combined with GC (GC–MS)	Good	Good
Chemiluminescence	Emission of radiation in chemical reaction; commonly applied for NO, NO_x, and O_3 detection	Good	—
Flame photometry detection (FPD)	Emission of radiation from excited molecules in flame; small concentration ranges	Good (for sulfur compounds)	—
Electrolytic conductivity	Ionization in appropriate solution; only for compounds containing sulfur, halogens, or nitrogen; simple setup	Good	Depends on specificity of ionization process
Coulometry	Stoichiometric reaction with appropriate, electrically generated reagent; only for compounds containing sulfur, nitrogen, or chloride	Good	Depends on specificity of ionization process
Colorimetry	Change of color by reaction with specific reagent in solution	Good	—
Thermal detection	Based on thermal conductivity or heat of combustion; limited application	—	Poor

recently been reviewed by several authors (Guiochon, 1990; Jennings, 1987; Roedel and Woelm, 1987). Gas chromatography is usually combined with a nonselective yet sensitive flame ionization detector (FID). The combined system represents a powerful tool for air monitoring. It offers excellent sensitivity and selectivity yet is restricted to organic species. A further improvement of selectivity can be achieved in combination with additional instrumentation, particularly mass spectrometry (GC–MS). Hitherto, the sensitivity and selectivity in the multicomponent analysis of air samples of such devices has been unequaled by other methods. However, the method only offers low time resolution, and owing to the complexity of the apparatus, it is only recently that field studies have been performed with compact GC devices.

It should be mentioned that although FID is commonly used in combination with GC it can also be applied alone as a sensitive yet nonselective analyzer for the determination of total hydrocarbon concentrations. Several methods besides FID exist that are based on ionization, such as photoionization detection (PID), electron capture detection (ECD), and mass spectrometry (MS). As already mentioned, MS is often combined with GC if the identification of individual peaks in the chromatogram represents a problem.

A second category of detection methods is based on chemical and electrochemical effects like chemiluminescence, flame photometry, electrolytic conductivity, coulometry, and colorimetry. All these schemes generally exhibit good sensitivity but only for certain species, as indicated in Table 1.3. For electrochemical devices (electrolytical conductivity and coulometry) the detection selectivity critically depends on the specificity of the involved chemical reaction that produces the ions. Chemiluminescence, flame photometry, and colorimetry can also be attributed to optical techniques since they involve either the detection of emitted photons or of transmitted light through the air sample in a specific solution (as in the case of colorimetry).

By far the most widely used detection scheme of this category is based on the process of chemiluminescence (e.g., see Kroon, 1978). This luminescence occurs in a chemical reaction in which excited molecules are produced that relax by emitting radiation whose intensity and frequency depends on the kind and concentration of compounds involved. Monitoring devices based on chemiluminescence are both sensitive and specific yet rather limited in their application. In practice they are used for the determination of the concentrations of O_3, NO, and NO_2 in air. Nitrogen dioxide (NO_2) can only be detected indirectly by previous conversion of NO_2 to NO via UV photodissociation. The subsequent detection by chemiluminescence yields the total content of nitric oxides (NO_x). The NO_2 concentration is then obtained as the difference between the NO_x and the NO concentration on a cyclic basis. Commercial NO/NO_x and O_3 monitors are commonly employed for routine in situ air control measurements.

A final category concerns methods based on thermal effects. The phenomena used are essentially thermal conductivity or heat of combustion. In the latter case only combustible gases or vapors, e.g., explosives at sufficiently high concentrations, are accessible to detection (Jungclaus et al., 1984). In general, thermal detectors exhibit poor selectivity, which impedes their application as direct air monitors. They are thus used as nonselective detectors in combination with a separation device, mainly GC.

With the exception of GC the nonspectroscopic detection schemes are practically only suited for the detection of single species or for whole groups of species like the hydrocarbons in the case of the FID. Furthermore, as Table 1.3 implies, they cannot meet all the challenging requirements expected from a universal instrument as discussed above. However, it should be emphasized that many of them (including others not mentioned here) are very useful and practical and may be applied also in harsh environments (e.g., in airplanes), thus playing an important role in today's air pollution control.

1.2.4. Spectroscopic Techniques for Air Monitoring

In recent years, spectroscopic techniques have attracted much interest because they offer some unique features that render such schemes very versatile tools for trace gas monitoring. The development can be pursued on the basis of some selected reviews (Christian and Callis, 1986; Grisar et al., 1992; Hanst, 1971; Hinkley, 1976; Killinger and Mooradian, 1983; Measures, 1988; Patel, 1978; Schiff and Platt, 1993).

The spectroscopic methods can be separated into those that use conventional light sources and those that use laser-based devices. The latter have profited from great progress in the development of tunable lasers. The basis of all spectroscopic techniques is some kind of interaction process between radiation and matter. It is thus convenient to differentiate between the essential processes involved, namely, scattering and absorption. The main schemes and their important features are summarized in Table 1.4 and briefly outlined in the following. Most of them are discussed in detail within the respective chapters of this book.

1.2.4.1. Scattering Processes

Figure 1.5 illustrates different types of scattering processes. The solid lines indicate real atomic or molecular levels, while dashed lines refer to virtual levels. Initial and final levels are denoted by i and f, respectively, while ν_0 and ν_s represent the frequency of the incident and scattered radiation, respectively. In the following, the characteristics of the various processes and their relevance to air monitoring are briefly outlined.

Table 1.4. Spectroscopic Techniques for Trace Gas Detection

Detection Scheme	Main Features	Sensitivity	Selectivity
Scattering processes			
Raman scattering	Inelastic interaction between photons and irradiated molecules; small cross sections; selectivity by frequency shift; only one wavelength needed; better for small wavelengths	Poor	Excellent
Laser-induced fluorescence (LIF)	Fluorescence of iradiated species small cross sections; only suitable for detection of atoms or radicals	Poor	Good
Absorption processes			
Infrared spectroscopy (incl. FTIR, matrix isolation, long path)	Absorption in fundamental IR range; large molecular cross sections; molecule specific	Good	Excellent
Differential optical absorption spectroscopy (DOAS)	Broadband source, separated from receiver by open path in the atmosphere; dual-wavelength operation; integrated response; mainly applied to UV (NO_2, SO_2, O_3)	Good	Good
Long-path absorption spectroscopy with lasers	Long single-pass or multipass cells; single or double-ended schemes in open atmosphere; integrated response; UV-vis-IR range	Good	Good
Light detection and ranging (lidar)/ differential absorption lidar (DIAL)	Atmospheric backscattering of laser pulses; three-dimensional profiling of pollutants; complex; mainly applied to UV and vis (SO_2, NO_2, O_3)	Good	Good

Table 1.4. *(Continued)*

Detection Scheme	Main Features	Sensitivity	Selectivity
Photoacoustic spectroscopy (PAS) with lasers	Measurement of absorbed energy in cell; short path length; mainly applied to IR; selective monitoring of numerous pollutants	Good	Good

Figure 1.5a refers to elastic scattering (Rayleigh and Mie) where $\nu_s = \nu_0$, i.e., no frequency shift occurs. Elastic scattering is important for atmospheric research on aerosols, clouds, and particulates, e.g., by light detection and ranging (lidar) techniques (see Chapter 3). Since elastic scattering per se is not molecule specific, it is only relevant for trace gas monitoring in conjunction with absorption processes to be discussed below.

The remaining scattering processes illustrated in Figs. 1.5b–1.5b, f involve specific molecular or atomic energy levels and thus intrinsically offer the potential for species-selective detection and for quantitative analyses in multi-component gas mixtures. Resonant scattering as shown in Fig. 1.5b again does not imply a frequency shift between ν_0 and ν_s yet involves specific energy levels. This scheme thus requires a tunable radiation source. It has been applied successfully for observations in the upper atmosphere. An example is the measurement of the layer of free atomic sodium in the mesosphere at an altitude of 90 km with a tunable dye laser lidar system (Hake et al., 1972).

Figure 1.5c refers to nonresonant Raman scattering, which involves a shift of the incident frequency ν_0 due to inelastic interaction of the photons with the irradiated molecules. This frequency shift $\pm\nu_r$ corresponds to vibrational-rotational frequencies that are characteristic of the molecules under study. The scattered frequencies ν_s are given by $\nu_S = \nu_0 - \nu_r$ and $\nu_{AS} = \nu_0 + \nu_r$, where ν_S and ν_{AS} correspond to Stokes (S) and Anti-Stokes (AS) processes, respectively. In Fig. 1.6 below frequency shifts ν_r of Q-branches of vibrational–rotational Raman spectra of typical atmospheric trace gases are summarized (Inaba, 1976). This demonstrates the great advantage of Raman scattering, namely, that a laser emission at only one (untuned) wavelength is necessary for detecting numerous gases since the selectivity is given by the characteristic

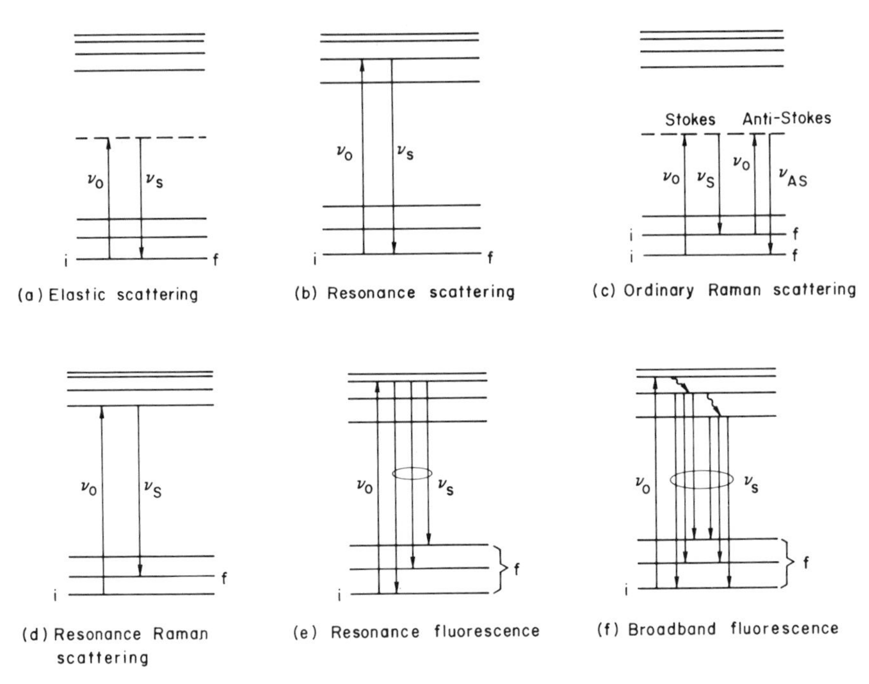

Figure 1.5. (a–f) Schematic atomic or molecular energy level diagrams with real (—) and virtual (---) levels. Various interaction processes associated with scattering and fluorescence are shown.

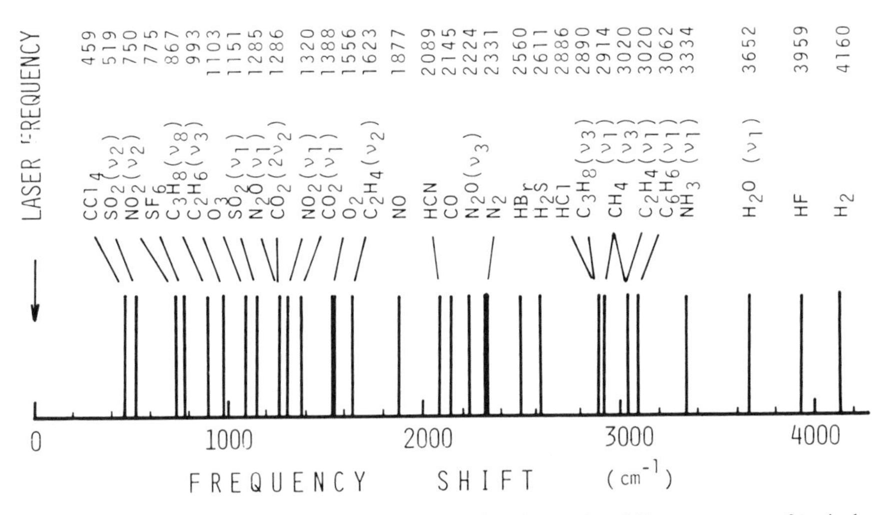

Figure 1.6. Frequency shifts of Q-branches of vibrational–rotational Raman spectra of typical molecules of interest in pollution monitoring. After Inaba (1976).

frequency shifts. The main disadvantages are the low scattering cross sections and possible interferences with fluorescence. As in Rayleigh scattering, the Raman cross section varies with ν_0^4 so that lasers with short wavelength [UV to visible (vis)] are preferable. The typically 10^3 times lower cross section of Raman scattering compared to Rayleigh scattering restricts the former technique to the monitoring of rather high molecular concentrations and/or short ranges.

When the exciting frequency ν_0 is tuned close to a proper resonance of the atom or molecule, resonance Raman scattering occurs as shown in Fig. 1.5d. This again requires a tunable source yet enhances the cross section over the ordinary Raman process by 3–6 orders of magnitude. Applications of Raman lidar systems to air monitoring are discussed in Chapter 3.

The remaining scattering processes, illustrated in Fig. 1.5e,f, are related to resonance and broadband fluorescence. This phenomenon refers to the spontaneous emission of a photon following the excitation into an excited state by absorption of the incident radiation at ν_0. The excited level can decay either by reemitting photons via transitions to lower levels resulting in discrete emission peaks (Fig. 1.5e) or by a combination of broad fluorescence and radiationless transitions caused by collisions with other atoms or molecules (Fig. 1.5f). As implied by Fig. 1.5e,f, the excitation of fluorescence always requires a tunable laser to match the proper resonance frequency. On the other hand, the measured fluorescence spectrum yields information on the atomic and molecular species responsible for the fluorescence.

The cross sections for fluorescence emission are small and comparable to those for Raman scattering (Inaba, 1976). Furthermore, the fluorescence is quenched in the atmosphere because of collisions with air molecules. The elegant technique of laser-induced fluorescence thus plays a minor role with respect to pollution monitoring in the troposphere. It has, however, been applied successfully for studying mesospheric atoms, as has already been mentioned above for the special case of resonance scattering (also referred to as resonance fluorescence).

Another interesting application of laser-induced fluorescence concerns hydrospheric pollution monitoring with airborne sensors. An early example of the monitoring of an oil spill on the ocean surface during an overflight is presented in Fig. 1.7 (Hoge and Swift, 1983). Figure 1.7a represents the depression of the water Raman backscatter signal as the oil slick was overflown. At the nominal speed of the airplane, 1 s corresponds to 85 m distance. The data in Fig. 1.7b indicate the fluorescence signal originating from the oil layer. The increased fluorescence clearly overlaps with the reduced Raman backscattering signal. The thickness of the oil layer shown in Fig. 1.7c was computed from the Raman data and oil extinction coefficients at the wavelengths used. The wavelength dependence of the fluorescence could also

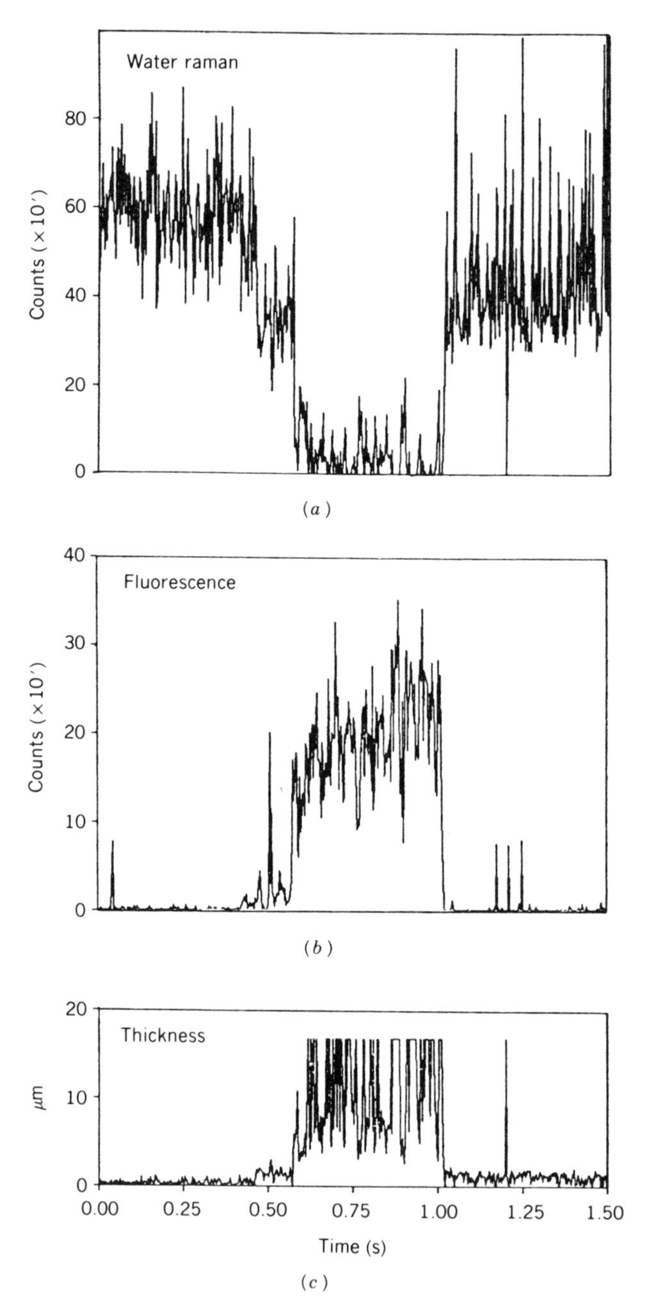

Figure 1.7. Water Raman backscatter (a) and oil fluorescence signals (b) obtained during the overflight of an oil slick on the ocean surface; (c) plot of the computed oil layer thickness. After Hoge and Swift (1983).

be used to differentiate between different oil types. In a very recent study, Hoge and co-workers examined chlorophyll pigments in ocean water by gathering fluorescence data during flight series, which demonstrates the great potential of this technique (see Hobbs, 1992).

Further examples of the different types of scattering applied to environmental monitoring are discussed in Chapter 3.

1.2.4.2. Absorption Processes

The most important tool for spectroscopic detection of gaseous air pollutants is based on measurement of the absorption of radiation by trace species. This approach inherently permits the identification and determination of concentrations of different species because the absorption features and strengths are molecule specific. The spectra occur over a broad portion of the electromagnetic spectrum, from the microwave region, where rotational transitions are responsible for the absorption, to the vacuum ultraviolet (VUV), where outler-shell electronic transitions cause the characteristic absorption (Hinkley et al., 1976). However, for chemical analysis and spectroscopic detection of gases the main wavelength ranges used are the middle or fundamental IR between, say, 2.5 and 25 μm and the visible-to-UV range (700 to 250 nm). In the former range, the so-called fingerprint region, the absorption is due to fundamental (as well as some overtone) and combination vibrational–rotational bands, whereas in the latter case electronic transitions with vibrational–rotational structure contribute to the absorption. The presence of absorption bands of typical pollutants in these wavelength regions has been outlined previously (Meyer and Sigrist, 1990) and is also illustrated in the central part of Fig. 3.2 (in Chapter 3). Molecular absorption cross sections in the mid-IR are typically on the order of 10^{-18} cm^2 and thus 6–8 orders of magnitude higher than Raman cross sections. This results in a considerable increase in detection sensitivity, which further enhances the attractiveness of absorption schemes for trace gas monitoring, particularly in the IR range (cf. Chapter 6).

Absorption spectroscopy on *trace* gases involves the measurement of small absorption coefficients owing to the low concentrations. This often requires long path lengths, especially for remote detection in the free atmosphere. This restricts the applicable wavelength ranges to atmospheric windows. The main windows occur for wavelengths shorter than 2.5 μm, from 3 to 5 μm, and from 8 to 14 μm (see Fig. 3.2). Atmospheric H_2O vapor and CO_2 drastically reduce the transmission beyond these spectral windows.

The quantitative identification of air constituents is limited by the spectral resolution and by the minimum detectable absorption coefficient α_{min}, i.e., by the sensitivity of the particular technique. A high spectral resolution is of primary concern for the proper identification of species, i.e., for detection selectivity. On the other hand, the sensitivity determines the minimum concentrations that can be measured reliably. Both aspects, sensitivity and selectivity, are related to the radiation source and detector, particularly to the tunability and spectral brightness of the source as well as to the spectral resolution and responsivity of the detector. Although incoherent radiation sources are successfully used for air pollution monitoring, as is demonstrated in Chapters 2, 6, and 7 of this book, the development of tunable lasers has given an impetus to the application of optical methods in air pollution control thanks to the high spectral brightness and low divergence of these sources. Various laser types are of interest including solid state, diode, dye, and gas lasers or laser systems in combination with nonlinear optical processes. The performance and applications of these sources are discussed in the relevant chapters of this book.

Different measurement schemes are applied to measure the small absorptions encountered in trace detection. Most are based on the well-known Beer–Lambert absorption law according to which the transmitted radiation intensity $I_2(\lambda)$ at wavelength λ is related to the incident intensity $I_1(\lambda)$ by

$$I_2(\lambda) = I_1(\lambda) \exp[-\alpha(\lambda) \cdot L] \tag{1.1}$$

where $\alpha(\lambda)$ indicates the common absorption coefficient (in cm^{-1}), and L is the absorption path length. Since straightforward differential absorption measurements are limited to absorbances αL of $\sim 10^{-4}$ (Patel, 1978), long path lengths L are employed to achieve sufficient sensitivity. This can be accomplished either by choosing an appropriate path through the (transparent) atmosphere with separate sender and receiver yielding path integrated data or by using a special cell design such as a White cell (White, 1942) in combination with a spectrometer (see Chapter 6).

Most commercial spectrometers are operated with conventional broadband sources in conjunction with monochromators or interferometers to establish the desired spectral resolution. The latter device is known as Fourier transform (FT) spectrometer. In comparison to dispersive spectrometers, FT systems possess a higher optical efficiency and a greater observation time efficiency because all spectral elements are simultaneously observed and measured. This unique feature and the available broad infrared range renders FTIR spectrometers very attractive for chemical analyses of multicomponent mixtures (Griffiths, 1975; Theophanides, 1984). However, the sensitivity is not always sufficient, e.g., for measurements of trace gases in clean air.

More recently, matrix isolation in combination with FTIR has been put forward (e.g., see Almond and Downs, 1989). The lack of spectral congestion compared to gas phase spectra avoids the problem of interferences and makes this technique very attractive for laboratory analyses of the composition of sampled air. It can be applied to a wide range of stable and moderately labile atmospheric trace gases with detection limits in the ppt range (cf. Chapter 7).

In *differential* absorption measurements the transmitted intensity is recorded either as the radiation wavelength is tuned across an absorption line of the species of interest or as it is switched from a wavelength with peak absorption ("on" position) to a nearby wavelength with negligible absorption ("off" position).

Today, commercial instruments based on differential optical absorption spectroscopy (DOAS) are available. They consist of a broadband radiation source like a xenon lamp and a receiver combined with a high-resolution spectrometer located some 100 m or even kilometers apart. Two adjacent wavelengths are used for the detection of a specific compound. These devices are mainly operated in the UV and visible range and are thus somewhat limited in the number of measureable species. However, they have successfully been applied to the monitoring of various gases, including highly reactive species such as the important OH radicals (Dorn et al., 1988) with detection limits in the sub-ppb range, as is discussed in Chapter 2.

Long-path absorption spectroscopy with *lasers* has profited from recent developments of tunable IR lasers, particularly lead salt diode lasers (Kuritsyn, 1985; Grisar et al., 1992). The technique of tunable diode laser absorption spectroscopy (TDLAS) offers very high spectral resolution, sensitivity, and time response. TDLAS has been performed both with open atmospheric paths and with confined air samples in long-path cells. Owing to the limited tuning range of a single diode laser, the technique is not as universal as FTIR spectroscopy because generally a different diode must be used for each compound being measured (Schiff, 1992). Nevertheless, the method has been applied to in situ measurements of a great variety of atmospheric trace gases, and detection limits in the ppt range have been achieved (as outlined in Chapter 5).

The sensitivity of long-path absorption measurements can be further improved by several orders of magnitude if the straightforward differential measurement is replaced by a first- or second-order *derivative* absorption measurement (Hinkley, 1976; Patel, 1978). In this scheme the wavelength of the incident radiation, generally of a tunable laser, is periodically modulated at a frequency ω_m across a small fraction of the line width of the molecular transition being measured. The absorbed as well as the transmitted power thus contain a time-varying component both at ω_m and $2\omega_m$. Hence, the problem arising from amplitude fluctuations that limit the sensitivity of the straightforward technique of differential absorption is avoided.

An interesting scheme that makes use of both scattering and absorption processes is differential absorption lidar, called DIAL. Its operation principle and versatile application for air monitoring are presented in Chapter 3. The double-ended version uses a topographic target or a retroreflector and thus yields an average concentration of specific substances along the laser beam path between source and reflecting target. The single-ended version relies on effective backscattering in the atmosphere and is thus most useful in the UV and visible spectral range, where Rayleigh and Mie scattering are strong. The unique feature is range-resolved monitoring, i.e., remote three-dimensional profiling of specific atmospheric trace species.

Finally, another method used for in situ absorption measurements that differs in various respects from the schemes previously discussed is photoacoustic spectroscopy (PAS). It is primarily a calorimetric technique whereby the absorbed energy is determined directly and not via the measurement of transmitted or backscattered radiation (see Chapter 4). Instead of radiation detectors, acoustic sensors are usually employed that measure the pressure modulation in the sample caused by the absorption of modulated radiation. In combination with a tunable laser as radiation source, this scheme permits the measurement of absorbances $\alpha L \sim 10^{-9}$ or even lower (Patel, 1978). Thus, small absorptions, i.e., small gas concentrations, can be measured even at short path lengths. Therefore, radiation wavelengths beyond atmospheric windows, e.g., in regions of strong water vapor absorption, can also be employed for the measurement. A further aspect is the wide dynamic range of PAS, which spans at least 5 orders of magnitude (Sigrist et al., 1989) and is thus of special interest for air monitoring since the gamut of pollutant concentrations is large. It should be mentioned that photoacoustics is a rather universal technique that has been applied successfully to many other fields (Mandelis, 1992).

1.3. CONCLUSIONS AND OUTLOOK

Selected global issues of air pollution have been illustrated by means of a short discussion of the "hot" topics of global warming, stratospheric ozone depletion, and photochemical smog formation. A profound understanding of these and related phenomena as well as the validation of model predictions require reliable methods suited for in situ measurements of numerous atmospheric trace species. In this respect spectroscopic techniques offer some unique features that are relevant to air monitoring. They are nonintrusive, do not require any sample preparation, and yield real-time data. They can be operated locally in situ for point monitoring or remote for determining the load of pollutants over a certain area, or even range-resolved for three-dimensional profiling. A further important aspect concerns the detection sensitivity, where excellent limits as low as 1 part in 10^{13} have been reported. The dynamic range spans up to 8 orders of magnitude in concentration, which means that in general the same instrument can be applied to measurements in heavily polluted areas as well as in rural regions. Since air masses contain numerous different trace constituents, the selectivity of detection is often more important than the sensitivity. This aspect represents a general strength of spectroscopic schemes. Since each molecule has its characteristic spectrum, it can easily be distinguished from other molecules if the spectral range and resolution of the instrument are sufficient. This feature enables monitoring of a number of species simultaneously, to detect selected compounds in complex mixtures or to perform chemical analyses of mixtures.

Despite these important advantages there are also some drawbacks that have so far delayed widespread application of spectroscopic devices for routine air monitoring. The instrumentation is often rather complex and thus requires special training of personnel. This is of course also true for other equipment used in pollution detection like gas chromatography. The costs for equipment are comparable to those for other devices if the great variety of species that can be detected with a single apparatus is taken into account. Another issue is the analysis of spectroscopic data. This analysis used to be rather complicated but has progressed greatly in recent years and is now easily practicable on a personal computer.

Further developments are still needed so that we can fully explore the great potential of spectroscopic schemes in air monitoring. These developments concern in particular laser and computer technology. The application of compact, widely tunable all-solid-state laser devices instead of complex systems with dye-, gas-, or cryogenically cooled diode lasers would mean a breakthrough toward moderate-cost, fully computerized instruments. The further improved system performance, the compactness and ruggedness, combined with high reliability and ease of operation will certainly favor the general applicability. Spectroscopic schemes will thus continue to yield valuable information and increase general knowledge of the global issues related to atmospheric trace gases.

REFERENCES

Almond, M. J., and Downs, A. J. (1989). Spectroscopy of matrix isolated species. *Adv. Spectrosc.* **17**.

Altshuller, A. P. (1986). The role of nitrogen oxides in nonurban ozone formation in the planetary boundary layer over North America, West Europe and adjacent areas of ocean. *Atmos. Environ.* **20**, 245–268.

Bolin, B., Döös, B. R., Jäger, J., and Warrick, R. A., Eds. (1986). *The Greenhouse Effect, Climatic Change, and Ecosystems.* Wiley, Chichester.

Bundesamt für Umwelt, Wald and Landschaft (BUWAL), Ed. (1989). *Ozon in der Schweiz*, Schriftenreihe Umweltschutz, No.101. BUWAL, Bern, Switzerland.

Christian, G., and Callis, J. B., Eds. (1986). *Trace Analysis: Spectroscopic Methods for Molecules*, Chem. Anal., Vol. 84. Wiley, New York.

Crutzen, P. J. (1988). Tropospheric ozone: An overview. In *Tropospheric Ozone: Regional and Global Scale Interactions* (I.S.A. Isaksen, Ed.), pp. 3–32. Reidel, Dordrecht, The Netherlands.

Cubasch, U. (1992). Das Klima der nächsten 100 Jahre. *Phys. Bl.* **48**, 85–89.

Dickinson, R. E., and Cicerone, R. J. (1986). Future global warming from atmospheric trace gases. *Nature (London)* **319**, 109–115.

Dorn, H. -P., Callies, J., Platt, U., and Ehhalt, D.H. (1988). Measurement of tropospheric OH concentrations by laser long-path absorption spectroscopy. *Tellus* **40B**, 437–445.

Ewing, G. W., Ed. (1990). *Analytical Instrumentation Handbook.* Dekker, New York.

Finlayson-Pitts, B. J., and Pitts, J. N. (1986). *Atmospheric Chemistry, Fundamentals and Experimental Techniques*, Wiley, New York.

Fox, D. L. (1985). Air pollution. *Anal. Chem.* **57**, 223R–238R.

Fox, D. L. (1987). Air pollution. *Anal. Chem.* **59**, 280R–294R.

Frei, R. W., and Albaiges, J., Eds. (1986). *Air and Water Analysis*, Curr. Top. Environ. Toxicol. Chem., Vol. 9. Gordon & Breach, New York.

Graedel, T. E. (1978). *Chemical Compounds in the Atmosphere.* Academic Press, New York.

Graedel, T. E., and Crutzen, P. J. (1989). The changing atmosphere. *Sci. Am.* **261/9**, 28–36.

Griffiths, P. R. (1975). *Chemical Infrared Fourier Transform Spectroscopy*, Chem. Anal., Vol. 43. Wiley, New York.

Grisar, R., Böttner, H., Tacke, M., and Restelli, G., Eds. (1992). *Monitoring of Gaseous Pollutants by Tunable Diode Lasers*, Proc. Int. Symp., Freiburg/Germany, October 17–18, 1991. Kluwer Academic, Dordrecht, The Netherlands.

Guiochon, G. (1990). Gas chromatography. *Rev. Sci. Instrum.* **61**, 3317–3339.

Hake, R. D., Jr., Arnold, D. E., Jackson, D. W., Evans, W. E., Ficklin, B. P., and Long, R. A. (1972). Dye-laser observations of the nighttime atomic sodium layer. *J. Geophys. Res.* **77**, 6839–6848.

Hanst, P. L. (1971). Spectroscopic methods for air pollution measurement. *Adv. Environ. Sci. Technol.* **2**, 91–213.

Harrison, R. M. (1984). Recent advances in air pollution analysis. *CRC Crit. Rev. Anal. Chem.* **15**, 1–61.

Harrison, R. M., and Perry, R., Eds. (1986). *Handbook of Air Pollution Analysis*, 2nd ed. Chapman & Hall, London.

Hinkley, E. D., Ed. (1976). *Laser Monitoring of the Atmosphere*, Top. Appl. Phys., Vol. 14. Springer-Verlag, Berlin.

Hinkley, E. D., Ku, R. T., and Kelley, P. L., (1976). Techniques for detection of molecular pollutants by absorption of laser radiation. In *Laser Monitoring of the Atmosphere* (E. D. Hinkley, Ed.), Top. Appl. Phys. Vol. 14, Chapter 6. Springer-Verlag, Berlin.

Hobbs, J. R. (1992). Airborne lidar examines ocean plant growth. *Laser Focus World*, November, pp. 15–16.

Hoge, F. E., and Swift, R. N., (1983). Experimental feasibility of the airborne measurement of absolute oil fluorescence spectral conversion efficiency. *Appl. Opt.* **22**, 37–47.

Houghton, J. T., Jenkins, G. J., and Ephraums, J. J. (1990). *Climate Change: The Intergovernmental Panel on Climate Change (IPCC) Scientific Assessment*. Cambridge Univ. Press, Cambridge.

Houghton, R. A., and Woodwell, G. M. (1989). Global climatic change. *Sci. Am.* **260/4**, 18–26.

Inaba, H. (1976). Detection of atoms and molecules by Raman scattering and resonance fluorescence. In *Laser Monitoring of the Atmosphere* (E.D. Hinkley, Ed.), Top. Appl. Phys. Vol. 14, Chapter 5. Springer-Verlag, Berlin.

Jennings, W. (1987). *Analytical Gas Chromatography*. Academic Press, Orlando, FL.

Jungclaus, G., Swanson, S., and Gorman, P. (1984). Sampling and analysis methods used in a trial burn at a hazardous waste incinerator, in *Identification and Analysis of Organic Pollutants in Air*, (L. H. Keith, Ed.,) pp. 425–442. Butterworth, Boston.

Keith, L. H., Ed. (1984). *Identification and Analysis of Organic Pollutants in Air*. Butterworth, Boston.

Killinger, D. A., and Mooradian, A., Eds. (1983). *Optical and Laser Remote Sensing*, Springer Ser. Opt. Sci., Vol. 39. Springer-Verlag, Berlin.

Kroon, D. J. (1978). Instrumental methods for automatic air monitoring stations. In *Air Pollution Control. Part III: Measuring and Monitoring Air Pollutants* (W. Strauss, Ed.), Chapter 6. Wiley, New York.

Kuritsyn, Yu. A. (1985). Infrared absorption spectroscopy with tunable diode lasers. In *Laser Analytical Spectrochemistry* (V. S. Letokhov, Ed.), Chapter 4. Adam Hilger, Bristol.

Mandelis, A., Ed. (1992). *Principles and Perspectives of Photothermal and Photoacoustic Phenomena*, Vol. 1. Elsevier, New York.

Measures, R. M., Ed. (1988). *Laser Remote Chemical Analysis*. Chem. Anal., Vol. 94. Wiley, New York.

Meyer, P. L., and Sigrist M. W. (1990). Atmospheric pollution monitoring using CO_2-laser photoacoustic spectroscopy and other techniques. *Rev. Sci. Instrum.* **61**, 1779–1807.

Organization for Economic Co-operation and Development (OECD), Ed. (1982). *Photochemical Smog: Contribution of Volatile Organic Compounds*. OECD, Paris.

Patel, C.K.N. (1978). Laser detection of pollution. *Science* **202**, 157–173.

Ramanathan, V., Cess, R. D., Harrison, E. F., Minnis, P., Barkstrom, B. R., Ahmad, E., and Hartmann, D. (1989). Cloud-radiative forcing and climate: Results from the earth radiation budget experiment. *Science* **243**, 57–63.

Roedel, W., and Woelm, G. (1987). *A Guide to Gas Chromatography*. Huethig, Heidelberg.

Schiff, H. I. (1992). Ground based measurements of atmospheric gases by spectroscopic methods. *Ber. Bunsenges. Phys. Chem.* **96**, 296–306.

Schiff, H. I., and Platt, U., Eds. (1993). *Optical Methods in Atmospheric Chemistry*, SPIE **1715**.

Schneider, S. H. (1987). Climate modeling. *Sci. Am.* **256/5**, 72–81.

Schneider, S. H. (1989). The changing climate. *Sci. Am.* **261/9**, 38–47.

Schurath, U. (1988). Bildung von Photooxidantien durch homogene Transformation von Schadstoffen. In *Ges. Strahlen Umweltforsch. Munich, Rep.* **17/88**, 136–151.

Seinfeld, J. H. (1989). Urban air pollution: State of the science. *Science* **243**, 745–752.

Sigrist, M. W., Bernegger, S., and Meyer, P. L. (1989). Atmospheric and exhaust air monitoring by laser photoacoustic spectroscopy. In *Photoacoustic, Photothermal and Photochemical Processes in Gases* (P. Hess, Ed.), Top. Curr. Phys., Vol. 46, Chapter 7. Springer-Verlag, Berlin.

Stolarski, R. S. (1988). The antarctic ozone hole. *Sci. Am.* **258/1**, 30–36.

Stolarski, R. S. (1992). Observation of global stratospheric ozone change. *Ber. Bunsenges. Phys. Chem.* **96**, 257–296.

Thain, W. (1980). *Monitoring Toxic Gases in the Atmosphere for Hygiene and Pollution Control*. Pergamon, Oxford.

Theophanides, T., Ed. (1984). *Fourier Transform Infrared Spectroscopy* Reidel, Dordrecht, The Netherlands.

Tomkin, J. (1992). What gas lies behind greenhouse effect? *Phys. Today*, December, pp. 13–15.

van Ham, J. (1989). Global elemental cycles and ozone. In *Atmospheric Ozone Research and Its Policy Implications* (T. Schneider, S. D. Lee, G. J. R. Walters and D. Grant, Eds.), pp. 57–72. Proc. 3rd US-Dutch Int. Symp., 1988. Elsevier, Amsterdam.

Warneck, P. (1988). *Chemistry of the Natural Atmosphere*. Academic Press, San Diego, CA.

White, J. U. (1942). Long optical paths of large aperture. *J. Opt. Soc. Am.* **32**, 285–288.

CHAPTER

2

DIFFERENTIAL OPTICAL ABSORPTION SPECTROSCOPY (DOAS)

U. PLATT

Institute for Environmental Physics, University of Heidelberg, Heidelberg, Germany

2.1. INTRODUCTION

2.1.1. Requirements for Spectroscopic Air Monitoring Techniques

Useful measurement techniques for atmospheric trace species should fulfill two main requirements. First, they must be sufficiently sensitive to detect the species under consideration at their ambient concentration levels. This can be a very demanding criterion; for instance, species present at mixing ratios ranging from as low as 0.1 ppt (mixing ratio of 10^{-13}, equivalent to about 2×10^6 molecules/cm^3) to several ppb (1 ppb corresponds to a mixing ratio of 10^{-9}) can have a significant influence on the chemical processes in the atmosphere (Logan et al., 1981; Perner et al., 1987a). Thus, detection limits from below 0.1 ppt to the ppb range are required, depending on the application. Second, it is equally important for measurement techniques to be specific, which means that the result of the measurement of a particular species must be neither positively nor negatively influenced by any other trace species simultaneously present in the probed volume of air. Given the large number of different molecules present at the ppt and ppb level, even in clean air, this is also not a trivial condition.

Further desirable properties of a measurement technique include the following: simplicity of design and use of the instruments based on it; capability of real-time operation (as opposed to samples being taken for later analysis); and the possibility of unattended operation. Also to be considered are weight, portability, and dependence of the measurement on ambient conditions.

Air Monitoring by Spectroscopic Techniques, Edited by Markus W. Sigrist. Chemical Analysis Series, Vol. 127.
ISBN 0-471-55875-3

2.1.2. The Principle of Absorption Spectroscopy

This variety of spectroscopic techniques make use of the absorption of electromagnetic radiation by matter (Fig. 2.1). Quantitatively the absorption of radiation is expressed by the Beer–Lambert law:

$$I(\lambda) = I_0(\lambda)\exp(-L\sigma(\lambda)c) \tag{2.1}$$

where $I_0(\lambda)$ denotes the initial intensity emitted by some suitable source of radiation, while $I(\lambda)$ is the intensity of the radiation after it passes through a layer of thickness L, where the species to be measured is present at the concentration c; the quantity $\sigma(\lambda)$ denotes the absorption cross section at the wavelength λ. The absorption cross section is a characteristic property of any species; $\sigma(\lambda)$ can be measured in the laboratory, while determination of the light path length L is usually trivial. Once these quantities are known, the trace gas concentration c can be calculated from the measured ratio $I_0(\lambda)/I(\lambda)$:

$$c = \log(I_0(\lambda)/I(\lambda))/(\sigma(\lambda)\,L) \tag{2.2}$$

The expression

$$D = \log(I_0(\lambda)/I(\lambda)) \tag{2.2a}$$

is called the *optical density* of a layer of a given species. (Note that in the literature the decadic as well as the natural logarithm are used in the definition of the optical density; in this chapter we always use the natural logarithm.) With this definition Eq. (2.2) becomes

$$c = D/(\sigma(\lambda)L) \tag{2.2b}$$

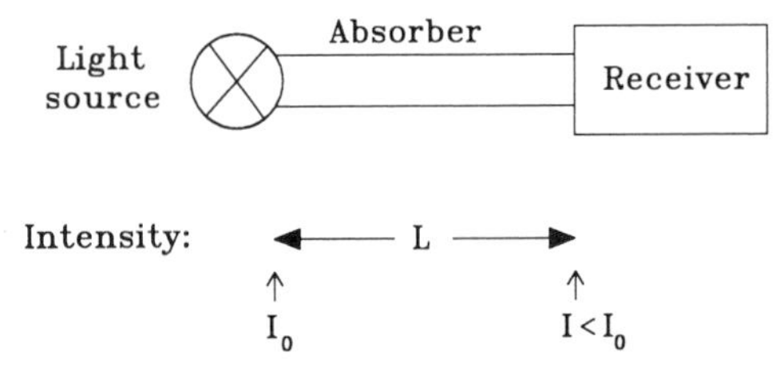

Figure 2.1. The basic principle of absorption spectroscopic trace gas detection.

2.1.3. Selection Criteria for Spectroscopic Techniques

To date no single measurement technique can fulfill all requirements, even nearly. Therefore in a particular application the selection of a technique will be based on the following particular requirements: What species are to be measured? Is the simultaneous determination of several species necessary? What is the required accuracy, time resolution, and spatial resolution? Also to be considered are logistic requirements like power consumption, mounting of light sources or retroreflectors (see below), or accommodation of the instrument on mobile platforms.

An important technical criterion of a spectroscopic instrument is the wavelength region used (see Fig. 2.2):

a. Microwave Spectroscopy. This spectral range is presently not used for tropospheric measurements, owing to the relatively large pressure broadening of the lines it would require measurements at reduced pressure.

b. Infrared (IR) Spectroscopy. This variety has been in use for several decades, initially developed for the detection of atmospheric CO_2 by nondispersive instruments [Ultrarot Absorptions-Spektrometrie (URAS)]. More modern instruments, based on Fourier transform techniques, can measure HNO_3, CH_2O, HCOOH, H_2O_2, and many other species by means of multiple-

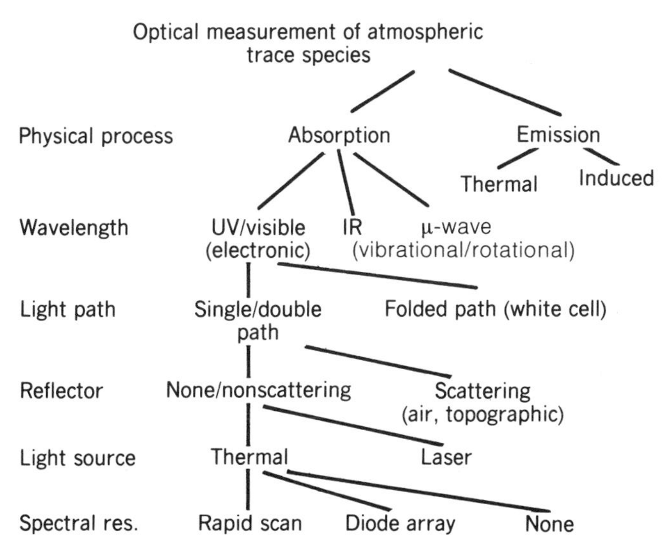

Figure 2.2. "Family tree" of spectroscopic measurement techniques for atmospheric trace species.

reflection cells with several-kilometer path lengths (Pitts et al., 1977; Tuazon et al., 1980, 1981). The sensitivity is in the low ppb range; thus such instruments appear to be best suited for studies of polluted air. In recent years tunable diode laser spectrometers (TDLS) have been developed to become field-usable instruments, successfully employed to measure HNO_3, NO, NO_2, CH_2O, and H_2O_2 at sub-ppb levels (Harris et al., 1989; Schiff et al., 1990; see also Chapter 5 of this volume). In the usual arrangement, coupled to a multiple-reflection cell, the strength of TDLS lies in the mobility of the instrument, allowing measurements on ships and airplanes, combined with high sensitivity. Limitations are due to the necessity of operating under low pressure (in most applications), thus introducing possible losses at the walls of the closed measurement cell. Also at present diode-laser technology still is quite complex.

c. Ultraviolet(UV)–Visible Absorption Spectroscopy. The most common optical arrangement is shown in Fig. 2.1. The light source and receiving system are typically separated by several kilometers. The strength of this technique lies in the absence of wall losses, good specificity, and the potential for real-time measurements. In particular the first property makes spectroscopic techniques especially well suited for the detection of unstable species like OH radicals (Dorn et al., 1988) or nitrate radicals (Platt et al., 1981). Limitations are due to logistic requirements (the need for electric power at two sites separated by several kilometers, but in sight of each other) in the case of unfolded path arrangements; also, conditions of poor atmospheric visibility can make measurements with this technique difficult.

2.2. LONG-PATH UV–VISIBLE ABSORPTION SPECTROSCOPY

In this section an overview will be given of the use of differential long-path UV–visible absorption spectroscopy to measure atmospheric trace gas concentrations.

2.2.1. Light Absorption in the Atmosphere

The attenuation of a light beam by absorption due to atmospheric constituents is in principle described by Eq. (2.1). For practical purposes of measurements in the atmosphere, however, Eq. (2.1) is oversimplified in that it neglects the presence of other causes of light extinction, which include:

a. Extinction Due to Rayleigh Scattering (i.e., scattering by air molecules). While this is not an absorption process, light scattered out of the probing light beam will normally not reach the detector; thus for our purposes it is justified to treat Rayleigh scattering as an absorption process. Simplified the "absorption" cross section can then be written as

$$\sigma_R(\lambda) \approx \sigma_{RO} \cdot \lambda^{-4} \qquad (\sigma_{RO} \approx 4.4 \times 10^{-16}\,\mathrm{cm}^2 \cdot \mathrm{nm}^4 \text{ for air}) \qquad (2.3a)$$

A more accurate treatment is found at Penndorf (1957). The Rayleigh extinction coefficient $\varepsilon_R(\lambda)$ is then given by

$$\varepsilon_R(\lambda) = \sigma_R(\lambda) \cdot c_{AIR} \qquad (2.3b)$$

where c_{AIR} denotes the concentration of air molecules (2.4×10^{19} cm^{-3} at 20 °C, 1 atm).

b. Extinction Due to Mie Scattering (i.e., scattering by atmospheric aerosol particles). This is only partly an absorption process, but by similar arguments as in the case of Rayleigh scattering it can be treated for practical purposes as an absorption process with the following extinction coefficient:

$$\varepsilon_M(\lambda) = \varepsilon_{MO} \cdot \lambda^{-n} \qquad (2.4)$$

with n being in the range 1...4 (Junge, 1963). Thus, a more comprehensive description of atmospheric absorption (in the presence of a single trace gas species) can be expressed as

$$I(\lambda) = I_0(\lambda) \exp[-L(\sigma(\lambda) c + \varepsilon_R(\lambda) + \varepsilon_M(\lambda))] \qquad (2.5)$$

Typical extinctions due to Rayleigh and Mie scattering at 300 nm are $1.3 \times 10^{-6}\,\mathrm{cm}^{-1}$ and $1\text{–}10 \times 10^{-6}\,\mathrm{cm}^{-1}$, respectively.

c. Absorption by Other Molecules Present in the Atmosphere. In the natural atmosphere many different molecular species will absorb light. Equation (2.5) must therefore be further extended:

$$I(\lambda) = I_0(\lambda) \exp\left[-L\left(\sum(\sigma_i(\lambda)\, c_i) + \varepsilon_R(\lambda) + \varepsilon_M(\lambda)\right)\right] \qquad (2.6)$$

where $\sigma_i(\lambda)$ and c_i denote the absorption cross section and the concentration of the ith species, respectively.

2.2.2. Differential Optical Absorption Spectroscopy (DOAS)

In contrast to laboratory spectroscopy, the true intensity, $I_0(\lambda)$, which would be received from the light source in the absence of any atmospheric absorption, is usually difficult to determine. It would involve removing the air from the open light path. While this seems to present a dilemma rendering long-path absorption spectroscopy useless in most cases, the solution lies in measuring the so-called differential absorption. This quantity can be defined as the part of the total absorption of any molecule "rapidly" varying with wavelength and is readily observable as will be shown below. Accordingly, the absorption cross section of a given molecule can be split into two portions:

$$\sigma_i(\lambda) = \sigma_{i0}(\lambda) + \sigma_i'(\lambda) \tag{2.7}$$

where $\sigma_{i0}(\lambda)$ varies only slowly with the wavelength λ, for instance, describing a general "slope," while $\sigma_i'(\lambda)$ shows rapid variations with λ, for instance, due to an absorption line (see Fig. 2.3). The meaning of "rapid" and "slow" variation of the absorption cross section as a function of wavelength is, of course, a question of the observed wavelength interval and the width of the absorption bands to be detected. Note that the extinction due to Rayleigh and Mie scattering can be assumed to be slowly varing with λ.

After introduction of Eqs. (2.7) into (2.6) we finally obtain

$$I(\lambda) = I_0(\lambda) \cdot \exp[-L\sum(\sigma_i'(\lambda)c_i)] \cdot \exp[-L(\sum(\sigma_{i0}(\lambda)\, c_i) + \varepsilon_R(\lambda) + \varepsilon_M(\lambda))] \cdot A(\lambda) \tag{2.8}$$

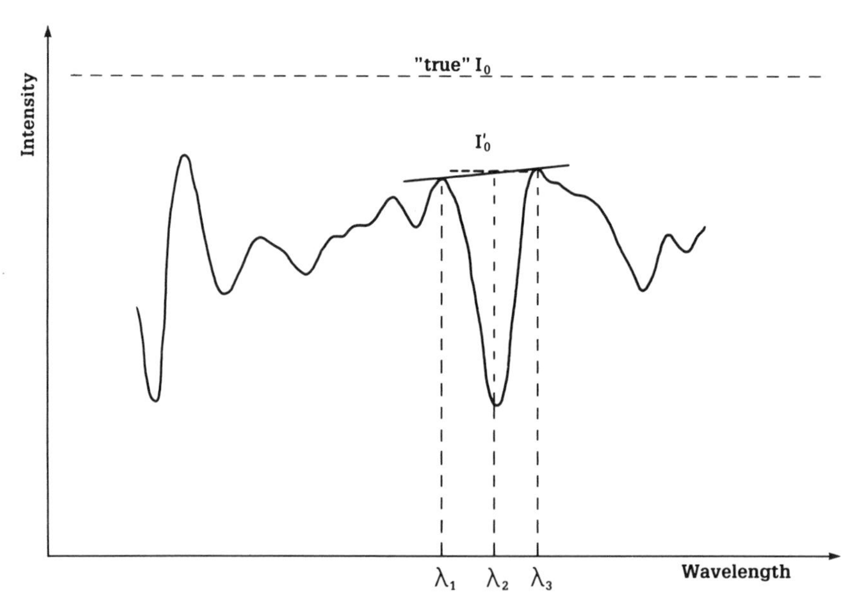

Figure 2.3. Differential absorption spectroscopy: practical example for the determination of I_0', D', and s'.

where the first exponential function describes the effect of the structured "differential" absorption of trace species, while the second exponential constitutes the slowly varying absorption of atmospheric trace gases as well as the influence of Mie and Rayleigh scattering. The attenuation factor $A(\lambda)$ describes the (slow) wavelength-dependent transmission of the optical system used.

Thus we can define a quantity I'_0 as the intensity in the absence of differential absorption:

$$I'_0(\lambda) = I_0(\lambda) \cdot \exp[-L(\sum(\sigma_{i0}(\lambda)c_i) + \varepsilon_R(\lambda) + \varepsilon_M(\lambda))] \cdot A(\lambda) \qquad (2.9)$$

As outlined in Fig. 2.4 the intensity I'_0 can, for instance, be interpolated from the light intensity at either side of a sufficiently narrow absorption line of the species:

$$I'_0 = I(\lambda_1) + [I(\lambda_3) - I(\lambda_1)] \cdot (\lambda_2 - \lambda_1)/(\lambda_3 - \lambda_1) \qquad (2.10)$$

The corresponding differential absorption cross section $\sigma'(\lambda)$ is then substituted for $\sigma(\lambda)$ in Eqs. (2.1) and (2.2); $\sigma'(\lambda)$ is determined in the laboratory (i.e., taken from literature data), just like $\sigma(\lambda)$. Likewise a differential optical density D' can be defined in analogy to Eq. (2.2) as the logarithm of the quotient of the intensities I'_0 and I_0 [as defined in Eqs. (2.8) and (2.9), respectively]:

$$D' = \log(I'_0(\lambda)/I(\lambda)) = L\sum(\sigma'_i(\lambda)c_i) \qquad (2.11)$$

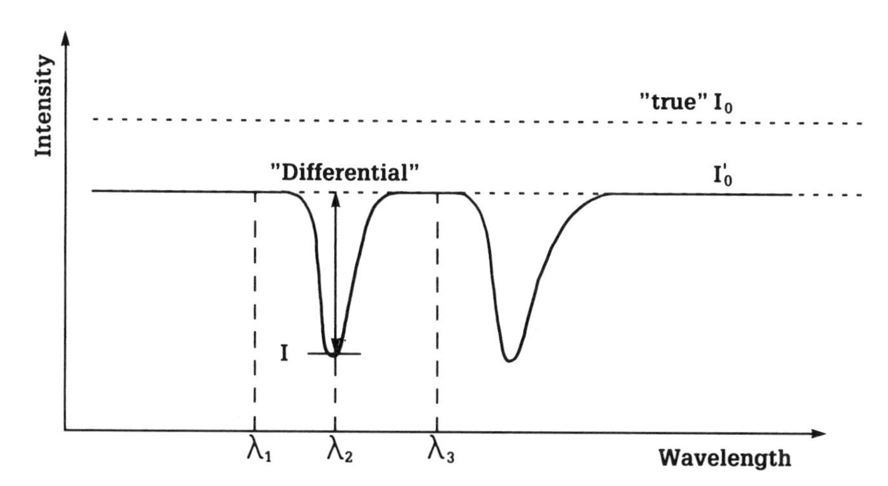

Figure 2.4. Differential absorption spectroscopy. The intensity I'_0 is interpolated from the light intensity at either side of a sufficiently narrow absorption line of the species. The "true" I_0, i.e., the light intensity at the detector in the absence of any broad-band attenuation process, is not important.

The species concentration can then be calculated according to Eq. (2.2b) with the differential optical density D' and differential absorption cross section $\sigma'(\lambda)$ substituted for the total quantities D and $\sigma(\lambda)$, respectively. Figure 2.3 illustrates the relationship of $\sigma(\lambda)$ and $\sigma'(\lambda)$ and the determination of D' from an actually measured spectrum. For practical measurements the usually finite resolution of the spectrometer used to measure D' must also be taken into account, resulting in an effective $\sigma'(\lambda)$ somewhat smaller than the value obtained at infinite spectral resolution, as described in Section 2.6.2.

Note that this procedure can only be applied to species the spectrum of which contains reasonably narrow absorption features, thus limiting the number of molecules detectable by this technique. A possible continuous absorption of the trace gas will be neglected by this procedure. In many cases, for instance, for the free radicals OH and NO_3 (see below), total and effective cross sections are nearly identical. On the other hand, this technique of observing "differential" absorptions is insensitive to extinction processes, which vary only slowly with wavelength, like Mie scattering by aerosol, dust, or haze particles. Since such processes attenuate the total available light intensity, however, they have an influence on the detection limit, as will be discussed in Section 2.6. Likewise slow variations in the spectral intensity of the light source or in the transmission of the optical system (telescope, spectrometer, etc.) are also eliminated.

2.3. USABLE SPECTRAL RANGES: SPECIES MEASURABLE WITH DOAS

Usable spectral ranges for the various trace gases are summarized in Fig. 2.5, which also gives an indication of the strength of the absorption and the corresponding detection limits that can be expected. Table 2.1 gives an overview of typical detection limits for the various species discussed here.

At short wavelengths the usable spectral range of differential optical absorption spectrometers is limited by rapidly increasing Rayleigh scattering (see above) and O_2 absorption. Those effects limit the maximum light path length to a few hundred meters in the wavelength range from 200 to 230 nm, where the sole usable absorption features of species like NO (Tajime et al., 1978) and NH_3 are located (see Fig. 2.5 and Table 2.1). Strong absorption features of SO_2 (Brand and Srikameswaran, 1972; Brand et al., 1973; Wu and Judge, 1981) are also found in this spectral region.

In the range from 230 to 260 nm absorption structures due to O_3 (Hartley band) and benzene ($A^1B_{2u} \leftarrow X^1A_{1g}$ transition) and its derivatives dominate,

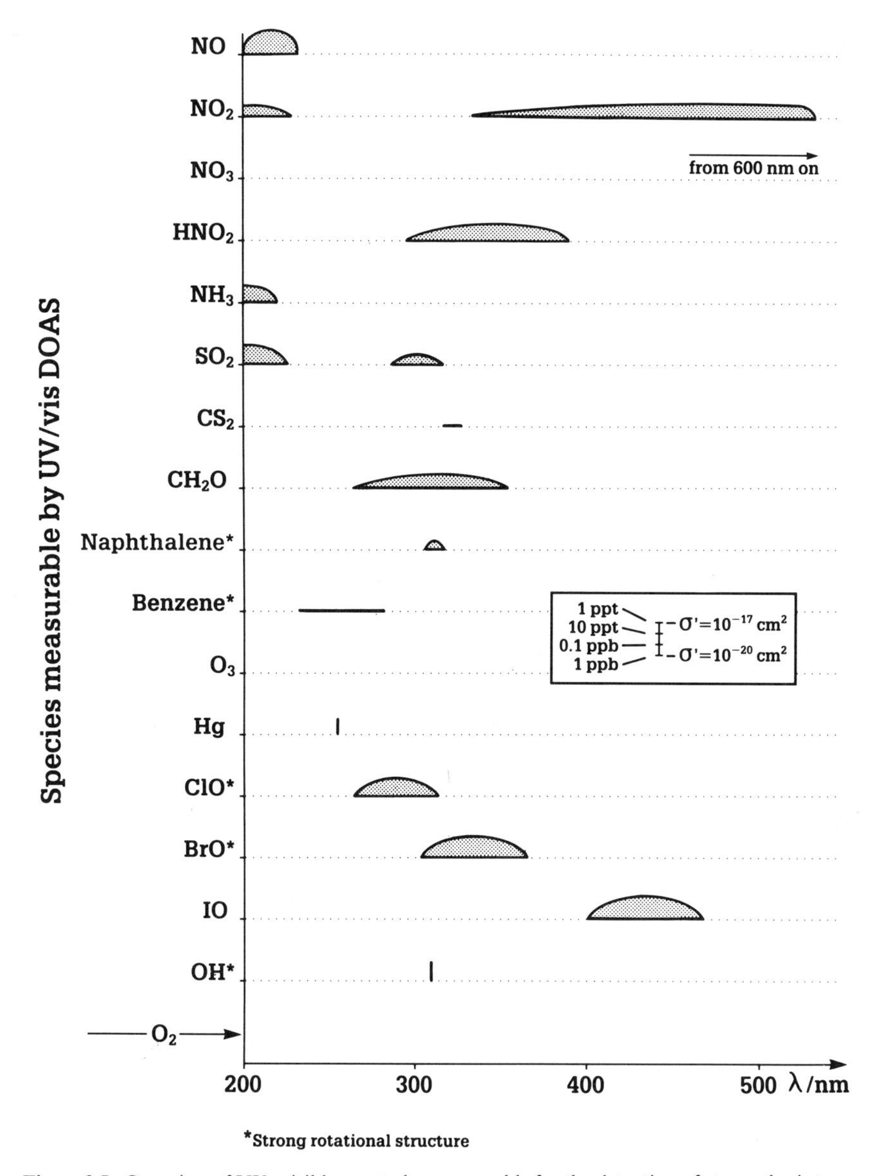

Figure 2.5. Overview of UV–visible spectral ranges usable for the detection of atmospheric trace gases. Vertical scale: log of absorption cross section of the molecule (10^{-20} to 10^{-17} cm^2/molecules); also approximate detection limit at 10 km light path length (1 ppt to 1 ppb, see insert). Molecules exhibiting strong rotational structure at atmospheric pressure are denoted by an asterisk (*).

Table 2.1. Substances Detectable by UV–Visible Absorption Spectroscopy

Species	Wave-length Interval (nm)	Differential Absorption Cross Section (cm^2/molecule)	Detection Limit[a] (5 km light path)
SO_2	200–230	6.5×10^{-18}	240[b]
	290–310	5.7×10^{-19}	17
CS_2	200–220		
	320–340	4.0×10^{-20}	500
NO	200–230	2.4×10^{-18}	240[b]
NO_2	200–230	2.6×10^{-20}	10^{5}[b]
NO_2	330–500	2.5×10^{-19}	80
NO_3	600–670	2.0×10^{-17}	2
NH_3	200–230	1.8×10^{-18}	800[b]
HNO_2	330–380	5.1×10^{-19}	40
O_3	300–330	4.5×10^{-21}	4000
CH_2O	300–360	4.8×10^{-20}	400

[a] Detection limits in pt (1ppt = 1 part in 10^{12}).
[b] 200-m light path.

interferences can be caused by forbidden oxygen band structures. Also polycyclic aromatic compounds like naphthalene (George and Morris, 1968; Neuroth et al., 1991) strongly absorb in this spectral range.

In the spectral range from about 290 to 310 nm the well-known SO_2 $A^1B_2 \leftarrow X^1A_1$ transition (Hamada and Merer, 1974; Brand and Nanes, 1973; Marx et al., 1980; Brassington, 1981; Brassington et al., 1984) dominates the absorption structure; also OH radicals absorb near 308 nm (Hübler et al., 1984). Note that the structures of the O_3, Huggins bands (Daumont et al., 1989) are very similar to those of SO_2 in spectral width (though about 2 orders of magnitude weaker) and thus may cause interferences in clean air if not taken into account. Also in that range are absorptions due to formaldehyde [$A^1A_2 \leftarrow X^1A_1$ transition from ca. 260 to 360 nm (Cantrell et al., 1990)].

At longer UV wavelengths between 300 and 400 nm absorptions due to CH_2O, NO_2 (Hall and Blacet, 1952; Schneider et al., 1987; Davidson et al., 1988), and HNO_2 (Newitt and Outridge, 1938; Perner and Platt, 1979; Harris et al., 1982, 1983; Bongartz et al., 1991) are usually visible; BrO bands (Wahner et al., 1988) are also located in this region.

Beyond 400 nm NO_2 is the dominating absorber (Fig. 2.5). Absorptions of IO, NO_3, H_2O, and $(O_2)_2$ (Perner and Platt, 1980) will also be found in this spectral region.

2.4. THE OPTICAL DESIGN OF LONG-PATH SPECTROMETERS

Instruments based on spectroscopic techniques have been designed in a variety of ways (Noxon, 1975, 1983; Davies et al., 1975; Perner et al., 1976; Bonafe et al., 1976; Kuznetzov and Nigmatullina, 1977; Millan, 1980; Millan and Hoff, 1978; Noxon et al., 1978, 1980; Platt, 1977; Platt et al., 1979; Platt and Perner, 1983; Johnston and McKenzie, 1984; Ender et al., 1986, 1990; Dorn and Platt, 1986; Axelsson et al., 1990; Plane and Nien, 1992). In the following subsections some options are discussed for the practical design and selection of key parameters like the length of the light path and the spectral resolution.

2.4.1. Key Components of DOAS Systems

Depending on the particular type of measurement desired, the elements of an optical long path spectrometer systems are selected:

a. Light Sources. An important requirement for DOAS light sources is a minimum of spectral variation, in particular at the scale of molecular band-widths (see Section 2.2.2). In other words, the emitted radiation intensity $I_0(\lambda)$ should only vary slowly with λ, i.e., the light sources should emit "white" light. In use are the following types:

- *Thermal light sources such as incandescent lamps or arc lamps.* Criteria are spectral brightness (watts/unit of radiating area and wavelength interval; see Section 2.6.2). This quantity is essentially given by the Planck function and the temperature of the radiating area (around 3000 K for incandescent lamps and 6000–10 000 K for Xe arc lamps). The spectrum of incandescent lamps is very similar to a black body radiator. Xenon arc lamps emit an essentially smooth spectrum in the spectral range discussed in Section 2.3, with the exception of a group of emission lines between 400 and 450 nm (Canrad-Hanovia 1986; Larche, 1953).
- *Lasers are usually more complex devices, but compared to thermal light sources generally give a much lower beam divergence and always a higher spectral intensity.* Problems here are due to the narrow spectral emission bandwidths or tuning range, which makes it usually difficult to observe molecular vibrational bands.
- *Light sources outside the atmosphere i.e., sun- or moon- or even starlight.* The largest problem with these light sources is due to their structured spectrum containing many Fraunhofer lines (see Sections 2.4.6 and 2.7.6).

b. Optical Elements Adapting the Absorption Path to Light Source and Spectrometer. The instrument may average the trace gas concentration over a 1-

(possibly several) km-long path stretching between a searchlight-type light source and the spectrometer. A variation of this design uses a reflector in the field to return the light of a source located next to the spectrometer, thus doubling the light path. Alternatively the path may be folded in a multiple reflection cell (White, 1942, 1976; Horn and Pimental, 1971; see also Chapter 6). Also range-resolving [light detection and ranging (lidar)] techniques may be employed. In the latter case the length L of the effective light path is calculated from the traveling time of the light; lidar techniques will be discussed elsewhere in this volume (see Chapter 3).

c. Measurement Pressure. The light path may be operated at ambient pressure or lower than ambient pressure. In the latter case, which will not be discussed here, the light path must run in a closed vessel.

d. Spectrometer and Radiation Detector Used to Measure the Spectra $I(\lambda)$ and $I_0(\lambda)$. Depending on the wavelength and sensitivity requirements a wide variety of detector designs may be used, ranging from nondispersive semiconductor or photomultiplier detectors to spectrograph–photodiode array combinations (see also Section 2.5). The photoacoustic detectors are a special case (see Chapter 4), directly measuring the difference $I_0 - I$ (being approximately proportional to $D \cdot I_0$ and thus the trace gas concentration) rather than determining I and I_0 separately.

A number of spectrometer designs are in use:

- *Czerny–Turner spectrometers (and variants).* This popular spectrometer design has been traditionally used for DOAS instruments (Platt et al., 1979; Platt and Perner, 1983). Besides the grating it requires at least two additional reflecting surfaces (collimating and camera mirrors). The additional reflection losses (reflectivity typically 0.8 in the near UV) are usually offset by high-efficiency (60–80%) plane blazed gratings. Since most of the incident light is directed into the desired diffraction order, stray light caused by reflection of unwanted orders off the spectrometer housing is minimized.
- *Flat-field spectrometers.* Modern holographic grating technology (Lerner and Thevenon, 1988) allows the design of spectrometers only consisting of entrance slit, grating, and detector (i.e., optomechanical scanning device or diode array). The presently low efficiency of the holographic gratings (20–40%) is approximately compensated by the reduced losses due to the low number of optical surfaces. While the stray light produced by holographic gratings itself is lower compared to ruled gratings, the higher intensity of unwanted orders may increase the stray light level of the whole spectrometer.
- *Fourier transform spectrometers.* While this type of instrument has been in use for IR absorption spectroscopy for a long time (see, e.g., Tuazon

et al., 1980), its application for DOAS in the UV–visible spectral range has been reported only recently (Colin et al., 1991; Vandaele et al., 1992).

As in most spectroscopic systems stray light requires careful consideration when one is selecting or designing a spectrometer. Sources of stray light include light scattered from optical elements (grating and mirrors), reflection of unused diffraction orders off the spectrometer walls, reflection of unused portions of the spectrum from walls near the focal plane, and reflections from the detector surface (Pierson and Goldstein, 1989). Note that stray light tends to be comparatively high in spectrometers filtering a relatively broad wavelength interval from a continuous spectrum, as in DOAS applications. To illustrate this point we consider a typical stray light level of 10^{-5} for a Czerny–Turner spectrometer (Pierson and Goldstein, 1989). This level gives the fractional light intensity found anywhere in the spectrum, when a single light beam (laser) of very narrow spectral width enters the spectrometer. The actual width of the line seen in the focal plane will then just be equal to the spectral resolution of the instrument. Continuous light entering the spectrometer can be thought of as being composed of a series of lines spaced at center-to-center distances equal to their width (i.e., the spectral resolution of the instrument). Since the total spectral range of light entering the spectrometer (for instance, from 300 to 600 nm) is on the order of 1000 times larger than the spectral resolution (of typically 0.3 nm for a low-resolution DOAS instrument), the level of stray light (per wavelength interval) is roughly 3 orders of magnitude larger than would be expected from the single line definition usually considered. Thus, in DOAS applications stray light levels I_{SL}/I can be expected to be closer to 10^{-2} than to 10^{-5}.

The effect of stray light is to reduce the recorded optical density, in that the expression given in Eq. (2.2a), $D = \ln(I_0(\lambda)/I(\lambda))$, must be replaced by

$$\begin{aligned} D_{SL} &= \ln\big((I_0(\lambda) + I_{SL})/(I(\lambda) + I_{SL})\big) \\ &= \ln\big(I_0(\lambda)/I(\lambda)\cdot(I_{SL}/I_0 + 1)/(I_{SL}/I + 1)\big) \\ &= D + \ln\big((I_{SL}/I_0 + 1)/(I_{SL}/I + 1)\big) \end{aligned} \tag{2.12}$$

For example, a stray light level of 10^{-2} would reduce a true optical density of $D = 1.00 \times 10^{-3}$ to an apparent $D_{SL} = 0.99 \times 10^{-3}$.

2.4.2. Practical Design of a Low-Resolution DOAS

Figure 2.6 shows the typical setup of a DOAS instrument (Platt and Perner, 1983, 1984; Dorn and Platt, 1986). The light source is designed to emit a beam

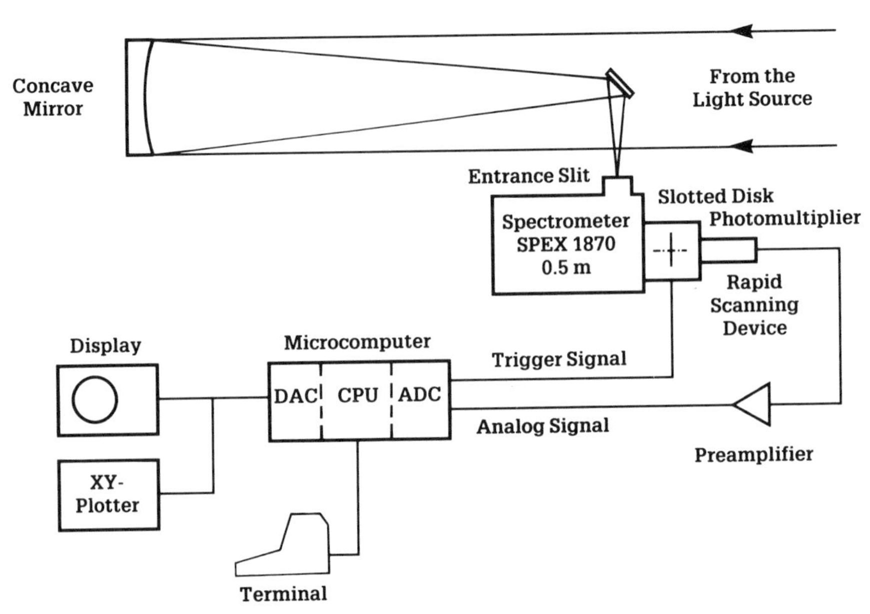

Figure 2.6. Typical setup of a low-resolution differential optical absorption spectrometer (DOAS); the searchlight-type light source is not shown. *Key*: ADC, analog-to-digital converter; DAC, digital-to-analog converter; CPU, central processing unit.

of light with minimum divergence. After passing a distance of several 100 m to several kilometers in the open atmosphere, the light is collected by a telescope and dispersed by a spectrometer. The spectrum projected into the focal plane of the spectrometer is then recorded either by a mechanical scanning device in combination with a photomultiplier tube (Platt and Perner, 1983, 1984; Dorn and Platt, 1986) or by a photodiode array (Dorn et al., 1988). Finally the recorded spectrum is digitized and transferred to a computer, which analyzes the spectral features due to trace gas absorptions along the light path and calculates the corresponding concentrations.

A more advanced design (Fig. 2.7) pioneered by Axelsson et al. (1990) uses a coaxial merge of a transmitting and a receiving telescope in conjunction with an array of corner-cube retroreflectors. Compared to folding the light path by a plane mirror [a technique used by Perner et al. (1976, 1987a)], this approach has the advantage of canceling the effect of atmospheric turbulence on the light beam. This is due to the capacity of corner-cube retroreflectors to return the incident light exactly (although with some lateral offset) into the direction of incidence (see Fig. 2.7).

During daytime, sunlight scattered within the measurement volume (not to

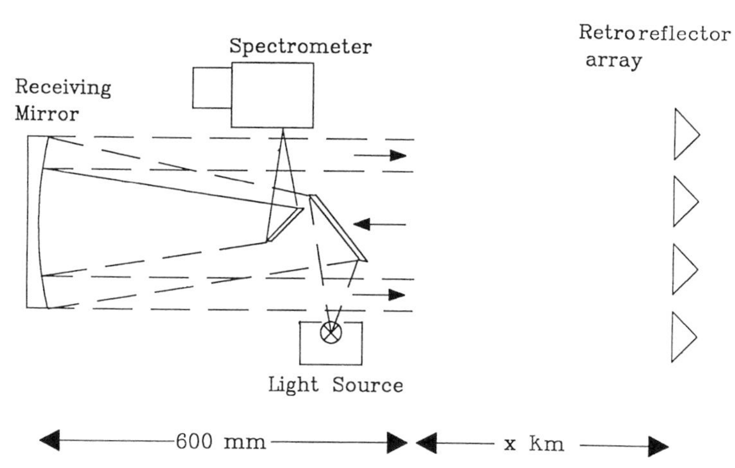

Figure 2.7. An advanced design pioneered by Axelsson et al. (1990) uses a coaxial merge of a transmitting and a receiving telescope. An outer ring of the receiving mirror is used as light source mirror, thus emitting a "hollow beam." Owing to beam divergence, the center hole is filled after a short distance. The high-precision retroreflector array returns the light exactly into the direction toward the source; however, because of the lateral offset of the reflected rays as well as atmospheric turbulence, a fraction of the reflected light also reaches the center of the mirror acting as primary of the receiving telescope.

be confused with stray light inside the spectrometer) may also reach the spectrometer. In contrast to a more or less continuous offset caused by stray light, scattered sunlight exhibits a strong Fraunhofer structure, which can disturb measurements even at stray light levels as low as a fraction of a percent of the received light source intensity. Fortunately, owing to the very narrow field of view of the instrument [on the order of 10^{-9} sr (steradians)], levels of scattered sunlight have been found to be extremely low, except when very hazy conditions prevail.

In such cases, however, the scattered light can easily be canceled by frequently interrupting the measurement and taking a "scattered light spectrum" with the telescope pointed beside the light source. Since only an offset of a fraction of a degree is required, this does not change the amount of scattered light entering the system, which can then be subtracted from the total spectrum.

Care must be taken that the transmission of the instrument and the spectral response curve of the detector [as summarized in $A(\lambda)$ in Eq. (2.9)] are only slowly varying functions of the wavelength. Then remaining undesired influences on the shape of the spectrum are eliminated by the mathematical treatment of the spectrum, as discussed in Section 2.6.

2.4.3. Light Utilization in a Long-Path Spectrometer

In the following discussion an attempt is made to obtain an idea of the light throughput of a typical long-path spectrometer. We take the dimension of a single-path DOAS system as outlined in Fig. 2.8a. Then the fraction x_1 of light emitted from the radiating surface (the ribbon of an incandescent lamp or the arc of a Xe high-pressure lamp) of the light source actually transformed into a light beam can be expressed as follows:

$$x_1 = r^2\pi/4\pi f^2 = 0.25(r/f)^2 \tag{2.13}$$

where r denotes the radius of the light-source (searchlight) mirror, and f its focal length. Assuming the radiating surface of the lamp to be circular (which is usually not the case but may be done for simplicity) with the radius r_1 (and thus the surface $r_1^2\pi$) the divergence of the beam (half angle) is given by

$$\alpha = r_1/f \tag{2.14}$$

and the radius R_b of the light beam at the receiving telescope located at the distance L from the searchlight (assuming $R_b \gg r$ and $\alpha = \tan\alpha$)

$$R_b = \alpha L = (r_1 L)/f \tag{2.15}$$

The fraction of light reaching the receiving telescope is given by $R^2\pi/R_b^2\pi$; thus

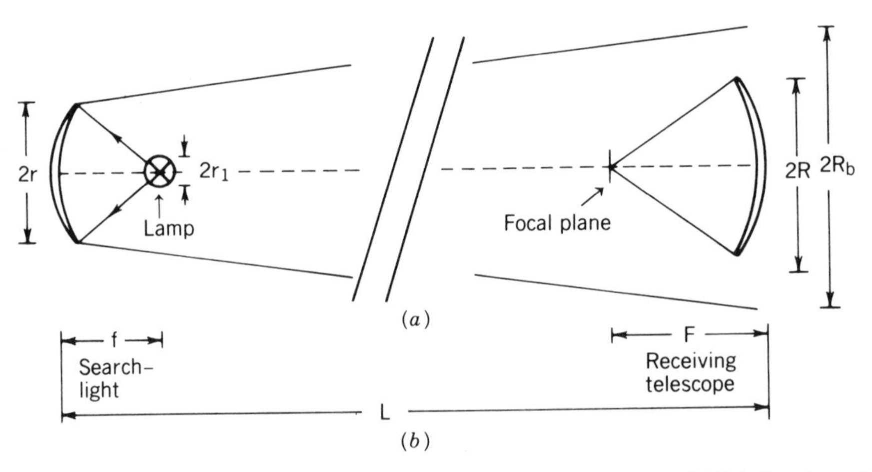

Figure 2.8. (a) Outline of the optics of a typical long-path spectrometer. (b) Relative sizes of receiving telescope focal spot and spectrometer entrance slit.

we obtain the fraction x_2 of the emitted light beam actually collected by the telescope:

$$x_2 = (R/R_b)^2 = [(Rf)/(r_1 L)]^2 \tag{2.16}$$

Combining Eq. (2.16) with Eq. (2.13) gives the fraction x_e of the total lamp output received by the telescope:

$$x_e = x_1 \cdot x_2 = 0.25[(Rr)/(r_1 L)]^2 \tag{2.17}$$

Note that x_e is independent of the focal length of the searchlight mirror (for a given radius r) and thus of its aperture. On the other hand, x_e is proportional to the product of the areas of the two mirrors and inversely proportinal to the radiating area of the lamp and thus its luminous intensity (radiated energy per unit of arc aera).

The important quantity is the illumination intensity D_b in the focal plane of the receiving telescope (i.e., the brightness of the focal spot). It is given by

$$D_b = P_1 \cdot x_e \cdot [(rF/L)^2 \pi]^{-1} \tag{2.18}$$

where P_1 denotes the lamp output power, and $(rF/L)^2\pi = R_f^2\pi$ gives the area of the focal spot. Substituting x_e from Eq. (2.17) yields

$$D_b = P_1 \cdot 1/(4\pi) \cdot [R/(r_1 F)]^2 \tag{2.19}$$

The illumination intensity of the focal spot depends neither on the length of the light path L (if light extinction is neglected) nor on the properties of the searchlight. Of course, the size of the focal spot (i.e., its radius R_f) will be proportional to the radius (r) of the searchlight mirror and inversely proportional to the light path length:

$$R_f = rF/L \tag{2.20}$$

From Eq. (2.19) it appears as if the light throughput of the system could be maximized by choosing the largest possible aperture $A_r = 2R/F$ of the receiving telescope. However, the aperture A_r must not be larger than the aperture A_s of the spectrometer (if $A_r > A_s$, a portion of the light received by the telescope will simply not be used). Thus, for a given value of A_s the brightness of the focal spot has a maximum value given by substituting $R = A_s F/2$ into Eq. (2.19):

$$D_b = P_1 \cdot 1/(16\pi) \cdot (A_s/r_1)^2 \tag{2.21}$$

The result is that for a long-path spectrometer the size of the optical system does not matter, provided that the aperture of the receiving system matches that of the spectrometer and the dimensions of the optical components exceed certain "minimum sizes" given by Eq. (2.20) as follows:

Depending on the relative sizes of the focus R_f and the entrance slit $(w \cdot h)$, three regimes can be distinguished (see Fig. 2.8b):

1. $w, h < R_f$. The amount of light entering the spectrometer will be independent of L (except for extinction). This is due to the independence of the illumination intensity from L stated in Eq. (2.19).
2. $w < R_f < h$. The amount of light entering the spectrometer will vary nearly in proportion to L^{-1}.
3. $R_f < w, h$. The amount of light entering the spectrometer will vary in proportion to L^{-2}.

Thus it can be concluded that the product of the size of the searchlight mirror and the focal length of the receiving system (the aperture of which matches the aperture of the spectrometer used) must be chosen sufficiently large to allow the system to be in regime 1.

Table 2.2 gives examples for the quantities derived above for a typical DOAS system with the following dimensions:

$$r = 0.15\,\text{m}$$
$$f = 0.3\,\text{m}$$
$$R = 0.15\,\text{m}$$
$$F = 1.5\,\text{m}$$

Further assuming entrance slit dimensions of (width w by height h) of 0.05 by 0.2 mm, the lightpaths for regimes 1 to 3 are obtained:

Regime 1: $L <$ about 2000 m
Regime 2: $2000 < L <$ about 9000 m
Regime 3: $L >$ about 9000 m

As can be seen from Table 2.2, the total loss factor from lamp output to the detector is close to 10^{-12} in the case of an optomechanical scanning device with a photomultiplier detector. The total number of photons per second and spectral interval (assuming a division of the spectrum into 400 intervals) reaching the detector can be estimated as follows: a 450-W xenon arc lamp emits about 2.5 W, corresponding to $dN_0/dt \approx 6 \times 10^{18}$ photons per second in the form of light in the spectral range from 440–460 nm (Canrad-Hanovia,

Table 2.2. Light Losses in a Typical Long-Path Spectrometer System[a]

Optical Element	Loss Factor
Lamp into beam, x_1[b]	0.08
Geometric losses, x_2[c] ($a = 10^{-3}$ rad; $R_b = 17$ m)	3.0×10^{-4}
Atmospheric attenuation (Mie and Rayleigh scattering)	0.2
Entrance slit	1.0
Spectrograph losses	0.2
Dispersion	5.0×10^{-3}
Multiplex losses (400 channels)	2.5×10^{-3}
PMT[d] quantum efficiency	0.2
Total loss factor	1.2×10^{-12}

[a] 10-km light path.
[b] See Eq. (2.13) in text.
[c] See Eq. (2.16) in text.
[d] PMT = photomultiplier tube.

1986). If we take the above loss factor from Table 2.2, about $dN/dt \approx 6 \times 10^6$ photoelectrons per second and spectral interval (of 0.05-nm width in this example) are registered. This number corresponds to a shot noise limit of about $4 \times 10^{-4}\,s^{-1/2}$ (1σ).

2.4.4. Arrangement of Light Paths

Common arrangements of DOAS light paths can be divided into three groups:

a. One (several kilometer) long path stretching between a searchlight-type light source and the spectrometer (see Figs. 2.1 and 2.8) can be used. The instrument will, of course, average the concentration over this extended volume of air. This is advantageous in applications where average concentration levels are desired. In particular local influences like small emission sources (for instance, near the instrument site) have little effect on the result of the measurement. On the other hand, investigations of fast chemical processes in the atmosphere (as in OH photochemistry; see Section 2.7.3) often require the assumption that all measured trace gases are homogeneously distributed over the measurement volume. This assumption is less likely to be fulfilled for large, extended measurement volumes.

b. A variation of this design uses a reflector [either a plane mirror (see Fig. 2.9) or a corner-cube retroreflector (as in Fig. 2.7)] in the field to return the light of a source located next to the spectrometer; thus the total light path length is twice the distance between the light source/spectrometer and the reflector. Advantages and limitations are similar to those specified in arrangement *a*, above.

c. Alternatively the path may be folded in a multiple-reflection cell (White, 1942, 1976; Horn and Pimentel, 1971; Hanst et al., 1973; Hanst, 1978; Ritz et al., 1992). This approach has two main advantages. First, the measurement volume is greatly reduced, thus making the assumption of homogeneous trace gas distribution much more likely to be fulfilled. Secondly the length of the light path can easily be changed (in certain increments of small multiples of the base path, depending on the particular design of the cell). Therefore it is

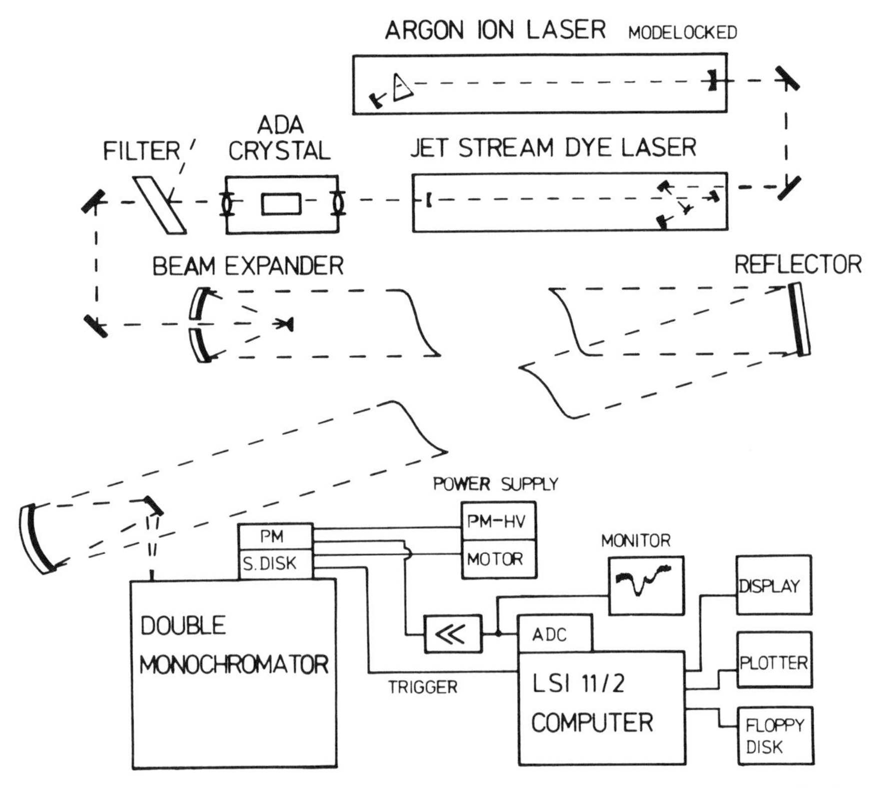

Figure 2.9. Optical setup of a high-resolution long-path spectrometer. A laser system supplies the high spectral intensity required because of the high resolution (0.002 nm instead of about 0.5 nm for a low-resolution DOAS). *Key*: ADC, analog-to-digital converter; PM, photomultiplier; HV, high voltage. Adapted from Platt et al. (1987).

possible to keep the length of the light path close to its optimum (see Section 2.6.4) for the atmospheric conditions. Disadvantages are due to the more complicated optical design and to reflection losses (high reflectivity of the cell mirrors is required); also, the high radiation intensity inside the folded light path can cause a photochemical reaction that may affect the accuracy of the measurement.

2.4.5. High-Resolution DOAS for the Detection of Atmospheric OH Radicals

In the case of OH radicals the narrow absorptions lines (see Table 2.4 in Section 2.7.3) dictate the use of a light source with high spectral intensity. A laser can be used with emission bandwidths that are either broad or narrow compared to the OH lines. The former approach has a fixed laser emission wavelength but requires combination with a high-resolution spectrometer (Hübler et al., 1984; Perner et al., 1987a; Platt et al., 1987, 1988; Dorn et al., 1988; Hofzumahaus et al., 1991; Mount, 1992). The latter variety of the technique uses a narrow-band laser scanning the OH line (Zellner and Hägele, 1985; Amerding et al., 1990, 1992). As an example, an experiment following the broad-band approach is shown in Fig. 2.9. A frequency-doubled dye laser coupled to a beam-expanding telescope serves as light source. The laser emission bandwidth [full width at half-maximum (fwhm)] of about 0.15 nm is large compared to the width of the OH line (0.0017 nm). The received light is dispersed by a 0.85-m double spectrometer (0.0018-nm resolution) (Perner et al., 1987a; Hofzumahaus et al., 1991). Figure 2.16 (see Section 2.6.3) shows a portion of the OH absorption spectrum together with the spectra of several interfering species (Hübler et al., 1984; Neuroth et al., 1991).

2.4.6. Optical Setup for the Detection of Stratospheric Trace Gases

DOAS using sunlight scattered in the zenith has proven to be a valuable tool for the determination of several stratospheric trace gases. To date the technique has been applied to the determination of vertical column densities of O_3, NO_2 (Brewer et al., 1973; Noxon, 1975; Noxon et al., 1978; Pommereau, 1982; Kerr et al., 1982; McKenzie and Johnston, 1984; Mount et al., 1987; Pommereau and Goutail, 1987, 1988; Johnston and McKenzie, 1989; Wahner et al., 1990; McKenzie et al., 1991; Johnston et al., 1992), NO_3 (Sanders et al., 1987; Solomon et al., 1989a), OClO (Solomon et al., 1987b, 1988, 1989b; Schiller et al., 1990; Perner et al., 1991; Fiedler et al., 1993) and $(O_2)_2$ (Sarkissian et al., 1991; Fiedler et al., 1990, 1992).

A spectrograph directed to the zenith (Fig. 2.10) observes the optical density of trace gas absorption bands indirectly via the scattering process of

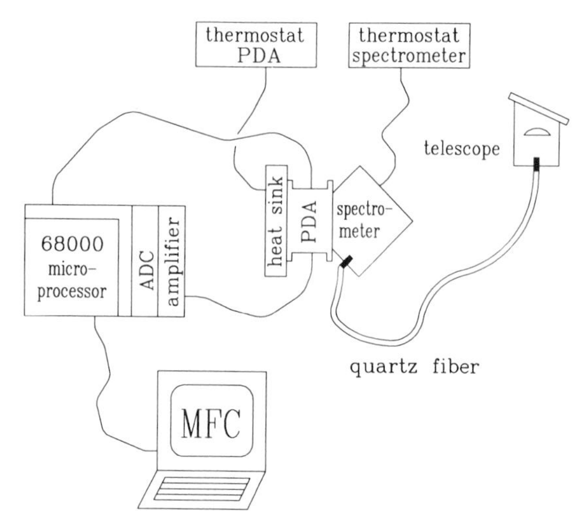

Figure 2.10. Optical setup of a zenith-scattered light observing spectrometer. Adapted from Kreher (1991). *Key*: PDA, photodiode array; MFC, software package.

the sunlight at different layers of the atmosphere. At large solar zenith angles this involves averaging over the contribution of nearly horizontal rays that reach the zenith at different altitudes with different intensities and information content about the trace gas layer. For more details on atmospheric radiation transport, see Solomon et al. (1987a) and Frank and Platt (1990). Simplified, the detected light from the zenith can be represented by a most probable light path through the atmosphere, defined by the most likely scattering height in the zenith. Typical heights of the most probable beam at 90° are 26 km (327 nm) down to 11 km (505 nm). Thus, in principle the wavelength-dependent beam height allows a certain degree of height resolution. Those slant light paths through the atmosphere are quite long under twilight conditions: At 90° zenith angle (and 327 nm) a horizontal light path running 600 km through the stratosphere integrates 90% of the trace gas absorption signature. From the optical density D as derived from the recorded spectra only effective column densities $\int c(z)\,dz$ [equivalent to $c_0 \cdot L$ if the concentration is independent of z, that is, if $c(z) = c_0$] can be derived. The term *effective* indicates that the light reaching the spectrometer is an average over several rays actually taking somewhat different routes through the atmosphere.

Modeling the physical process of radiation transport, including Rayleigh scattering, refraction, and Mie extinction, enables determination of the *air mass factor* (AMF), which is defined as the ratio of the effective column and the vertical column of a given trace gas. Typically the AMF reaches values of around 20 at 90° zenith angle. This AMF is then applied to conversion of the measured slant columns into vertical columns. It is a function of zenith angle and wavelength, and depends to some extent on assumptions of the relative vertical trace gas profiles and the air density.

2.5. DETECTORS FOR UV–VISIBLE SPECTROMETERS

Two different detector designs are currently being used for DOAS instruments: (a) optomechanical scanning devices of the "slotted disk" type, as shown in Fig. 2.11, and (b) solid state device photodetectors (photodiode arrays), as shown in Fig. 2.12.

2.5.1. Optomechanical Scanning Devices

In this design a segment of the dispersed spectrum (typically a few nanometers up to 100 nm) produced in the exit focal plane of the spectrograph is scanned by a series of moving exit slits etched radially in a thin metal disk ("sloted disk") rotating in the focal plane. At a given time, one particular slit serves as an exit slit. The light passing through this slit is received by a photomultiplier tube (PMT) mounted behind the focal plane. The output signal of the PMT is digitized by a high-speed analog-to-digital (A/D) converter and recorded by a computer. During one scan (i.e., one sweep of an exit slit over the spectral interval of interest), several hundred digitized signal samples are taken.

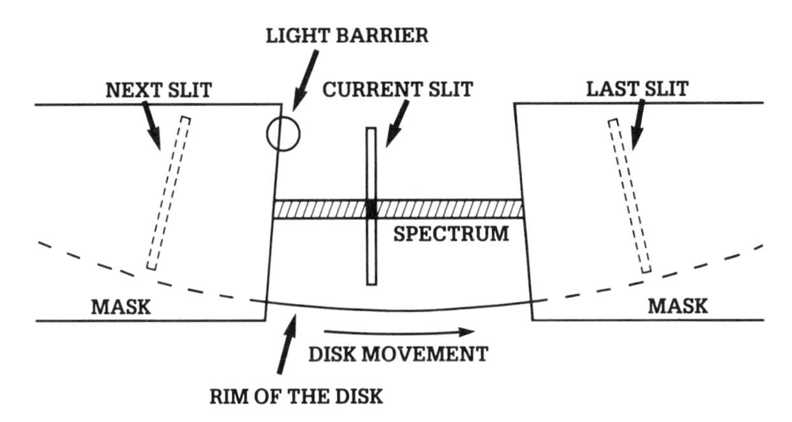

Figure 2.11. Optomechanical rapid scanning device. At any given moment one of the radial slits etched into a thin metal disk ("slotted disk") rotating in the focal plane of the spectrometer acts as an exit slit sweeping over the spectrum. A photomultiplier receives the portion of the spectrum falling through that slit.

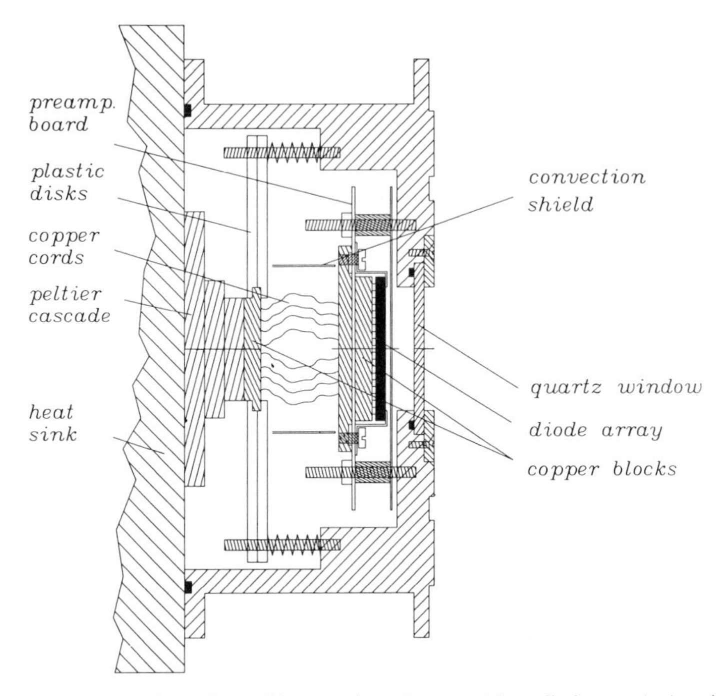

Figure 2.12. Cross section of a solid state photodetector (photodiode array). An air-cooled three-state Peltier device allows cooling to about −40 °C. Adapted from Stutz and Platt (1992).

Consecutive scans are typically performed at a rate of 100 scans per second and are signal averaged by the software.

The process of taking one spectral scan and the interaction of hardware and software thereby is best illustrated in conjunction with Fig. 2.11. The central wavelength of the scan is selected by the spectrograph setting, and the width of the scan region by a mask located very close to the slotted disk. The distance between the slits along the rim of the disk is slightly larger than the aperture of the mask, so that at any time no more than one slit is irradiated. As a slit becomes visible at the left edge of the mask, it is detected by an infrared light barrier located there (see Fig. 2.11), and a trigger signal is sent to the computer. As the slit then sweeps over the spectrum, the computer continuously takes digitized samples of the light intensity at the current position of the slit. Thus, one sweep of a slit is divided into several hundred channels, each associated with a wavelength interval several times narrower than the resolution of the spectrograph. During each scan, the digitized samples are added to the corresponding channels in the computer memory; all consecutive scans are thereby superimposed in the computer memory (signal averaged).

After finishing one scan, the computer waits for the next trigger pulse to indicate the next slit (shown by dashed lines in Fig. 2.11) approaching the left edge of the mask. In order to preserve the spectral resolution while superimposing a large number of individual scans, the rotational speed of the slotted disk is kept constant to within $\pm 0.1\%$. The signal variations seen by the computer during a single spectral scan can be caused by several effects:

1. The light absorption or extinction by atmospheric constituents varies with wavelength.
2. The light losses due to mirror reflectivities and the like may vary with wavelength.
3. The output of the light source may vary with wavelength and time.
4. Owing to turbulence, atmospheric refraction may change with time.
5. Random noise is added to the signal by the photomultiplier, preamplifier, and A/D converter.

Since a single scan takes less than 10 ms, the effect of atmospheric scintillations is very small because the frequency spectrum of atmospheric turbulence close to the ground peaks around 0.1–1 Hz and contains very little energy at frequencies above 10 Hz (Haugen, 1973). In addition, typical spectra obtained during several minutes of integration time represent an average over some 10,000–40,000 individual scans. Thus, effects of noise and temporal signal variations are very effectively suppressed. In fact, even momentarily blocking the light beam entirely (e.g., owing to a vehicle driving through the beam) has no noticeable effect on the spectrum.

2.5.2. Photodiode Array Detectors

The main disadvantage of the optomechanical scanning of the absorption spectra as desribed in Section 2.5.1 lies in the multiplex losses. Only the section of the spectrum passing through the exit slit is reaching the detector. Thus typically only on the order of 1% of the light is utilized, making relatively long integration times necessary.

Therefore a solid state detector consisting of several hundred individual photodetectors arranged in a linear row is—in principle—a far superior device for recording absorption spectra (Jones, 1985 a, b; Talmi and Simpson, 1980; Vogt et al., 1978). Additional advantages are that there are no moving parts, no high voltage is required, and the quantum efficiency (at least in the red and near-IR spectral regions) is better than that of most photomultiplier cathodes (Yates and Kuwana, 1976). Figure 2.12 shows the cross section of a typical silicon photodiode array detector system (using a Reticon RL 1024 R detector chip); it consists of 1024 diodes, each measuring about 0.025 by 2.5 mm integrated on a single silicon chip.

Unfortunately experience shows that photodiode array detectors pose some problems when used to record weak absorption features in DOAS instruments. For one thing; the dark current of the photodiodes can cause problems, since it is "signal dependent." This apparent paradox is due to the different average voltages across the diode junctions during exposures, when the array is darkened or irradiated, respectively. A detailed explanation of this effect is given by Stutz and Platt (1992). Since the dark current decreases exponentially with temperature, it can always be reduced to negligible levels by cooling the array. However, additional complexity is added to the simple design by cooling the array to very low temperatures (− 40 to − 80 °C) in order to reduce the dark current to acceptable levels. In addition this often makes vacuum insulation necessary, further complicating design and operation of the device. An even more severe problem is the presence of Fabry–Pérot etalon structures in the recorded spectra. Those structures are caused by the protective overcoat of the array and also by vapor deposits on the array, which are very difficult to avoid (Mount et al., 1992; Stutz and Platt, 1992). Particular problems are due to changes in the thickness of the layer of vapor deposit, causing continuous changes in the etalon modulation of the array sensitivity (Mount et al., 1992).

Presently diode array detectors seem to have great advantages in applications where strong absorptions ($D > 0.01$) occur or only low light levels are available (as in zenith scattered light observation of stratospheric species; see Section 2.4.6). For the observation of very low optical densities ($D \approx 10^{-2}$–10^{-4}) optomechanical scanning devices still give much better results, despite their inferior light utilization compared to that of diode arrays.

2.6. EVALUATION OF SPECTRA, SENSITIVITY, AND DETECTION LIMITS

In principle the concentration of the absorbing species is determined according to Eq. (2.2) by using the differential optical density $D' = \log(I'_0(\lambda)/I(\lambda))$ as defined in Eq. (2.11):

$$c = D'/(\sigma'(\lambda)L) \tag{2.2'}$$

2.6.1. Evaluation Procedures

Since the differential absorption cross section $\sigma'(\lambda)$ and the light path L in Eq. 2.2′ are known, in practice two additional problems arise: (1) the $I'_0(\lambda)$

spectrum has to be determined (see Section 2.2); (2) in certain spectral regions overlapping absorptions by several trace species may occur (see Fig. 2.13, here; and Figs. 2.15 and 2.21, below), which must be separated. The latter problem is discussed in Section 2.6.3. The first problem is solved by (digitally) high pass filtering the spectra during the evaluation procedure. The cutoff frequency of the filter in wavelength space is chosen so as to minimize attenuation of the molecular absorption bands to be detected. Since the widths of those bands is roughly around 1 nm, this suggests cutoff frequencies on the order of a fraction of a nm^{-1}. On the other hand, the slowly varying slope of the spectra should be removed as completely as possible. There are many possible choices of suitable filtering techniques:

a. Experience has shown that the slowly varying light extinction factors $\exp(-\Sigma(\sigma_{i0}(\lambda)c_i))$, $\exp(-\varepsilon_M(\lambda))$, and $A(\lambda)$ of Eqs. (2.8) and (2.9) can be quite well approximated by a least square fitted polynomial $P(\lambda) = \Sigma a_n \lambda^n$ of suitable order [usually fifth (see Platt et al., 1979)]. Thus the expression for $I'_0(\lambda)$ given in Eq. (2.9) becomes

$$I'_0(\lambda) = I_0(\lambda) \cdot \exp[-L(\Sigma(\sigma_{i0}(\lambda)c_i + \varepsilon_R(\lambda) + \varepsilon_M(\lambda))] \cdot A(\lambda) \approx P(\lambda) \quad (2.22)$$

Dividing $I(\lambda)$ from Eq. 2.8 by this approximation for $I'_0(\lambda)$ yields:

$$I(\lambda)/I'_o(\lambda) = \exp[-L(\Sigma(\sigma'_i(\lambda)c_i))] \approx I(\lambda)/P(\lambda) \quad (2.23)$$

In the case of only one absorbing species this exactly corresponds to Eq. (2.1), with $P(\lambda)$ as an approximation for $I_0(\lambda)$. Figure 2.14 shows an example of a measured original atmospheric spectrum (lower trace) and the version divided by a least square fitted polynomial (upper trace). Here the optical density D' can then be determined from the light intensities at either side one of the bands as well as in the center of the band (see Fig. 2.3).

b. Division by a digitally smoothed copy $S(\lambda)$ of the spectrum can be used instead of dividing by $P(\lambda)$.

c. Also Fourier transform techniques have been applied to obtain suitable filtering of the spectra (Colin et al., 1991).

Additional work needs to be done on the issue of optimum filtering of DOAS spectra. Note that the reference spectra must be subjected to the same filtering process as the measured atmospheric spectra before the differential absorption cross section σ' can be determined.

Generally, in order to determine correct differential absorption cross sections σ', reference spectra taken at a high resolution compared to the resolution of the DOAS instrument $\sigma_{hr}(\lambda)$ must be folded with the instrument

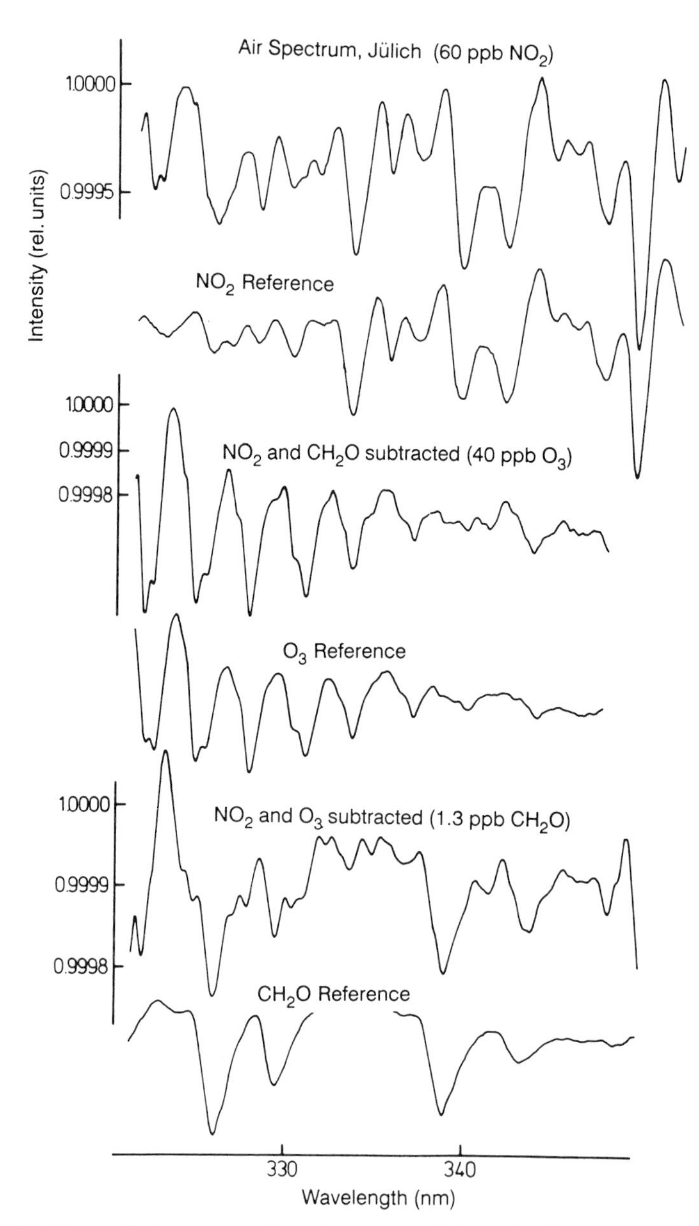

Figure 2.13. Deconvolution process of an air spectrum with overlapping NO_2, O_3, and CH_2O absorptions. This example shows an atmospheric absorption spectrum from 325 to 350 nm taken at Jülich, Germany (top trace). The most dominant absorption features are due to NO_2 (NO_2 reference spectrum, second trace from top). Three reference spectra due to NO_2, O_3, and CH_2O, respectively, were simultaneously fitted to the atmospheric spectrum. Using the coefficients thus determined, one can subtract the suitable scaled logarithms of the NO_2 and CH_2O reference spectra, leaving a residual spectrum clearly showing O_3 absorption features (two center traces). An analogous operation with the NO_2 and O_3 reference spectra shows that CH_2O absorption features are also present in the spectrum (two bottom traces). Adapted from Platt and Perner (1983).

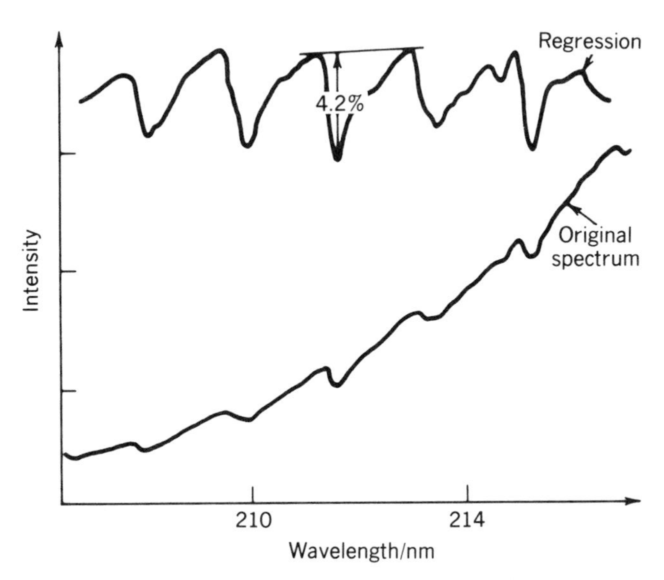

Figure 2.14. UV spectrum of SO_2 and NO (206–216 nm) taken near Jülich, Germany, at 270-m light path. Lower trace: original air spectrum. Upper trace: spectrum after division by least square fitted polynomial. The regular band structure due to SO_2 now becomes much more visible; some features near 214 nm are due to atmospheric NO.

function $W(\lambda)$ including any digital filtering done to the spectra) to obtain the effective absorption cross section as seen by the DOAS instrument [see also Eq. (2.34)]:

$$\sigma(\lambda) = \int \sigma_{\text{hr}}(\lambda)\, W(\lambda - \lambda')\, d\lambda'$$

From the spectrum $\sigma(\lambda)$ thus calculated, $\sigma'(\lambda)$ is determined, as described above.

2.6.2. Sensitivity and Detection Limits

The detection limit for a particular substance can be calculated according to Eq. (2.2) if the differential absorption cross section, the minimum detectable optical density D_0, and the length of the light path are known.

In general D_0 is determined by photoelectron statistics (shot noise), which is a function of light intensity, and detector-noise-induced "background absorption structures," here denoted as B, being on the order of 10^{-4} for a typical DOAS system with an optomechanical scanning device (see Section 2.5).

Photoelectron shot noise is proportional to $N^{-1/2}$, where N is the total number of photons recorded around the center of the absorption line during the time interval t of the measurement:

$$D_0 = (1/N + B^2)^{1/2} \tag{2.24}$$

If sufficiently light is available, of course, D_0 is limited by B (Platt et al., 1979; Platt and Perner, 1983). When N from Section 2.4.3 is used, ideally an integration time on the order of minutes is sufficient.

2.6.3. Interferences

If the absorption features of several species overlap in the same spectral region, usually a simultaneous least square fit of the respective absorption spectra is used to separate the contributions of the individual species. The fit coefficients will then be a measure for the $\sigma_i' \cdot c_i$ products of the absorbing species. Since Eq. (2.23) is a nonlinear equation usually the logarithm of $I(\lambda)/I_0'(\lambda)$ is formed, thus allowing a (mathematically simpler) linear fit. Examples of the deconvolution process are shown in Figs. 2.13, 2.15, and 2.16. The example in Fig. 2.13 shows an atmospheric absorption spectrum from 325 to 350 nm taken at Jülich Germany (top trace). The most dominant absorption features are due to NO_2 (NO_2 reference spectrum, second trace from the top). Three reference

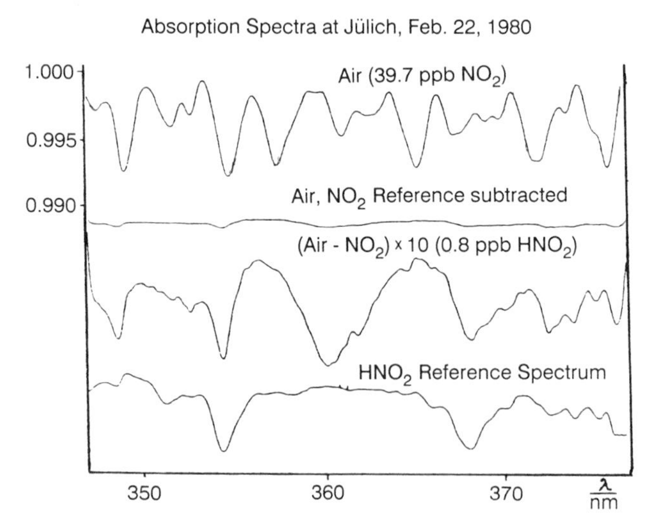

Figure 2.15. Deconvolution process of overlapping features of NO_2 and HONO spectrum (taken at Jülich, Germany, February 22, 1980). Adapted from Platt and Perner (1983).

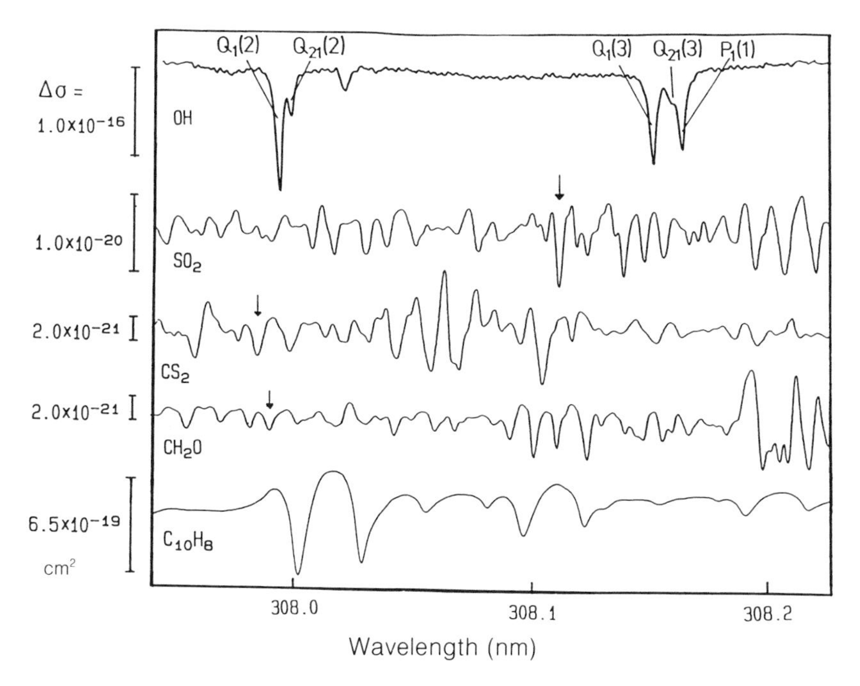

Figure 2.16. OH absorption spectrum (uppermost trace) together with the spectra of several interfering species (SO_2, CS_2, CH_2O, and naphthalene). Vertical bars indicate the differential absorption cross sections of the various species. Note that the cross sections of the interfering species are several orders of magnitude lower than that of OH.

spectra due to NO_2, O_3, and CH_2O, respectively, were simultaneously fitted to the atmospheric spectrum. After the coefficients are thus determined the suitable scaled logarithms of the NO_2 and CH_2O reference spectra are substracted, leaving a residual spectrum clearly showing O_3 absorption features (two center traces in Fig. 2.13). An analogous operation with the NO_2 and O_3 reference spectra indicates that CH_2O absorption features are also present in the spectrum (two bottom traces in Fig. 2.13). A similar deconvolution process of overlapping features of NO_2 and HONO is illustrated by Fig. 2.15 (Platt and Perner, 1983). Another example, a section of the OH absorption spectrum, is presented in Fig. 2.16 together with the spectra of several interfering species (Hübler et al., 1984; Neuroth et al., 1991). It can be seen that the interference by these species can be readily eliminated, since the OH spectrum consists of several isolated groups of lines whereas the interfering absorption features are more or less evenly distributed over the observed spectral interval. Therefore the strengths of the interfering spectra can be quite accurately measured in the "gaps" between the OH lines. Once known in strength the interfering spectra can easily be substrated from the observed spectra.

2.6.4. Optimum Light Path

While for a given minimum detectable optical density the detection limit improves proportionally to the length of the light path, the actual detection limit will not always improve with a longer light path. This is because with a longer light path the received light intensity $I(\lambda)$ tends to be lower, increasing the noise associated with the measurements of $I(\lambda)$ and thus increasing the minimum detectable optical density. In the following discussion we consider a few simple points that may clarify this relationship and also enable us to derive the detection limits.

For a typical DOAS system, as described above, there are no geometrical light losses up to path lengths of about 2 km (for longer light paths, the size of the light source mirror should be increased). However, in addition to geometrical light losses, light attenuations due to atmospheric absorption as well as to scattering affect the magnitude of the received light signal:

$$I_{\text{received}} = I_{\text{source}} \exp(-L/L_0) \tag{2.25}$$

where the absorption length L_0 reflects the combined effects of broad-band atmospheric absorption and Mie and Rayleigh scattering as given in the second term of Eq. (2.8):

$$L_0 = 1/[\Sigma(\sigma_{i0}(\lambda)c_i) + \varepsilon_R(\lambda) + \varepsilon_M(\lambda)] \tag{2.26}$$

Since the noise level of a photomultiplier signal is essentially dependent on the number of photons received and thus is proportional to the square root of the intensity, the signal-to-noise ratio of a DOAS system as a function of the light path length L can be expressed as

$$D'/N \approx L \exp(-L/2L_0) \cdot G(L) \tag{2.27}$$

where $G(L)$ is constant (regime 1), $G(L) = L^{-1}$ (regime 2), or $G(L) = L^{-2}$ (regime 3), respectively (see Section 2.4.3). Relationship (2.27) yields optimum D'/N ratios for light path lengths of $L = 2L_0$ (regime 1), $L = L_0$ (regime 2), or a boundary maximum at the shortest light path (regime 3). While there are no lower limits for L_0 in the atmosphere (fog), the upper limits are given by Rayleigh scattering and the scattering by atmospheric background aerosol, which indicate optimum light path lengths ($2L_0$) in excess of 10 km for wavelengths above 300 nm.

Another point to consider is the width Γ of the absorption line (or band). At a given spectral intensity of the light source $N(\lambda)$, the light intensity around the line center will be $N(\lambda)\,d\lambda \approx N(\lambda)\Gamma$ (in photons per nanometer and second); thus D_0 becomes

$$D_0 \approx [1/(N(\lambda)\Gamma t) + B^2]^{1/2} \tag{2.28}$$

From Eq. (2.28) it becomes clear that the detection of a spectrally narrow absorption line requires a proportionally higher spectral intensity of the light source (see Section 2.4.5).

2.6.5. Optimum Resolution

For a given (grating) spectrometer the light throughput varies in proportion to the square of the spectral resolution (here expressed as the width of the instrument function Γ_0 in nanometers); thus the light intensity at the output becomes

$$I \approx I_i/\Gamma_0^2 \tag{2.29}$$

The "signal" in the spectrum is the differential optical density D', which (for a given species, wavelength, and light path) is proportional to the concentration. In the case of shot noise limitation (see Section 2.6.4) the minimum detectable optical density D_0 in the spectrum is inversely proportional to the square root of I:

$$D_0 \alpha I^{-1/2} \propto 1/\Gamma_0 \tag{2.30}$$

In general the signal as a function of the spectral resolution will vary in proportion to the differential absorption cross section σ' (R) as a function of resolution. Figures 2.17 and 2.18 show two typical bands of NO_2. While $\sigma'(\Gamma_0)$ in principle can be an arbitrary function of Γ_0 (decreasing with increasing Γ_0), in the example of Fig. 2.18 $\sigma'(\Gamma_0)$ can be simply approximated as a linear function:

$$\sigma'(\Gamma_0) = f(\Gamma_0) \approx \sigma_0'(1 - a\Gamma_0) \tag{2.31}$$

where a is a constant (in the example of Fig. 2.18, $a \approx 2.4\ \text{nm}^{-1}$). Thus in this special case the signal-to-noise ratio becomes

$$D'/N = \sigma_0'(1 - a\Gamma_0)\cdot\Gamma_0 \propto \Gamma_0 - a\Gamma_0^2 \tag{2.32}$$

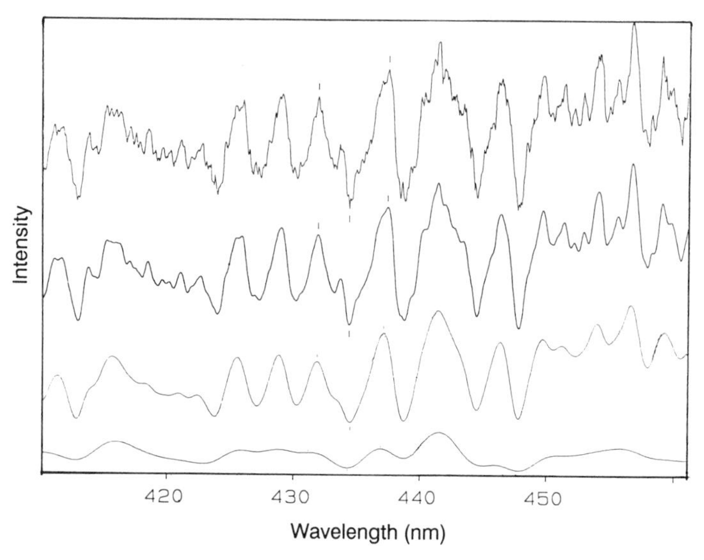

Figure 2.17. NO_2 spectra at different spectral resolution for the same NO_2 column density. Traces from top to bottom: 0.01-, 0.3-, 1.0-, and 3.0-nm resolution. The apparent absorption cross section decreases by almost one order of magnitude, when the resolution is reduced from 0.3 to 3 nm.

The optimum D'/N (in this case) is thus obtained at the resolution

$$d/d\Gamma_0(D'/N) = 1 - 2a\Gamma_0 = 0 \tag{2.33a}$$

or

$$\Gamma_{0\text{opt}} = 1/(2a) \approx 1.2\,\text{nm} \tag{2.33b}$$

where $a \approx 2.4\,\text{nm}^{-1}$ (see Fig. 2.18). Note that in general the instrument function $W(\lambda)$ will not be of Gaussian shape; the spectrum $D_W(\lambda)$ of a given trace species (expressed as optical density) as seen by a particular spectrometer is obtained by folding the true spectrum $D_T(\lambda)$ with the instrument function:

$$D_W(\lambda) = \int D_T(\lambda) \cdot W(\lambda - \lambda')d\lambda' \tag{2.34}$$

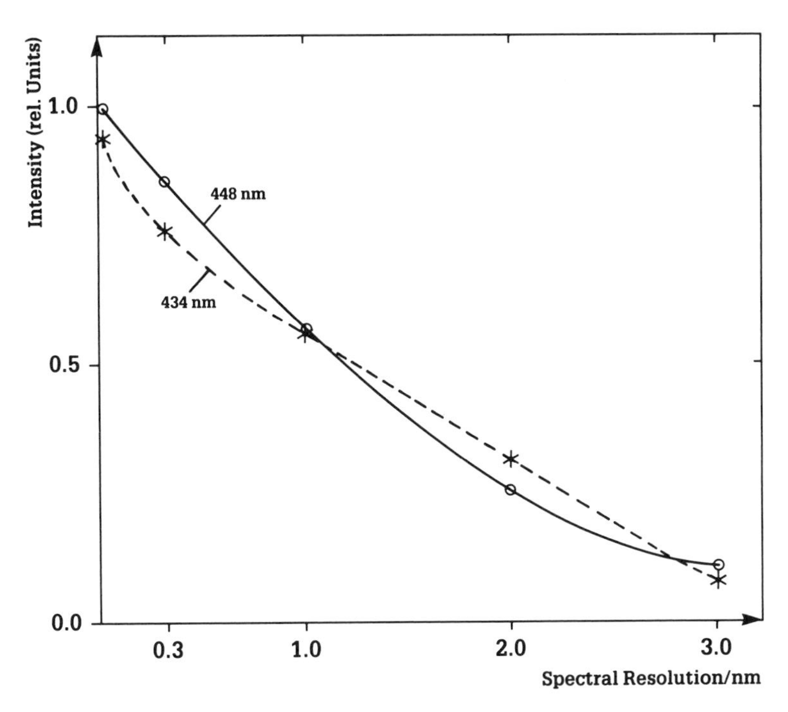

Figure 2.18. Optical density of the spectra shown in Fig. 2.17 as a function of spectral resolution. In this example $D'(\Gamma_0)$, proportional to $\sigma'(\Gamma_0)$, can be simply approximated as a linear function $\sigma'(\Gamma_0) = f(\Gamma_0) \approx \sigma'_0(1 - a\Gamma_0)$, where a is a constant (here $a \approx 2.4\,\text{nm}^{-1}$).

2.7. SAMPLE APPLICATIONS

Applications of DOAS measurements of atmospheric trace species range from air pollution control to the detection of very rare atmospheric species. In particular, the absence of wall losses makes the techniques especially well suited for the detection of unstable species that could be lost at the inner surfaces of tubes used to bring sample air into analysis instrumentation. The first reliable detections of species like nitrous acid [HONO (Perner and Platt, 1979; Platt et al., 1980b; Harris et al., 1982; Atkinson et al., 1986; Pitts et al., 1984a, b, 1985; Perner et al., 1987b; Appel et al., 1990; Winer and Biermann, 1991)], NO_3 radicals (see Section 2.7.4), and OH radicals (see Section 2.7.5) were made by DOAS techniques. Further applications include the determination of vertical gradients of SO_2 (Platt, 1978) as well as of NO_2, O_3, and NO_3 (Brauers et al., 1989) using light paths running at different height ranges above the ground. Also vertical fluxes of atmospheric species could be determined by employing DOAS sensors in eddy-accumulation-type arrangements (Nestlen et al., 1981).

2.7.1. Observation of Air Pollution

Applications of DOAS instruments include the measurement of SO_2, NO_2, O_3, and formaldehyde (CH_2O) in polluted air masses. Figure 2.19 shows a synopsis of the absorption spectra of the above species in the spectral region from 280 nm to 485 nm. Examples of atmospheric spectra taken at light path lengths of several kilometers in the near-UV are given in Figs. 2.15 and 2.20. In the spectral range from 200 to 230 nm the absorption features of NO and NH_3 (together with the stronger $X \rightarrow C^1B_2$ system of SO_2) becomes accessible (Figs. 2.14 and 2.21). Owing to the strong attenuation (mostly by Rayleigh scattering and O_2 and O_3 absorption) of short-wavelength UV light, only light path lengths on the order of a few 100 m are possible.

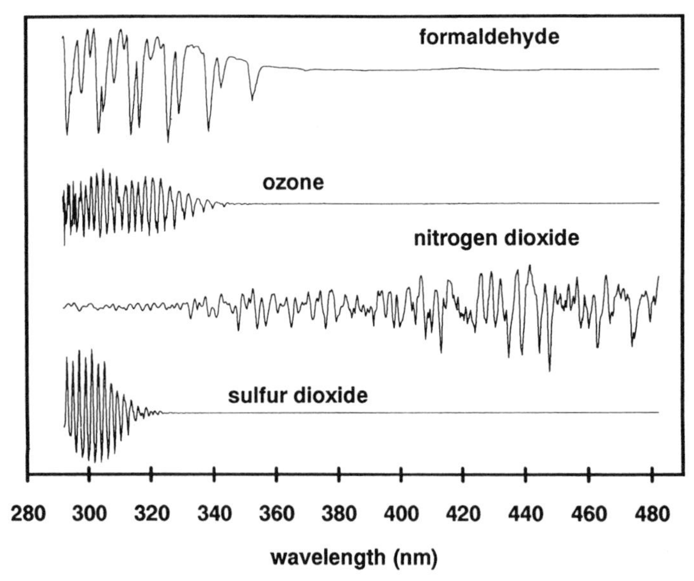

Figure 2.19. Spectra of CH_2O, O_3, NO_2, and SO_2 in the spectral range of 280–485 nm. Strengths of absorptions are not to scale. Note the strong, virtually interference-free features of NO_2 at wavelengths above 360 nm. From Brauers, unpublished results (1992).

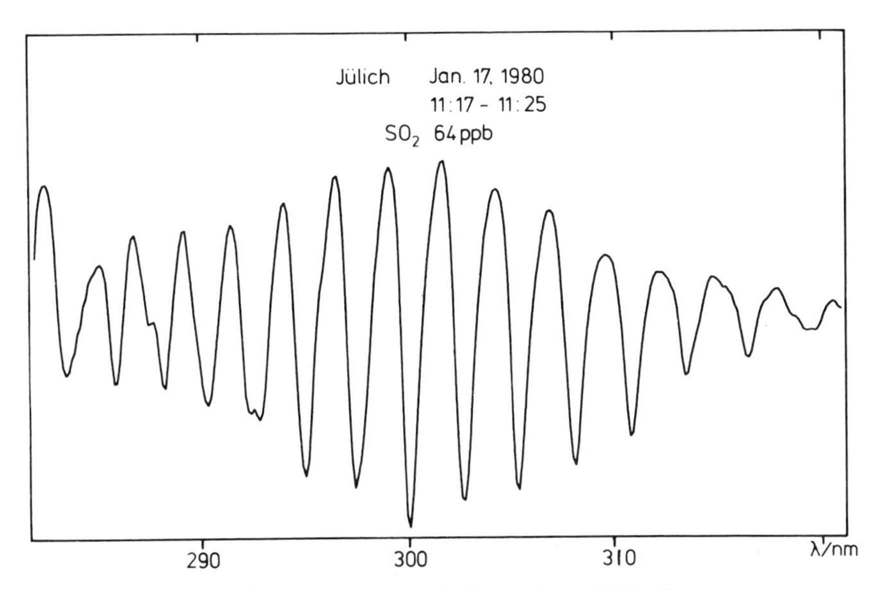

Figure 2.20. SO_2 spectrum around 300 nm taken at Jülich, Germany.

2.7.2. Observation of the NO_x–O_3 Photostationary State

Simultaneous observation in both above spectral regions would, for instance, allow the determination of the NO_x–O_3 photostationary state as mediated by the following reactions with a single instrument:

$$NO + O_3 \longrightarrow NO_2 + O_2 \qquad (R2.1)$$

$$NO_2 + h\nu \longrightarrow NO + O \qquad (R2.2)$$

$$O + O_2 + M \longrightarrow O_3 + M \qquad (R2.3)$$

where M indicates a chemically inert constituent like N_2, which acts mainly as an energy-transfer agent. Since all participating species NO, NO_2, and O_3 are measurable by DOAS (see Table 2.1) in the same volume of air, deviations from the photostationary state can give information about other processes oxidizing NO to NO_2 (Parrish et al., 1986).

2.7.3. Investigating Free Radical Reactions in the Atmosphere

Free radicals play a central role in many chemical systems such as flames, living cells, or the atmosphere. Owing to their high chemical reactivity,

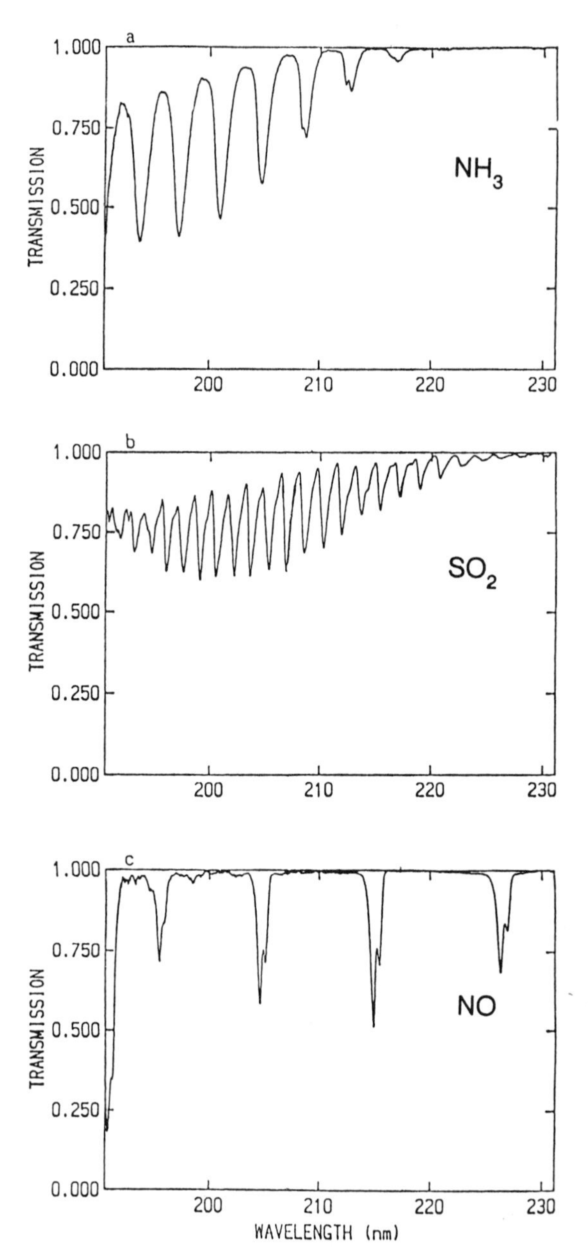

Figure 2.21. Laboratory spectra of NH_3, SO_2, and NO: 200–230 nm. Adapted from Edner et al. (1990).

radicals are the driving force for most chemical processes in the atmosphere. In particular they initiate and carry reaction chains. Thus, knowledge of the concentration of those species in the atmosphere is a key requirement for the investigation of atmospheric chemistry. For instance, degradation of most oxidizable trace gases in the atmosphere, like hydrocarbons, carbon monoxide, or chlorofluorohydrocarbons (CFCs) is initiated by free radicals. The degradation products frequently give rise to further radical production (chain branching) and the formation of secondary oxidants like ozone and peroxides (Logan et al., 1981; Platt et al., 1990). The following free radicals are relevant to chemical processes in the lower atmosphere (troposphere): OH, HO_2, RO_2, (R standing for a organic radical like CH_3), NO_3, and possibly the halogen oxide radicals IO and BrO. Table 2.3 summarizes data on concentration and reactivity of some of these radicals.

As can be seen from Table 2.3 the radicals OH and NO_3 exhibit the highest reactivity (product of typical concentration and the reaction rate constant) toward atmospheric species. As a consequence of their high reactivity the steady state concentration of free radicals in the atmosphere is generally quite low, even compared to atmospheric trace gas standards. For instance, the atmospheric lifetime of the OH radical never exceeds 1 s. On the other hand, a consequence of this short lifetime is that transport processes can be largely neglected when one is considering the budgets of most free radicals. Therefore, measurement of the instantaneous concentration of free radicals together with

Table 2.3. Typical Atmospheric Concentrations and Reaction Rate Constants of Some Radicals[a]

	Radical			
	OH	HO_2	CH_3O_2	NO_3
Typical Atmospheric Concentration (molecules/cm^3)				
Average	5×10^5	4×10^7	2×10^7	10^7
Maximum	1×10^7	5×10^8	1×10^8	1×10^{10}[b]
Reaction with Trace Species (rate constants, cm$^3 \cdot$ molecules$^{-1} \cdot s^{-1}$)				
CO	2.4×10^{-13}[c]	≪	≪	$< 4 \times 10^{-16}$[d]
n-Butane	2.4×10^{-12}	≪	≪	6.5×10^{-17}
Isoprene	1.0×10^{-10}	≪	≪	8.2×10^{-13}
α-Pinene	5.4×10^{-11}	≪	≪	5.8×10^{-12}

[a] Data from Atkinson (1990), if not otherwise noted.
[b] Observed values.
[c] Atkinson et al. (1989).
[d] Burrows et al. (1985).

Table 2.4 Data Pertinent to the Spectroscopic Detection of OH Radicals and NO_3 Radicals

Parameter	Radical	
	OH	NO_3
Wavelength of strongest band or line (λ/nm)	308	662
Line width (Γ/nm)	0.0017	3.2
Effective absorption cross section (σ/cm^2)	1.3×10^{-6}	1.8×10^{-17}
Minimum detectable optical density (D_0)	10^{-4}	10^{-4}
Detection limit [C_0, molecules/cm^3 ($l = 5$ km)]	1.5×10^6	1.1×10^7
Interfering molecules	SO_2, CH_2O, CS_2, $C_{10}H_8$	H_2O

a comprehensive set of other trace gases linked with the radical's chemical cycles enables to study in situ of chemical reactions in the atmosphere (Perner et al., 1987a; Platt et al., 1987, 1988; Dorn et al., 1988; Mount, 1992).

To date the results obtained by absorption spectroscopy appear to be more convincing than the data obtained by other techniques. In the case of NO_3 radicals visible absorption spectroscopy also is the most commonly used technique to measure the atmospheric concentration of this species (Noxon et al., 1978, 1980; Noxon, 1983; Platt et al., 1980a, 1981, 1984, 1989; Plane and Nien, 1992).

As discussed above, the design of a spectroscopic system for the detection of atmospheric trace species is strongly influenced by the nature of the absorption spectrum of the particular molecules of interest. Table 2.4 illustrates this point, showing for instance that the width of the absorption bands of OH and NO_3 differ by roughly 3 orders of magnitude. Consequently the design of the optical setup is quite different for both molecules.

2.7.4. Measurement of NO_3 Radicals by DOAS

Owing to the large width of the NO_3 absorption bands, a thermal light source in conjunction with a parabolic mirror gives sufficient spectral intensity to reach a minimum detectable optical density of $D_0 = 10^{-4}$ (see also Section 2.6.2). Also a relatively small spectrometer with a spectral resolution of about 0.6 nm can be used (Platt et al., 1981, 1984). Figures 2.22 and 2.23 give examples of NO_3 absorption bands observed in clean and polluted air, respectively. Possible interference by absorption lines of water vapor are easily eliminated because the water lines are much narrower than the NO_3 band. In particular "pure" interference spectra can be obtained during daytime, when the concentration of NO_3 radicals is far below the detection limit, owing to

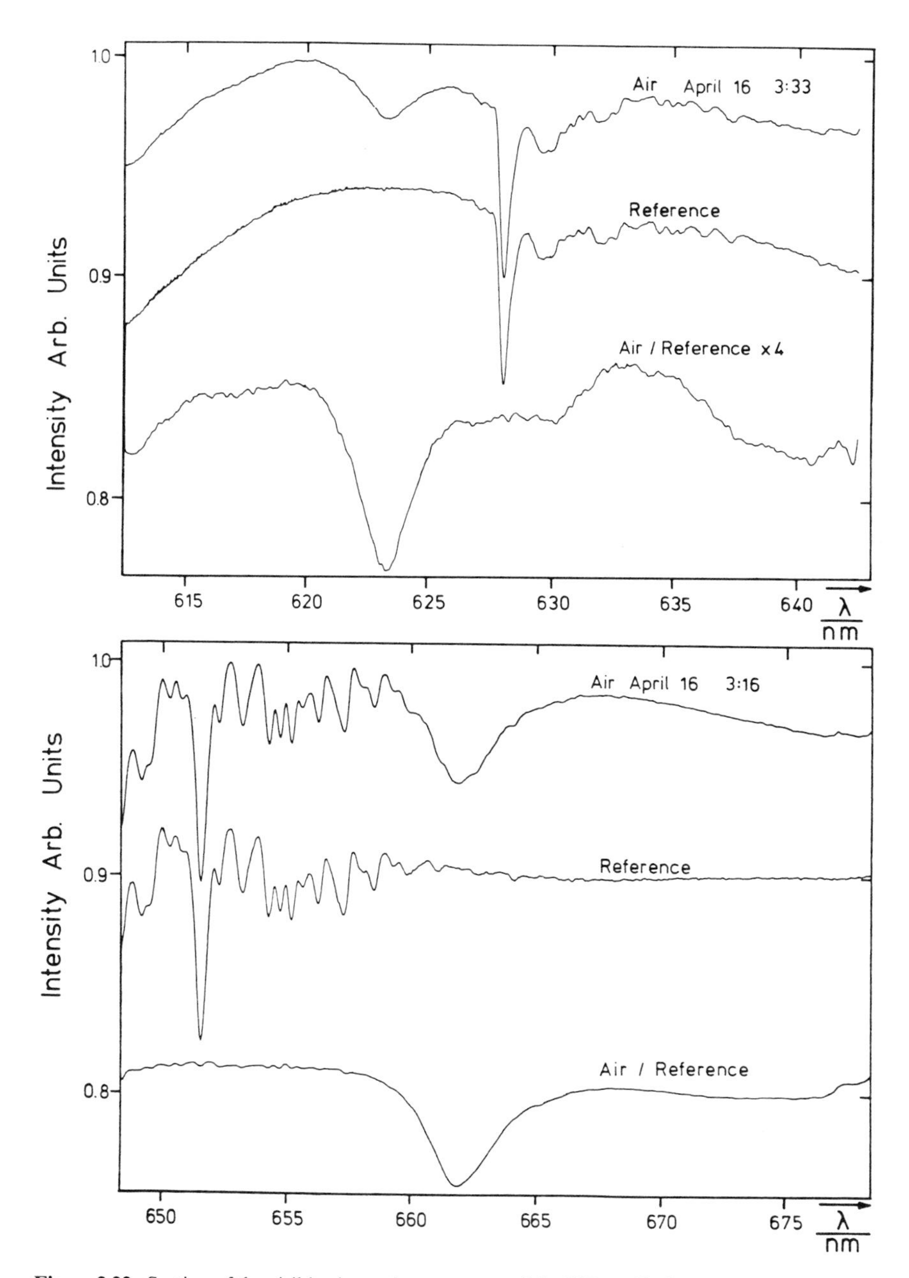

Figure 2.22. Section of the visible absorption spectrum of the NO_3 radical (upper panel: 613–642 nm; lower panel: 649–678 nm); the absorption corresponds to about 1.5×10^9 molecules/cm^3 (60 ppt) at a 4.8-km light path in Deuselbach/Hunsrück, Germany. Uppermost traces: spectrum as measured in the atmosphere (April 16, 1980). Center traces: "reference spectrum" taken the following day; owing to rapid photolysis, the NO_3 concentration is well below 10^7 molecules/cm^3, and thus only absorption structures of H_2O remain. Lowest traces: ratio of above spectra; water lines are removed. Adapted from Platt et al. (1981).

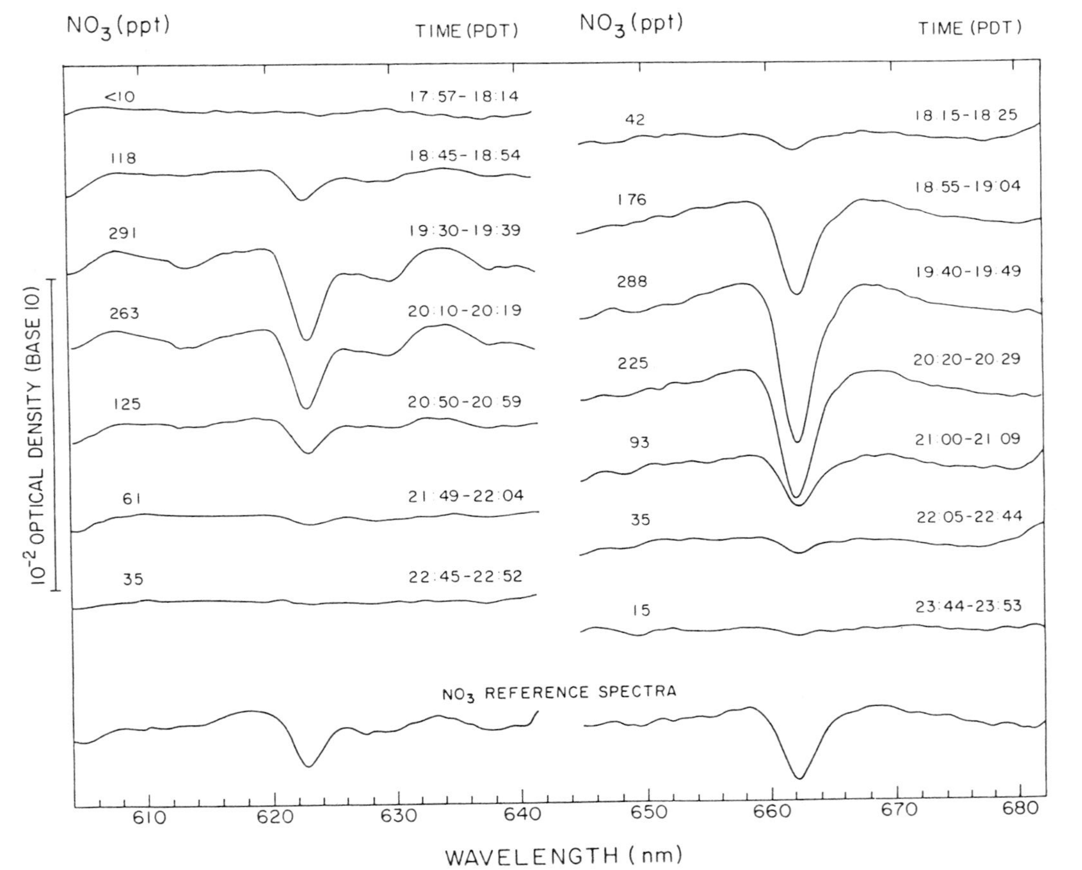

Figure 2.23. Development of the visible NO_3 absorption bands at 623 and 662 nm during the evening of Sept. 12, 1979. PDT = Pacific Daylight Time. Adapted from Platt et al. (1980a).

their rapid photolytic destruction (see R 2.5a,b, below). With this setup, routine measurements of NO_3 radical concentrations have been made yielding nighttime concentrations of up to 400 ppt ($\approx 10^{10}$ molecules/cm^3) (Platt et al., 1980a, 1981, 1982, 1989; Pitts et al., 1984c; Biermann et al., 1988). Figures 2.24 and 2.25 give typical nighttime NO_3 concentration profiles.

The ultimate goal of the development of measurement techniques for atmospheric trace species is, of course, the investigation of chemical reaction systems. In order to examine the type of problem that might be solved by observation of the atmospheric concentration of free radicals, an overview of the sources and sinks of the two radicals described earlier (in Section 2.7.3) will now be given.

The prerequisite for nitrate radical production is the simultaneous presence of nitrogen dioxide and ozone in the same air mass (Platt et al., 1990; Wayne et al., 1991). Reaction of NO_2 with ozone will then form NO_3 radicals:

$$NO_2 + O_3 \longrightarrow NO_3 + O_2 \qquad \text{(R2.4)}$$

During daytime NO_3 radicals are rapidly destroyed by photolysis.

$$NO_3 + h\nu \longrightarrow NO + O_2 \qquad \text{(R2.5a)}$$
$$\longrightarrow NO_2 + O \qquad \text{(R2.5b)}$$

The photolytic lifetime is about 5 s. Another daytime loss mechanism of NO_3 is reaction with NO. At night the NO_3 lifetimes and thus concentrations often

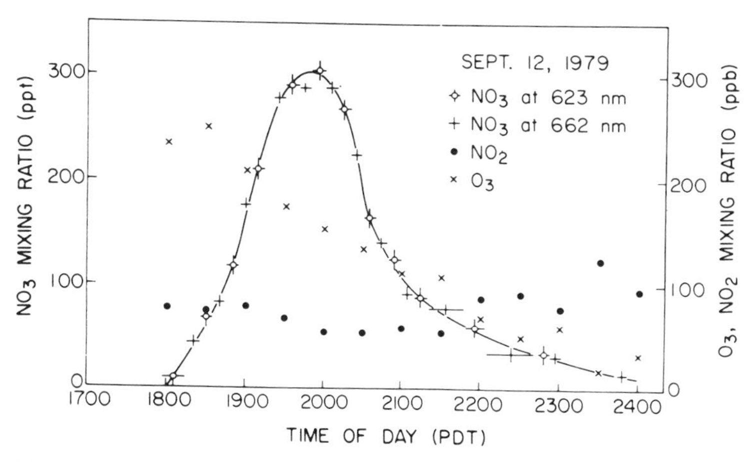

Figure 2.24. Development of the NO_3 concentration as determined from the spectra shown in Fig. 2.23. PDT = Pacific Daylight Time. Adapted from Platt et al. (1980a).

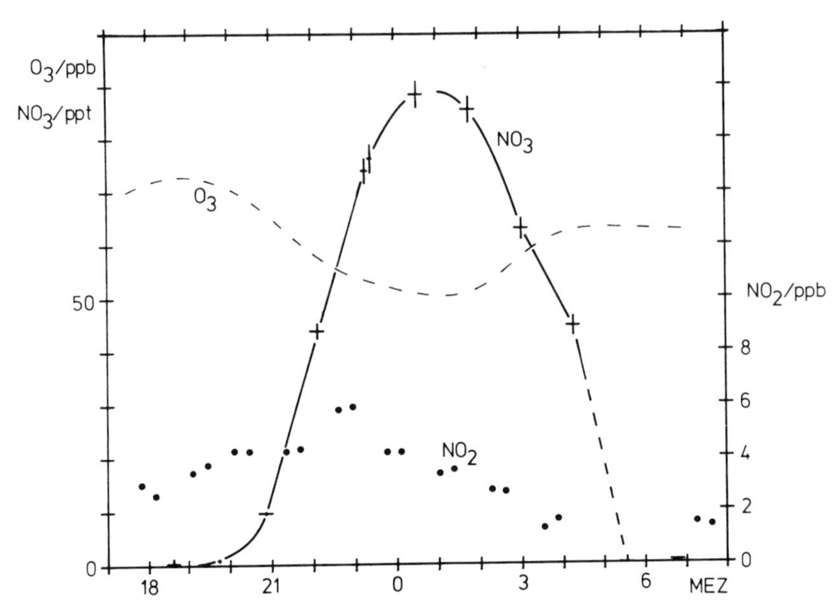

Figure 2.25. Typical nighttime NO_3 concentration profiles measured by low-resolution DOAS at Deuselbach/Hunsrück, Germany, on June 7/8, 1983; light path, 4.8 km. *Key*: MEZ, Central European time.

become much larger; the main destruction mechanisms are then reactions with organic species (see Table 2.3) or heterogeneous losses (i.e., reactions at the ground or at the surface of aerosol or cloud particles) of either NO_3 or N_2O_5, which is in equilibrium with NO_3 and NO_2:

$$NO_2 + NO_3 \longleftrightarrow N_2O_5 \tag{R2.6}$$

Under "moderately polluted" conditions typical for rural areas in industrialized countries NO_3 and N_2O_5 concentrations are of the same order of magnitude. A heterogeneous reaction (het) with water will convert N_2O_5 to nitric acid:

$$N_2O_5 + H_2O \xrightarrow{\text{het}} HNO_3 \tag{R2.7}$$

From measured concentrations of O_3, NO_2, and NO_3 the atmospheric lifetime of NO_3 as limited by the combination of any first-order loss process

$$NO_3 + X \longrightarrow \text{products} \tag{R2.8}$$

can be calculated:

$$d/dt\,[NO_3] = [O_3]\cdot[NO_2]\cdot k_5 - [NO_3]\cdot\Sigma([X_i]\cdot k_i)$$

Assuming stationary state conditions with respect to NO_3, the NO_3 lifetime τ_{NO_3} becomes (Platt et al., 1980a, 1989)

$$\tau_{NO_3} = [NO_3]/([O_3]\cdot[NO_2]\cdot k_5) \quad (2.35)$$

Since k_5 is known from laboratory measurements and the concentrations of O_3, NO_2, and NO_3 can be simultaneously measured by the DOAS technique, τ_{NO_3} can readily be calculated for various atmospheric conditions.

2.7.5. Measurement of OH Radicals by DOAS

Hydroxyl radicals are mainly produced by photolysis of ozone to yield excited oxygen atoms [$O(^1D)$], which in turn react with water vapor:

$$O(^1D) + H_2O \longrightarrow OH + OH \quad (R2.9)$$

OH radicals rapidly react with most organic species, the products being further radicals (peroxy radicals), which in turn reproduce HO_2 or OH radicals. While those reaction sequences are generally quite complex, only two reactions account for most of the "final" loss of HO_x (sum of OH and HO_2 radicals): the recombination of HO_2—

$$HO_2 + HO_2 \longrightarrow H_2O_2 + O_2 \quad (R2.10)$$

and the reaction of OH with NO_2—

$$OH + NO_2 + M \longrightarrow HNO_3 + M \quad (R2.11)$$

where M is as in R 2.3. A series of OH measurement campaigns were performed with the DOAS technique using optomechanical scanning of the spectra (see Section 2.5.1; also Perner et al., 1987a; Platt et al., 1988) as well as (in later versions of the instrument) diode array detectors (Dorn et al., 1988; Hofzumahaus et al., 1991). Observed levels of OH radicals ranged from the detection limit ($1-2\times10^6$ molecules/cm^3) up to about 10^7 molecules/cm^3. Figure 2.26 shows a typical diurnal profile of the OH concentration.

Note that NO_3 and OH are in a sense complementary to each other: OH radicals are directly produced by photochemical reactions; thus their concentration will be highest at noontime (see Fig. 2.26) and very low at night.

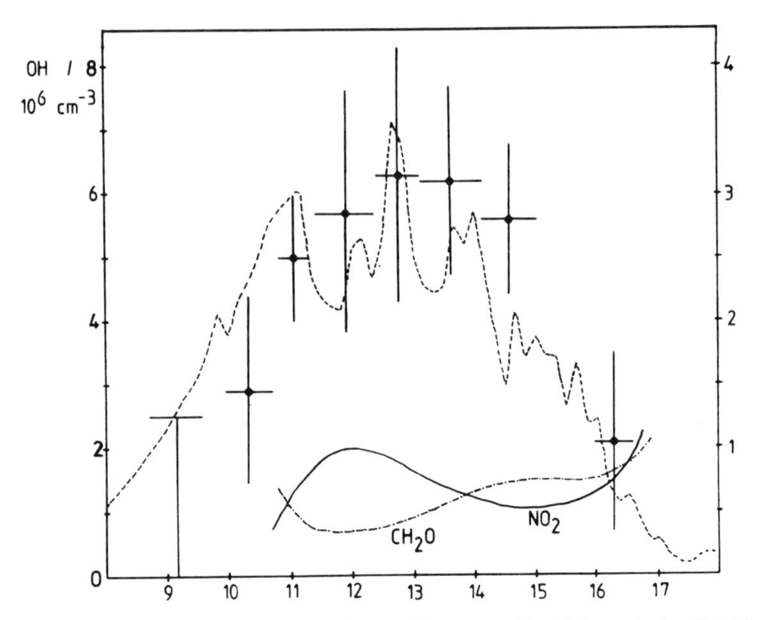

Figure 2.26. Typical daytime OH concentration profile measured by high-resolution DOAS with a frequency-doubled dye laser as light source in the Schauinsland, Black Forest, Germany, on June 25, 1984; light path, 4.8 km. Adapted from Platt et al. (1988).

Conversely nitrate radicals are destroyed by daylight; thus they will reach high concentrations only at night (see Fig. 2.25) whereas their daytime levels are negligible. On the other hand, important pathways for the removal of both NO_3 and OH radicals lead to the formation of nitric acid. Owing to the stability of the HNO_3 molecule and its good solubility in water, nitric acid is the "final fate" of nitrogen oxides in the atmosphere. Measurement of the concentration of both radicals enables us, for instance, to compare the relative efficiency of the two radicals in converting NO_x to nitric acid (and thus initiating the removal of NO_x from the atmosphere). Figure 2.27 shows a comparison of the rates of NO_2 to HNO_3 conversion (in percent per hour) as a consequence of R 2.4 followed by R 2.6/R 2.7 and R 2.10, respectively. Whereas the OH mechanism shows strong diurnal and seasonal variations, the NO_3–N_2O_5 mechanism is relatively independent of season. It is also interesting to note that the temperature dependence of the reaction of O_3 with NO_2 (R 2.4) is compensated by the longer nights in the winter.

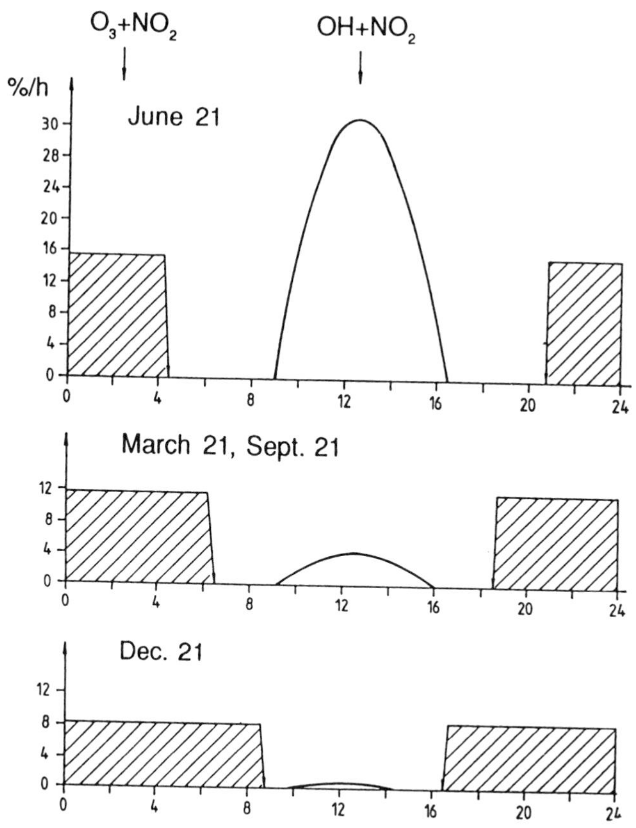

Figure 2.27. Comparison of the calculated NO_2-to-nitric acid conversion rates due to the NO_3–N_2O_5 mechanism (R2.4 followed by R2.6 and R2.7) and the OH reaction (R2.10), respectively. Note that the NO_3–N_2O_5 mechanism is relatively independent of the season. In the case of the NO_3–N_2O_5 mechanism the temperature dependence of the reaction of $O_3$with NO_2 (R2.4) is compensated by the longer nights in the winter.

2.7.6. Determination of Total Column Densities by Means of Extraterrestrial Light Sources

Raw spectra of zenith-scattered sunlight (or moon- or starlight in UV–visible range are dominated by strong Fraunhofer lines (optical densities of up to 0.5). In order to remove the Fraunhofer structure spectra taken at different zenith angles (of the observed celestial body) are always ratioed. In Fig. 2.28a sample spectra of zenith-scattered light taken at 75° and 90° solar zenith angles are shown. Only after the spectra of Fig. 2.28a are ratioed do the much weaker

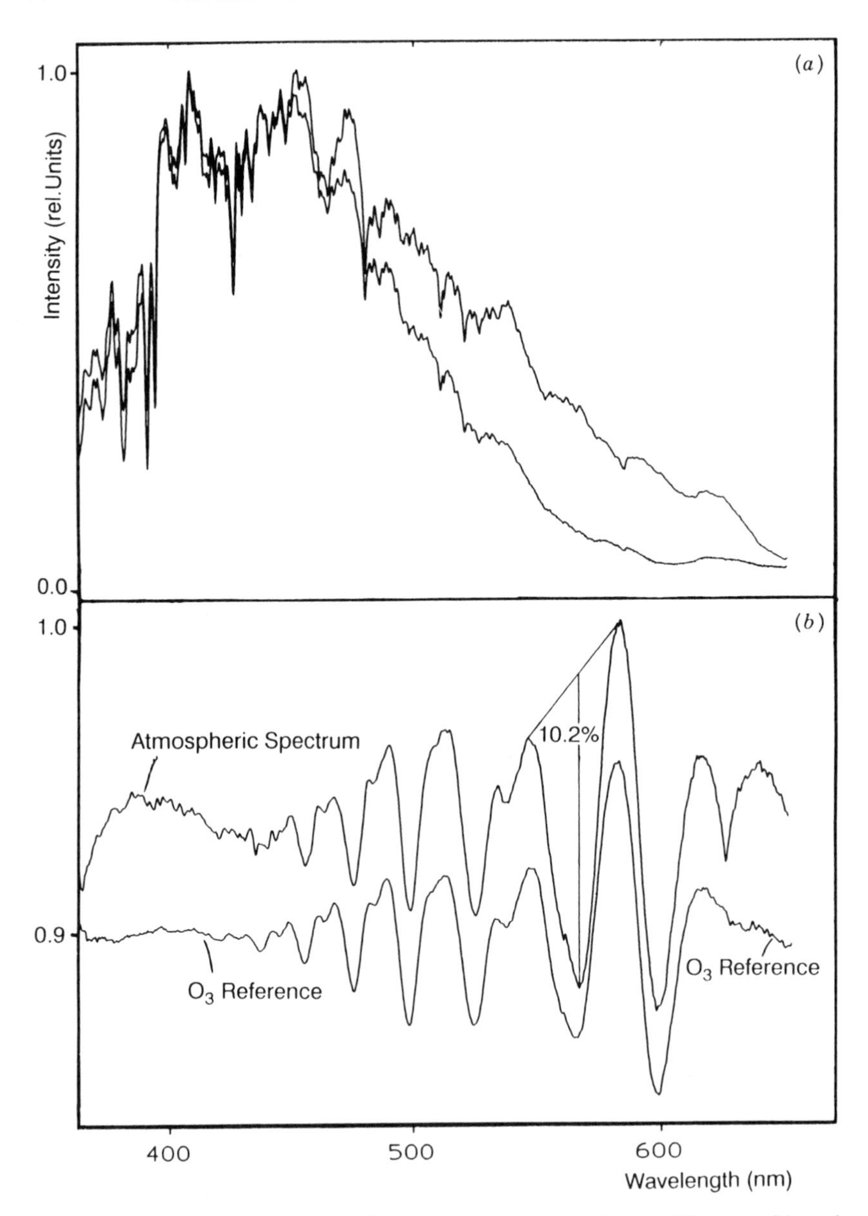

Figure 2.28. Zenith-scattered sunlight light spectra. (a) Spectra taken at different zenith angles; note the dominating Fraunhofer structure. (b) Ratio of spectra taken at large and small zenith angles; atmospheric absorption features due to O_3 (strong Chappuis bands), NO_2 (weak), and OClO (very weak) become visible.

atmospheric absorption structures become visible (Fig. 2.28b). Here strong O_3 Chappuis bands, weaker NO_2 visible bands, and very weak absorptions due to OClO are discernible. Figure 2.29 shows an example of vertical column data extracted from zenith sky spectra taken in Kiruna, Sweden, at Feb. 4–8, 1990, using the optical arrangement described in Section 2.4.6. The data are plotted together with the solar zenith angle, which varied between 84° and 95°. The (3σ) errors of the O_3 columns shown result only to a minor extent from the error in the measured slant column, but mainly from the error of the air mass factor (AMF).

A potential source of systematic errors in calculating the AMF is the influence of Mie scattering by stratospheric particles [polar stratospheric clouds (PSCs) or background aerosol]. Mie extinction in the stratosphere alters the relative intensities of light beams, reaching the zenith at different heights and therefore carrying different amounts of information about the trace gas layer. The relative contributions of these beams, given by their

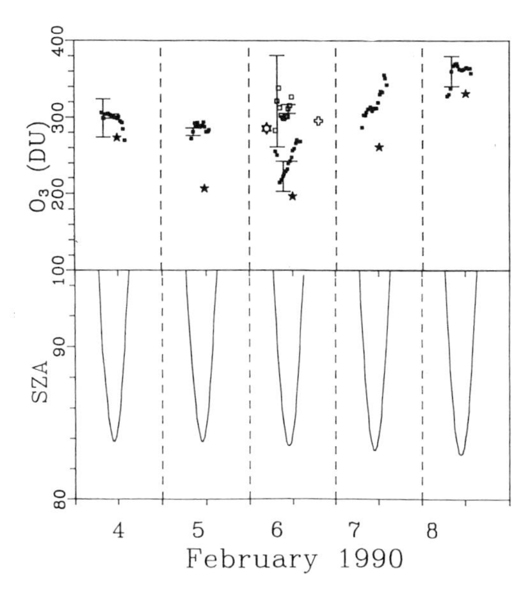

Figure 2.29. Vertical column densities for O_3 evaluated with standard AMF neglecting Mie extinction (filled squares, upper panel) and the corresponding solar zenith angle (lower panel) vs. universal time for Kiruna, Sweden (68 °N), Feb. 4–8, 1990. Error bars show typical 3σ uncertainties. *Key*: open squares; ozone columns calculated using AMFs including Mie scattering; filled stars, Total Ozone Mapping Spectrometer (TOMS) data; open star, integrated ozone sonde data from Hofmann and Deshler (1990); open cross, integrated O_3 sonde value (CHEOPS ozone pool); SZA, solar zenith angle; DU, Dobson units. Adapted from Fiedler et al. (1992).

intensities, determines the average optical density observed by a zenith-observing spectrograph.

2.8. CONCLUSIONS

Long-path differential optical absorption spectroscopy has proven to be an extremely useful technique for the detection of atmospheric trace gases and still has much development potential.

The strengths of the DOAS technique include the absence of wall losses, good specificity, and the potential for real-time measurements. Quite unique are its "inherent calibration" properties. The required differential absorption cross sections of the molecules of interest can be measured in the laboratory; therefore no calibration in the field, for instance, by preparing calibration gases, is required. The influence of the instrument function (see Sections 2.4 and 2.6.5) is corrected mathematically.

In particular the absence of wall losses makes spectroscopic techniques especially well suited for the detection of unstable species like OH radicals.

Although a large number of trace gases important for atmospheric chemistry and pollution control (see Table 2.1) can be sensitively measured by DOAS, many species do not exhibit suitable UV–visible absorption structures and thus are not detectable by DOAS. Another problem can be the limitation of DOAS measurements by conditions of poor visibility.

In the case of unfolded path arrangements the large measurement volume can, for certain applications, complicate the interpretation of the measurement results. Also, logistic requirements, (the need to set up instrumentation at two sites separated by several kilometers but in sight of each other) are sometimes cumbersome.

Future developments will no doubt include measurement at shorter wavelengths (below 300 nm) and possibly at low pressure. Folded path arrangements will likely gain importance. Also, the potential of diode array detectors is not yet fully explored.

REFERENCES

Amerding, W., Herbert, A., Schindler, T., Spiekermann, M., and Comes, F. J. (1990). In situ measurements of tropospheric OH radicals—A challenge for the experimentalist. *Ber. Bunsenges. Phys. Chem.* **94**, 776–781.

Amerding, W., Herbert, A., Spiekermann, M., Walter, J., and Comes, F. J. (1992). Ein schnell durchstimmbares Laserspektrometer. *Ber. Bunsenges. Phys. Chem.* **96**, 314.

Appel, B. R., Winer, A. M., Tokiwa, Y., and Biermann, H. W. (1990). Comparison of atmospheric nitrous acid measurements by annular denuder and differential optical absorption systems. *Atmos. Environ.* **24A**, 611–616.

Atkinson, R. (1990). Gas-phase tropospheric chemistry of organic compounds: A review. *Atmos. Environ.* **24A**, 1–41.

Atkinson, R., Carter, W. P. L., Pitts, J. N., Jr., and Winer, A. M. (1986). Measurements of nitrous acid in an urban area. *Atmos. Environ.* **20**, 408–409.

Atkinson, R., Baulch, D. L., Cox, R. A. Hampson, R. F., Kerr, J. A., and Troe, J. (1989). Evaluated kinetic and photochemical data for atmospheric chemistry: Supplement III. *J. Phys. Chem. Ref. Data* **18**, 881–1095.

Axelsson, H., Galle, B., Gustavsson, K., Regnarsson, P., and Rudin, M. (1990). A transmitting/receiving telescope for DOAS-measurements using retroreflector technique. *Dig. Top. Meet. Opt. Remote Sens. Atmos., OSA*, **4**, 641–644.

Biermann, H. W., Tuazon, E. C., Winer, A. M., Wallington, T. J., and Pitts, J. N. (1988). Simultaneous absolute measurements of gaseous nitrogen species in urban ambient air by long pathlength infrared and ultraviolet-visible spectroscopy. *Atmos. Environ* **22**, 1545–1554.

Bonafe, U., Cesari, G., Giovanelli, G., Tirabassi, T., and Vittori, O. (1976). Mask correlation spectrophotometry advanced methodology for atmospheric measurements. *Atmos. Environ.* **10**, 469–474.

Bongartz, A., Kames, J., Welter, F., and Schurath, U. (1991). Near–UV absorption cross sections and trans/cis equilibrium of nitrous acid. *J. Phys. Chem.* **95**, 1076–1082.

Brand, J. C. D., and Nanes, R. (1973). The 3400–3000 Å absorption of sulfur dioxide. *J. Mol. Spectrosc.* **46**, 194–199.

Brand, J. C. D., and Srikameswaran, K. (1972). The $\Pi^*-\Pi$ (2350A) band system of sulphur dioxide. *Chem. Phys. Lett.* **15**, 130–132.

Brand, J. C. D., Jones, V. T., and DiLauro, C. (1973). The $^3B_1-^1A_1$ band system of sulfur dioxide: Rotational analysis of the (010), (100), and (110) bands. *J. Mol. Spectrosc.* **45**, 404, 411.

Brassington, D. J. (1981). Sulfur dioxide absorption cross-section measurements from 290 nm to 317 nm. *Appl. Opt.* **20**, 3774–3779.

Brassington, D. J., Felton, R. C., Jolliffe, B. W., Marx, B. R., and Moncrieff, J. T. M. (1984). Errors in spectroscopic measurements of SO_2 due to nonexponential absorption of laser radiation, with application to the remote monitoring of atmospheric pollution. *Appl. Opt.* **23**, 469–475.

Brauers, T., Dorn, H. -P., and Platt, U. (1989). Spectroscopic measurements of NO_2, O_3, SO_2, IO and NO_3 in maritime air. *Proc. COST(Coopération européenne dens le domaine de lao Recherche Scientifique et Technique)*. 611 Meet., Varese, Italia, pp. 237–242.

Brewer, A. W., McElroy, C. T., and Kerr, J. B. (1973). Nitrogen dioxide concentrations in the atmosphere. *Nature (London)* **246**, 129–133.

Burrows, J. P., Tyndall, G. S., and Moortgat, G. K. (1985). Absorption spectrum of NO_3 and kinetics of the reactions of NO_3, with NO_2, Cl, and several stable atmospheric species at 298 K. *J. Phys. Chem.* **89**, 4848–4856.

Canrad-Hanovia (1986). "Compact Arc Lamps," data sheet. Canrad-Hanovia Inc., Newark, NJ.

Cantrell, C. A., Davidson, J. A., McDaniel, A. H., Shetter, R. E., and Calvert, J. G. (1990). Temperature-dependent formaldehyde cross section in the near-ultraviolet spectral region. *J. Phys. Chem.* **94**, 3902–3908.

Colin, R., Carleer, M., Simon, P. C., Vandaele, A. C., Dufour, P., and Fayt, C. (1991). "Atmospheric Absorption Measurement by Fourier Transform DOAS," EUROTRAC annual report 1991 (TOPAS subproject), pp. 14–46. Int. Sci. Secr. Garmisch-Partenkirchen, Germany.

Daumont, D., Barbe, D. A., Brion, J. and Malicet, J. (1989). "In Ozone in the atmosphere." (R. D. Bojkov, and Fabian, P., Eds.), (Proc. Quadrennial Ozone Symp., 1988), pp. 710–712. Deepak Publishing, Hampton, VA.

Davidson, J. A., Cantrell, C. A., McDaniel, A. H., Shetters, R. E., Madronich, S., and Calvert, J. G. (1988). Visible-ultraviolet absorption cross sections for NO_2 as a function of temperature. *J. Geophys. Res.* **93D**, 7105–7112.

Davies, J. H., van Egmond, N. D., Wiens, R., and Zwick, H. (1975). Recent developments in environmental sensing with the Barringer correlation spectrometer. *Can. J. Remote Sens.* **1**, 85–94.

Dorn, H. -P., and Platt, U. (1986). Eine empfindliche optische Nachweismethode für Spurenstoffe in der Atmosphäre. *Elektrizitaetswirtschaft* **24**, 967–970.

Dorn, H. -P., Callies, J., Platt, U., and Ehhalt, D. H. (1988). Measurement of tropospheric OH concentrations by laser long-path absorption spectroscopy. *Tellus* **40B**, 437–445.

Edner, H., Sunesson, A., Svanberg, S., Uneus, L., and Wallin, S. (1986). Differential optical absorption spectroscopy system used for atmospheric mercury monitoring. *Appl. Opt.* **25**, 403–409.

Edner, H., Amer, R., Ragnarson, P., Rudin, M., and Svanberg, S. (1990). Atmospheric NH_3 monitoring by long-path UV absorption spectroscopy. *In Environment and Pollution Measurement Sensors and Systems* (H. O. Nielson, Ed.) *Proc. SPIE* **1269**, 14–20.

Fiedler, M., Frank, H., Gomer, T., Hausmann, M., Pfeilsticker, K., and Platt, U. (1990). Groundbased spectroscopic trace gas measurements in the Arctic stratosphere during winter 1989/90. *Proc. Eur. Ozone Workshop*, 1st, Schliersee, pp. 61–64.

Fiedler, M., Frank, H., Gomer, T., Hausmann, M., Pfeilsticker, K., and Platt, U. (1993). Groundbased spectroscopic measurements of stratospheric NO_2 and OClO in Arctic winter 1989/90. *Geophys. Res. Lett.* (in press).

Frank, H., and Platt, U. (1990). Advanced calculation procedures for the interpretation of skylight measurements. *Proc. Eur. Ozone Workshop 1st*, Schliersee, pp. 65–68.

George, G. A., and Morris, G. C. (1968). The intensity of absorption of naphthalene from 30 000 cm^{-1} to 53 000 cm^{-1}. *J. Mol. Spectrosc.* **26**, 67–71.

Hall, C. T., and Blacet, F. E. (1952). Separation of the absorption spectra of NO_2 and N_2O_4 in the range of 2400–5000 Å. *J. Chem. Phys.* **20**, 1745–1749.

Hamada, Y., and Merer, A. J. (1974). Rotational structure at the long wavelength end of the 2900 Å system of SO_2. *Can J. Phys.* **52**, 1443–1457.

Hanst, P. L. (1978). Air pollution measurement by Fourier transform spectroscopy. *Appl. Opt.* **17**, 1360–1366.

Hanst, P. L., Lefohn, A. S., and Gay, B. W. (1973). Detection of atmospheric pollutants at parts-per-billion levels by infrared spectroscopy. *Appl. Spectrosc.* **27**, 188–198.

Harris, G. W., Carter, W. P. L., Winter, A. M., Pitts, J. N., Platt, U., and Perner, D. (1982). Observations of nitrous acid in the Los Angeles atmosphere and implications for the predictions of ozone-precursor relationships. *Environ. Sci. Technol.* **16**, 414–419.

Harris, G. W., Winer, A. M., Pitts, J. N., Platt, U., and Perner, D. (1983). Measurement of HONO, NO_3 and NO_2 by long–path differential optical absorption spectroscopy in the Los Angeles basin: Optical and laser remote sensing. *Springer Ser. Opt. Sci.* **39**, 106–113.

Harris, G. W., Mackay, G. I., Iguchi, T., Mayne, L. K., and Schiff, H. I. (1989). Measurements of formaldehyde in the troposphere by tunable diode laser absorption spectroscopy. *J. Atmos. Chem.* **8**, 119–137.

Haugen (Ed.) (1973). "Workshop on Micrometeorology. " Am. Meteorol. Soc. Sci. Press, Ephrata, PA.

Hofmann, D. J., and Deshler, T. (1990). Balloonborne measurements of polar stratospheric clouds and ozone at −93 °C in the Arctic in February 1990. *Geophys. Res. Lett.* **17**, 2185–2188.

Hofzumahaus, A., Dorn, H. -P., Callies, J., Platt, U., and Ehhalt, D. H. (1991). Tropospheric OH concentration measurements by laser long-path absorption spectroscopy. *Atmos. Environ.* **25A**, 2017–2022.

Horn, D., and Pimentel, G. C. (1971). 2. 5 km low-temperature multiple-reflection cell. *Appl. Opt.* **10**, 1892–1898.

Hübler, G., Perner, D., Platt, U., Tönnissen, A., and Ehhalt, D. H. (1984). Groundlevel OH radical concentration: New measurements by optical absorption. *J. Geophys. Res.* **89D**, 1309–1319.

Johnston, P. V., and McKenzie, R. L. (1984). Long-path absorption measurements of NO_2 in rural New Zealand. *Geophys. Res. Lett.* **11**, 69–72.

Johnston, P. V., and McKenzie, R. V. (1989). NO_2 observations at 45S during the decreasing phase of solar cycle 21, from 1980 to 1987. *J. Geophys. Res.* **94D**, 3473–3486.

Johnston, P. V., McKenzie, R. L., Keys, J. G., and Matthews, W. A. (1992). Observations of depleted stratospheric NO_2 following the Pinatubo volcanic eruption. *Geophys. Res. Lett.* **19**, 211–213.

Jones, D. G. (1985a). Photodiode array detectors in UV–VIS spectroscopy: Part I. *Anal. Chem.* **57**, 1057–1073.

Jones, D. G. (1985b). Photodiode array detectors in UV–VIS spectroscopy: Part II. *Anal. Chem.* **57**, 1207–1214.

Junge, C. E. (1963). "Air Chemistry and Radioactivity," Int. Geophys. Ser., Vol. 4. Academic Press, New York and London.

Kerr, J. B., McElroy, C. T., and Evans, W. F. (1982). Mid-latitude summertime measurements of stratospheric NO_2. *Can J. Phys.* **60**, 196–200.

Kreher, K. (1991). Messung der Breitenverteilung (50 °N–70 °S) von stratosphärischem Ozon und Stickstoffdioxid mittels optischer Absorptionsspektroskopie. Diploma Thesis, University of Heidelberg.

Kuznetzov, B. I., and Nigmatullina, K. S. (1977). Optical determination of the nitrogen dioxide content in the atmosphere. *Atmos Ocean. Phys.* **13**, 614–617.

Larche, K. (1953). Die Strahlung des Xenon—Hochdruckbogens hoher Leistungsaufnahme. *Z. Phys.* **136**, 74–86.

Lerner, J. M., and Thevenon, A. (1988). "The Optics of Spectroscopy. " Jobin-Yvon Optical Systems/Instruments SA. Longjumeau, France.

Logan, J. A., Prather, M. J., Wofsy, S. C., and McElroy, M. B. (1981). Tropospheric chemistry: A global perspective. *J. Geophys. Res.* **86D**, 7210–7254.

Marx, B. R., Birch, K. P., Felton, R. C., Jolliffe, B. W., Rowley, W. R. C., and Woods, P. T. (1980). High-resolution spectroscopy of SO_2 using a frequency-doubled continuous-wave dye laser. *Opt. Commun.* **33**, 287–291.

McKenzie, R. L., and Johnston, P. V. (1984). Springtime stratospheric NO_2 in Antarctica. *Geophys. Res. Lett.* **11**, 73–75.

McKenzie, R. L., Johnston, P. V., McElroy, C. T., Kerr, J. B., and Solomon, S. (1991). Altitude distributions of stratospheric constituents from ground-based measurements at twilight. *J. Geophys. Res.* **96D,** 15,499–15,512.

Millan, M. M., and Hoff, R. M. (1978). Remote sensing of air pollutants by correlation spectroscopy—instrumental response characteristics. *Atmos. Environ.* **12**, 853–864.

Millan, M. M. (1980). Remote sensing of air pollutants. A study of some atmospheric scattering effects. *Atmos. Environ.* **14**, 1241–1253.

Mount, G. H. (1992). The measurement of tropospheric OH by long path absorption. I. Instrumentation. *J. Geophys. Res.* **97D**, 2427–2444.

Mount, G. H., Sanders, R. W., Schmeltekopf, A. L., and Solomon, S. (1987). Visible spectroscopy at McMurdo station, Antarctica: 1. Overview and daily variation of NO_2 and O_3, Austral. spring, 1986. *J. Geophys. Res.* **92D**, 8320–8328.

Mount, G. H. Sanders, R. W., and Brault, J. W. (1992). Interference effects in reticon photodiode array detectors. *Appl. Opt.* **31**, 851–858.

Nestlen, M., Fischer, C., Hartmann, R., Platt, U., Münnich, K. O., and Flothmann, D. (1981). "Austausch von Luftverunreinigungen an der Grenzfläche Atmosphäre/Erdoberfläche (Trockene Deposition)," Zwischenbericht für das Umweltbundesamt zum Teilprojekt 2: Experimentelle Arbeiten, Forschungsbericht 104 02 609.

Neuroth, R., Dorn, H. P., and Platt, U. (1991). High resolution spectral features of a series of aromatic hydrocarbons and BrO: Potential interferences with OH measurements. *J. Atmos. Chem.* **12**, 12,287–12,298.

Newitt, D. M., and Outridge, L. E. (1938). The ultraviolet absorption bands ascribed to HNO_2. *J. Chem. Phys.* **6**, 752–754.

Noxon, J. F. (1975). Nitrogen dioxide in the stratosphere and troposphere measured by ground-based absorption spectroscopy. *Science* **189**, 547–549.

Noxon, J. F. (1983). NO_3 and NO_2 in the mid-Pacific troposphere. *J. Geophys. Res.* **88D**, 11017–11021.

Noxon, J. F., Norton, R. B., and Henderson, W. R. (1978). Observation of atmospheric NO_3. *Geophys. Res. Lett.* **5**, 675–678.

Noxon, J. F., Norton, R. B., and Marovich, E. (1980). NO_3 in the troposphere. *Geophys. Res. Lett.* **7**, 125–128.

Parrish, D. D., Trainer, M., Williams, E. J., Fahey, D. W., Huebler, G., Eubank, C. S., Liu, S. C., Murphy, P. C., Albritton, D. L., and Fehsenfeld, F. C. (1986). Measurements of the $NO_x - O_3$ photostationary state at Niwot Ridge, Colorado. *J. Geophys. Res.* **91D**, 5361–5370.

Penndorf, R. J. (1957). Tables of refractive index for standard air and the Rayleigh scattering coefficient for the spectral region between 0. 2 and 20.0 μm and their application to atmospheric optics. *J. Opt. Soc. Am.* **47**, 176–182.

Perner, D., and Platt, U. (1979). Detection of nitrous acid in the atmosphere by differential optical absorption. *Geophys. Res. Lett.* **6**, 917–920.

Perner, D., and Platt, U. (1980). Absorption of light in the atmosphere by collision pairs of oxygen $(O_2)_2$. *Geophys. Res. Lett.* **7**, 1053–1056.

Perner, D., Ehhalt, D. H., Pätz, H. W., Platt, U., Röth, E. P., and Volz, A. (1976). OH-radicals in the lower troposphere. *Geophys. Res. Lett.* **3**, 466–468.

Perner, D., Platt, U., Trainer, M., Hübler, G., Drummond, J. W., Junkermann, W., Rudolph, J., Schubert, B., Volz, A., Ehhalt, D. H., Rumpel, K. J., and Helas, G. (1987a). Measurement of tropospheric OH concentrations: A comparison of field data with model predictions. *J. Atmos. Chem.* **5**, 185–216.

Perner, D., Kessler, C., and Platt, U. (1987b). HNO_2, NO_2, and NO measurements in automobile engine exhaust by optical absorption. In *Monitoring of Gaseous Pollutants by Tunable Diode Laser.* (R. Grisar, H. Preier, G. Schmidtke, and G. Restelli, Eds.), pp. 116–119. Reidel, Dordrecht, The Netherlands.

Perner, D., Klüpfel, T., Parchatka, U., Roth, A., and Jörgensen, T. (1991). Ground-based UV–vis spectroscopy: Diurnal OClO profiles during January 1990 above Söndre Strömfjord, Greenland. *Geophys. Res. Lett.* **18**, 787–790.

Pierson, A., and Goldstein, J. (1989). Stray light in spectrometers: Causes and cures. *Lasers Optronics* (September), pp. 67–74.

Pitts, J. N., Jr., Finlayson, B. J., and Winer, A. M. (1977). Optical systems unravel smog chemistry. *Environ. Sci. Technol.* **11**, 568–573.

Pitts, J. N., Jr., Biermann, H. W., Winer, A. M., and Tuazon, E. C. (1984a). Spectroscopic identification and measurement of gaseous nitrous acid in dilute auto exhaust. *Atmos. Environ.* **18**, 847–854.

Pitts, J. N., Jr., Sanhueza, E., Atkinson, R., Carter, W. P. L., Winter, A. M., Harris, G. W., and Plum, C. N. (1984b). An investigation of the dark formation of nitrous acid in environmental chambers. *Int. J. Chem. Kinet.* **16**, 919–939.

Pitts, J. N., Jr., Biermann, H. W., Atkinson, R., and Winer, A. M. (1984c). Atmospheric implications of simultaneous nighttime measurements of radical and HONO. *Geophys. Res. Lett.* **11**, 557–560.

Pitts, J. N., Jr., Wallington, T. J., Biermann, H. W., and Winer, A. M. (1985). Identification and measurement of nitrous acid in an indoor environment. *Atmos. Environ.* **19**, 763–767.

Plane, J. M. C., and Nien, C. F. (1992). Differential optical absorption spectrometer for measuring atmospheric trace gases. *Rev. Sci. Instrum.* **63**, 1867–1876.

Platt, U. (1977). Mikrometeorologische Bestimmung der SO_2-Abscheidung am Boden, PhD. thesis, University of Heidelberg.

Platt, U. (1978). Dry deposition of SO_2. *Atmos. Environ.* **12**, 363–367.

Platt, U., and Perner, D. (1983). Measurements of atmospheric trace gases by long path differential UV/visible absorption spectroscopy. *Springer Ser. Opt. Sci.* **39**, 95–105.

Platt, U., and Perner, D. (1984). Ein Instrument zur spektroskopischen Spurenstoffmessung in der Atmosphäre. *Fresenius' Z. Anal. Chem.* **317**, 309–313.

Platt, U., Perner, D., and Pätz, H. (1979). Simultaneous measurement of atmospheric CH_2O, O_3, and NO_2 and by differential optical absorption. *J. Geophys. Res.* **84D**, 6329–6335.

Platt, U., Perner, D., Harris, G. W., Winer, A. M., and Pitts, J. N. (1980a). Detection of NO_3 in the polluted troposphere by differential optical absorption. *Geophys. Res. Lett.* **7**, 89–92.

Platt, U., Perner, D., Harris, G. W., Winer, A. M., and Pitts, J. N. (1980b). Observations of nitrous acid in an urban atmosphere by differential optical absorption. *Nature (London)* **285**, 312–314.

Platt, U., Perner, D., Schröder, J., Kessler, C., and Tönnissen, A. (1981). The diurnal variation of NO_3. *J. Geophys. Res.* **86D**, 11,965–11,970.

Platt, U., Perner, D., and Kessler, C. (1982). The importance of NO_3 for the atmospheric NO_x cycle from experimental observations. *Proc. Symp. Comp. Nonurban Troposphere*, 2nd, Williamsburg, Virginia, 1982, pp. 25–28.

Platt, U., Winer, A. M., Biermann, H. W., Atkinson, R., and Pitts, J. N. (1984). Measurement of nitrate radical concentrations in continental air. *Environ. Sci. Technol.* **18**, 365–369.

Platt, U., Rateike, M., Junkermann, W., Hofzumahaus, A., and Ehhalt, D. H. (1987). Detection of atmospheric OH radicals. *Free Radical Res. Commun.* **3**, 165–172.

Platt, U., Rateike, M., Junkermann, W., Rudolph, J., and Ehhalt, D. H. (1988). New tropospheric OH measurements. *J. Geophys. Res.* **93D**, 5159–5166.

Platt, U., Perner, D., and Semke, S. (1989). Observation of nitrate radical concentrations and lifetimes in tropospheric air. In "Ozone in the Atmosphere" (R. D. Bojkov and P. Fabian, Eds.), Proc. Quadrenn. Ozone Symp., 1988, pp. 512–515. Deepak Publishing, Hampton, VA.

Platt, U., LeBras, G., Poulet, G., Burrows, J. P., and Moortgat, G. (1990). Peroxy radicals from night-time reaction of NO_3 with organic compounds. *Nature (London)* **348**, 147–149.

Pommereau, J. P. (1982). Observation of NO_2 diurnal variation in the stratosphere. *Geophys. Res. Lett.* **9**, 850–853.

Pommereau, J. P., and Goutail, F. (1987). An advanced visible–UV spectrometer for atmospheric composition measurements, *Eur. Space Agency* [*Spec. Publ.*] *ESA SP*. **ESA SP-270**, 197–200.

Pommereau, J. P., and Goutail, F. (1988). O_3 and NO_2 ground-based measurements by visible spectrometry during the arctic winter and spring 1988. *Geophys. Res. Lett.* **15**, 891–894.

Ritz, D., Hausmann, M., and Platt, U. (1993). An improved open path multi-reflection cell for the measurement of NO_2 and NO_3. In *Optical Methods in Atmospheric Chemistry* (H. I. Schiff and U. Platt, Eds.), Proc. SPIE **1715**, 200–211.

Sanders, R., Solomon, S., Mount, G., Bates, M. W., and Schmeltekopf, A. (1987). Visible spectroscopy at McMurdo station, Antarctica: 3. Observation of NO_3. *J. Geophys. Res.* **92D**, 8339–8342.

Sarkissian, A., Pommereau, J. P., and Goutail, F. (1991). Identification of polar stratospheric clouds from the ground by visible spectrometry. *Geophys. Res. Lett.* **18**, 779–782.

Schiff, H. I., Karecki, D. R., Harris, G. W., Hastie, D. R., and Mackay, G. I. (1990). A tunable diode laser system for aircraft measurements of trace gases. *J. Geophys. Res.* **95D**, 10, 147–10, 154.

Schiller, C., Wahner, A., Platt, U., Dorn, H.-P., Callies, J., and Ehhalt, D. H. (1990). Near UV atmospheric absorption measurements of column abundances during airborne Arctic stratospheric expedition, January–February 1989:2. OClO observations. *Geophys. Res. Lett.* **17**, 501–504.

Schneider, W., Moortgat, G. K., Tyndall, G. S., and Burrows, J. P. (1987). Absorption cross-sections of NO_2 in the UV and visible region (200–700 nm) at 298 K. *J. Photochem. Photobiol.* **40**, 195–217.

Solomon, S., Schmeltekopf, A. L., and Sanders, R. W. (1987a). On the interpretation of zenith sky absorption measurements. *J. Geophys. Res.* **92D**, 8311–8319.

Solomon, S., Sanders, R. W., Carroll, M. A., and Schmeltekopf, A. L. (1989b). Visible spectroscopy at McMurdo station, Antarctica: 2. Observation of OClO. *J. Geophys. Res.* **92D**, 8329–8338.

Solomon, S., Mount, G. H., Sanders, R. W., Jakoubek, R. O., and Schmeltekopf, A. L. (1988). Observation of the nighttime abundance of OClO in the winter stratosphere above Thule, Greenland. *Science* **242**, 550–555.

Solomon, S., Miller, H. L., Smith, J. P., Sanders, R. W., Mount, G. H., Schmeltekopf, A. L., and Noxon, J. F. (1989a). Atmospheric NO_3. 1. measurement technique and the annual cycle at 40 °N. *J. Geophys. Res.* **94D**, 11,041–11,048.

Solomon, S., Sanders, R. W., Carroll, M. A., and Schmeltekopf, A. L. (1989b). Visible and near-ultraviolet spectroscopy at McMurdo Station, Antarctica. 5. Observations of the diurnal variations of BrO and OClO. *J. Geophys. Res.* **94D**, 11,393–11,403.

Stutz, J., and Platt, U. (1993). Problems in using diode arrays for open path DOAS measurements of atmospheric species. In *Optical Methods in Atmospheric Chemistry*, (H. I. Schiff and U. Platt, Eds.), Proc. SPIE **1715**, 329–340.

Tajime, T., Saheki, T., and Ito, K. (1978). Absorption characteristics of the γ-0 band of nitric oxide. *Appl. Opt.* **17**, 1290–1294.

Talmi, Y., and Simpson, R. W. (1980). Self-scanned photodiode array: A multichannel spectrometric detector. *Appl. Opt.* **19**, 1401–1414.

Tuazon, E. C., Winer, A. M., Graham, R. A., and Pitts, J. N. (1980). Atmospheric measurements of trace pollutants by kilometer-pathlength FT-IR spectroscopy. *Environ. Sci. Technol.* **10**, 259–299.

Tuazon, E. C., Winer, A. M., and Pitts, J. N. (1981). Trace pollutant concentrations in a multiday smog episode in the California South Coast Air Basin by long path length Fourier transform IR spectroscopy. *Environ. Sci. Technol.* **15**, 1232–1237.

Vandaele, A. C., Carleer, M., Colin, R., and Simon, P. C. (1993). Detection of urban O_3, NO_2, H_2CO and SO_2 using Fourier Transform Spectroscopy. In *Optical Methods in Atmospheric Chemistry* (H. I. Schiff and U. Platt, Eds.), Proc. SPIE **1715**, 288–292.

Vogt, S. S., Tull, R. G., and Kelton, P. (1978). Self-scanned photodiode array: High performance operation in high dispersion astronomical spectrophotometry. *Appl. Opt.* **17**, 574–592.

Wahner, A., Ravishankara, A. R., Sander, S. P., and Friedl, R. R. (1988). Absorption cross section of BrO between 312 and 385 nm at 298 and 223 K. *Chem. Phys. Lett.* **152**, 507–512.

Wahner, A., Callies, J., Dorn, H. P., Platt, U., and Schiller, C. (1990). Near UV atmospheric absorption measurements of column abundances during airborne Arctic stratospheric expedition, January–February 1989: 1. Technique and NO_2 observations. *Geophys. Res. Lett.* **17**, 497–500.

Wayne, R. P., Barnes, I., Biggs, P., Borrows, J. P., Canosa-Mas, C. E., Hjorth, J., LeBras, G., Moortgat, G. K., Perner, D., Poulet, G., and Sidebottom, H. (1991). The nitrate radical: Physical, chemistry, and the atmosphere. *Atmos. Environ.* **25A**, 1–203.

White, J. U. (1942). Long optical paths of large aperture. *J. Opt. Soc. Am.* **32**, 285–288.

White, J. U. (1976). Very long optical paths in air. *J. Opt. Soc. Am.* **66**, 411–416.

Winer, A. M., and Biermann, H. W. (1991). Measurements of nitrous acid, nitrate radicals, formaldehyde, and nitrogen dioxide for the Southern California Air Quality Study by differential optical absorption spectroscopy. In *Measurement of Atmospheric Gases* (H. I. Schiff, Ed.), Proc. SPIE **1433**, 44–57.

Wu, C. Y. R., and Judge, D. L. (1981). SO_2 and CS_2 cross section data in the ultraviolet region. *Geophys. Res. Lett.* **8**, 769–771.

Yates, D. A., and Kuwana, T. (1976). Evaluation of a self-contained linear diode array detector for rapid scanning spectrophotometry. *Anal. Chem.* **48**, 510–514.

Zellner, R., and Hägele, J. (1985). A double-beam UV-laser differential method for monitoring tropospheric trace gases. *Opt. Laser Technol.* **4**, 79–82.

CHAPTER

3

DIFFERENTIAL ABSORPTION LIDAR (DIAL)

SUNE SVANBERG

Department of Physics,
Lund Institute of Technology, Lund, Sweden

3.1. INTRODUCTION

Advanced techniques are needed to monitor our threatened environment, i.e., to evaluate pollution levels and developmental trends. While measurements obviously do not improve the environmental conditions per se, they can provide the impetus for implementing appropriate environmental protection programs. Measurement can then confirm positive changes induced by these actions. In this way advanced measurement techniques can play a very constructive role in the process of improving the environment.

Measurements are required on a local, a regional, and also a global scale. Tropospheric pollution has obvious manifestations in terms of health problems, water and soil acidification, and forest damage. Human-induced stratospheric changes in the ozone layer, as evidenced by the occurrence of "ozone holes" at the polar caps, may have much more far-reaching consequences (Farman et al., 1985; Stolarski, 1988). The "greenhouse" effect, due to a global increase in infrared (IR)-absorbing gases such as CO_2, CH_4, and N_2O, is another process of paramount importance (Bach et al., 1979; Revelle, 1982; Mason, 1989). Gaseous pollutants injected into the atmosphere enter very complex atmospheric chemistry chains (Wayne, 1985; Seinfeld, 1986; Trush, 1988).

Optical spectroscopy based on the specific absorption properties of different gases provides sensitive and selective measurements of atmospheric constituents. Its nonintrusive nature and real-time data capability makes it particularly useful. It is well adapted to various remote-sensing approaches yielding data from large atmospheric volumes. Optical spectroscopic remote

Air Monitoring by Spectroscopic Techniques, Edited by Markus W. Sigrist. Chemical Analysis Series, Vol. 127.
ISBN 0-471-55875-3

sensing can be performed in a *passive* mode, employing natural radiation sources such as the sun or the blue sky, or in an *active* mode, using an artificial source such as a lamp or a laser. These approaches are illustrated in Fig. 3.1. In reflective passive monitoring, frequently performed from satellites, the modification in the infalling spectral distribution due to target absorption is studied. Passive atmospheric absorption—or, in the IR region, emission—can be utilized. Active monitoring in transmission frequently utilizes a CW (continuous wave) optical transmitter to sense the average absorption over an atmospheric path. Lidar (*li*ght *d*etection *a*nd *r*anging) provides active optical remote sensing in backscattering. A pulsed laser transmitter is used, and light, backscattered from molecules and particles, is collected by an optical telescope and is detected and range-resolved in a radar-like mode. If the wavelength of the laser is varied from an absorption line of a pollutant gas to a close-by position, the detected changes in the backscattered light intensity can be used to evaluate range-resolved concentration profiles. This is the *differential absorption* version of the lidar method, and this is the main subject of the present chapter.

In all optical monitoring of the atmospheric constituents it is important to avoid the spectral regions where major atmospheric species basically block out the transmission. The same limitations pertain to normal, earthbound astronomical observations. Below 200 nm the Schumann–Runge bands of molecular oxygen (Thompson et al., 1963) put a definitive halt to atmospheric

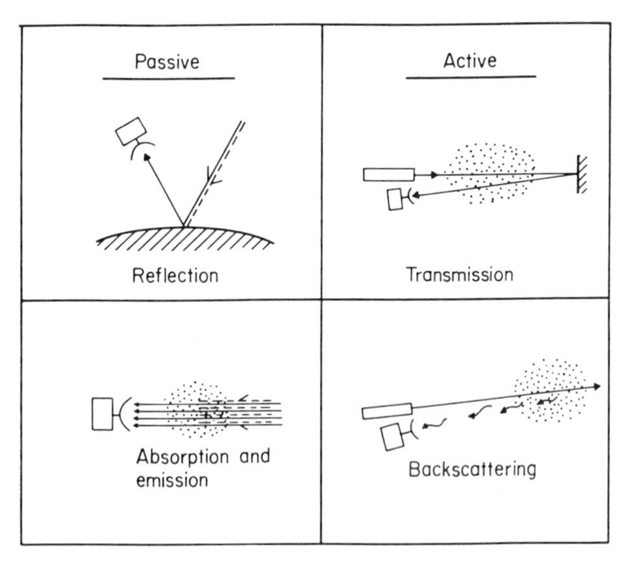

Figure 3.1. Schemes for optical remote sensing of the environment. From Svanberg (1980).

spectroscopy, leaving vacuum ultraviolet (VUV) spectroscopy as a laboratory discipline. Astronomical observations are effectively halted by stratospheric ozone absorption, which (presently) causes a cutoff at about 300 nm. The zero solar background below 300 nm makes the region down to 200 nm very attractive for atmospheric remote sensing with lidars, although the increasing Rayleigh and Mie scattering for short wavelengths imposes range limitations. In the visible region the atmosphere is obviously transparent, while substantial regions in the IR are completely blocked out by water and carbon dioxide absorption bands. An overview of the atmospheric transmission, pollutant absorption bands, and available laser sources is given in Fig. 3.2 (Grant and Menzies, 1983).

The absorption lines of the major atmospheric species and most important minor species including many pollutants are given in the extensive HITRAN compilation (Rothman et al., 1987). Very recently this material has become available on diskettes for easy personal-computer access (Killinger, 1992). The line widths of the molecular absorption lines are strongly pressure dependent and the relative strengths temperature dependent, which forms the basis for extractions of meteorological information (Korb and Weng, 1983). It also allows the deconvolution of line-of-sight passive observations of molecules present in the troposphere (broadened lines) as well as the stratosphere (sharp lines) (Menzies and Seals, 1977).

The purpose of the present chapter is to describe differential absorption lidar (DIAL) techniques for three-dimensional mapping of atmospheric pollutants. Such techniques allow the remote monitoring of ambient air, industrial emissions, natural emissions due to various geophysical phenomena (volcanoes or other geothermal sources), and monitoring of leaks on natural gas pipelines as well as warning systems for chemical warfare gases. The techniques can also be used for measurements of meteorological parameters such as temperature, humidity, and wind speed. Tropospheric as well as stratospheric studies can be performed. The present review is mainly focused on the description of DIAL monitoring of industrial pollution and geophysical emanations. A pedagogical rather than a comprehensive approach is taken, and no attempt is made to include all important references to the field. For the sake of convenience, examples are largely taken from work performed in the author's lidar groups at the Chalmers University of Technology and the Lund Institute of Technology over a time span of more than 15 years. Comprehensive references can be found in the books and review articles cited below.

The general topic of the present chapter has previously been treated in a few specialized monographs. Hinkley (1976) gives an early comprehensive review of laser monitoring of the atmosphere. In the proceedings of a conference edited by Killinger and Mooradian (1983), non-laser optical techniques such as DOAS (*d*ifferential *o*ptical *a*bsorption *s*pectroscopy) and FTIR (*F*ourier

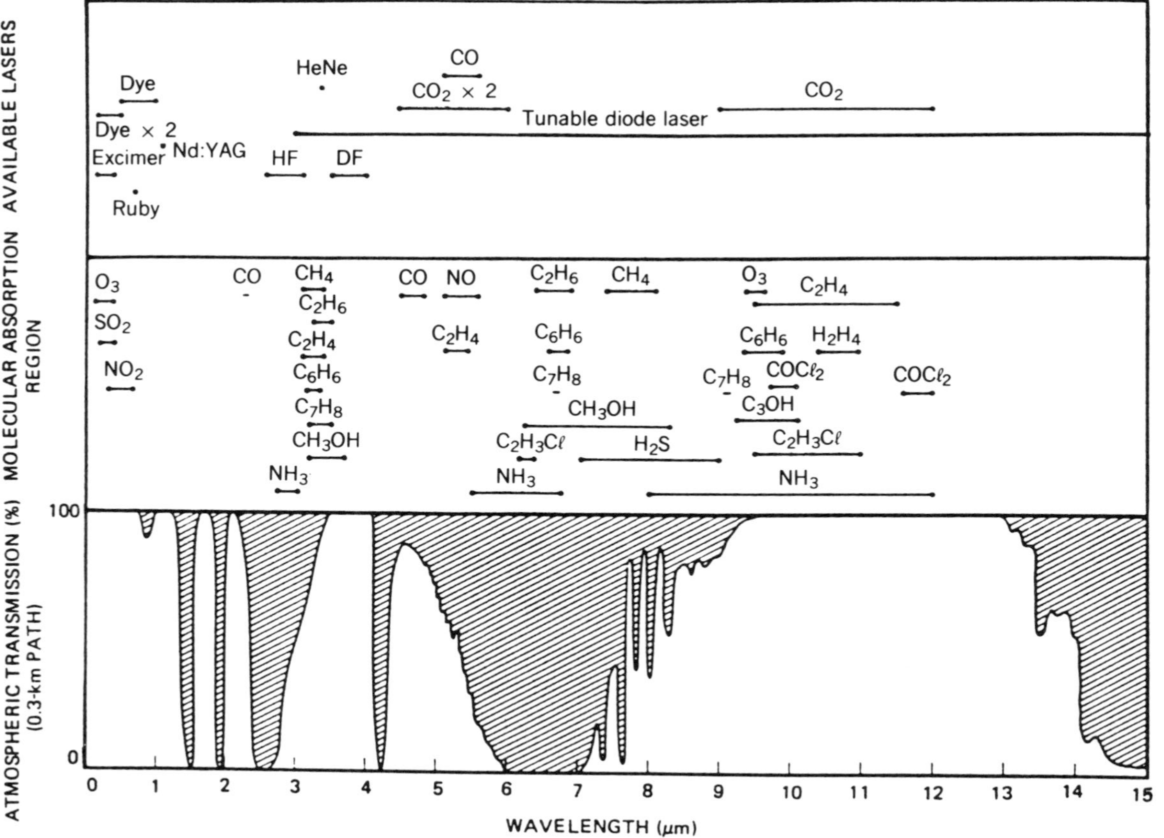

Figure 3.2. Atmospheric transmission, pollutant absorption bands, and available laser sources. From Grant and Menzies (1983).

*t*ransform *i*nfra*r*ed spectroscopy) are included. Measures (1984) has written the most thoroughgoing treatise on laser remote sensing. In a follow-up multi-author volume edited by Measures (1988), the general field of laser remote chemical analysis is treated.

In a series of international laser radar conferences the topic has been extensively illuminated through the years. The number of conferences in the series presently amounts to 17, and the latest ones are cited here (McCormick, 1982a, 1992; Megie, 1984; Carswell, 1986; Stefanutti, 1988; Zuev, 1990).

Several reviews of the field of laser remote sensing of the atmosphere have been published, e.g., by Carswell (1983), Grant and Menzies (1983), Grant (1987), Kobayashi (1987), Fredriksson (1988), and Zanzottera (1990). The present author has also written a few shorter reviews (Svanberg, 1978, 1980, 1985, 1991). Many aspects of practical optical spectroscopy as well as the atomic and molecular physics background are treated in a new monograph (Svanberg, 1992).

We begin our discussion in Section 3.2 with some basic considerations pertaining to optical atmospheric monitoring. The field is introduced by a discussion of long-path absorption measurements, in particular using DOAS. Then different varieties of lidar measurements are introduced and discussed. The use of topographic targets has a clear connection with the long-path absorption technique. Atmospheric backscattering (Mie, Raman, and fluorescence) enables us to perform range-resolved measurements that are particularly characteristic of the lidar technique. Next, Section 3.3 is devoted to a thorough description of the differential absorption lidar (DIAL) method, allowing detailed monitoring of important pollutants. We then discuss in Section 3.4, important laser types used in modern lidar systems and describe in Section 3.5 the basic elements of lidar system design. In Section 3.6, concerning DIAL measurements of tropospheric pollutant gases, some examples of practical monitoring of different important pollutants are given. A special section (3.7) is reserved for a novel DIAL application—that of atomic mercury monitoring, which apart from the pollution aspects contains elements of geophysical studies (mining, geothermal energy, volcanoes, etc). Finally, the outlook for the future of DIAL technology is given in Section 3.8.

3.2. BASIC CONSIDERATIONS

There are two major kinds of optical methods applicable in active remote sensing of the atmosphere:

- Long-path absorption monitoring
- Lidar (*l*ight *d*etection *a*nd *r*anging), with subdivisions:

Topographic target lidar
Mie scattering lidar
Fluorescence lidar
Raman scattering lidar
Differential absorption lidar (DIAL)

Since there are many elements in common between the two basic techniques, we will first consider long-path absorption measurements as an introduction to the various lidar schemes.

3.2.1. Long-Path Absorption Monitoring

Long-path absorption techniques are based on the same principles as spectrophotometry. However, by using well-collimated normal light beams or laser beams, it is possible to use a path length of several kilometers instead of the 1-cm cuvette typically used in the chemical laboratory. The principle is given in Fig. 3.3. A single-ended arrangement can be achieved by utilizing a corner cube retroreflector at the end of the light path and collecting the back-reflected light with a telescope. Since all detected photons have traveled the same path, no range resolution is obtained and only average concentrations can be determined. The light source can be a high-pressure xenon lamp, as in the case of DOAS. It is difficult to measure weak absorption lines in the presence of strong atmospheric turbulence because of strong scintillation. A fast scanning detection or parallel ccd (charge-coupled devices) spectral detection can then be used to overcome such difficulties. Tunable diode lasers and CW line-tunable CO_2 lasers are useful coherent sources for long-path absorption measurements. We shall come back to the scintillation problem later.

Although the topic of the present chapter is DIAL measurements, the

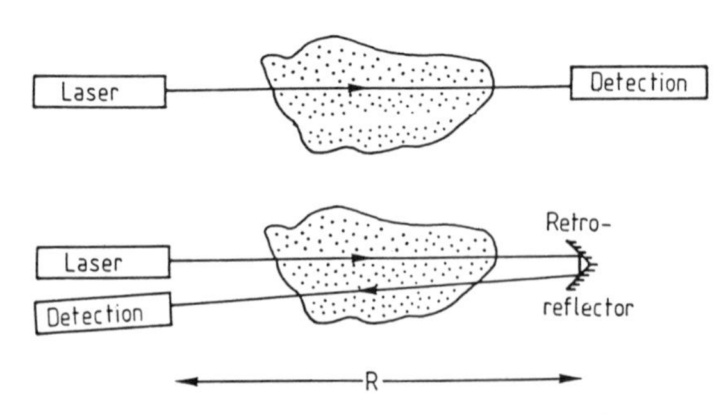

Figure 3.3. Principle of long-path absorption atmospheric spectroscopy.

pertinent principles are best illustrated by starting with a DOAS description. (DOAS is extensively treated by Platt in Chapter 2 of this book.) A DOAS setup is shown in Fig. 3.4 (Edner et al., 1993b). A high-pressure Xe lamp placed in the focus of a telescope is used to launch a well-collimated light beam over a distance of 100 m up to several kilometers. Light is reflected by a mirror and is directed into a Newtonian telescope. Light is focused into the entrance slit of a spectrometer and is detected by a photomultiplier tube (PMT). For fast scanning a number of slits are placed radially on a fast-rotating wheel in front of the PMT. The signals from a large number of scans are added in a computer. The sweep triggering is obtained from a photodiode observing the light from a light-emitting diode through the slotted wheel. In this way a chosen spectral region can repeatedly be swept during a time for which the atmosphere can be considered to be "frozen." An alternative way to reduce the influence of turbulence is to detect all wavelengths in a certain wavelength interval simultaneously by using a diode array detector.

An atmospheric recording of ambient SO_2 over a path length of 2000 m is shown in Fig. 3.5. In Fig. 3.5a the total light from the "white" lamp as seen by the detector is shown for the spectral region 280–320 nm covering the region of SO_2 absorption. Actually, a small absorptive structure can be observed. This can be enhanced by selecting the proper region and magnifying the structure as shown in Fig. 3.5b. A polynomial has been fitted to the general curve. This

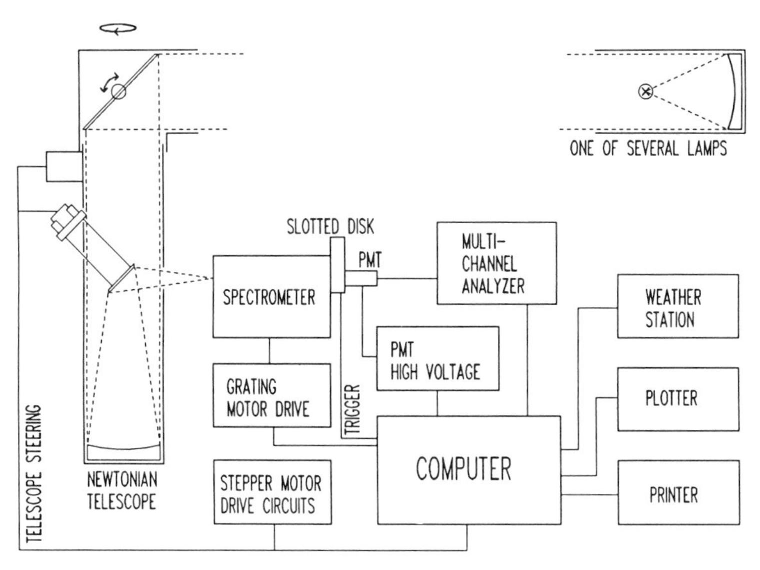

Figure 3.4. Setup for differential optical absorption spectroscopy (DOAS). From Edner et al. (1992a, 1993b).

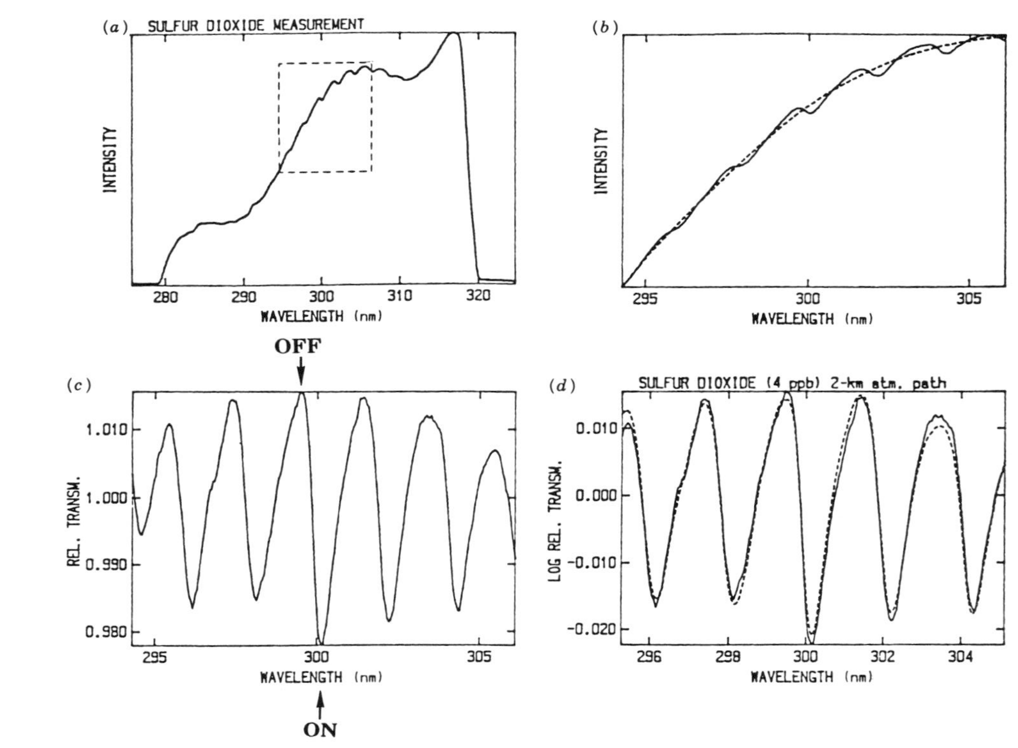

Figure 3.5. SO_2 DOAS measurement: (a) the raw recording of the lamp intensity observed over a 2000-m absorption path; (b) a magnified spectral region with a fitted polynomial (dashed line); (c) the recorded intensity divided by the polynomial; finally, in (d) the logarithm of the curve has been formed and a fit to a laboratory spectrum (dashed line) has been made, yielding a mean SO_2 concentration of 11 $\mu g/m^3$ (4 ppb). From Edner et al. (1992a, 1993b).

polynomial cannot reproduce the fast intensity variation, but rather provides a smooth curve for normalization. In Fig. 3.5c the intensity recording has been divided with the polynomial. Finally, in Fig. 3.5d the logarithm of the curve is formed for fitting to a weighted laboratory spectrum by employing the Beer–Lambert law. In this way an average concentration value of the pollutant is obtained for the measurement path. As can be seen, SO_2 provides a strongly modulated structure well suited also for DIAL. In the figure the two wavelengths normally used in DIAL are indicated. They have been chosen for maximum differential absorption within the smallest possible wavelength interval. Clearly, the "effective" absorption cross sections will vary with the spectral resolution, which is normally lower than the intrinsic line width owing to pressure (and Doppler) broadening. Thus it is important to determine the effective cross sections with the same resolution (the same apparatus!) in the laboratory as in the practical field measurements. It should also be noted that the Beer–Lambert law is not strictly valid when the instrument limits the resolution. Here laser sources have a great advantage over classical spectroscopic equipment.

Pollutant gas concentrations can be expressed as number densities (molecules/m^3), which can be related to volume or mass fractions (ppm—parts per million, 10^6; ppb—parts per billion, 10^9; ppt—parts per trillion, 10^{12}). Nowadays the most common way to state pollution levels is to indicate mass per volume, i.e., $\mu g/m^3$.

Further DOAS spectra—of NO_2 and O_3—are given in Fig. 3.6. For both molecules the *differential* absorption is lower, but there is a substantial *general* absorption for both gases. In particular, ozone does not exhibit much structure, and DIAL measurements require a rather large wavelength separation for achieving a large enough change in the broad and little-structured ozone absorption bond. (See also Fig. 3.23 in Section 3.5.4.) Recently, a renewed spectroscopic study of the ozone molecule suggested use of a particularly strong differential structure close to 283 nm for DOAS measurements, as shown in Fig. 3.6b (Axelsson et al., 1990).

3.2.2. The Lidar Method

In the lidar method, laser pulses are transmitted into the atmosphere and backscattered radiation is collected by an optical telescope and detected as illustrated in Fig. 3.7. The first lidar experiments were performed by Fiocco and Smullin (1963).

A particularly useful characteristic of the lidar method is its capability of remotely monitoring large areas. The size of the covered area is of course closely related to the platform arrangement for the lidar system. These aspects are also illustrated in Fig. 3.7. From a fixed laboratory an industrial area or a

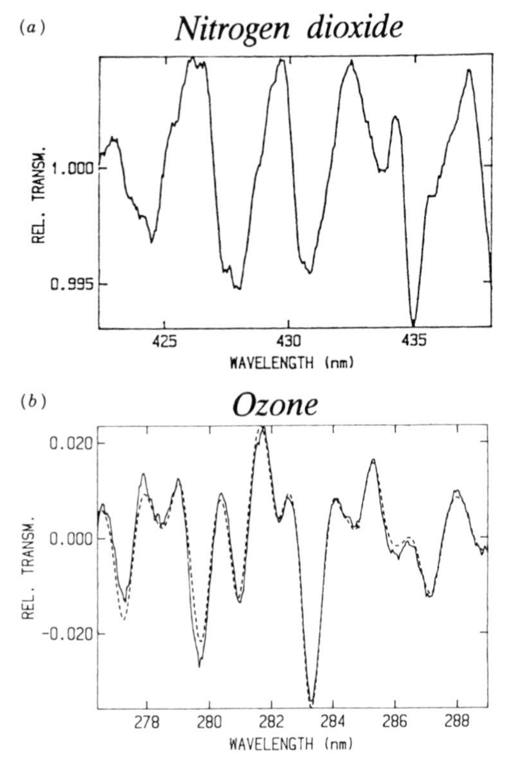

Figure 3.6. DOAS recordings for NO_2 and O_3 over an optical path of 2000 m. Measured intensities have been divided by fitted polynomials to enhance the structure. The NO_2 recording (a) yields a mean gas concentration of 29 $\mu g/m^3$ (12 ppb), whereas the O_3 recording (b) yields 161 $\mu g/m^3$. From Edner et al. (1992a, 1993b) and Axelsson et al. (1990), respectively.

section of a city can be covered by scanning. A mobile system, particularly if supplied with its own electric power generator, can conveniently be deployed for various measurement campaigns in urban or industrial areas. Airborne systems have also been constructed and used very successfully for regional measurements. Finally, satellites can provide platforms for future global coverage by lidar space systems now being planned.

3.2.2.1. Topographic Target Lidar

In the lidar approach, a laser pulse is transmitted into the atmosphere and backscattered radiation is detected as a function of time by an optical receiver in a radar-like fashion. If the laser beam is directed against a distant,

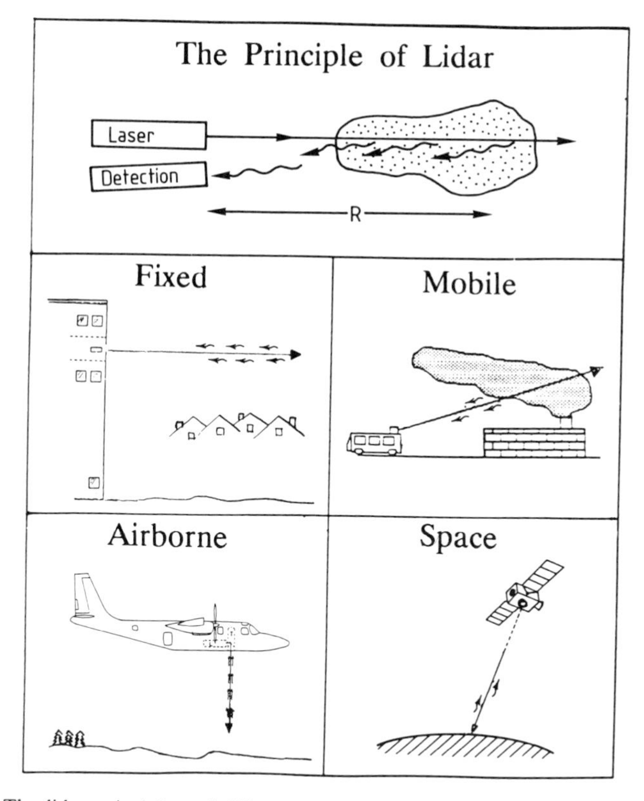

Figure 3.7. The lidar principle and different platform arrangements for lidar systems: Fixed laboratory, mobile system, airborne system, and spacecraft installation.

back-reflecting mirror (retroreflecting corner cube) at a distance R, a strong optical echo signal is received after a time

$$t = 2R/c \tag{3.1}$$

where c is the velocity of light. This is a principle of, e.g., lunar ranging against retroreflectors placed by *Apollo* astronauts (range precision: a few centimeters) or satellite tracking for geodetical applications. Even if a natural (topographic) target such as a brick wall, vegetation, or a mountainside is chosen, there will be a distinct but fainter echo signal. This is the principle of military range finders. Ranging is illustrated in Fig. 3.8, where a nitrogen-laser-pumped dye laser operating in the blue spectral region is fired against a mountainside at a distance of 5.5 km (Fredriksson et al., 1979). The backscattered light is received

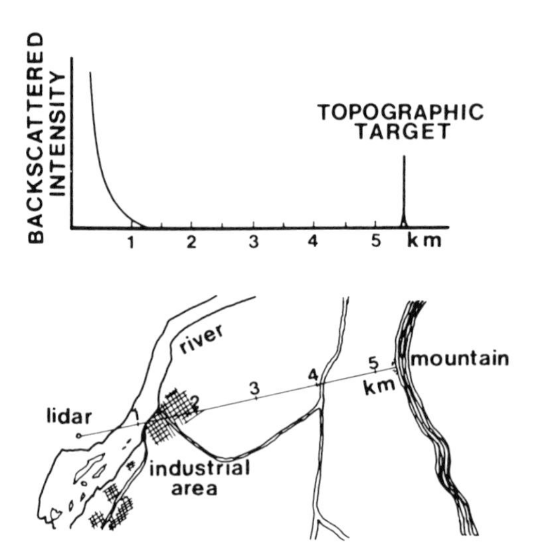

Figure 3.8. Illustration of NO_2 measurements against a topographic target. From Fredriksson et al. (1979).

by an optical telescope of diameter 30 cm and the optical echo is transformed into an electronic signal in a fast PMT. With light traveling at 300 m/μs the echo will be detected about 37 μs after the laser pulse. For a pulse length of δt, the minimum width of the echo signal expressed as a range δR will be

$$\delta R = \delta t/2c \tag{3.2}$$

Owing to the finite response time of the detection system, the width can frequently be larger. The intensity $P(\lambda)$ of the echo depends on many factors including laser pulse energy, range, telescope area, and the diffuse reflectance (albedo) of the target. It also depends on the wavelength-dependent absorption cross section $\sigma(\lambda)$ of the atmosphere. If only the absorption cross section for the pollutant gas varies when the wavelength of the laser is changed from λ_{on} (*on* the absorption line) to the nearby wavelength λ_{off} (*off* the absorption line) and no other gas constituents interfere (or absorb equally for the two wavelengths), the Beer–Lambert law yields

$$P(\lambda_{\text{on}})/P(\lambda_{\text{off}}) = \exp\left[-2RN(\sigma_{\text{on}} - \sigma_{\text{off}})\right] \tag{3.3}$$

where $[\sigma_{\text{on}} - \sigma_{\text{off}}]$ is called the *differential absorption cross section*, and N is the average concentration of the gas studied.

Equation (3.3) assumes that the diffuse reflectance of the target does not change for the small wavelength change $\Delta\lambda = \lambda_{on} - \lambda_{off}$. This is normally the case for solid materials interrogated in the UV or visible region. However, in the IR region, e.g., around 10 μm, where the CO_2 laser operates, a *differential albedo* can exist for the target material (Shumate et al., 1982; Grant, 1982; Englisch et al., 1983). This is due to the sharpness of the molecular vibration spectra even in solids or liquids. This phenomenon can then be used for remote characterization of the target, e.g., in airborne geophysical applications. In such measurements it is instead important to choose the wavelengths in such a way that a zero atmospheric differential absorption value is obtained.

If a gaseous pollutant with a nonzero differential absorption for the small wavelength change chosen is present in the air, the relative topographic echo heights will change as described by Eq. (3.3). The echo serves the role of a power meter placed at the end of the (doubled) path, with the considerable convenience that nobody needs to place the power meter at the remote location. For the case shown in Fig. 3.8, the laser wavelength was changed from 448.1 to 446.5 nm. A 4% change in echo height was observed. Based on the relevant differential absorption cross section for the wavelength pair used, it was concluded that the average concentration on NO_2 over the 5.5 km path from the lidar system to the mountain was about 2 ppb. In airborne lidar measurements the surface of the earth is available as a topographic target for mean concentration determinations.

3.2.2.2. Mie Scattering Lidar

Mie scattering from particles and Rayleigh scattering from molecules provide strong signals observed in backscattering in lidar sounding of the free atmosphere. These scattering processes are *elastic*, i.e., the scattered photons have the same frequency as the incoming light (apart from possible small Doppler shifts). This signal can clearly be seen in Fig. 3.8 for ranges up to 1300 m and can be considered to be due to a "distributed" topographic target, present at all distances. Rayleigh scattering has a strong $1/\lambda^4$ dependence on the wavelength and can yield a dominating signal at short UV wavelengths. Mie scattering has a slower wavelength dependence ($\approx 1/\lambda^2$) and increases in relative importance to be dominant in particle-rich air probed at visible and near-IR wavelengths. The Mie backscattering from particles thus allows a mapping of the relative distribution of particles over large areas if the lidar system is scanned (see, e.g., Shimizu et al., 1985). However, since Mie scattering theory (van de Hulst, 1957; Kerker, 1969) involves many normally inaccessible particle parameters, quantitative results are difficult to obtain. Mie scattering is extremely useful in providing the "distributed mirror" needed in DIAL, the main topic of this chapter. As discussed below, measurements are then made at two close-lying wavelengths exhibiting a nonzero differential absorption cross section for the gas of interest, and the two signals are divided to eliminate all unknown Mie scattering parameters.

From Fig. 3.8 it is obvious that the atmospheric backscattering exhibits a strong range dependence, basically reflecting the $1/R^2$ illumination law. We would now like to discuss the lidar signals more closely and introduce the general lidar equation, yielding the received laser radar intensity $P(\lambda, R)$ from a range R:

$$P(\lambda, R) = CWn_b(R)\sigma_b \frac{\Delta R}{R^2} \exp\left(-2\int_0^R [\sigma(\lambda)N(r) + K_{ext}(r)]\,dr\right) \quad (3.4)$$

Here C is a system constant; W is the transmitted pulse energy; and $n_b(R)$ is the number density of scattering objects with backscattering coefficient σ_b. The exponential factor describes the attenuation of the laser beam and the backscattered radiation due to the presence of absorbing molecules of concentration $N(r)$ and absorption cross section $\sigma(\lambda)$ and due to scattering particles with wavelength-independent extinction $K_{ext}(r)$.

There exists a complicated relation between σ_b and K_{ext} for particle scattering. Certain simplifying assumptions are frequently made to allow an evaluation of the particle distribution $n_b(R)$. Much work has been invested into inverting the lidar equation. One of the most commonly used technique is the Klett inversion (Klett, 1981, 1986). If the absorption due to the particles is very small (thin clouds), the extraction of relative values of $n_b(R)$ is very simple, with a constant σ_b value assumed. A close-range recording of industrial particulate pollution is shown in Fig. 3.9. Superimposed on a general $1/R^2$ signal due to a uniform background particle distribution, localized signals of increased backscattering are seen owing to the industrial plumes. The stepwise attenuation of the uniform backscattering when the main plume is being passed can also clearly be seen.

Pure Mie scattering is extensively used for studying stratospheric dust from volcanic eruptions (McCormick, 1982b; McCormick et al., 1984; Osborn et al., 1992; Stefanutti et al., 1992b). Such studies are important for assessing perturbations in the earth's radiation budget. The year 1816 became known as "the year without a summer" because of an extensive volcanic eruption. Recent major eruptions that have been much studied by vertically sounding Mie lidars were due to El Fuego, Guatemala (1974), El Chichon, Mexico (1982), and Mt. Pinatubo, the Philippines (1991). Lidar recordings of the development of the stratospheric dust layer from Mt. Pinatubo are shown in Fig. 3.10 (Osborn et al., 1992). The Mie-to-Rayleigh scattering ratio has been plotted, eliminating (e.g.) the $1/R^2$ distance dependence.

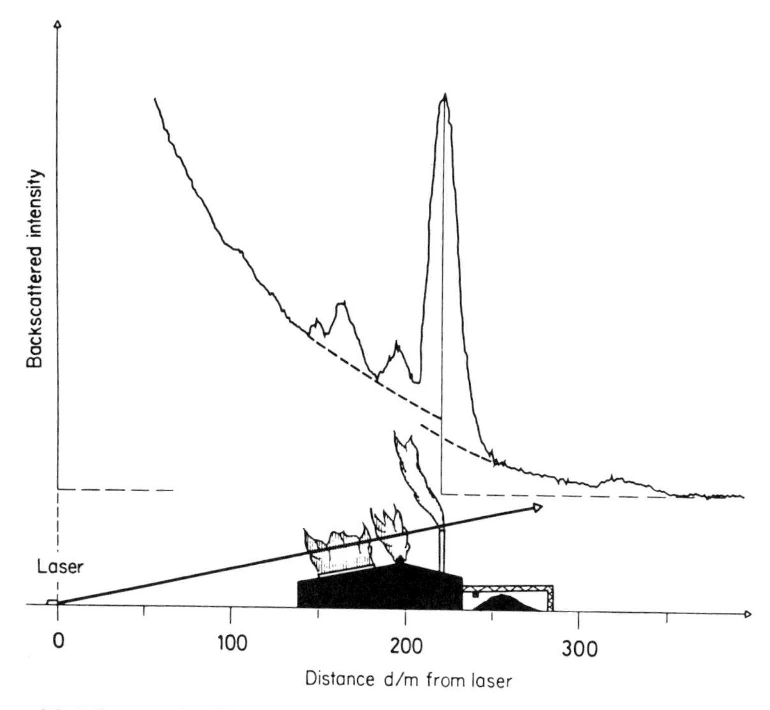

Figure 3.9. Mie scattering lidar particle monitoring at an iron-alloy plant. From Fredriksson et al. (1976).

3.2.2.3. Raman Lidar

Raman scattering is a much more well-defined scattering process, that can be used for quantitative measurements of gaseous constituents of the atmosphere. For the case of Raman lidar the constants in the lidar equation (3.4) have a new meaning, so that $n_b(R)$ is the number density of the Raman scattering molecules and σ_b is the Raman cross section. In contrast to the Mie and Rayleigh processes, Raman scattering is *inelastic*. In vibrational Raman scattering, a molecular vibrational quantum is picked up from the incoming photon, leaving the scattered quantum characteristically red-shifted (Stokes-shifted) with respect to the Rayleigh scattering by an amount corresponding to the molecular vibrational energy. Raman scattering from the major atmospheric constituents, O_2, N_2, and H_2O, is seen in the insert of Fig. 3.11, exhibiting the characteristic Raman shifts (1556, 2331, and 3652 cm^{-1}) for these molecules (Fredriksson et al., 1976). The main difficulty in using Raman scattering is its inherent weakness, typically 10^3 times weaker than the

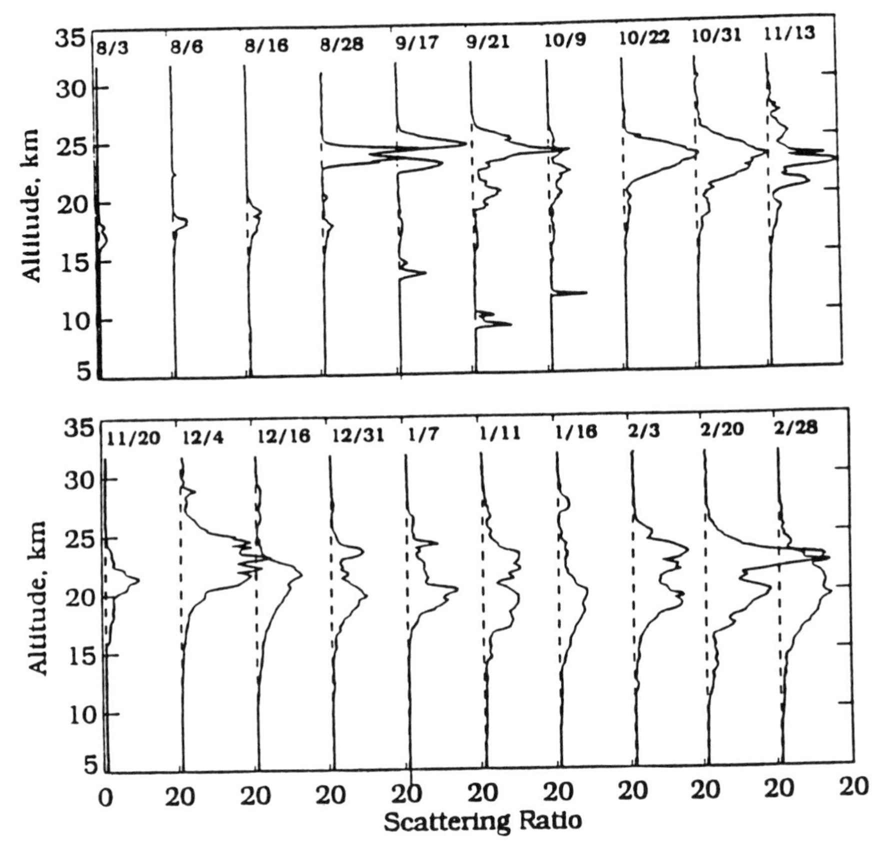

Figure 3.10. Lidar recordings of stratospheric dust due to the Mt. Pinatubo volcanic eruption. The Mie scattered intensity divided by the expected Rayleigh scattering intensity from a particle-free atmosphere has been plotted. Recordings taken at Hampton, Virginia, covering the period, Aug. 3, 1991, to Feb. 28, 1992. From Osborn et al. (1992).

Rayleigh scattering from the same molecule. This must be compensated by very high concentrations and short measuring ranges even when large lidar systems are used (Hirschfeld et al., 1973). Thus, the Raman technique is useful mainly for the major atmospheric constituents. First Raman lidar recordings were actually performed for N_2 and O_2 (Leonard, 1967).

A Raman lidar recording is obtained by suppressing the strong elastically backscattered light and instead centering the detection band (selected by a sharp interference filter with high out-of-band rejection) on the appropriate Raman-shifted wavelength. A nitrogen Raman lidar recording is shown in Fig. 3.11 together with a corresponding elastic recording (Fredriksson et al.,

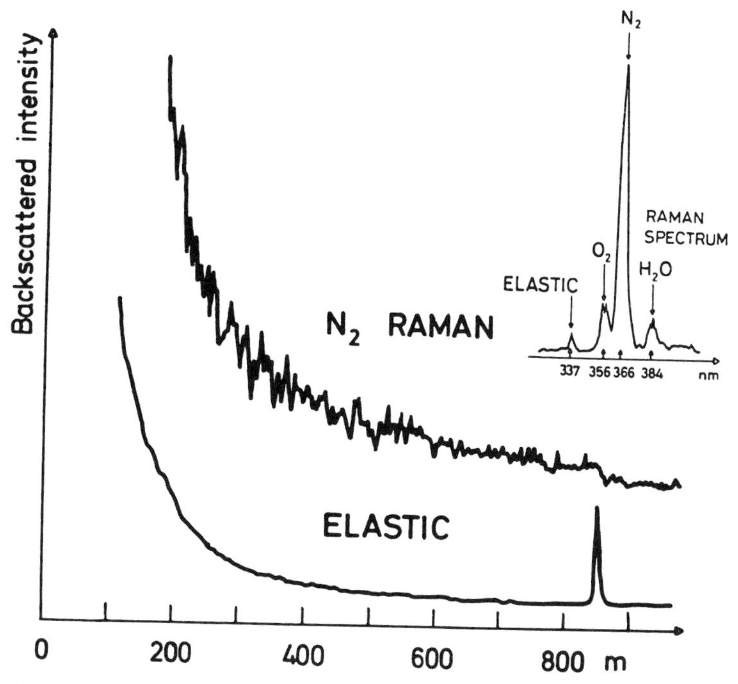

Figure 3.11. Elastic backscattering and N_2 Raman scattering recordings against a solid target at an 850-m distance. The measurements were performed with a nitrogen laser operating at 337 nm. Inset: a spectrally resolved recording of atmospheric backscattering is shown, featuring peaks due to oxygen, nitrogen, and water vapor. From Fredriksson et al. (1977).

1977). In this case the beam from a nitrogen laser ($\lambda = 337$ nm) was terminated against the bricks of a smoke stack, clearly seen as an echo at 850 m in the elastic signal. In contrast, the nitrogen Raman recording shows the uniformly decreasing $1/R^2$ falloff for the ambient nitrogen and a sudden disappearance of the signal at the encounter of the solid target.

Nitrogen Raman signals can be used for measuring atmospheric attenuation (nontrivial deviations from $1/R^2$ dependence), since the signal is not strongly influenced by the particle backscattering. Another major application is vertical sounding of water vapor profiles, which are of great meteorological importance and are useful in radiation budget assessments. Examples of water vapor mixing ratio recordings taken with a high-energy XeCl excimer laser ($\lambda = 308$ nm) are shown in Fig. 3.12 (Ansmann et al., 1992). Here the water Raman signal has been divided by the nitrogen Raman signal. Water vapor can also be measured by DIAL (Browell et al., 1978; 1981).

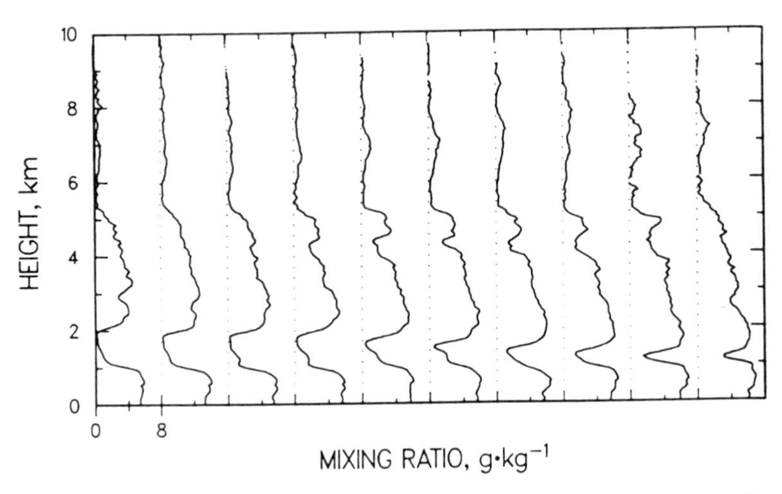

Figure 3.12. Water vapor Raman lidar curves, plotted as mixing ratios (water vapor to dry air). From Ansmann et al. (1992).

Rotational Raman scattering utilizes the extremely small wavelength shifts due to the absorption or emission of a quantum corresponding to a rotational transition. Since the population distribution on the different rotational levels is very temperature dependent, rotational Raman scattering can be employed for vertical temperature measurements using lidar signals from N_2 or O_2 (Hauchecorne et al., 1992).

3.2.2.4. Fluorescence Lidar

In fluorescence lidar a laser tuned to the absorption line of an atmospheric species is used and fluorescence light is detected in the subsequent decay. For this case $n_b(R)$ in Eq. (3.4) is the number density of fluorescing atoms and σ_b is the fluorescence cross section. Fluorescence lidar is a powerful technique for measurements at mesospheric heights, where the pressure is low and the fluorescence is not quenched by collisions. The collisions induce radiationless decays to the ground state, eliminating many atoms from contributing to the fluorescence light emission. At normal atmospheric pressures the quenching dominates by a factor of about 10^3 over fluorescence, effectively eliminating the light. After first demonstrations by Bowman et al. (1969), the technique has been used extensively to monitor layers of various alkali and alkaline earth atoms (Li, Na, K, Ca, Ca^+, and Fe) at a height of about 100 km (Chanin, 1983; Fricke and von Zahn, 1985; Hansen and von Zahn, 1990; Collins et al., 1992; Kane et al., 1992). The mesospheric atoms are formed from vaporization of micrometeorites. By monitoring the movements of the layers the atmospheric dynamics can be studied, including the formation of gravitational waves.

An example of a recording of the mesospheric sodium layer is shown in Fig. 3.13 (Collins et al., 1992). A narrow-band laser was tuned to the sodium resonance line. In the same recording, taken at the South Pole, the Mie scattering return from polar stratospheric clouds (PSCs) is shown. Such clouds are very important in the polar ozone destruction process.

A novel application of fluorescence lidar is the creation of "guide stars" of fluorescing sodium atoms in the mesosphere for rectifying images of large-scale astronomical telescopes. The technique of adaptive imaging strongly enhances the performance of ground-based telescopes and is now being implemented widely (Thompson and Gardner, 1987; Collins, 1992). Recently, it was realized that the sodium wavelength can conveniently be generated by mixing the two Nd:YAG laser wavelengths 1.064 mm and 1.319 μm (Jeys et al., 1989).

3.2.2.5. *Solid Target Fluorescence Lidar*

In Fig. 3.8 the typographic target echo was recorded by using the elastically backscattered light. However, fluorescence is induced in the target by the laser pulse, and if the signal is strong enough the wavelength contents of the echo can be analyzed at the site of the receiver system. An optical multichannel analyzer with an image-intensified array detector is well suited for such measurements. Then the full fluorescence spectrum can be captured for every laser shot. The image intensifier can also be gated down to 5 ns to accept light

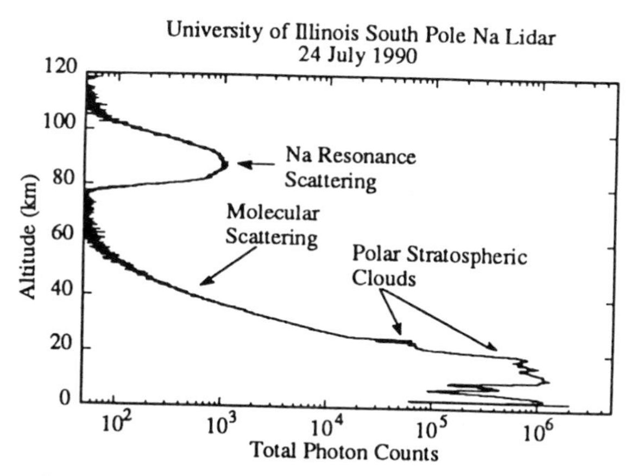

Figure 3.13. Recording at the South Pole of the mesospheric sodium layer (seen in resonance fluorescence) and polar stratospheric clouds (seen in Mie backscatter). From Collins et al. (1992).

arriving only at the right delay. In this way ambient light can be suppressed.

Fluorescence spectroscopy has long been used for analytical and diagnostic purposes (Udenfriend, 1962, 1969; Wehry, 1976; Hercules, 1966; Lakowicz, 1983). Laser-induced fluorescence (LIF) has an interesting potential for remote sensing of environmental parameters. For quite some time, hydrospheric pollution monitoring has been performed with airborne laser-based fluorosensors. Different kinds of oil can be identified by their fluorescence properties. Other pollutants and algal bloom patches can similarly be studied (O'Neill et al., 1980; Capelle et al., 1983; Hoge et al., 1986; Reuter, 1991). By use of the blue-green transmission window of water, bathymetric measurements of sea depths can be performed as well (Kim, 1977; Hoge et al., 1980). The field of laser-based hydrospheric monitoring is covered by Measures (1984). LIF has also been used by American (Hoge et al., 1983; Hoge, 1988; Chappelle et al., 1985), Italian (Cecchi and Pantani, 1991) and Swedish groups (Svanberg, 1990; Edner et al., 1992d) for studies of land vegetation.

Examples of laboratory spectra for various oil products that might appear in the aquatic environment are shown in Fig. 3.14 (Celander et al., 1978). As a rule, light petroleum fractions exhibit blue-shifted, intense fluorescence whereas heavier fractions also have longer wavelength components and fluoresce more weakly. At short UV wavelengths the penetration of the exciting light into the oil is limited to micrometers. For longer wavelengths the penetration depth is larger. Thus, in order to assess the thickness of an oil film the choice of excitation wavelength is important. In the assessment of marine oil spills, the fluorescence characteristics of different oil products play an important role in airborne measurements and in the decision regarding the correct oil-fighting countermeasures to be implemented.

Algae fluorescence monitoring can be important for measuring the total marine productivity, which originates in the conversion of solar energy, CO_2, and nutrients into organic matter by microscopic phytoplankton. Recently, huge algal blooms, for instance, of *Chrysochromulina polylepis*, leading to devastating consequences for most other marine life forms, have occurred owing to eutrophication of coastal waters. Some classes of algae exhibit LIF spectra with certain characteristic features (Celander et al., 1978) in addition to the dominating peak at 685 nm due to chlorophyll *a*. A blue-green fluorescence is normally observed for water even in the absence of oil spills. This fluorescence is due to organic material (*Gelbstoff*). Measurement scenarios for fluorescence lidar monitoring of water and land vegetation are shown in Fig. 3.15 with examples of remotely recorded spectra.

In aquatic monitoring a strong Raman signal due to the water molecules is observed at a Raman shift of about $3400\,cm^{-1}$ (O—H stretch vibration), as shown in Fig. 3.15a. This signal is very useful, since it is possible to utilize it as a reference to normalize algal and *Gelbstoff* signals to the same effective water

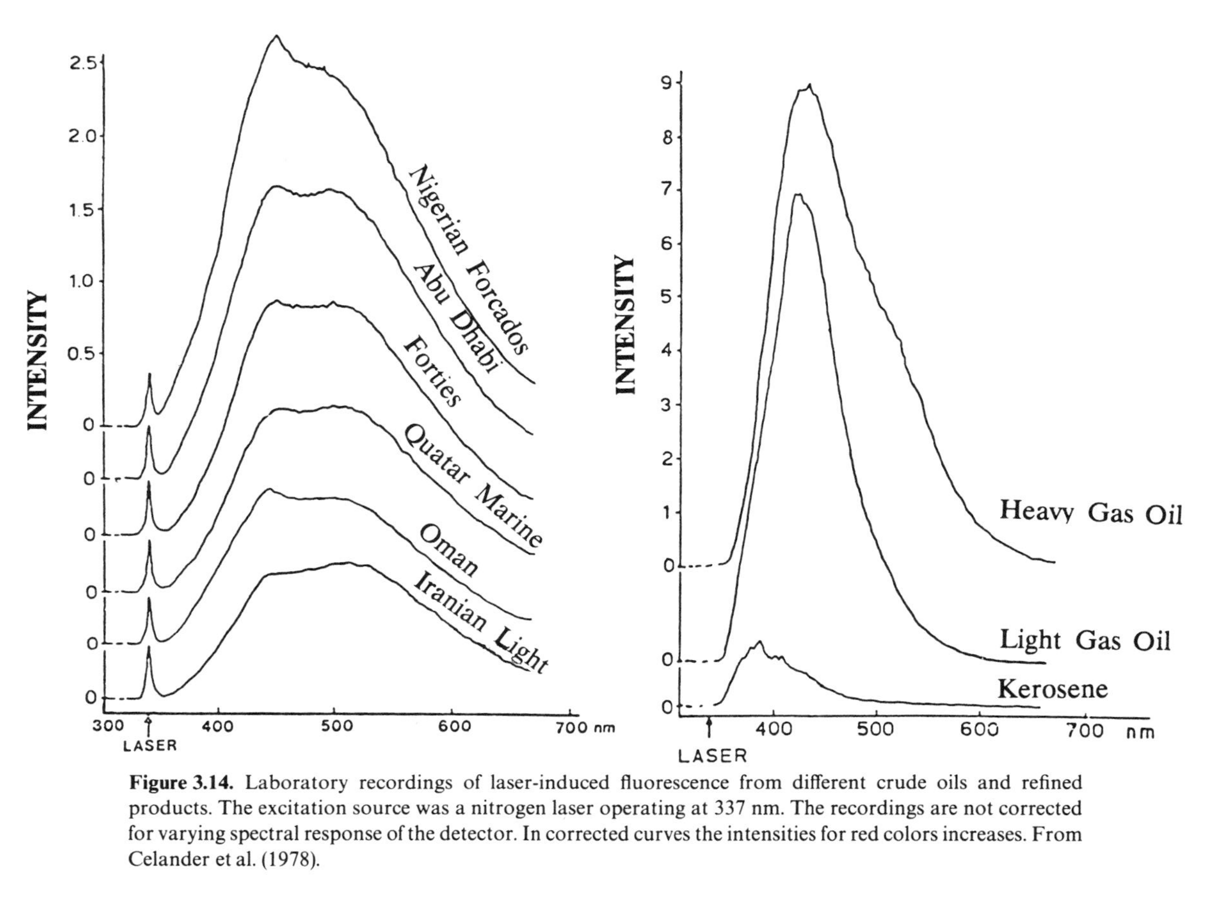

Figure 3.14. Laboratory recordings of laser-induced fluorescence from different crude oils and refined products. The excitation source was a nitrogen laser operating at 337 nm. The recordings are not corrected for varying spectral response of the detector. In corrected curves the intensities for red colors increases. From Celander et al. (1978).

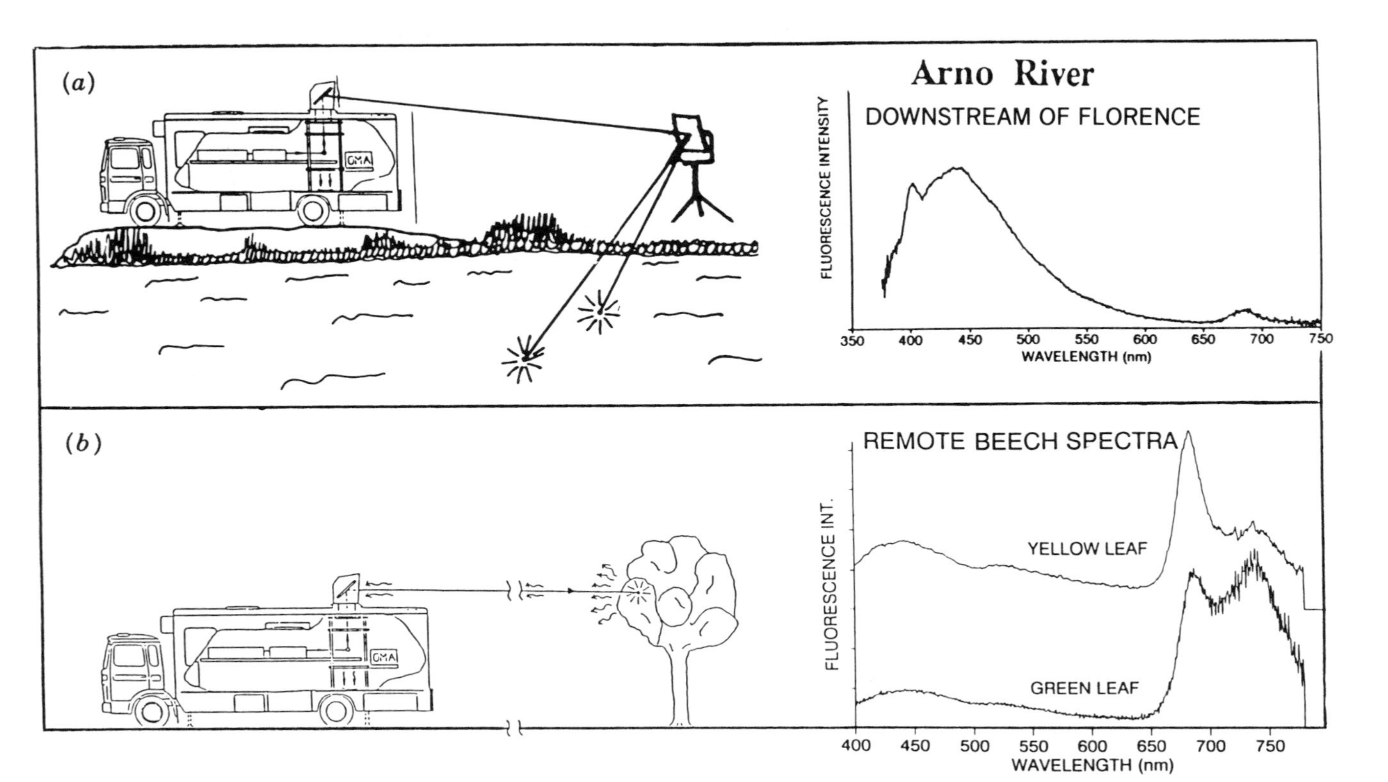

Figure 3.15. Fluorescence lidar measurements of (a) water and (b) terrestrial vegetation. A frequency-tripled Nd:YAG laser operating at 355 nm was used. Remote laser-induced fluorescence spectra are shown for a measurement distance of about 50 m. Adapted from Edner et al. (1992d,e).

measuring volume. A careful analysis of the shape of the water Raman signal provides information on the water temperature. Water molecules form aggregates of different sizes with slightly different Raman shifts. The relative occurrence of mono-, di-, and poly-water molecules is temperature dependent (Leonard et al., 1979; Breschi et al., 1992).

Land vegetation monitoring by fluorescence lidar is shown in Fig. 3.15b, illustrating different signatures for green and lightly yellow maple leaves. The ratio of the two chlorophyll peaks at 690 and 735 nm relates to different plant physiological conditions, as discussed by Lichtenthaler and Rinderle (1988). Land vegetation can be efficiently characterized by multispectral reflectance measurements using spaceborne sensors in satellites such as LANDSAT, SPOT, and ERS-1 (Chen, 1885). An *active* remote sensing technique such as LIF might, in certain circumstances, complement *passive* reflectance monitoring.

3.3. PRINCIPLES OF DIFFERENTIAL ABSORPTION LIDAR (DIAL)

The principles of DIAL are schematically represented in Fig. 3.16, where a measurement scenario in an industrial area is illustrated. Laser pulses are transmitted into the atmosphere, which for the time being is (unrealistically) assumed to contain a uniform distribution of Mie scattering particles. A gaseous pollutant is emitted from the industry as indicated and penetrates in between the uniformly distributed particles, which provide "topographic targets" located everywhere. The industrial effluent is assumed to be free of additional particles. Laser light is alternatingly transmitted at a wavelength (λ_{on}), where the species under investigation absorbs, and at a neighboring, off-resonant wavelength (λ_{off}). In the presence of an absorbing gas cloud the on-resonance signal is attenuated through the cloud and the off-resonant one is not. If the particle concentration is low, the off-resonant signal will exhibit a pure $1/R^2$ dependence as shown in Fig. 3.16, whereas additional intensity losses occur for the on-resonance signal when the gas is encountered. The differences between the two curves is best visualized by dividing them for each range interval. For identical curves a ratio value of unity is obtained for all ranges. This is true only if the Mie scattering from the particles does not change when the wavelength is changed from the on to the off value. For spherical particles of uniform size (monodisperse particles) the Mie scattering cross section exhibits a strong oscillatory behavior as a function of wavelength. However, for natural atmospheres the size and shape of the particles vary continuously and the oscillations are smeared out. Thus, for practical cases the on- and off-wavelength curve division is unaffected if the wavelength separation is small. The resulting curve is called the DIAL curve.

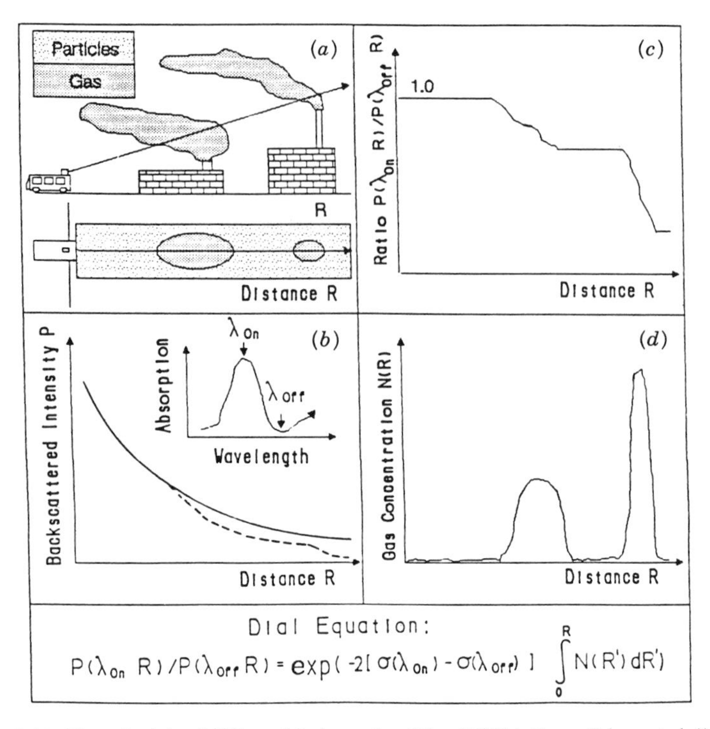

Figure 3.16. The principle of differential absorption lidar (DIAL). From Edner et al. (1987b).

When a gas cloud is encountered there is a downward slope on this ratio curve, which after the cloud passes resumes its horizontal direction but at a lower level. From the DIAL curve it is clear that the gas concentration as a function of range can be calculated basically by employing the Beer–Lambert law and using the differential absorption cross section for the gas. Such a calculated concentration curve is also included in Fig. 3.16.

At this point we note that we really do not need our initial pedagogical (and unrealistic) assumption that there is a uniform particle distribution. Even if there would have been a localized cloud of additional particles increasing the backscattering strongly for this particular range, the resulting upward bump would be equally present in the on- and off-resonance curves and would thus not show up in the ratio (DIAL) curve. Even most other troublesome and unknown parameters are eliminated in the division, and the gas concentration as a function of the range along the beam can be evaluated with knowledge of the differential absorption cross section only.

What we have found here using simplified arguments can of course be put on a rigorous mathematical foundation. The DIAL curve is obtained by forming the lidar equation (3.4) first for the *on* wavelength and then for the *off* wavelength. Then the two equations are divided by each other, yielding

$$\frac{P(\lambda_{\text{on}}, R)}{P(\lambda_{\text{off}}, R)} = \exp[-2(\sigma_{\text{on}} - \sigma_{\text{off}}) \int_0^R N(r)dr] \tag{3.5}$$

The assumptions we have made are that σ_b and $K_{\text{ext}}(r)$ are wavelength independent for a sufficiently small wavelength change. If we want to calculate the average concentration value $N_{\text{av}}(R, R + \Delta R)$ for a certain range interval ΔR, this can easily be made using Eq. (3.5). We obtain

$$N_{\text{av}}(R, R + \Delta R) = \frac{1}{2(\Delta R)[\sigma_{\text{on}} - \sigma_{\text{off}}]} \cdot \ln \frac{P(\lambda_{\text{off}}, R + \Delta R) P(\lambda_{\text{on}}, R)}{P(\lambda_{\text{on}}, R + \Delta R) P(\lambda_{\text{off}}, R)} \tag{3.6}$$

From the hand-waving description as well as from Eq. (3.6) it is obvious that a sufficiently large range interval ΔR must be chosen to allow a significant average concentration value N_{av} for the corresponding interval to be evaluated. The larger the concentration values are, the easier the expression following the "ln" in the equation obtains a significant value for a given value of ΔR. This is easily understood by observing that this expression is the DIAL curve value at R divided by the DIAL curve value at $R + \Delta R$. A large slope on a noise-free DIAL curve yields high-quality data for the range-resolved concentration curve. For high concentrations a high spatial resolution can be used while still producing accurate data. To enable the numerical evaluation to be performed in the presence of the noise in the experimental data, the DIAL curve is normally "smoothed" before the concentration calculation by forming a sliding average over a number of digital range channels in the electronic detection system.

The DIAL technique was pioneered by Shotland (1966), who studied water vapor. First DIAL measurements on an atmospheric pollutant were reported by Rothe et al. (1974a, b) and Grant et al. (1974) for the case of NO_2.

In this section we shall present two simple applications of the DIAL technique: the case of a strong pollutant plume in an otherwise clean atmosphere, and the case of a uniform gas distribution in the atmosphere. Later, many examples of the more common situation of nonuniform distributions will be given.

In Fig. 3.17 we demonstrate a remote measurement of the NO_2 contents in an industrial plume. The laser beam was directed just above the top of the smokestack. To the left in the figure are shown the on- and off-resonance lidar recordings for NO_2. A logarithmic representation is used to decrease the large signal differences at close and far-off ranges. In the elastic backscatter curves a

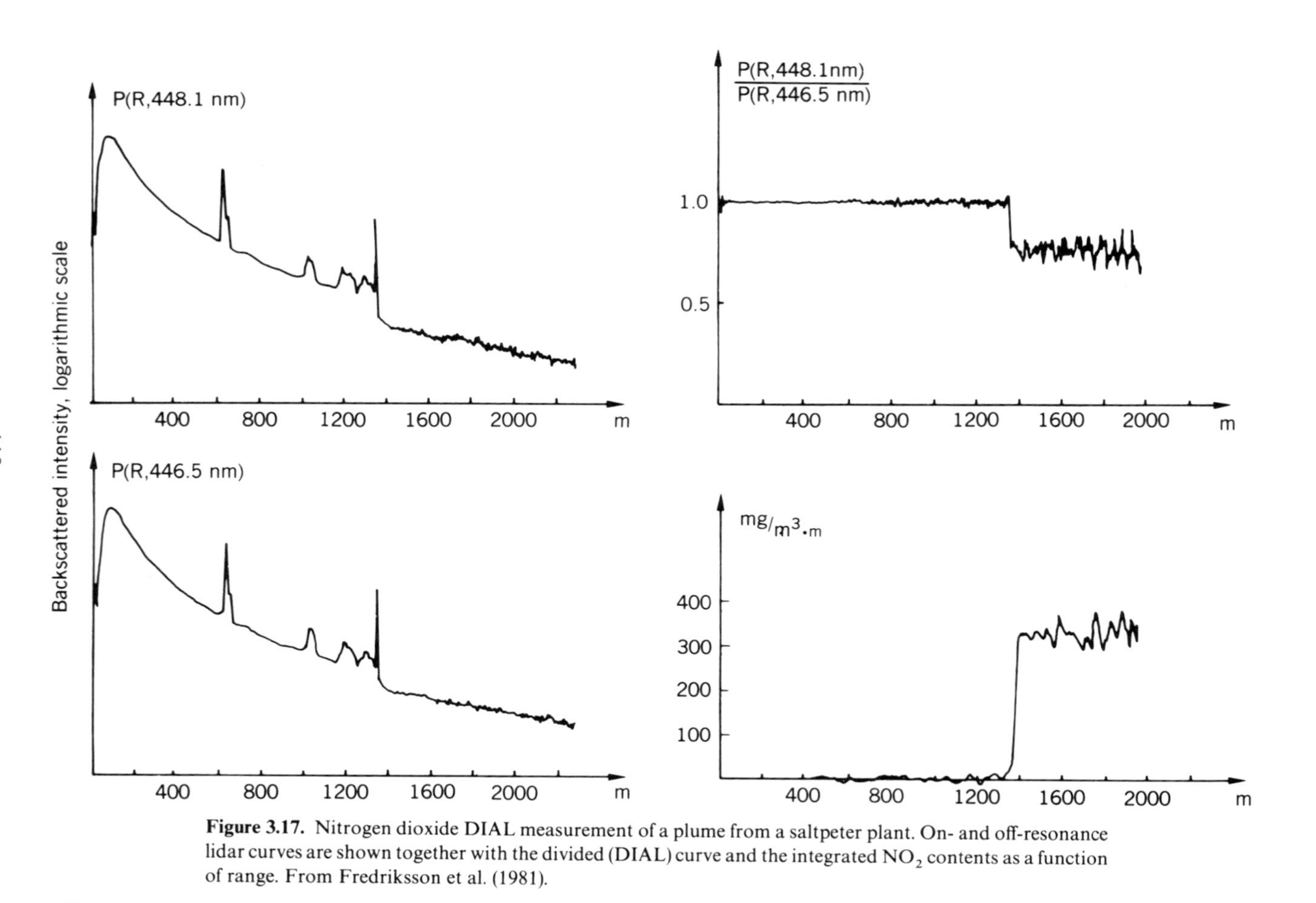

Figure 3.17. Nitrogen dioxide DIAL measurement of a plume from a saltpeter plant. On- and off-resonance lidar curves are shown together with the divided (DIAL) curve and the integrated NO_2 contents as a function of range. From Fredriksson et al. (1981).

number of particle-containing plumes accidentally drifting into the laser beam are clearly seen in addition to the one that is actively aimed at (at a 1350-m distance). The backscattered radiation is attenuated only by this plume, as can be clearly seen in the curves. It can also be seen that the attenuation is larger in the 448.1 nm curve because of the strong NO_2 absorption at this on-resonance wavelength. In the upper-right part of the figure the divided (DIAL) curve is shown. It can then clearly be seen that the effect of the particle plumes is eliminated. The only effect "surviving" the ratio formation is the step at 1350 m. Using Eq. (3.6) and the differential cross section for the NO_2 absorption, we can calculate the integrated NO_2 concentration as shown in the lower right corner. Knowing the gas flow rate from the stack it is possible to conclude that the emission amounted to about 40 kg/h.

The case of a uniform gas distribution is normally not a realistic one. However, the permanent gases such as N_2 and O_2 of course show this behavior, and in Fig. 3.18 we show the case of atmospheric oxygen in order to illustrate the application of the DIAL equation. This measurement was performed in the very weak oxygen absorption band surrounding the atomic mercury line at 254 nm. In the top part of the figure individual lidar curves for on- and off-resonance laser tuning are shown. At about a 1200-m distance the laser beam hits a hill, resulting in sharp echoes in both curves. The close-range intensity has been reduced by ramping up the amplification of the system to reach its full constant value only at a range of 600 m. In the lower part of the figure the DIAL curve is shown. For the case of a uniform distribution, Eq. (3.5) takes the very simple form of a pure exponential:

$$\frac{P(\lambda_{\rm on}, R)}{P(\lambda_{\rm off}, R)} = \exp[-2NR(\sigma_{\rm on} - \sigma_{\rm off})] \tag{3.7}$$

which is clearly shown in Fig. 3.18, Equations (3.7) and (3.3) are identical, as expected.

3.4. LASERS FOR DIAL

Since the introduction of the laser in 1960 a large number of laser types have been developed. Some of these are more practical and useful than others, and a few systems have emerged as the DIAL lasers of choice. Important requirements for DIAL use are tunability, high pulse energy, and sufficiently short pulse length. Since the lasers need to be used in operational field equipment they must be sufficiently rugged and practical. The most useful DIAL lasers are shown in Fig. 3.19. Two of the laser systems shown, the Nd:YAG laser and

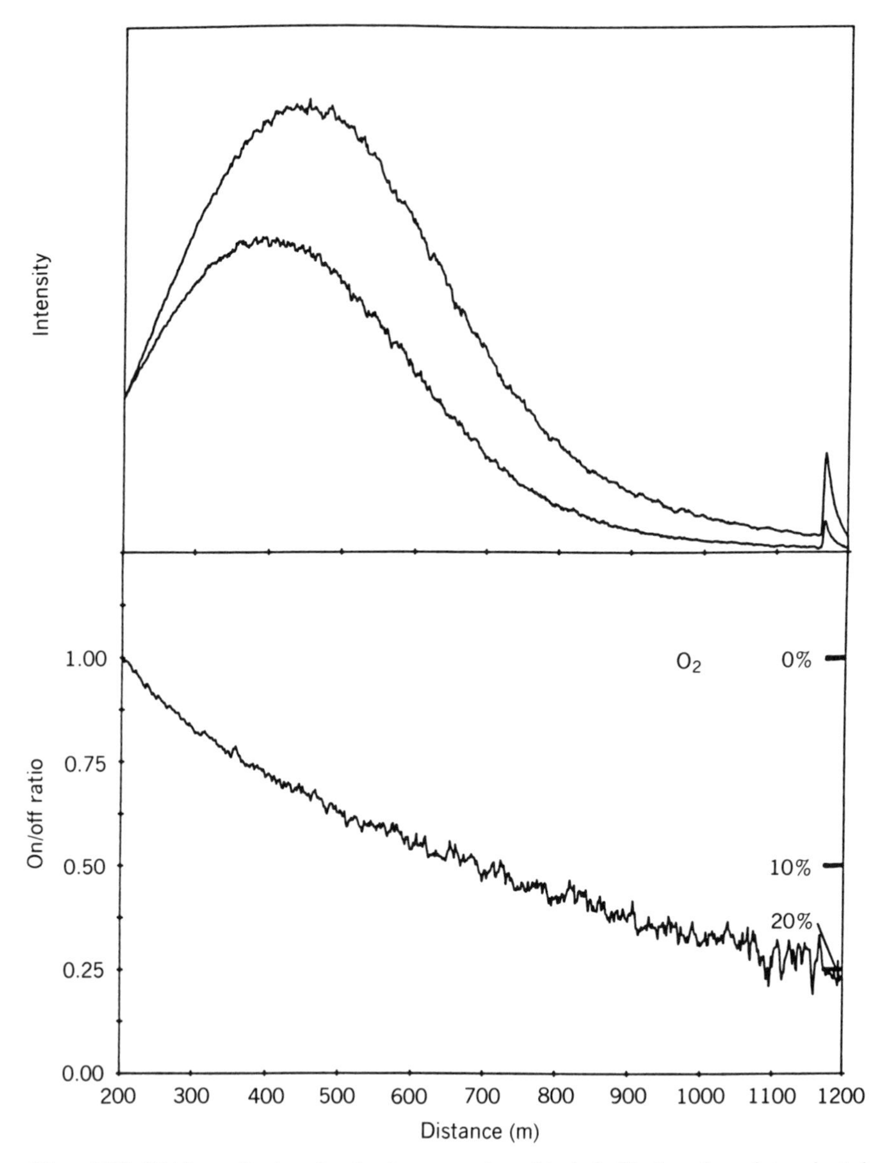

Figure 3.18. DIAL monitoring of molecular oxygen in ambient air. The laser beam is terminated against a hillside at about 1200-m distance. In the upper part of the figure on- and off-resonance curves are shown for O_2. The photomultiplier voltage has been ramped. In the lower part of the figure the DIAL curve is shown, following an exponential as expected for a uniform gas distribution. From Edner et al. (1991c).

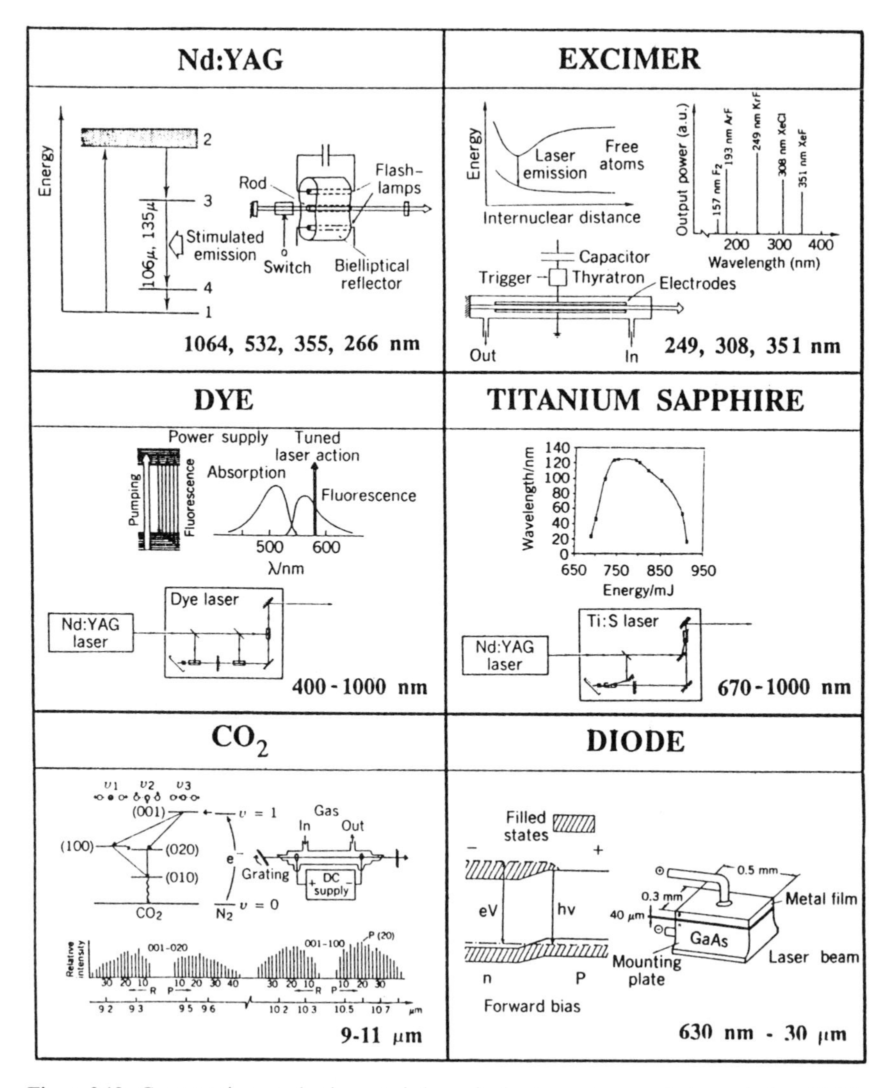

Figure 3.19. Construction and characteristics of six laser types of interest for DIAL measurements.

the excimer laser, are fixed-frequency lasers normally used as pumping sources for dye or titanium–sapphire lasers, which are widely tunable.

Laser sources are described in a number of textbooks (see, e.g., Siegman, 1986; Svelto, 1989; Svanberg, 1992).

3.4.1. The Nd: YAG Laser

The neodymium: YAG (yttrium aluminum garnet) laser is a highly efficient solid state laser using Nd^{3+} ions in a crystalline matrix. The Nd: YAG laser is a four-level laser with a broad absorption band that permits broad-band flashlamps to be used for pumping. The upper laser transition level has a long lifetime and permits storage of excess population. If a shutter is placed in the cavity, lasing can be prevented although population inversion has been reached. By opening the shutter when most of the pumping capacity of the lamp has been utilized, a giant pulse at 1.064 μm can be generated with megawatt power and about 10-ns pulse duration (Q-switching technique). In that way pulses with an energy of tens of millijoules can be formed. In order to achieve higher pulse energies the beam is passed through flashlamp-pumped amplifiers with laser rods of increasing diameter. In commercial Nd: YAG lasers used for lidar systems, one or two amplifiers are used to achieve pump energies well above 1 J. Repetition rates are 10–30 Hz.

By using phase-matched KDP (potassium dihydrogen phosphate) crystals, efficient frequency doubling, tripling, or quadrupling to 532, 355, or 266 nm, respectively, can be obtained. The doubled and tripled outputs are frequently used for pumping rhodamine and coumarin dyes in a dye laser for DIAL applications. The quadrupled output can be useful for ozone lidar systems, frequently in conjunction with stimulated Raman scattering in deuterium or hydrogen, producing outputs at 289 or 299 nm, respectively.

During the last few years high-power Nd: YAG lasers have become much more compact than they used to be, making them even more attractive for lidar use. Flashlamp quality has also greatly improved, allowing typically 20 million shots before replacement. Diode laser pumping of Nd: YAG lasers is quickly developing. Here pulsed diode lasers at 808 nm are used, matching a strong absorption line of the active medium. Small, very reliable units employing this all-solid-state technology are available. As the price of diode lasers continues to decline rapidly, high-energy lidar transmitters with strongly reduced power and cooling requirements will undoubtedly soon be available.

3.4.2. The Excimer Laser

Excimer lasers use noble-gas halogenides as the lasing medium. Such molecules can be formed in excited states in a fast discharge (thyratron switching) in a mixture of Ar, Kr, or Xe with HCl or HF. Since the ground state is immediately dissociating, excimer molecules are ideal as laser media. In lidar contexts the XeCl excimer laser at 308 nm is important for pumping dye lasers for DIAL. The KrF laser at 249 nm is common for tropospheric ozone lidar systems, and is then frequently used in connection with stimulated Raman scattering in deuterium or hydrogen to produce 268-, 277-, 292-, or 313-nm output. For stratospheric ozone monitoring, less absorbed, longer wavelengths must be used, e.g., 308 nm from XeCl or 351 nm from XeF. In direct use of excimer lasers as lidar transmitters, a specially designed laser resonator (unstable resonator) is utilized to yield a much better beam quality than is available from excimer lasers for dye-laser pumping.

Excimer lasers yield pulse energies of hundreds of millijoules. An especially attractive feature is their high repetition rates (up to several hundred hertz). On the other hand, special installations and precautions are needed in the handling of very corrosive and toxic gases.

3.4.3. The Dye Laser

Dye lasers have long been the standard source in DIAL systems. The active medium consists of strongly fluorescing dye molecules, normally dissolved in methanol or ethanol. Flashlamp pumping in the broad absorption band of the dye can be used, but more frequently the harmonics of a Nd:YAG laser or a XeCl excimer laser is employed. The fire hazard in using direct flashlamp pumping of the flammable dye liquid as well as the comparatively long pulses are reasons for infrequent use of the direct pumping modality. In the dye laser oscillator normally a grating preceded by a prism beam expander is used for narrowing down the bandwidth and for tuning. For DIAL it is important to be able to change the wavelength between adjacent shots. It is not practical to turn the grating by using the normal sine-drive mechanics of the laser. Instead, a dual-wavelength option is frequently used in the laser. One possibility is to use a beamsplitter and direct half the laser mode toward one grating and the other half toward the second grating. The two arms are then alternatingly blocked for adjacent shots. Another, more convenient option is to have two gratings covering half of the expanded beam each. By alternatingly blocking the beam path in front of the gratings, every second shot can be fired at the preselected *on* and *off* wavelengths.

In order to achieve a sufficient pulse energy for long-range monitoring, one or two dye amplifier cells are normally used after the oscillator, as shown in Fig. 3.19. The dye amplifiers are pumped by the same pump laser, the output of which is split up in beams of suitable pulse energies. Rhodamine dyes are used to reach SO_2 and O_3 wavelengths around 300 nm after frequency doubling. Output powers of up to 200 mJ can be achieved at the dye's fundamental wavelengths when green Nd: YAG laser pumping is employed. Blue and green dyes can be pumped by the 355 nm output from a tripled Nd: YAG laser or conveniently directly and efficiently by a XeCl excimer laser. In this way 450 nm for NO_2 or 226 and 254 nm after frequency doubling for NO and Hg, respectively, can be generated.

3.4.4. The Titanium–Sapphire Laser

Titanium–sapphire is a new tunable solid state laser material covering the wavelength region from 670 nm to about 1 μm. Since the upper state lifetime is only 3 μs, flashlamp pumping, although possible, is not very efficient. Frequently, a frequency-doubled Nd: YAG laser is preferred for pumping. In order to get a high pulse energy, an end-pumped amplifier stage is frequently used. Commercial lasers of this kind have just become available and give considerable promise. NO_2 wavelengths are achieved by frequency doubling; SO_2, O_3, and Hg wavelengths, by frequency tripling; and NO wavelengths, by frequency quadrupling. Output pulse energies in oscillator/amplifier units can be similar to those achievable with dye lasers.

3.4.5. The CO_2 Laser

The carbon dioxide laser is a gas laser, emitting on a variety of vibrational–rotational lines of the CO_2 molecule in the 9–12 μm region. In the gas discharge tube nitrogen is added and pumping is provided by collisional exchange between ground-state CO_2 molecules and vibrational excited N_2 molecules. At the normal gas pressure of few torr the Doppler-broadened lines allow a very limited tuning of only 50 MHz and one has to rely on the accidental coinciding of pollution molecular absorption lines and the grating-selected laser lines. Many pollutants are accessible because of the richness of lines in the IR region. Frequently a CW seeder laser is employed when heterodyne detection is used in IR lidar systems.

The HF (hydrogen-fluoride) and DF (deuterium-fluoride) lasers are other pulsed gas lasers operating in the IR region. These lasers cover the 2.7–3.0 μm and 3.7–4.0 μm regions, respectively.

3.4.6. The Diode Laser

The diode laser is a highly efficient tunable solid state laser with very small dimensions. It consists of a highly doped *p–n* junction that is operated biased in the forward direction. The lasing occurs over the bandgap, the size of which varies for different materials. Normally, a resonator is provided by polishing the semiconductor chip surfaces. Better laser control is achieved by using an external resonator and coating the diode itself with an antireflection layer. We have discussed earlier the use of powerful 808-nm diode lasers for solid state laser pumping. Pulsed diode lasers are used in certain cloud height meters and range finders. CW units are used for seeding pulsed IR lasers and as local oscillators in heterodyne detection systems.

3.5. LIDAR SYSTEM DESIGN

A good DIAL system should be designed to be able to detect low gas concentrations at large ranges. Thus, signal intensity and signal noise are major concerns. Obviously, the detected backscattered intensity will increase with laser pulse energy, telescope area, and detector efficiency as indicated by the lidar equation. However, a lidar system can additionally strongly improve in sensitivity (particularly in the IR region) by introduction of heterodyne detection schemes instead of direct photon detection. A further aspect of major concern is the influence of the atmospheric turbulence (scintillation) in the lidar signals. Another technical consideration in laser radar system design is the need to be able to handle the large dynamic range typical for lidar signals. Before describing some DIAL systems in more detail we shall next address these general points.

3.5.1. Dynamic Range Reduction

A special signal-handling problem in lidar systems is that the backscattered intensity has a basic $1/R^2$ dependence. Thus, the detection system must be able to handle strong close-range signals at almost the same time as weak signals from afar. This calls for a very large dynamic-range capability that is hard to provide. Thus various methods to reduce the dynamic-range requirements have been developed.

By separating the axis of transmission from the detection telescope axis, the close-range signal is reduced since the transmission and detection lobes do not spatially overlap until a certain range. Even in a coaxial system a strong "geometrical compression" can be achieved (Harms et al., 1978; Harms, 1979) by introducing an aperture defining the telescope field of view in the far-field image plane of the telescope. The closer-range scattering volume will be imaged at larger distance from the telescope mirror and will thus be out of focus in the aperture plane. Only a fraction of the totally backscattered light can thus pass the aperture, reducing the close-range signal.

There are also electronic means to reduce the dynamic-range requirements for the detection system. One possibility is to use a logarithmic amplifier. A logarithmic representation was used above in Fig. 3.17. A further possibility is to ramp the high voltage supplied to the photomultiplier dynode chain, triggering the ramp at the time of the pulse release. This was illustrated in Fig. 3.18. Maximum and constant amplification is provided after a time delay when the signal has decreased to a level that can be handled by the transient digitizer, which has a limited number of bits for the digitizing of the signal.

Geometrical compression, logarithmic amplification, and dynode-chain ramping all modify the recorded signal with regard to the lidar equation (3.4). However, since all curves are influenced in the same way, this problem is divided away when one is forming the DIAL ratio between the signals.

3.5.2. Heterodyne Detection

Heterodyne, or *coherent*, detection is a valuable scheme particularly in the IR region, where PMTs do not exist to provide strong and noise-free amplification (Menzies, 1976). Infrared detectors have a high efficiency but yield very low signal levels, calling for electronic amplification. This can easily induce additional noise. This can be circumvented in the heterodyne scheme, where the incoming signal amplitude A_S is mixed with the radiation of a narrow-band local oscillator with amplitude A_L on the detector, as shown Fig. 3.20. The detected signal S from the detector is the square of the resulting amplitude:

$$\begin{aligned} S &= (A_S \cos \omega_S + A_L \cos \omega_L)^2 \\ &= A_S A_L \cos(\omega_S - \omega_L) + (\text{high-frequency terms}) \end{aligned} \tag{3.8}$$

Most terms in Eq. (3.8) oscillate at high optical frequencies (10^{13}–10^{14}). If the signal is passed through an electronic filter centered at the difference (intermediate) frequency $\omega_{IF} = \omega_S - \omega_L$, only the slowly oscillating component can pass. We note that this signal is proportional to the original signal amplitude A_S as well as to the local oscillator amplitude A_L, which can be increased arbitrarily for low-noise amplification. Since the signal is proportional to the square root of the normal lidar signal (A_S^2), heterodyne detection also handles

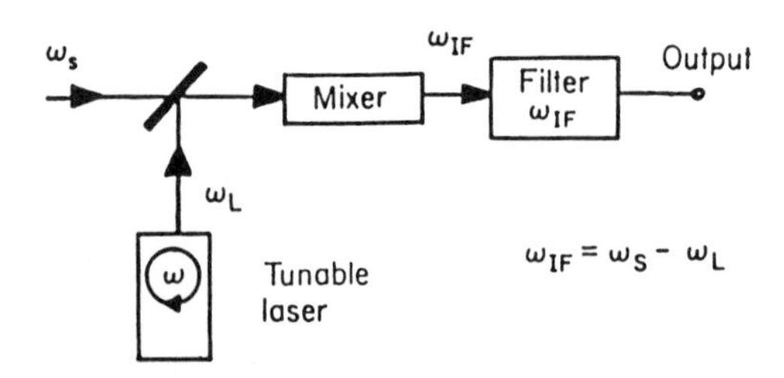

Figure 3.20. The principle of heterodyne detection.

the problems with dynamic range in an efficient way. Because coherence phenomena constitute an integrated part of heterodyne detection, it is important to adequately handle speckle noise through appropriate coherent lidar design.

By varying the local oscillator frequency or by analyzing the intermediate frequency spectrum, it is possible to detect Doppler shifts induced by movements of the backscattering particles. In this way it is possible to measure wind velocity with such a Doppler lidar system (Huffaker et al., 1984; Menzies, 1991). A wind speed of 1 m/s along the line of sight of the laser beam roughly corresponds to a frequency shift of 200 kHz at 10 μm wavelength.

An alternative way of measuring stratospheric wind speeds is to directly observe the Doppler shift in Rayleigh scattered narrow-band laser light. For this purpose Fabry–Pérot interferometers (Garnier and Chanin, 1992) or sharp atomic edge filters (Korb et al., 1992) are used on the detection side.

3.5.3. Atmospheric Turbulence

The useful range and the accuracy of a DIAL system will be limited by the number of backscattered photons detected by the system (the shot noise). However, even if a strong signal is obtained, the result is not necessarily accurate. This is related to the fact that lidar recordings have to be performed at both the *on* and *off* wavelengths (at least one and in general many lidar returns at each wavelength) in order to calculate the concentration from the DIAL curve. It is then very important that the atmospheric conditions remain the same to provide identical backscattering from particles and molecules. Because of atmospheric turbulence, as well as more macroscopic changes induced by winds (cloud and particle plume movements), such changes occur if the recordings for *on* and *off* wavelengths are not performed simultaneously. Then the curves are different for other reasons than for the presence of the gas to be studied, and erratic concentration values (even negative!) can be obtained. Simultaneous recording can in principle be performed by using two individually tuned lasers with overlapping beams. However, since the wavelength difference is normally chosen very small to avoid differential Mie scattering, it can be difficult to optically filter the two return signals in separate detection channels. By delaying one laser about 100 μs with respect to the other, the same detector and electronic transient digitizer can be used to detect the two laser returns. Experiments have shown (Killinger and Menyuk, 1981) that the time structure of turbulence is such that the atmosphere is practically "frozen" if the recordings are made within 1 ms. Since in many DIAL systems, many on- and off-resonance shots are averaged in separate memories, atmospheric movements are averaged out and no systematic error is made. Still the noise level increases owing to the turbulence. However, since the lidar curves are frequently evaluated out to maximum range, the signal becomes photon limited and the turbulence is in practice found to be no serious limitation even if a single laser is used at a repetition rate of 10–100 Hz, firing on, on- and off-resonance wavelengths every second shot. For such systems, changes in the DIAL curve larger than 1% can be considered significant. Thus, this number for differential absorption can form a basis for a calculation of the DIAL detectivity for different pollutants, as further discussed below (see Table 3.1 in Section 3.6.6). In order to test the proper performance of a DIAL system, a DIAL measurement and data evaluation can be performed without actually changing the wavelength of the laser between the shots. Clearly, a zero concentration value should be obtained for all range intervals. The size of nonzero concentration readings (positive or negative!) indicates the uncertainty in the particular measurement situation.

The problem of spectrally separating the on- and off-resonance signals can be solved in the gas correlation lidar technique (Edner et al., 1984). Gas correlation techniques for passive atmospheric monitoring have been described by Ward and Zwick (1975) and Lee and Zwick (1985). The principle of gas correlation lidar is shown in Fig. 3.21. A rather crude laser system with a comparatively broad line width is utilized. Since the laser wavelength is not sharp, it covers both on- and off-resonance wavelengths at the same time. However, the information for on- and off-resonance wavelengths can be separated on the detection side by splitting the received radiation into two parts. One part is detected directly, whereas the other part is first passed through a cell filled with an optically thick sample of the gas to be studied. In this way all the on-resonance radiation is filtered away, leaving only the off-resonance radiation to be detected. In the direct channel the sum of the on- and off-resonance radiation is detected. Unknown factors are eliminated by dividing the signals. The simultaneous detection of the two signals eliminates influences due to atmospheric turbulence and fluctuations due to changing reflectivity in airborne measurements using the ground as a topographic target.

3.5.4. Fixed DIAL System Design

Differential absorption lidar systems can be arranged with different degrees of versatility and complexity. Fixed systems frequently can have a simpler design than mobile systems. In particular, electricity and cooling demands can be met using the normal utilities available in laboratories. We will start with a description of a vertically looking ozone lidar system with a layout as shown in Fig. 3.22 (Edner et al., 1991a). Since the ozone molecule does not exhibit a strongly structured spectrum but rather only a gradually increasing absorption from 320 to 260 nm, it is necessary to place the on- and off-resonance

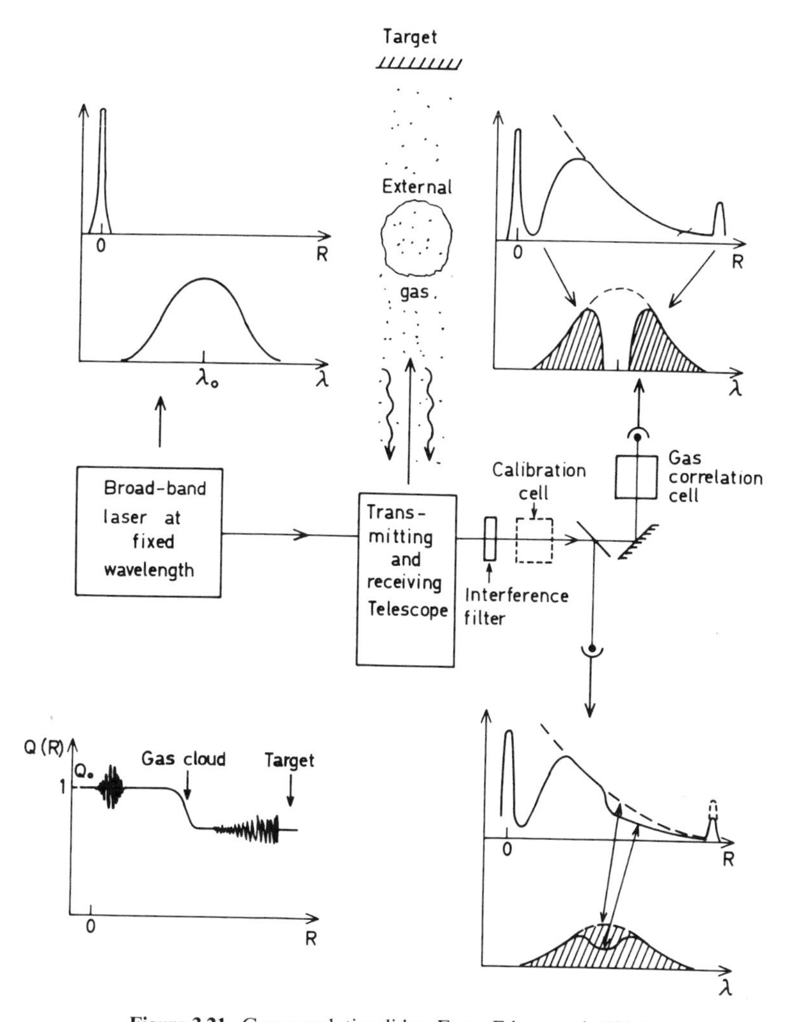

Figure 3.21. Gas correlation lidar. From Edner et al. (1984).

wavelengths much further apart than would normally be the case for DIAL. The absorption spectrum of ozone is shown in Fig. 3.23. Since there are no sharp structures, a continuously tunable laser is not needed but rather a step-tunable system. This can be achieved by an excimer laser or a frequency-quadrupled Nd:YAG laser in combination with stimulated Raman shifting. When a KrF laser is used the primary emission occurs at 249 nm whereas the quadrupled YAG emission occurs at 266 nm. The positions of the Stokes-

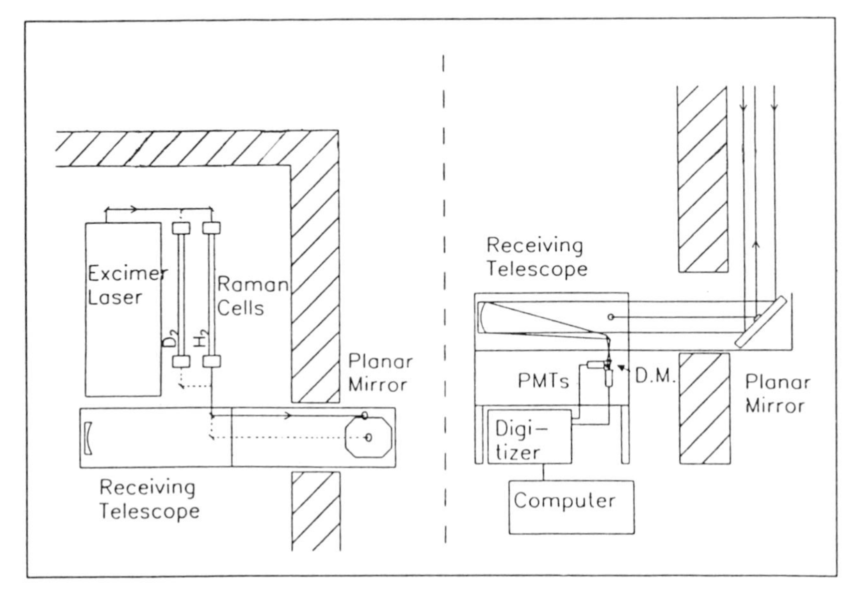

Figure 3.22. A fixed ozone lidar system based on an excimer laser. Left: top view; Right: side view. From Edner et al. (1991a).

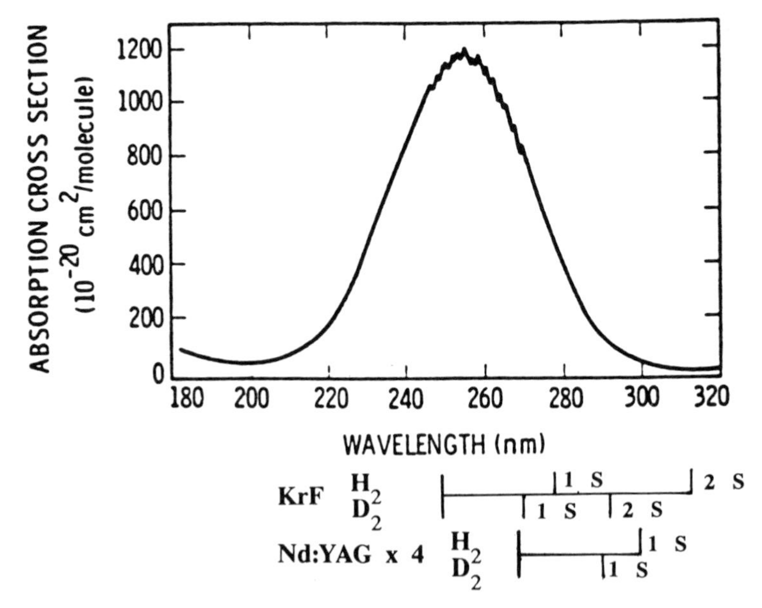

Figure 3.23. The ozone absorption cross section in the short-wavelength region, with fixed-frequency laser lines indicated. From Molina and Molina (1986).

shifted component obtained with hydrogen or deuterium in the Raman converting cell are indicated in Fig. 3.23.

The system shown in Fig. 3.22 uses a KrF laser (pulse energy, 300 mJ; repetition rate, 100 Hz) and incorporates two Raman cells, one filled with hydrogen and one filled with deuterium (Raman shifts 4155 and 2987 cm^{-1} for H_2 and D_2, respectively). In the system the first and second Stokes components from H_2 at 277 and 313 nm are normally utilized and are transmitted together with the primary excimer radiation via a first surface aluminized mirror vertically into the atmosphere. Backscattered radiation is reflected via the same mirror into a horizontally looking telescope using a 30-cm-diameter off-axis parabolic aluminized mirror. A dichroic mirror is used to direct the two Stokes components into separate photomultiplier tubes, mounted behind narrow-band interference filters. In this system the on- and off-resonance wavelengths can be transmitted and recorded separately because of the large wavelength separation. Each photomultiplier is connected to an input channel of a digital oscilloscope where the transients are averaged for many laser pulses. The data are transferred into a personal computer, where the DIAL curve is formed. A practical range of about 2 km is obtained.

As a further example of a fixed lidar system a CO_2-laser-based system operated by ENEA (Comitato Nationale per la Ricerca e per lo Sviluppo dell' *E*nergia *N*ucleare e delle *E*nergie *A*lternative) at Frascati, Italy; is chosen (Barbini et al., 1991). A schematic diagram of the system is shown in Fig. 3.24. Two line-tuned pulsed CO_2 lasers generating 80-ns-long pulses with energy up to 4 J are used, tuned to the on- and off-resonance wavelengths, respectively. The lasers are fired close in time for freezing the atmospheric conditions as discussed above. The two laser beams are sent along the rotation axes of the telescope, which is mounted in a dome at the top of a 5-m-high tower. By this arrangement it is possible to keep the overlap between the transmission and detection lobes. The radiation is detected by a HgCdTe detector cooled to 77 K. This system is used for mapping of H_2O and O_3 and has a range of about 5 km for spatially resolved measurements.

3.5.5. Mobile DIAL System Design

Mobile DIAL systems are very versatile for research and operational measurements. A number of efficient systems have been constructed (see e.g., Hawley et al., 1983; Jolliffe et al., 1987; Staer et al., 1984; Fredriksson et al., 1981; Edner et al., 1987b; Wolf et al., 1990; Zanzottera, 1990). In Fig. 3.25 two Swedish systems of similar layout are shown, one constructed at the Chalmers Institute of Technology (Fredriksson et al., 1981) and the other one at the Lund Institute of Technology (Edner et al., 1987b). A photograph of the newer and larger system during measurements in an industrial area is shown in Fig. 3.26.

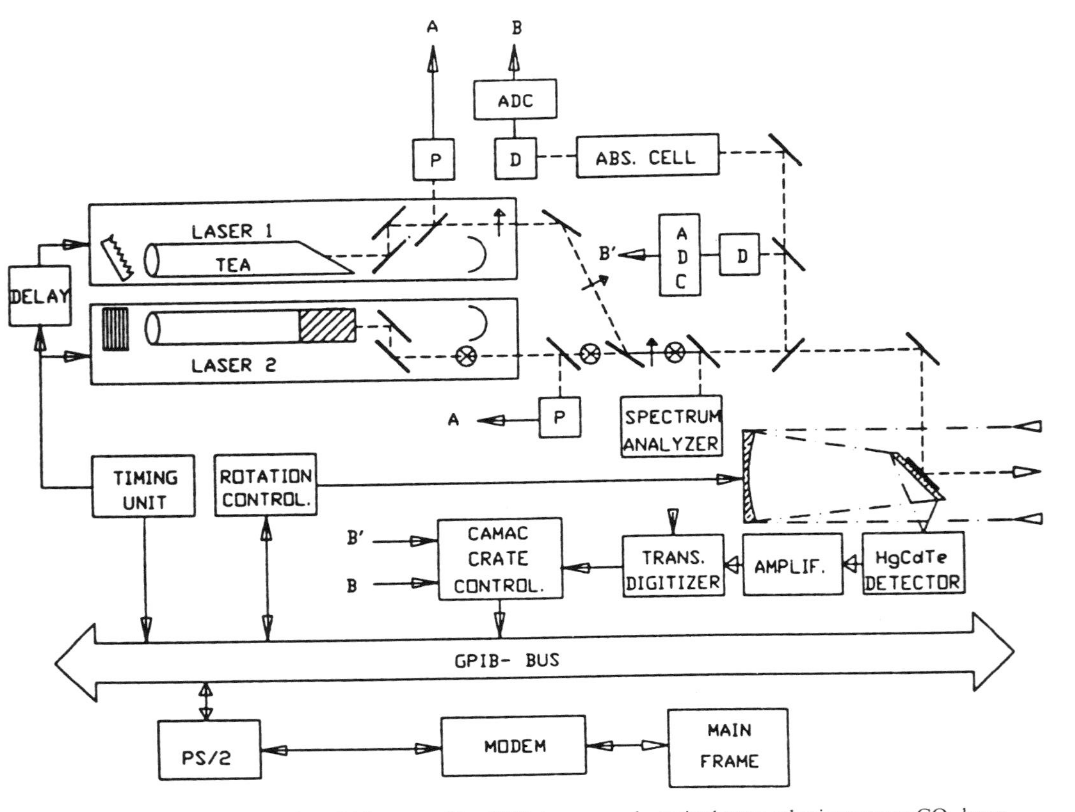

Figure 3.24. Layout of a CO_2 DIAL system *Key*: TEA, transversaly excited atmospheric pressure CO_2 laser; ADC, analog-to-digital converter; D, detector; P, power meter; GPIB, general purpose interface bus; PS/2, computer. From Barbini et al. (1991).

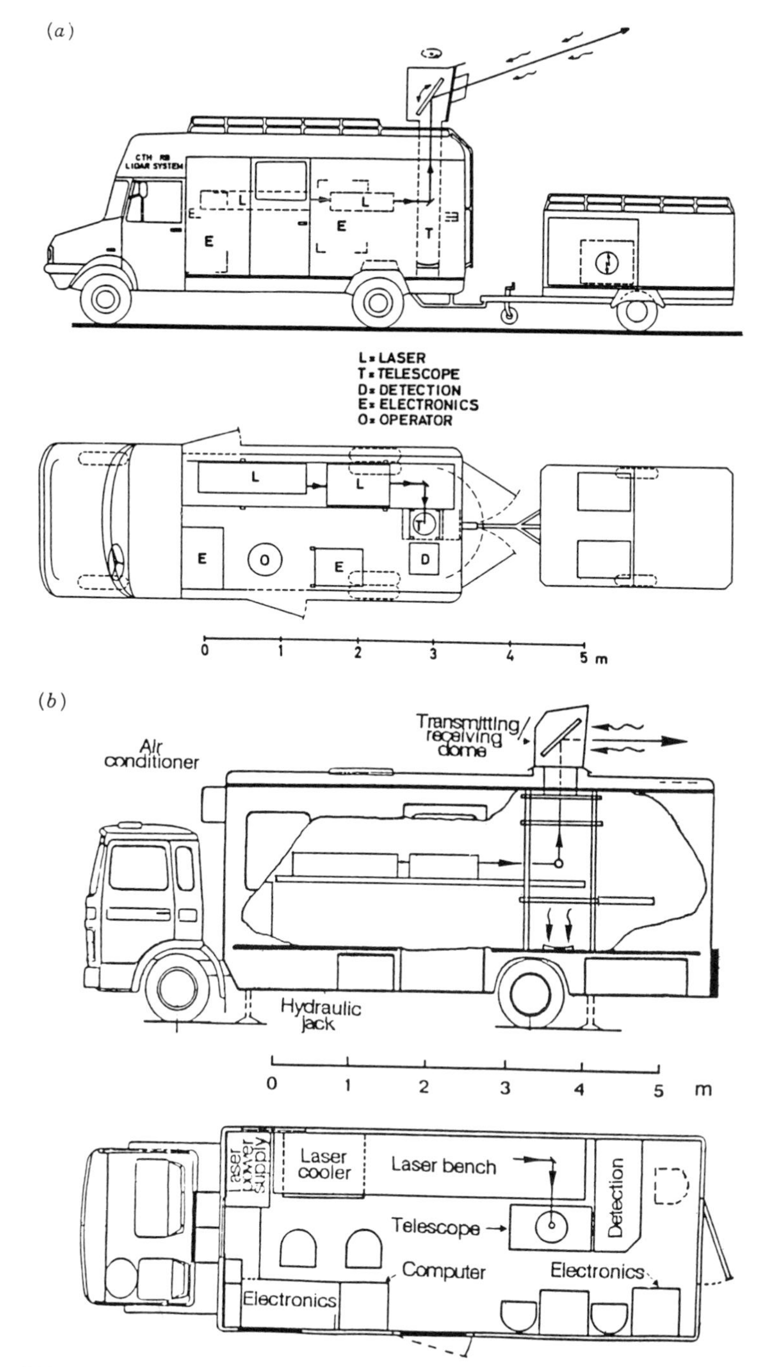

Figure 3.25. Two Swedish mobile DIAL systems (a,b). From Fredriksson et al. (1981) and Edner et al. (1987b), respectively.

Figure 3.26. Photograph of the newer Swedish mobile lidar system.

A schematic of the optical arrangements in the older system is shown in Fig. 3.27. The new system is similar but uses updated optics and electronics.

Nd:YAG lasers are used as transmitters in the Swedish DIAL systems. The pulse energy at 10 Hz at the fundamental wavelength was 250 mJ in the original system of 1981. By using frequency conversion in nonlinear crystals, 100-mJ pulses at 532 nm (the second harmonic) and 50-mJ pulses at 355 nm (the third harmonic) could then be generated. Since that time, laser technology has been greatly improved. In the new system operating at 20 Hz the fundamental pulse energy is presently 1.200 J, with 500 mJ at 532 nm and 200 mJ at 355 nm. The frequency-doubled output is used to pump yellow and red rhodamine dyes in a dye laser in order to reach UV wavelengths around 300 nm for DIAL measurements of SO_2 and O_3. The frequency-tripled radiation is used to pump blue stilbene and coumarin dyes for reaching NO_2 wavelengths, and after frequency doubling NO and Hg wavelengths. Arrangements are made to provide fast wavelength switching from on- to off-resonance wavelengths every second laser shot. In the older system this is made by rocking the laser oscillator grating by employing a stepper-motor-driven eccentric wheel acting on a cam on the grating mount. In the newer system two independently set gratings are used in the oscillator and a rotating quartz block in used to engage one or the other grating for alternate laser shots.

For high pulse energies the laser beam is expanded in a Galilean telescope with one negative and one positive lens to reduce the beam power density before hitting the system steering mirror. (A Newtonian telescope cannot be used since air breakdown is obtained in the focus between the two positive lenses.) The laser beam is sent vertically and coaxially with the receiving telescope toward a large first surface aluminum folding mirror used to direct

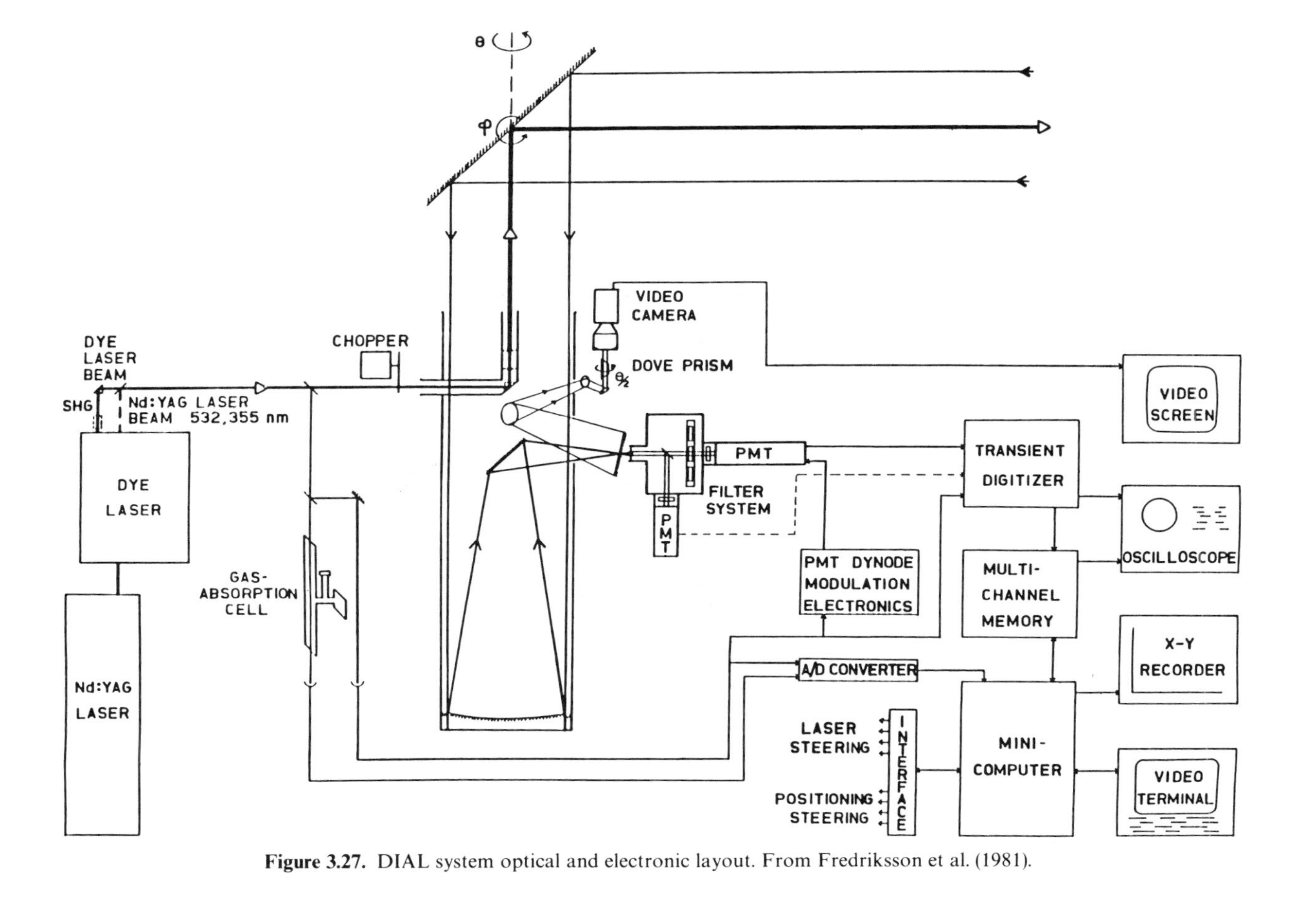

Figure 3.27. DIAL system optical and electronic layout. From Fredriksson et al. (1981).

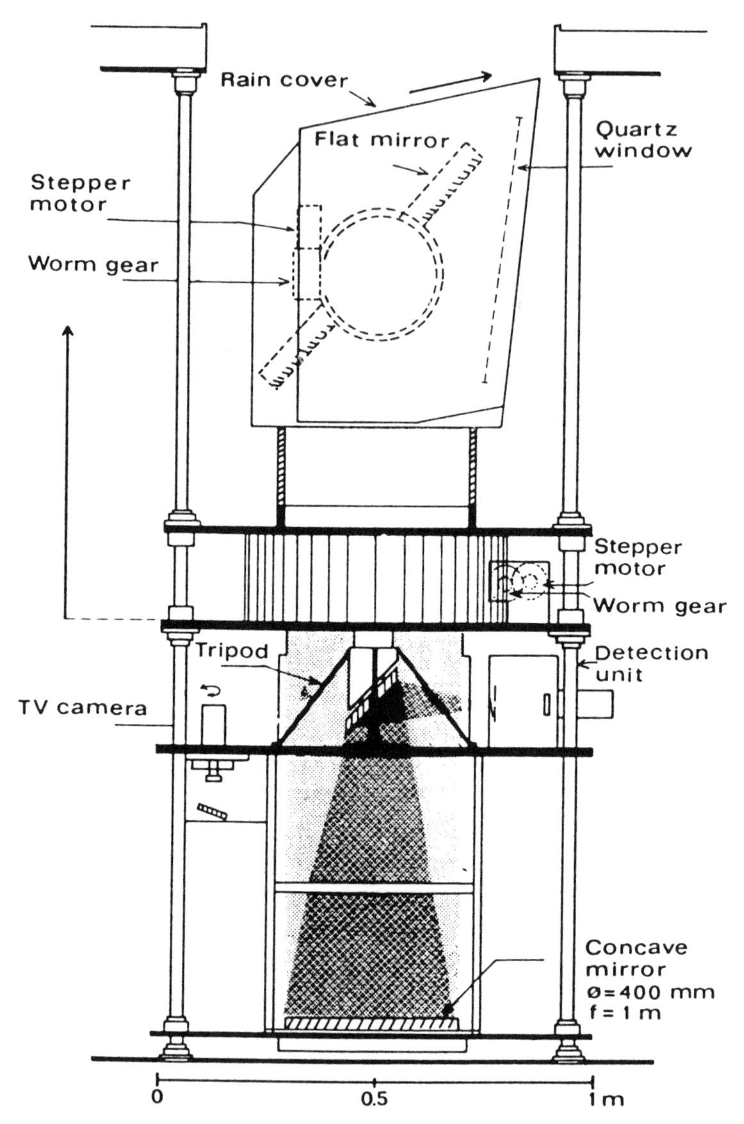

Figure 3.28. Arrangement with receiver telescope and retractable mirror dome in mobile DIAL system. The dome is shown in its retracted position. From Edner et al. (1987b).

the laser beam into the measurement direction. The mirror is placed in a dome equipped with a large quartz window, sealing off the lidar system toward the outdoor weather conditions. The dome rotation and the mirror vertical positioning are computer controlled via powerful stepper motors. In Fig. 3.28 the layout of the telescope/dome arrangement of the newer system is shown. In order to reduce the overall height of the mobile system during transport, the dome can be retracted into the laboratory as shown in the figure. This also improves the serviceability of the dome construction. A motorized cover can be moved forward for rain protection of the quartz window once the dome has been hoisted out of the roof orifice into operating position.

Light backscattered from the atmosphere or from topographic targets passes into the system via the quartz window and is directed via the folding mirror down into a vertical Newtonian telescope that has a diameter of 30 cm in the older system and 40 cm in the newer system. The telescope mirrors are spherical (for cost reasons) rather than parabolic. Via a secondary mirror the light is focused toward an image plane, where an aperture is placed defining the telescope field of view. With a telescope focal length of 1 m an aperture diameter of 2 mm corresponds to a telescope field of view of 2 mrad, matching a typical laser beam divergence. A small beam divergence and a correspondingly small telescope field of view is desirable in order to achieve rejection of ambient light, competing with the laser-induced signal during daytime operation. This is not an important consideration below 300 nm, where the stratospheric ozone layer provides an efficient cutoff of background radiation. The vertical telescope arrangement with a rooftop steering mirror has the following important advantages over most other constructions:

- 360° horizontal scanning capability
- Eye-safe near-field operation, with the laser beam leaving system high above the ground
- Easy optical alignment
- Space-saving arrangement in the mobile laboratory
- Easy weather protection

On the other hand, high elevation angles cannot be used. For vertical sounding the whole dome can be dismantled or, more conveniently, a fixed 45° mirror be used in front of the quartz window.

The aperture is arranged as a hole in a polished metal mirror that directs all light, except the signal photons passing the aperture, toward a ccd camera that displays sharply the target area except a black spot at the position of the laser beam. Such an arrangement is very useful for swift and accurate positioning of the measuring beam. When the dome is rotated, the TV image will also rotate unless counter measures are taken. In the older system the TV camera is preceded by a Dove prism that is rotated with half the angular speed relative to the dome rotation. In the newer system the whole TV camera is rotated synchronously with the dome rotation.

When the stepper motors are activated, 10-turn precision potentiometers are also turned to generate analog voltages that are compared with set voltages defining the limits of allowed firing directions. If these limits are accidently passed, e.g., owing to computer or operator failure, the laser firing is inhibited. Such a system is an additional security measure when non-eye-safe wavelengths are used (see also Section 3.5.7).

The light that passes the aperture is collimated by means of a lens. A narrow-band high-transmission interference filter centered at the laser wavelength is used to reject daylight. A number of interference filters corresponding to different pollutants (wavelengths) are mounted on a filter wheel. The optical transient is converted into an electrical transient in a PMT. In order to reduce the dynamic range requirements the PMT voltage is ramped as discussed in Section 3.5.1. In this way the close-range intensity is reduced. Beyond about 600 m the true signal shape is obtained.

The transients from the PMT are fed to a transient digitizer, which is a critical component in a DIAL system. The transient digitizer normally has an 8-bit resolution (256 intensity levels), a very limited value, which is the origin of the dynamic range reduction needs. For tropospheric air pollution monitoring the separation between the temporal channels in the transient digitizer is 10 ns, which corresponds to a range interval of 1.5 m, matching a commonly used laser pulse length of 10 ns. The data string of numbers from the transient digitizer is fed to a personal computer. The on- and off-resonance transients are added in separate memories. In the older system a separate hardware signal averager was used since the computers were not fast enough to directly handle the data stream. Before one forms the DIAL curve by division of the on- and off-resonance curves, it is important to subtract the background intensity due to the ambient light and the PMT dark current. This is done by automatically blocking the laser beam every ninth shot and subtracting this signal multiplied by 8 from the intensity collected during the eight previous shots. This is done for the on- as well as the off-resonance wavelengths.

For a given measurement direction normally 50 data collection cycles as described above are used, corresponding to 400 transients for each wavelength. Then the DIAL curve as the ratio of the on- and off-resonance curves is formed. From the DIAL curve the range-resolved concentration curve is then calculated for the given measurement direction using Eq. (3.6). Examples of this process were already given in Figs. 3.17 and 3.18. A new measurement direction can then be chosen, and the procedure is repeated. After the final measurement direction is finished, the procedure can be repeated if it is desired to form more representative averages. Finally, a concentration map can be calculated from all the collected data. We shall return to this process in Section 3.6.

To make the mobile system fully self-contained and operational, a motor generator on a trailer is brought along. The new system is equipped with a 20-kV · A diesel-powered unit, amply covering the demands of the system. Apart from the laser, particularly energy-consuming units are a closed-loop cooler for the laser and the laboratory air-conditioning system. The new system is also equipped with four hydraulically operated supporting legs that, when in use, fully stabilize the mobile laboratory for accurate beam pointing.

3.5.6. Airborne DIAL System Design

Airborne DIAL systems provide monitoring of gas constituents over wide areas. Such systems are normally looking downward or upward in a fixed position, but scanning systems utilizing a laterally moving mirror can also be used. Because of the swift movement, special requirements as to high repetition rates pertain. Furthermore, it is necessary that the on- and off-resonance pulses be fired simultaneously or with very small temporal spacing, since the atmospheric backscattering conditions otherwise certainly would be changed. For airborne operation, weight and power consumption are normally also critical factors.

The layout of an airborne DIAL system operated by NASA is shown in Fig. 3.29 (Browell et al., 1983; Browell, 1991). The system utilizes two individually Nd:YAG pumped dye lasers, firing with a temporal separation of 100 μs. The laser beams are split into two equal parts, one transmitted toward the nadir and one transmitted toward the zenith. Correspondingly, two receiving telescopes are operated back to back. The system is capable of large area mapping of particles, ozone, and water vapor.

3.5.7. Eye Safety

DIAL systems are operated in the open environment, and it is necessary that the measurements be performed in such a way that eye damage will not occur to persons in the operating zone of the system. The safe use of lasers is regulated by American standards, which have also been adopted in most other countries (ANSI, 1986; Sliney and Wolbarsht, 1980). Of special concern is the spectral region from 400 to 1400 nm, where the cornea is transparent and the radiation is focused onto a small spot on the retina. Outside this region the radiation is absorbed before reaching the retina and the irradiation threshold values are relaxed by a factor of about 10^4. With a normal laser beam divergence of a few milliradians the UV laser beam for NO, Hg, O_3, or SO_2 measurements is

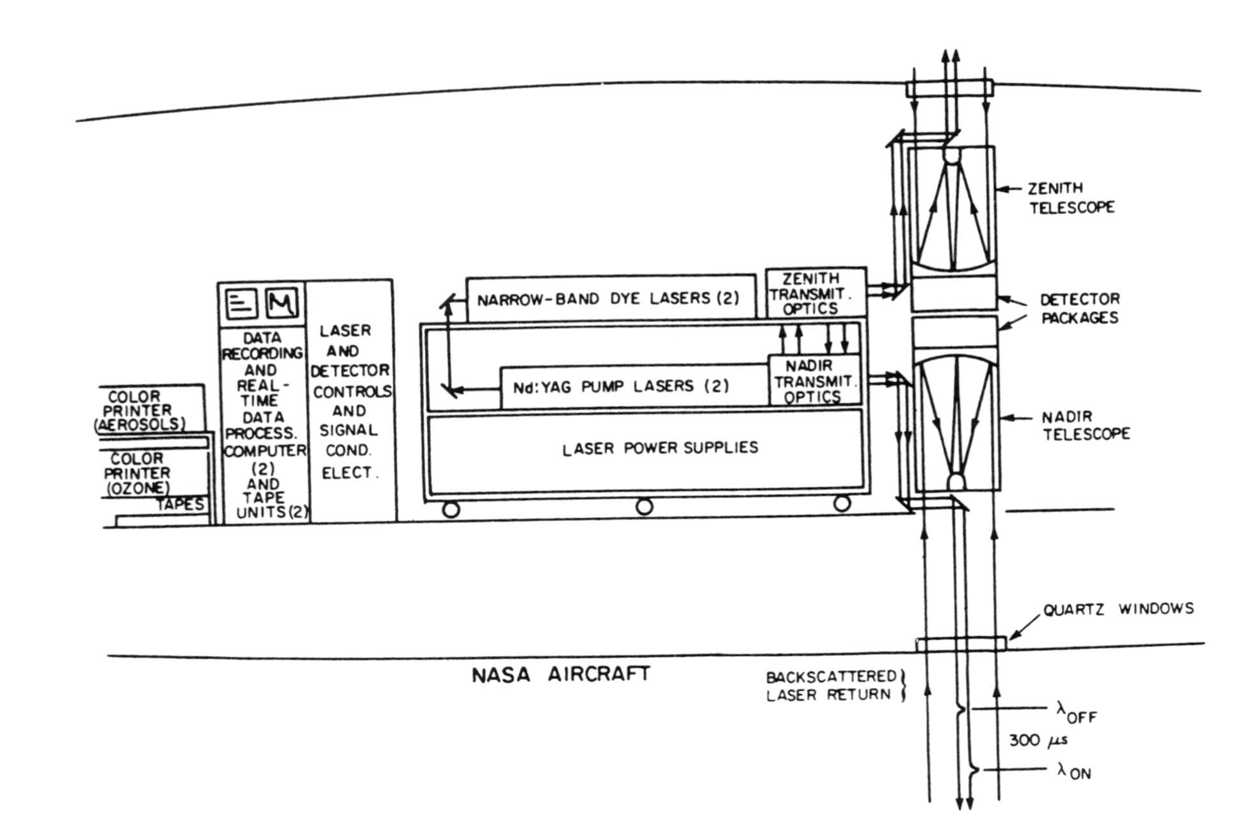

Figure 3.29. The NASA [National Aeronautical and Space Administration (U.S.)] airborne DIAL system. From Browell (1991).

eye-safe for distances beyond 100 m or shorter. Combined with rooftop transmission from mobile systems, UV DIAL measurements give little safety concern. For NO_2 monitoring in the blue region the eye safety distance is several kilometers and special precautions must be taken with continuous beam-path monitoring during measurements. Special consideration must be given to the possible use of binoculars by ground observers of a lidar airborne system.

3.6. DIAL MONITORING OF TROPOSPHERIC GASES

In this section we shall give several examples of DIAL monitoring of atmospheric gases. Most of the examples concern measurements of industrial pollution. The measurements are performed with DIAL systems as described in the previous section.

3.6.1. SO_2 Monitoring

Sulfur dioxide is one of the most important pollutants produced in the burning of fossil fuels. The amount of SO_2 produced is directly related to the sulfur contents in the fuel. In the atmosphere the gas is converted into sulfuric acid and sulfate particles largely responsible for acidification of certain areas. The absorption spectrum of SO_2 is prominent in the region around 300 nm, as shown in Fig. 3.5. There also the most common wavelength pair utilized in DIAL measurements is indicated.

The DIAL technique is very powerful for determining the total flux of a pollutant from an industrial plant. This can be done by making a vertical scan with the lidar system beam through the atmosphere downwind from the plant. In this way not only the stack emissions are captured but also the diffuse emissions from ventilators and leaking valves. Such a downwind recording at a Swedish pulp mill is shown in Fig. 3.30. The DIAL curves recorded in different directions are automatically evaluated on the system computer, and the concentrations are represented on a type of gray scale. Data were collected for a total time of 20 min. The area integrated concentration value N_A can be calculated from the data in the figure. By multiplying this value with the wind velocity component perpendicular to the measurement plane, the total flux F_{tot} from the industry can be determined:

$$F_{tot}\,[\mathrm{kg/s}] = N_A\,[(\mathrm{kg/m^3})\cdot m^2]\cdot v_\perp\,[\mathrm{m/s}] = \int_A N\,da\cdot v\cos\phi\,[\mathrm{kg/s}] \quad (3.9)$$

where ϕ is the angle between the measurement direction and the direction

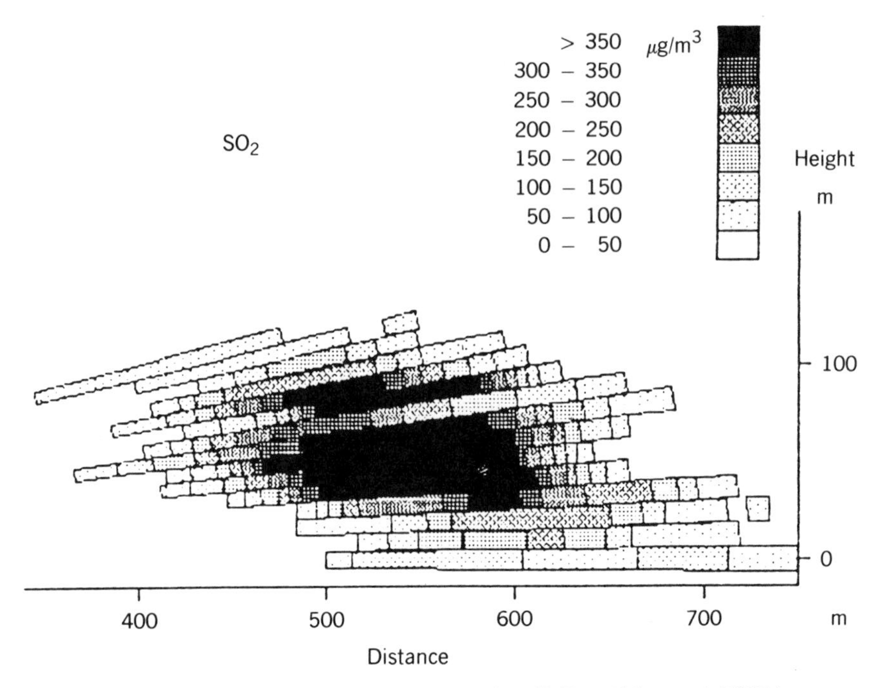

Figure 3.30. SO_2 plume scan at a Swedish pulp mill. From Edner et al. (1987b).

perpendicular to the plume. For accurate flux determinations it is obviously as important to make a correct wind velocity measurement as to perform the DIAL measurements correctly. For the measurement situation shown in Fig. 3.30 an hourly flux of 230 kg is estimated.

Since wavelengths for DIAL SO_2 monitoring are obtained by frequency doubling of very efficient rhodamine dyes, high-energy laser pulses can be obtained and thus a substantial range is available. This is illustrated in Fig. 3.31, where on- and off-resonance curves and the corresponding DIAL curves are shown for a horizontal range out to 4 km. At a distance of about 3 km thin clouds were encountered, but the signal was retrieved also after the clouds were passed. The recording was performed as part of a study of a possible conversion of H_2S into SO_2 downwind from an Icelandic geothermal field. The SO_2 concentration was found to be very low and did not increase for larger ranges and correspondingly longer times available for atmospheric chemistry to take place.

Various aspects of practical SO_2 monitoring have been discussed by Egebäck et al. (1984).

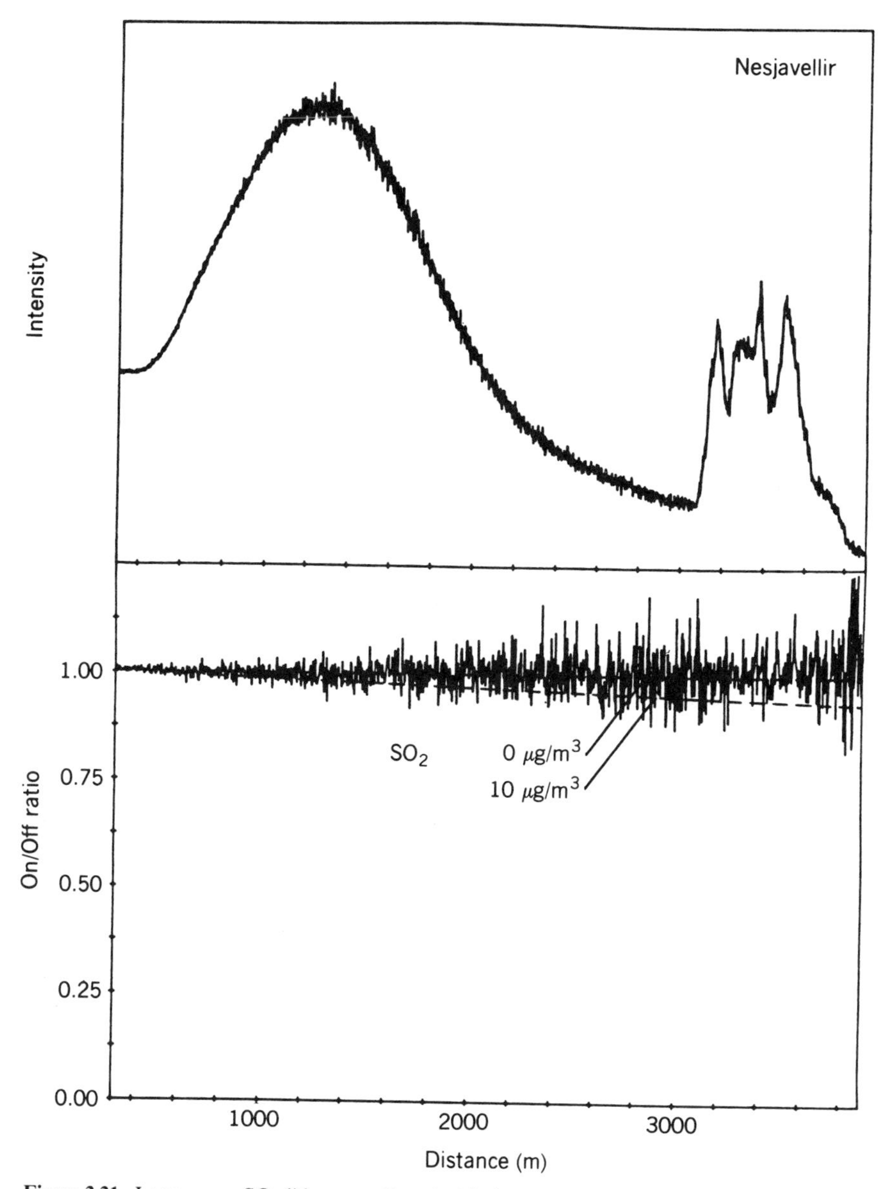

Figure 3.31. Long-range SO_2 lidar recordings (at Nesjavellir, Iceland) for on- and off-resonance wavelengths and corresponding divided (DIAL) curve. From Edner et al. (1991c).

3.6.2. O_3 Monitoring

Great attention is presently being focused on the ozone molecule. A steadily increasing concentration of tropospheric ozone is thought to be at least partially related to the increasing damage to forests observed throughout Europe. The stratospheric ozone layer is being depleted, most likely because of chemical reactions with fluorinated hydrocarbons (Freons). Measurements of stratospheric ozone have been performed by many groups (Werner et al., 1983; Uchino et al., 1983; Megie et al., 1985; Browell et al., 1990; Stefanutti et al., 1992a). In order to reach stratospheric heights, lidar systems with high pulse energies ($\simeq$ 1 J), large telescope diameters ($\simeq$ 1 m), and photon-counting detection electronics are needed. For stratospheric measurements it is necessary to use wavelengths that are only weakly absorbed by ozone (long wavelengths) in order to avoid excessive absorption already at lower heights. As discussed in Section 3.2.1 a special problem with O_3 monitoring is the need for a wide wavelength separation, necessitating a correction to be applied for different Mie scattering at the different wavelengths (Browell, 1985). This correction is particularly critical when particle layers are being passed, e.g., at the tropopause. DIAL data for tropospheric ozone are shown in Fig. 3.32 (Edner et al., 1992b). Here the new Swedish mobile DIAL system was used at high elevation angles. The range- and angle-resolved data can be converted into a vertical profile as shown in the figure.

3.6.3. NO_2 Monitoring

Nitrogen oxide (NO) is formed in all high-temperature combustion and is an important pollutant from industrial activities and, in particular, from automotive traffic. Shortly after the emission of NO into the atmosphere this molecule is oxidized into NO_2 and further on to HNO_3, which contributes to water and soil acidification. NO_2 absorbs in the blue spectral region and was the first pollutant to be measured by the DIAL technique (Rothe et al., 1974a, b; Grant et al., 1974). Recordings for NO_2 have already been shown in Fig. 3.17.

Practical NO_2 monitoring by DIAL techniques has been discussed by Fredriksson and Hertz (1984).

3.6.4. NO Monitoring

The NO molecule has a strong absorption band, the γ-band at short UV wavelengths, as shown in Fig. 3.33. First atmospheric UV lidar measurements were reported by Aldén et al. (1982b), who used stimulated Raman scattering to produce the required radiation. For DIAL plume mapping of NO frequency mixing was employed, first generating 575 nm by dye radiation frequency

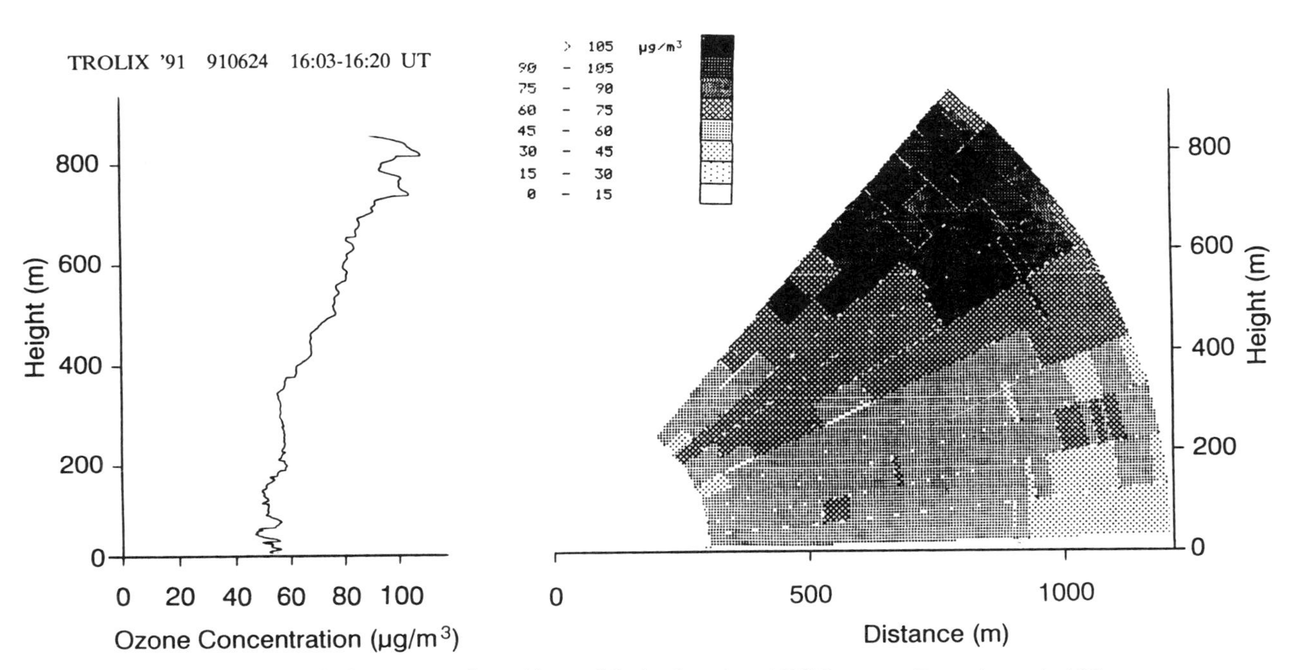

Figure 3.32. Vertical ozone sounding with a mobile dye-laser-based DIAL system. From the vertical lidar sweep a vertical ozone profile has been evaluated. From Edner et al. (1992b).

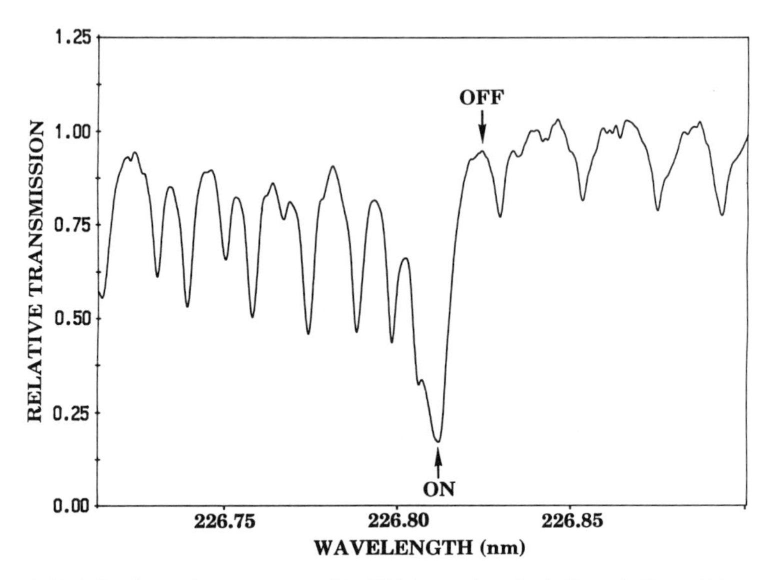

Figure 3.33. NO absorption spectrum with DIAL wavelengths indicated. From Edner et al. (1988).

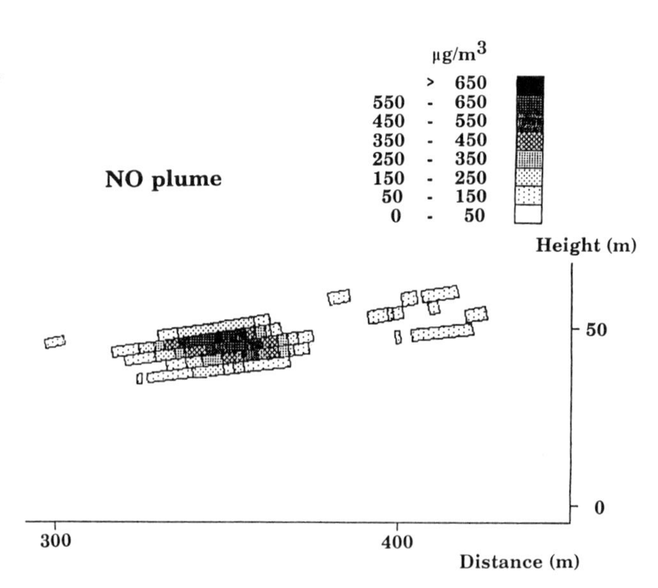

Figure 3.34. Vertical NO scan through the plume from a small heating plant. From Edner et al. (1988).

doubling and then mixing with residual fundamental Nd:YAG radiation in a second KDP crystal to reach 226 nm (Edner et al., 1988; Jolliffe et al., 1987). The result from a vertical scan downwind from a smokestack is seen in Fig. 3.34.

Since NO absorbs at approximately half the wavelength on NO_2, it is possible to measure the two gases simultaneously using the same frequency-doubled laser (Nickolov and Svanberg, 1986). The low conversion efficiency of the KPB (potassium pentaborate) frequency-doubling crystal was initially a drawback but with the occurrence of the new nonlinear material BBO (β-barium borate) double species monitoring became very realistic. Such measurements have been reported by Kölsch et al. (1989).

3.6.5. Monitoring of Other Gases

DIAL monitoring of a number of other pollutant gases has also been demonstrated, although the techniques may be less operational than for SO_2, O_3, NO_2, NO, and Hg. DIAL monitoring of toluene and benzene at 267 and 253 nm, respectively, has recently been reported (Milton et al., 1992). Molecular chlorine (Cl_2) exhibits a broad absorption spectrum in the UV region. Like for ozone it is necessary to have a sufficiently large wavelength separation in the DIAL measurements. A demonstration of DIAL measurements on an artificial chlorine cloud has been made by Edner et al. (1987a). HCl emitted from incineration ships has been monitored by Weitkamp (1981) using a DF laser transmitter at 3.6 μm. By using frequency mixing techniques, wavelengths around 5 μm could be produced by CO_2 lasers, allowing DIAL monitoring of CO (Killinger et al., 1980), NO (Menyuk et al., 1980), and hydrazine and other fuels (Menyuk et al., 1982). Hydrocarbons can be measured at the CH-stretch wavelength 3.4 μm, and practical DIAL measurements have been performed by Milton et al. (1988). Special, simplified systems have also been constructed for CH_4 detection, i.e., from leaking natural gas pipelines. In one system the gas correlation lidar technique (discussed in Section 3.5.3) was employed for methane leak detection (Galetti, 1987). Several demonstrations of direct and heterodyne DIAL monitoring at CO_2 laser wavelengths ($\approx 10\ \mu$m) have been made covering species such as Freon-12, ethylene, ozone, ammonia, and sulfur hexafluoride (see, e.g., Grant, 1989).

3.6.6. Sensitivity and Detection Limits of DIAL

Preferred wavelength regions, differential absorption cross sections, and detection limits in ppb for a 100-m measuring path are given in Table 3.1 for pollutants for which realistic DIAL measurements can be performed. The detection limits correspond to a 1% change in the DIAL curve over the range

Table 3.1. Important DIAL Detected Pollutants with Detection Limits for a 100-m Measurement Interval (200-m Absorption Path)[a]

Gas	Wavelength	Differential Cross ($atm^{-1} \cdot cm^{-1}$)	Detection Limit (ppb)
NO	226 nm	100	5
Benzene	253 nm	61	8
Hg	254 nm	670,000	0.001
Toluene	267 nm	30	17
O_3	280 nm	30	9
SO_2	299 nm	25	20
NO_2	450 nm	10	50
HCl	3.6 μm	6	90
C_2H_4	10 μm	31	16

[a] Only rough values are given. Differential cross-sections depend on details in wavelength-pair choice, laser line width, etc.

interval chosen. Many other gases can be studied, but the experience with such measurements might be less extensive. The table is given to provide the reader with a realistic assessment of the practical applicability of the DIAL technique.

3.7. SPECIAL CASE STUDY: ATOMIC MERCURY MONITORING

Mercury is a troublesome pollutant that is unique in the atmosphere, as it is mainly present in atomic form (Jepsen, 1973). All other pollutants are molecules, for which each electronic transition is accompanied by thousands of vibrational–rotational transitions, giving rise to a distributed band spectrum. For an atom the electronic transition probability (oscillator strength) is instead concentrated in a single line, or at least in a low number of isotopic and hyperfine-structure lines, which because of Doppler and collisional broadening appear as a single line. Because of this the DIAL detection limit for Hg is about 3 orders of magnitude lower than for other pollutants such as SO_2 or NO_2. When these latter pollutants can be detected on the ppb level, mercury can be detected at ppt concentrations. That is exactly what is needed, since atomic mercury has an Atlantic background concentration of about 0.25 ppt or 2 ng/m^3 (Slemr et al., 1981).

Locally increased amounts of mercury in the air can be caused by industrial activities such as chlorine-alkali plants, where liquid mercury electrodes are used. Mercury is also emitted from coal-fired power plants as well as from incineration plants. Recently attention has been given to strong mercury emissions from crematoria, which in some countries give rise to more mercury pollution than normal incineration plants. As a toxic heavy metal, mercury and its cycle in the environment have been much studied (see, e.g., Ngriau, 1979; Mitra, 1986; Lindqvist, 1991).

Apart from being a pollutant caused by human activities, mercury is a very interesting geophysical tracer gas associated with ore deposits (Bristow and Jonasson, 1972; McCarthy, 1972), geothermal energy (Robertson et al., 1977; Varekamp and Buseck, 1983), and seismic and volcanic phenomena (Varekamp and Buseck, 1981). These latter aspects are illustrated in Fig. 3.35 (Svanberg, 1991).

The mercury resonance line ($6s^1S_0$–$6p^3P_1$) occurs at 253.65 nm. Its width and shape at ambient atmospheric conditions are shown in the cell absorption spectrum in Fig. 3.36. There also the isotopic contributions to the Hg line shape are indicated and neighboring "forbidden" absorption lines of molecular oxygen are shown (see also Fig. 3.18). First attempts to remotely monitor mercury were made using Anti-Stokes Raman shifting of frequency-doubled dye laser radiation (Aldén et al., 1982a). On- and off-resonance curves and a

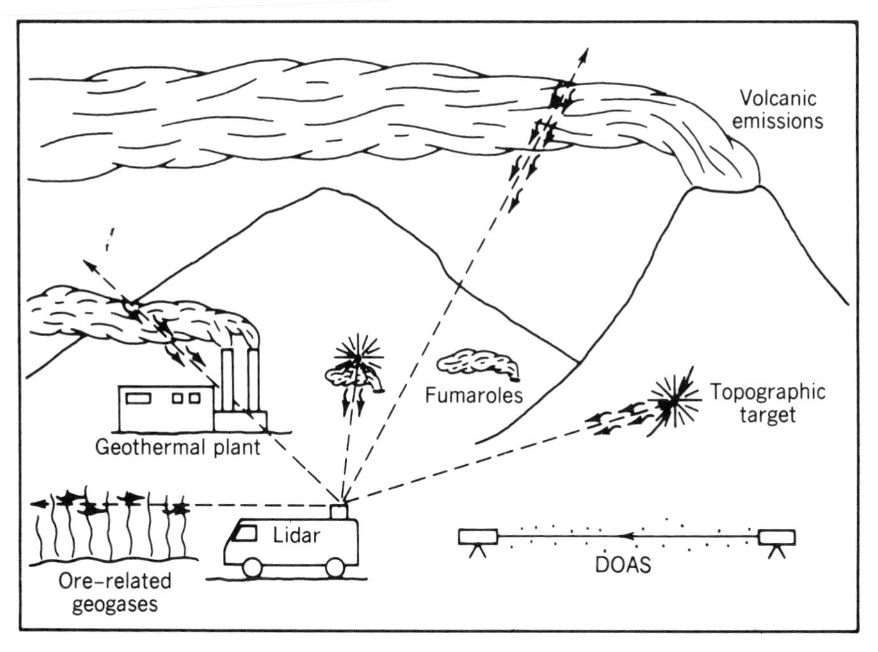

Figure 3.35. Measurement scenarios for atomic mercury of geophysical origin. From Svanberg (1991).

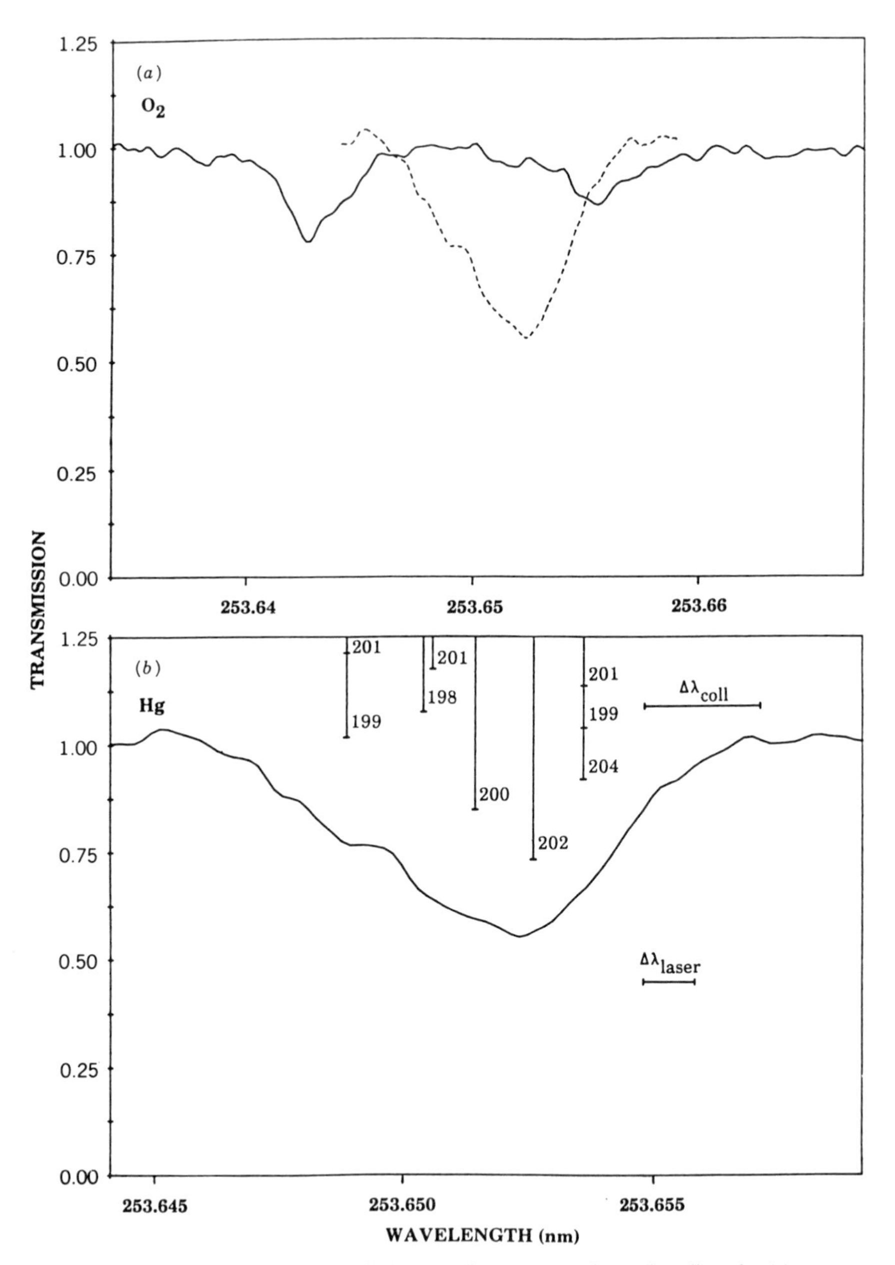

Figure 3.36. Laboratory recording of the atomic mercury absorption line: in (a) oxygen "forbidden" lines are also shown: in (b) the isotopic contributions to the line shape are shown. From Edner et al. (1989).

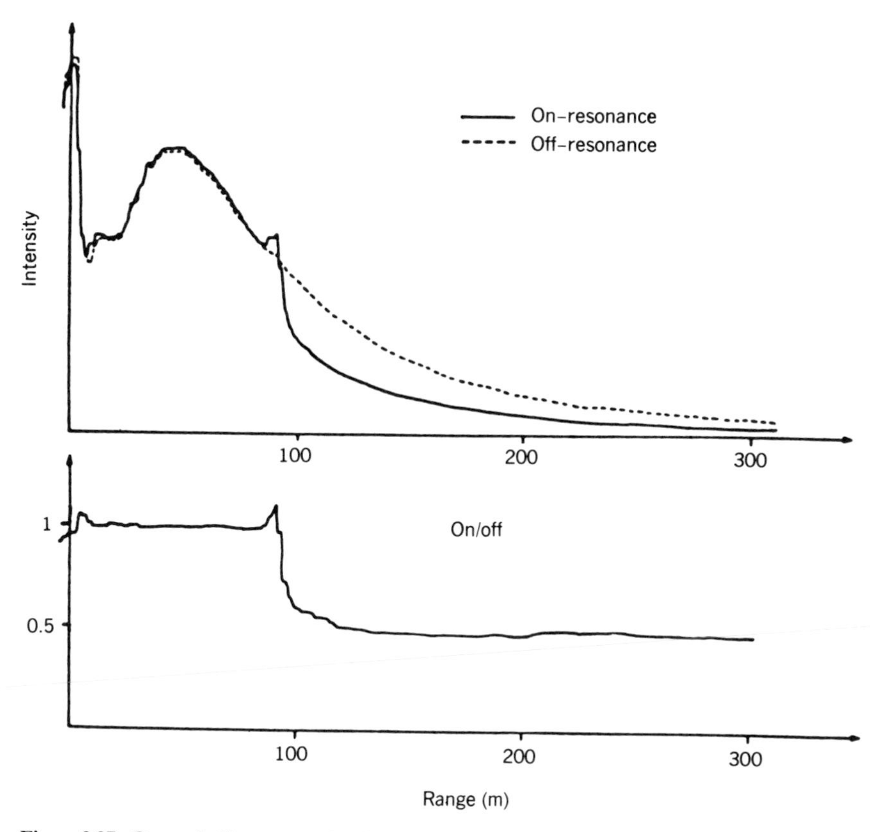

Figure 3.37. On- and off-resonance Hg lidar curves for an artifical mercury cloud at about 100-m distance. Note the fluorescence emission peak in the on-resonance curve. In the lower part of the figure the divided (DIAL) curve is shown. From Aldén et al. (1982a).

resulting DIAL curve from an artificial mercury cloud at a distance of about 100 m are shown in Fig. 3.37. However, the laser power and line width did not allow practical monitoring. In the on-resonance curve it is possible to see a fluorescence signal associated with the dense mercury cloud. It occurs weakly even at atmospheric pressure because the 254 nm line is the only radiative decay channel available. The occurrence of fluorescence does not induce any problems in the evaluation of mercury concentrations, as demonstrated by Edner et al. (1989). Real measurements of industrial mercury pollution became possible when narrow-band coumarin dye laser emission at 508 nm could be efficiently directly doubled using the new nonlinear crystal BBO. Figure 3.38 shows the result of a horizontal scan around the cell house of a Swedish chlorine-alkali plant (Edner et al., 1989), performed with the new Swedish

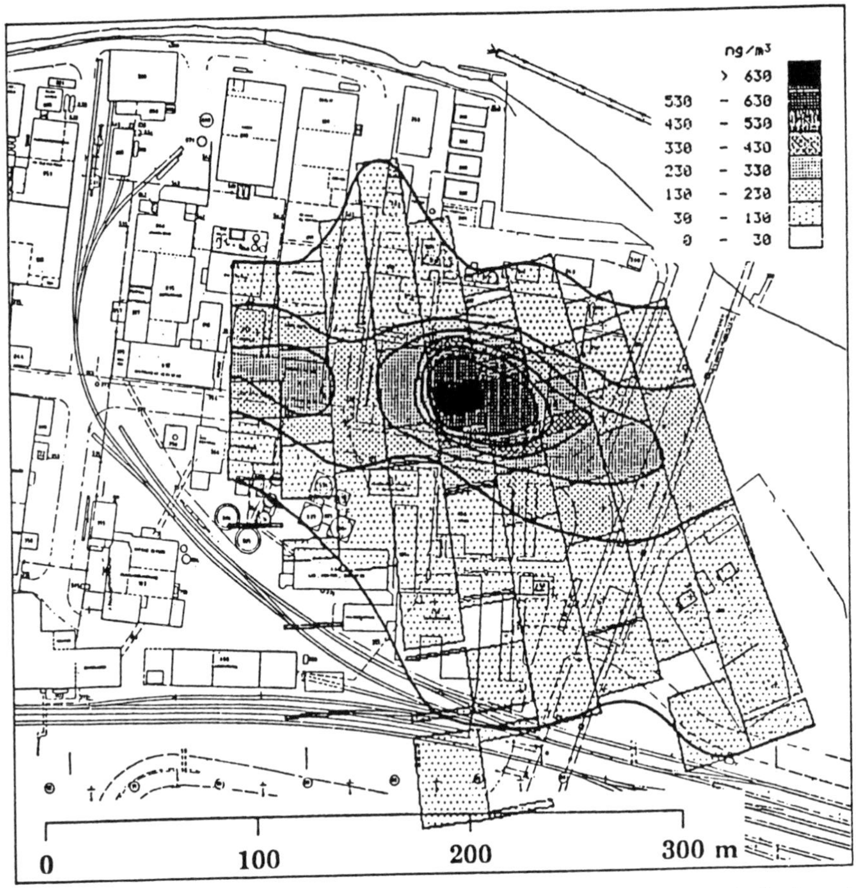

Figure 3.38. Horizontal lidar scan at a Swedish chlorine-alkali plant showing the horizontal distribution of atomic mercury. From Edner et al. (1989).

DIAL system. From data in vertical scans downwind from the cell house and available wind data, an emission of 30 g/h is obtained. Measurements at a similar Italian plant yielded similar emission values and an estimated yearly mercury emission of 500 kg (Ferrara et al., 1992). A vertical scan featuring a strong plume and a weaker one is shown in Fig. 3.39. The mercury emission from the stack of a crematorium is shown in Fig. 3.40 during the cremation process (Edner et al., 1991b). In order to alleviate environmental problems in this connection, techniques like selenium addition at the cremation or gas filtering are being tested.

Lidar measurements of mercury have been performed in Icelandic and Italian geothermal fields by the Swedish lidar group. Although no enhanced

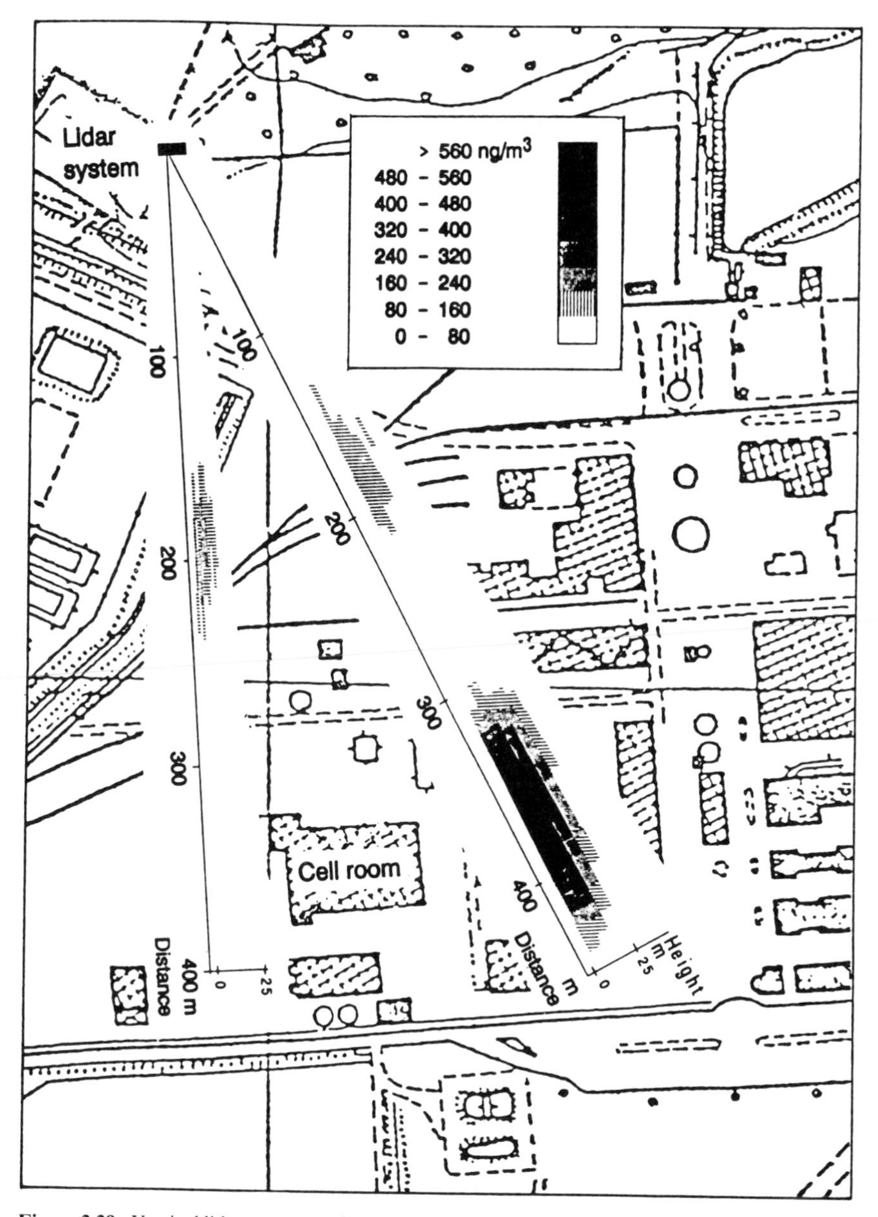

Figure 3.39. Vertical lidar scans, upwind and downwind from the cell house at the chlorine-alkali plant at Rosignano Solvay, Italy. The vertical Hg concentration plots are shown in diagrams placed along the lines of measurement. The weak plume is related to a mercury waste deposit. From Ferrara et al. (1992).

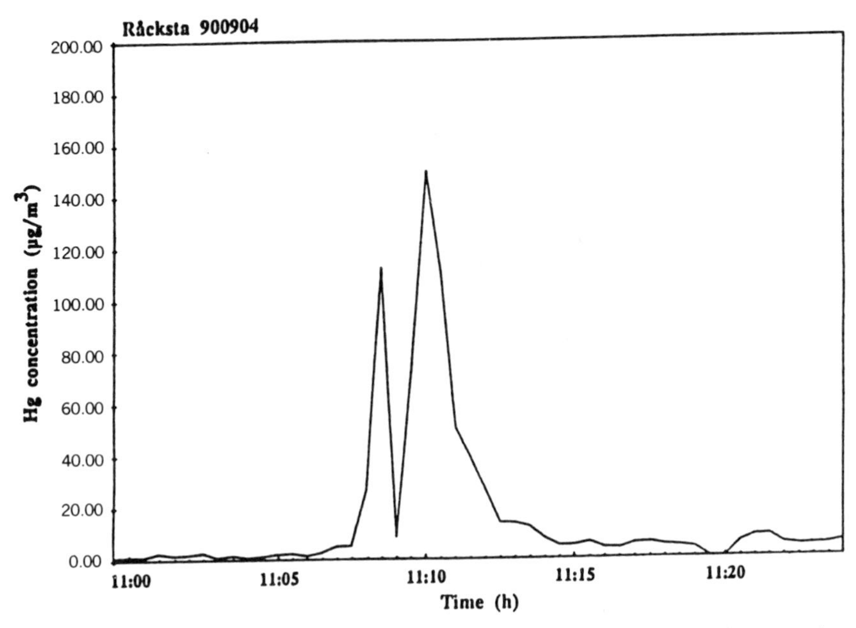

Figure 3.40. The concentration of atomic mercury in the stack effluents of a Swedish crematorium during a single cremation. The recording was taken with the Swedish mobile DIAL system, placed 200 m from the crematorium. From Edner et al. (1991b).

Figure 3.41. Photograph of the Swedish mobile DIAL system during atomic mercury measurements at the geothermal plant at Castelnuovo di Val di Cecina, Italy. From Edner et al. (1992c).

atomic mercury emissions could be detected in three Icelandic geothermal fields (Edner et al., 1991c), strongly elevated values were found in Italy (Edner et al., 1992c). A photograph of the new Swedish mobile DIAL system during measurement at a geothermal power station in Tuscany, Italy, is shown in Fig. 3.41. As can be seen in Fig. 3.42, substantial emissions are recorded. The

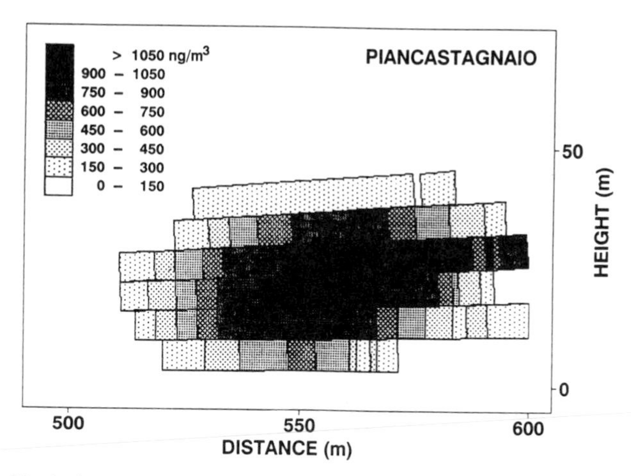

Figure 3.42. Vertical mercury lidar scan through the plume from the geothermal power plant at Piancastagnaio, Italy (20 MW). From Edner et al., 1992c).

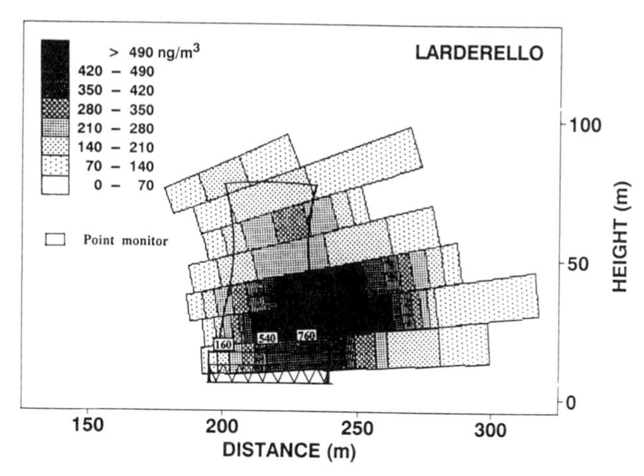

Figure 3.43. Vertical lidar scan at the geothermal power plant at Larderello, Italy. The DIAL data for atomic mercury concentration are compared with the data from three point monitors placed close to the cooling tower. From Edner et al. (1992c).

yearly total flux of mercury from the plant is estimated to be about 200 kg, combining DIAL scans with wind velocity data. Data from a vertical scan close to the cooling towers at the largest geothermal power station in Europe are shown in Fig. 3.43. In the figure data obtained using point monitors based on mercury amalgamation in gold and subsequent atomic absorption monitoring are inserted. As can be seen there is a good agreement between the two measurement techniques.

The aforementioned absence of elevated concentrations of atomic mercury in the air over geothermal fields in Iceland presents a puzzle. There can be no

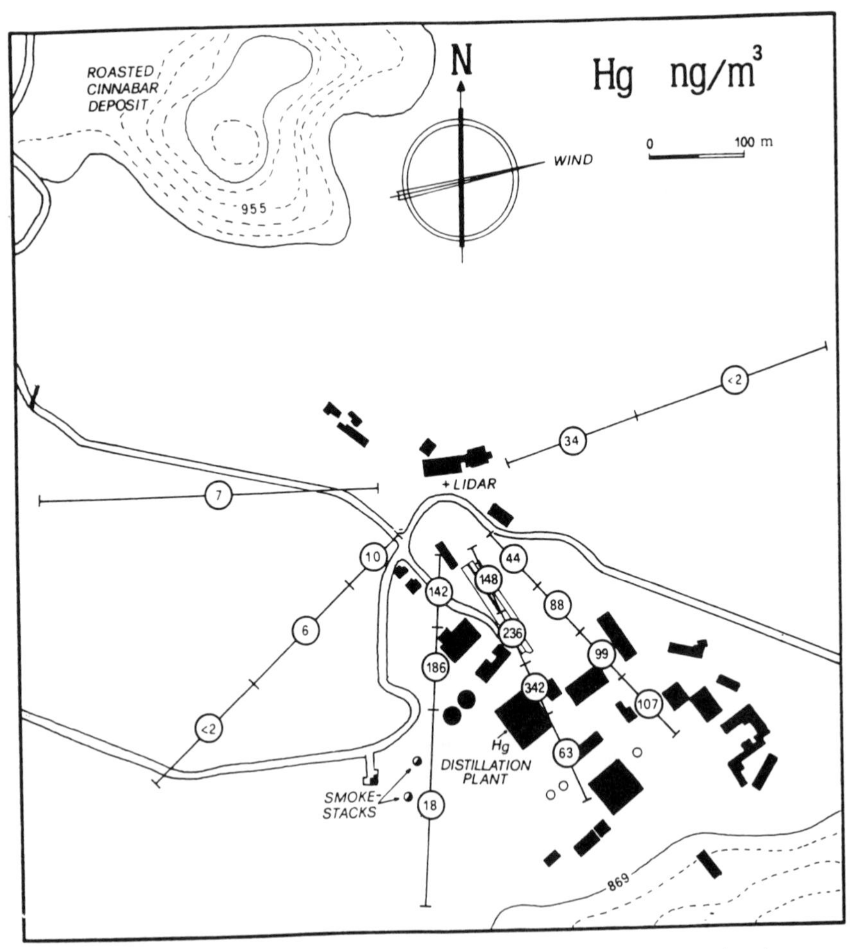

Figure 3.44. Mercury concentration data from a horizontal lidar scan at the abandoned mercury mine at Abbadia S. Salvatore, Italy. From Edner et al. (1993a).

doubt about the presence of significant amounts of mercury in well fluids in these areas. The lidar technique is sensitive only to atomic mercury. Mercury in a form other than elemental vapor would therefore not have been revealed in our search. The occurrence of the mercury in some form other than elemental vapor seems to be the most probable explanation for the failure of the lidar search to detect significant concentrations. Mercury in Californian geothermal fields is known to be present mainly in elemental form (Robertson et al., 1977), as is the case in Italy.

In connection with mercury mining there is a substantial emission of mercury to the atmosphere. We have performed DIAL monitoring of the air over the now abandoned mercury mine at Abbadia S. Salvatore in Tuscany, Italy (Edner et al., 1993a). A horizontal map over the mining area is shown in Fig. 3.44, with mean concentrations given in circles for selected path lengths. As can be seen strongly elevated concentrations are found, particularly in connection with a distillation plant. On the other hand, the concentration quickly reaches the background value (2 ng/m^3) outside the central area.

In connection with the mine there are large deposits of roasted cinnabar ore still containing about 0.2% of mercury. There is a substantial outgassing from these cinnabar banks. The DIAL technique can be used to measure the atomic mercury gradient due to the outgassing, which is particularly strong at elevated temperature. Comparisons with point monitoring distributed along a

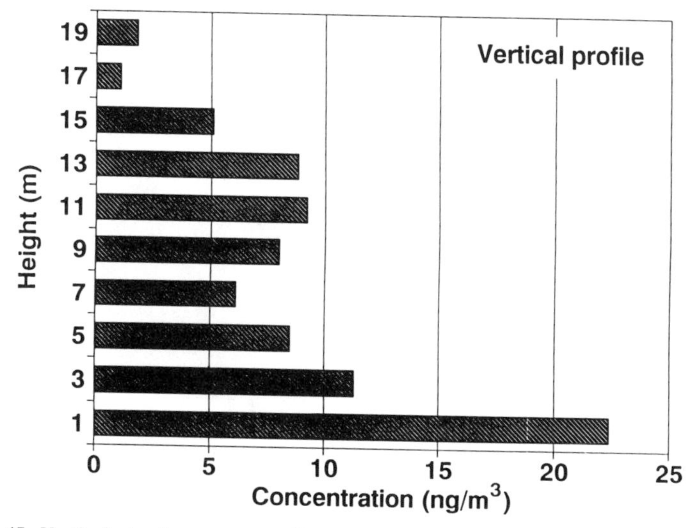

Figure 3.45. Vertical atomic mercury gradient over roasted cinnabar ore deposits at Abbadia S. Salvatore. Italy, measured with a mercury DIAL system. From Edner et al. (1993a).

vertical wire suspended from a crane showed good agreement (Ferrara et al., 1991). Data are shown in Fig. 3.45. Attempts have also been made to measure the very low mercury gradient above lake surfaces by comparing data obtained in long paths chosen at low heights above the surface (Edner and Svanberg, 1991). It has been postulated that this may account for missing mercury in the full environmental cycle of this element. In these measurements the lidar beam was directed onto large first surface mirrors mounted to provide measurement paths across a lake at 0.5 and 2.0 m above the surface. In initial measurements the concentration gradient over the lake surface was not high enough to be detected.

We conclude this section on atmospheric mercury monitoring by showing an example of a background mercury concentration measurement illustrating the sensitivity of the technique. The DIAL curve in Fig. 3.46 allows a determination of the ambient air atomic mercury concentration over the city of Lund, Sweden, which was found to be $1.5 \pm 1.0\,\mathrm{ng/m^3}$ Hg over a 1-km path.

As shown by the foregoing examples, DIAL monitoring of atomic mercury is a very powerful technique that can be employed not only for traditional pollution monitoring but also for studying geophysical processes.

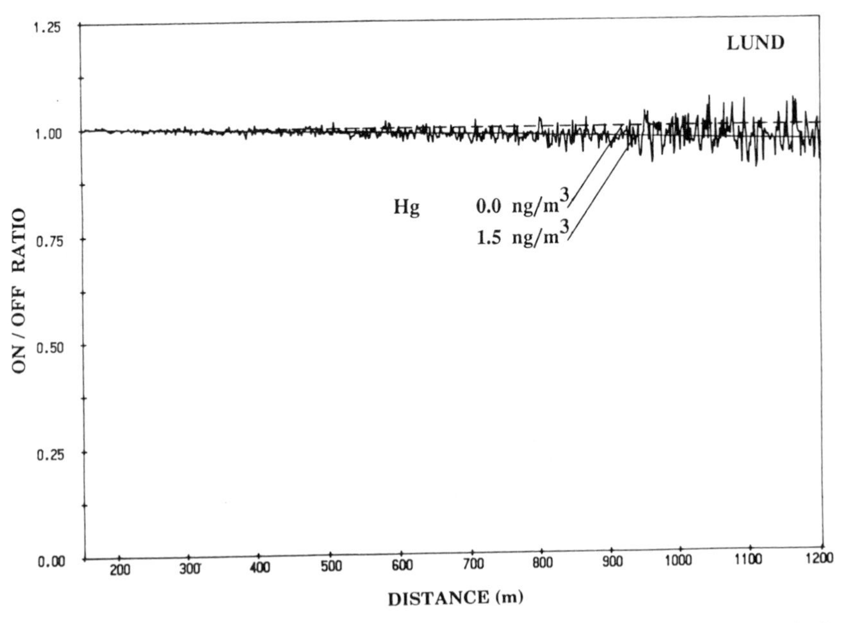

Figure 3.46. DIAL curve from a background atomic mercury concentration measurement in the city of Lund, Sweden. From Edner and Svanberg (1991).

3.8. OUTLOOK

The DIAL technique for monitoring of atmospheric pollutants is reaching a substantial level of general applicability and maturity. It provides unique possibilities for three-dimensional mapping of the atmosphere and for measuring total fluxes from industrial and urban areas. The technique is starting to be applied for routine monitoring. The largest obstacles to the widespread application of the powerful DIAL technique is system complexity. This is particularly true for the laser part. A very interesting development is the emergence of diode-pumped Nd: YAG lasers that can be used for pumping of titanium–sapphire lasers. Moderate-cost systems using this all-solid-state technology would mean a breakthrough for DIAL technology: high reliability and ease of operation could be anticipated. Very powerful and low-cost computers are already available for near-real-time data processing into understandable graphs. The increased compactness possible with the new technology will allow installation in small vehicles that can easily be moved from one measuring site to another without tedious setup procedures. The same technology will also enable construction of realistic DIAL airborne systems.

The ultimate application of the DIAL technique is global monitoring of meteorological parameters and atmospheric pollution from space. Increasing system performance, compactness, and reliability in conjunction with the planned availability of serviceable space platforms should make execution of such a scenario possible. Both NASA and ESA (the European Space Agency) are planning space lidar systems providing global wind, temperature, and pollution monitoring (see, e.g., Couch et al., 1991).

ACKNOWLEDGMENTS

The author gratefully acknowledges fruitful cooperation with a large number of present and previous co-workers and graduate students in the field of environmental remote sensing. This work was supported by the Swedish Space Board, the National Swedish Environment Protection Board, the Swedish Natural Science Research Council, the Swedish Space Corporation, and the Knut and Alice Wallenberg Foundation.

REFERENCES

Aldén, M., Edner, H., and Svanberg, S. (1982a). Remote measurements of atmospheric mercury using differential absorption lidar. *Opt. Lett.* **7**, 221–223.

Aldén, M., Edner, H., and Svanberg, S. (1982b). Laser monitoring of atmospheric NO using ultraviolet differential-absorption techniques. *Opt. Lett* **7**, 543–545.

American National Standards Institute (ANSI), (1986). "American National Standards for the Safe Use of Lasers,"ANSI-Rep. Z136.1. ANSI, Washington, DC.

Ansmann, A., Riebesell, M., Wandinger, U., Weitkamp, C., Voss, E., Lahmann, W., and Michaelis, W. (1992). Combined Raman elastic-backscatter LIDAR for vertical profiling of moisture, aerosol extinction, backscatter, and LIDAR ratio. *Appl. Phys.* **B55**, 18–28.

Axelsson, H., Edner, H., Galle, B., Ragnarson, P., and Rudin, M. (1990). Differential absorption spectroscopy (DOAS) measurements of ozone in the 280–290 nm wavelength region. *Appl. Spectrosc.* **44**, 1654–1658.

Bach, W., Pankrath, J., and Kellogg, W., Eds. (1979). *Man's Impact on Climate.* Elsevier, Amsterdam.

Barbini, R., Colao, F., Palucci, A., Ribezzo, S., Hermsen, T., and Orlando, S. (1991). "Ozone Measurements from the ENEA Lidar/DIAL System," EUROTRAC Annual Report 1990, Part 7, pp. 46–51. EUROTRAC, Garmisch-Partenkirchen, Germany.

Bowman, M. R., Gibson, A. J., and Sandford, M. C. W. (1969). Atmospheric sodium measured by a tuned laser radar. *Nature* (*London*) **221**, 456–457.

Breschi, B., Cecchi, G., Pantani, L., Raimondi, V., Tirelli, D., Valmori, G., Mazzinghi, P., and Zoppi, M. (1992). Measurement of water column temperature by Raman scattering. EARSeL (Eur. Assoc. Remote Sens. Lab.) *Adv. Remote Sens.* **1**, 131–134.

Bristow, Q., and Jonasson, I. R. (1972). Vapour sensing for mineral exploration. *Can. Min. J.* **93**, 39–47.

Browell, E. V. (1985). Ultraviolet DIAL measurements of O_3 profiles in regions of spatially inhomogeneous aerosols. *Appl. Opt.* **24**, 2827–2836.

Browell, E. V. (1991). Differential absorption lidar detection of ozone in the troposphere and lower stratosphere. In *Optoelectronics for Environmental Science* (S. Martellucci and A. N. Chester, Eds.), pp. 77–89. Plenum, New York.

Browell, E. V., Wilkerson, T. D., and McIllrath, T. J. (1978). Water vapor differential-absorption lidar development and evaluation. *Appl. Opt.* **18**, 3474–3483.

Browell, E. V., Carter, A. F., and Wilkerson, T. D. (1981). Airborne differential absorption lidar system for water vapor investigation. *Opt. Eng.* **20**, 84–90.

Browell, E. V., Carter, A. F., Jr., Shipley, S. T., Allen, R. J., Butler, C. F., Mayo, M. N., Siviter, J. H., and Hall, W. M. (1983). Airborne LIDAR system and measurements of ozone and aerosol profile. *Appl. Opt.* **22**, 522–532.

Browell, E. V., Butler, C. F., Ismail, S., Fenn, M. A., Kooi, S. A., Carter, A. F., Tuck, A. F., Toon, O. B., Proffitt, M. H., Loewenstein, M., Schoeberl, M. R., Isaksen, I., and Braathen, G. (1990). Airborne lidar observations in the wintertime arctic stratosphere: Ozone. *Geophys. Res. Lett.* **17**, 325–328.

Capelle, G. A., Franks, L. A., and Jessup, D. A. (1983). Aerial testing of a KrF laser-based fluorosensor. *Appl. Opt.* **22**, 3382–3387.

Carswell, A. I. (1983). Lidar measurement of the atmosphere. *Can. J. Phys.* **61**, 378–395.

Carswell, A. I., Chairman (1986). *Thirteenth International Laser Radar Conference*, NASA Conf. Publ. 2431. NASA, Toronto.

Cecchi, G., and Pantani, L. (1991). Vegetation monitoring by means of spectral resolved fluorescence lidar. *Eur. Space Agency [Spec. Publ.]* **ESA SP-319**, p. 687.

Celander, L., Fredriksson, K., Galle, B., and Svanberg, S. (1978). "Investigation of laser-induced fluorescence with applications to remote sensing of environmental parameters," Rep. GIPR-149. Göteborg Institute of Physics, CTH, Göteborg, Sweden.

Chanin, M. L. (1983). Rayleigh and resonance sounding of the stratosphere and mesosphere. *Springer Ser. Opt. Sci.* **39**, 192–198.

Chappelle, E. W., Wood, F. M., Newcomb, W. W., and McMurtrey, J. E., III (1985). Laser-induced fluorescence of green plants. 3. LIF spectral studies of five major plant types. *Appl. Opt.* **24**, 74–80.

Chen, H. S. (1985). *Space Remote Sensing Systems.* Academic Press, Orlando, FL.

Collins, G. P. (1992). Making stars to see stars: DOD adaptive optics work is declassified. *Phys. Today* **45**, (2), 17–21.

Collins, R. L., Bowman, K. P., and Gardner, C. S. (1992). Lidar observations of polar stratospheric clouds and stratospheric temperatures at the South Pole. *NASA Conf. Publ.* **3158**, 293–295.

Couch, R. H., Rowland, C. W., Ellis, K. S., Blythe, M. P., Regan, C. R., Koch, M. R., Antill, C. W., Cox, J. W., DeLorme, J. F., Crockett, S. K., Remus, R. W., Casas, J. C., and Hunt, W. H. (1991). Lidar-in-space technology experiment: NASA's first in-space lidar system for atmospheric research. *Opt. Eng.* **30**, 88–95.

Edner, H., and Svanberg, S. (1991). Differential lidar measurements of atmospheric mercury. *Water, Air, Soil Pollut.* **56**, 131–139.

Edner, H., Svanberg, S., Unéus, L., and Wendt, W. (1984). Gas correlation lidar. *Opt. Lett.* **9**, 493–495.

Edner, H., Fredriksson, K., Sunesson, A., and Wendt, W. (1987a). Monitoring Cl_2 using a differential absorption lidar system. *Appl. Opt.* **26**, 3183–3185.

Edner, H., Fredriksson, K., Sunesson, A., Svanberg, S., Unéus, L., and Wendt, W. (1987b). Mobile remote sensing system for atmospheric monitoring. *Appl. Opt.* **26**, 4330–4338.

Edner, H., Sunesson, A., and Svanberg, S. (1988). NO plume mapping using laser radar techniques. *Opt. Lett.* **12**, 704–706.

Edner, H., Faris, G. W., Sunesson, A., and Svanberg, S. (1989). Atmospheric atomic mercury monitoring using differential absorption lidar techniques. *Appl. Opt.* **28**, 921–930.

Edner, H., Ragnarson, P., Svanberg, S., and Wallinder, E. (1991a). "Vertical Lidar Probing of Ozone and Related Trace Species," EUROTRAC Annual Report 1990, Part 7, pp. 37–40. EUROTRAC Secretariat, Garmisch-Partenkirchen, Germany.

Edner, H., Olsson, P., and Wallinder, E. (1991b). "Measurements of Mercury with Optical Techniques at the Råcksta Crematorium," Lund Rep. At. Phys. LRAP-118. Lund Institute of Technology, Lund, Sweden.

Edner, H., Faris, G. W., Sunesson, A., Svanberg, S., Bjarnason, J.Ö., Kristmansdòttir, H., and Sigurdsson, K.H. (1991c). Lidar search for atomic mercury in Icelandic geothermal fields. *J. Geophys. Res.* **94D**, 2977–2985.

Edner, H., Ragnarson, P., Spännare, S., and Svanberg, S. (1992a). "A differential absorption spectroscopy (DOAS) system for urban atmospheric pollution monitoring," Lund Rep. At. Phys. LRAP-133. Lund University, Lund, Sweden.

Edner, H., Svanberg S., and Wallinder, E. (1992b). Photo-oxidants: precursors and products, Evaluation of DIAL systems for tropospheric ozone. In *EUROTRAC Symposium '92*. (P. M. Borrell, P. Borrell, T. Civitas, and W. Seiler, Eds.) SPB Acad. Publishing, The Hague, Netherlands.

Edner, H., Ragnarson, P., Svanberg, S., Wallinder, E., De Liso, A., Ferrara, R., and Maserti, B. E. (1992c). Differential absorption lidar mapping of atmospheric atomic mercury in Italian geothermal fields. *J. Geophys. Res.* **97D**, 3779–3786.

Edner, H., Johansson, J., Svanberg, S., Wallinder, E., Bazzani, M., Breschi, B., Cecchi, G., Pantani, L., Radicati, B., Raimondi, V., Tirelli, D., and Valmori, G. (1992d). Laser-induced fluorescence monitoring of vegetation in Tuscany. EARSeL (Eur. Assoc. Remote Sens. Lab.). *Adv. Remote Sens.* **1**, 119–130.

Edner, H., Johansson, J., Svanberg, S., Wallinder E., Cecchi, G., and Pantani, L. (1992e). Fluorescence lidar monitoring of the Arno River. EARSeL (Eur. Assoc. Remote Sens. Lab.). *Adv. Remote Sens.* **1**, 42–45.

Edner, H., Ragnarson, P., Svanberg, S., Wallinder, E., Bargagli, R., Ferrara, R., and Maserti, B.E. (1993a). Atmospheric mercury mapping in a Cinnabar mining area. *Sci. Total Environ.* **133**, 1–15.

Edner, H., Ragnarson, P., Spännare, S., and Svanberg, S. (1993b). Differential optical absorption spectroscopy (DOAS) system for urban atmospheric pollution monitoring. *Appl. Opt.* **32**, 327–333.

Egebäck, A. L., Fredriksson, K., and Hertz, H. M. (1984). DIAL techniques for the control of sulfur dioxide emission. *Appl. Opt.* **23**, 722–729.

Englisch, W., Wieseman, W., Boscher, J., and Rother, M. (1983). Laser remote sensing measurements of atmospheric species and natural target reflectivities. *Springer Ser. Opt. Sci.* **39**, 38–43.

Farman, J. C., Gardiner, B. G., and Shanklin, J. D. (1985). Large losses of total ozone in Antarctica reveal seasonal ClO_x/NO_x interaction. *Nature (London)* **315**, 207–210.

Ferrara, R., Maserti, B. E., Morelli, M., Edner, H., Ragnarson, P., Svanberg, S., and Wallinder, E. (1991). Vertical distribution of atmospheric mercury concentration over a Cinnabar deposit. In *Heavy Metals in the Environment* (J.G. Farmer, Ed.), C.E.P. Consultants, Edinburgh. pp. 247–250.

Ferrara, R., Maserti, B. E., Edner, H., Ragnarson, P., Svanberg, S., and Wallinder, E. (1992). Mercury emissions to the atmosphere from a chloralkali complex measured by the lidar technique. *Atmos. Environ.* **26A**, 1253–1258.

Fioccio, G., and Smullin, L. D. (1963). Detection of scattering layers in the upper atmosphere (60–140 km) by optical radars. *Nature (London)* **199**, 1275–1276.

Fredriksson, K. (1988). Differential absorption lidar for pollution mapping. In *Laser Remote Chemical Analysis* (R. M. Measures, Ed.), pp. 273–331. Wiley (Interscience), New York.

Fredriksson, K., and Hertz, H. M. (1984). Evaluation of the DIAL technique for studies on NO_2 using a mobile lidar system. *Appl. Opt.* **23**, 1403–1411.

Fredriksson, K., Lindgren, I., Svanberg, S., and Weibull, G. (1976). "Measurements of the emission from industrial smoke-stacks using laser radar techniques," Rep. GIPR-121. Göteborg Institute of Physics, CTH, Göteborg, Sweden.

Fredriksson, K., Galle, B., Linder, A., Nyström, K., and Svanberg, S. (1977). "Laser radar measurements of air pollutants at an oil-burning power station," Rep. GIPR-150. Göteborg Institute of Physics, CTH, Göteborg, Sweden.

Fredriksson, K., Galle, B., Nyström K., and Svanberg, S. (1979). Lidar system applied in atmospheric pollution monitoring. *Appl. Opt.* **18**, 2998–3003.

Fredriksson, K., Galle, B., Nyström, K., and Svanberg, S. (1981). Mobile lidar system for environmental probing. *Appl. Opt.* **20**, 4181–4189.

Fricke, K. H., and von Zahn, U. (1985). Mesopause temperatures derived from probing the hyperfine structure of the D_2 resonance line of sodium by lidar. *J. Atmos. Terr. Phys.* **47**, 499–512.

Galetti, E. (1987). Detection of methane leaks with a correlation lidar. *Opt. Soc. Am., Tech. Dig. Ser.* **18**, 80–83.

Garnier, A., and Chanin, M.L. (1992). Description of a Doppler Rayleigh LIDAR for measuring winds in the middle atmosphere. *Appl. Phys.* **B55**, 35–40.

Grant, W. B. (1982). Effect of differential special reflectance of DIAL measurement using topographic targets. *Appl. Opt.* **21**, 2390–2394.

Grant, W. B. (1987). Laser remote sensing techniques. In *Laser Spectroscopy and Its Applications* (L. J. Radziemski, R.W. Solarz, and J.A. Paisner, Eds.), pp. 565–621. Dekker, New York.

Grant, W. B. (1989). The mobile atmospheric pollutant mapping (MAPM) system: A coherent CO_2 DIAL system. In *Laser Application in Meteorology and Earth and Atmospheric Remote Sensing* (M. M. Sokoloski, Ed.), SPIE **1062**, 172–190.

Grant, W. B., and Menzies, R. T. (1983). A survey of laser and selected optical systems for remote measurement of pollutant gas concentration. *APCA J.* **33**, 187–194.

Grant, W. B., Hake, R. D., Jr., Liston, E. M., Robbins, R. C., and Proctor, E.K., Jr. (1974). Calibrated remote measurement of NO_2 using differential absorption backscattering technique. *Appl. Phys. Lett.* **24**, 550–552.

Hansen, G., and von Zahn, U. (1990). Sudden sodium layers in polar latitudes. *J. Atmos. Terr. Phys.* **52**, 585–608.

Harms, J. (1979). Lidar return signals for coaxial and noncoaxial systems with central obstruction. *Appl. Opt.* **18**, 1559–1566.

Harms, J., Lahmann, W., and Weitkamp, C. (1978). Geometrical compression of lidar return signals. *Appl. Opt.* **17**, 1131–1135.

Hauchecorne, A., Chanin, M. L., Keckhut, P., and Nedeljkovic, D. (1992). LIDAR monitoring of the temperature in the middle and lower atmosphere. *Appl. Phys.* **B55**, 29–34.

Hawley, J. G., Fletcher, D., and Wallace, G. F. (1983). Ground-based ultraviolet differential absorption lidar (DIAL) system and measurements. *Springer Ser. Opt. Sci.* **39**, 128–137.

Hercules, D. H., Ed. (1966). *Fluorescence and Phosphorescence Analysis.* Wiley (Interscience), New York.

Hinkley, E. D. Ed. (1976). *Laser Monitoring of the Atmosphere*, Top. Appl. Phys., Vol. 14 Springer-Verlag, Heidelberg.

Hirschfeld, T., Schildkraut, E. R., Tannenbaum, H., and Tanenbaum, D. (1973). Remote spectroscopic analysis of ppm-level air pollutants by Raman spectroscopy. *Appl. Phys. Lett.* **22**, 38–40.

Hoge, F. E. (1988). Ocean and terrestrial lidar measurements. In *Laser Remote Chemical Analysis* (R. M. Measures, Ed.), pp. 409–503. Wiley (Interscience), New York.

Hoge, F. E., Swift, R. N., and Frederick, E. B. (1980). Water depth measurement using an airborne pulsed neon laser system. *Appl. Opt.* **19**, 871–883.

Hoge, F. E., Swift, R. N., and Yungel, J. K. (1983). Feasibility of airborne detection of laser-induced fluorescence of green terrestrial plants. *Appl. Opt.* **22**, 2991–3000.

Hoge, F. E., Swift, R. N., and Yungel, J. K. (1986). Active-passive airborne ocean color measurement. 2: Applications. *Appl. Opt.* **25**, 48–57.

Huffaker, R. M., Lawrence, T. R., Post, M. J., Priestley, J. T., Hall, F. F., Jr., Richter, R. A., and Keeler, R. J. (1984). Feasibility studies for a global wind measuring satellite system. (Windsat): Analysis of simulated performance. *Appl. Opt.* **23**, 2523–2536.

Jepsen, A. F. (1973). Measurements of mercury vapor in the atmosphere. *Adv. Chem. Ser.* **123**, 80–94.

Jeys, T. H., Brailove, A. A., and Mooradian, A. (1989). Sum frequency generation of sodium resonance radiation. *Appl. Opt.* **28**, 2588–2591.

Jolliffe, B. W., Michelson, E., Swann, N. R. W., and Woods, P. T. (1987). A differential absorption lidar system for measurements of tropospheric NO, NO_2, SO_2, and O_3. *Opt. Soc. Am., Tech. Dig. Ser.* **18**, 26–30.

Kane, T. J., Mui, P., and Gardner, C. S. (1992). Evidence for substantial seasonal variations in the structure of the mesospheric iron layer. *Geophys. Res. Lett.* **19**, 405–408.

Kerker, M. (1969). *The Scattering of Light*, Academic Press, New York.

Killinger, D. A. (1992). USF HITRAN-PC. University of South Florida, Tampa.

Killinger, D. A., and Menyuk, N. (1981). Effects of turbulence-induced correlation on laser remote sensing errors. *Appl. Phys. Lett.* **38**, 968–970.

Killinger, D. A., and Mooradian, A., Eds. (1983). *Optical and Laser Remote Sensing*, Springer Ser. Opt. Sci. Vol. 39. Springer-Verlag. Heidelberg.

Killinger, D. A., Menyuk, N., and DeFoe, W.E. (1980). Remote sensing of CO_2 using frequency-doubled CO_2 laser radiation. *Appl. Phys. Lett.* **36**, 402–405.

Kim, H. H. (1977). Airborne bathymetric charting using pulsed blue-green lasers. *Appl. Opt.* **16**, 46–56.

Klett, J. D. (1981). Stable analytical inversion for processing lidar returns. *Appl. Opt.* **20**, 211–220.

Klett, J. D. (1986). Extinction boundary value algorithms for lidar inversion. *Appl. Opt.* **25**, 2462–2464.

Kobayashi, T. (1987). Techniques for laser remote sensing of the environment. *Remote Sens. Rev.* **3**, 1–56.

Kölsch, H. J., Rairoux, P., Wolf, J. P., and Wöste, L. (1989). Simultaneous NO and NO_2 DIAL measurements using BBO crystal. *Appl. Opt.* **28**, 2052–2056.

Korb, C. L. and Weng, C. Y. (1983). Differential absorption Lidar technique for measurement of the atmospheric pressure profile. *Appl. Opt.* **22**, 3759–3770.

Korb, C. L., Gentry, B. M., and Weng, C. Y. (1992). The edge technique: Theory and application to the lidar measurement of atmospheric winds. *Appl. Opt.* **31**, 4202–4213.

Lakowicz, J. R. (1983). *Principles of Fluorescence Spectroscopy*, Plenum, New York.

Lee, H. S., and Zwick, H. H. (1985). Gas filter correlation instrument for the remote sensing of gas leaks. *Rev. Sci. Instrum.* **56**, 1812–1819.

Leonard, D. A. (1967). Observation of Raman scattering from the atmosphere using a pulsed nitrogen ultraviolet laser. *Nature* (*London*) **216**, 142–143.

Leonard, D. A., Caputo, B., and Hoge, F. E. (1979). Remote sensing of subsurface water temperature by Raman scattering. *Appl. Opt.* **18**, 1732–1745.

Lichtenthaler, H. K., and Rinderle, U. (1988). The role of chlorophyll fluorescence in the detection of stress conditions is plants. *CRC Crit. Rev. Anal. Chem.* **19**, Suppl. 1, S29–S88.

Lindqvist, O., Ed. (1991). *Mercury as an Environmental Pollutant*, *Water*, *Air*, *Soil Pollut.*, Vol. 56, Reidel, Dordrecht, The Netherlands.

Mason, B. J. (1989). The greenhouse effect. *Contemp. Phys.* **30**, 417–432.

McCarthy, J. H., Jr. (1972). Mercury vapor and other volatile components in the air as guides to ore deposits. *J. Geochem. Explor.* **1**, 143–162.

McCormick, M. P., Chairman (1982a). *Eleventh International Laser Radar Conference*, NASA Conf. Publ. 2228, Abst. Pap. University of Wisconsin–Madison.

McCormick, M. P. (1982b). Lidar measurements of Mt. St. Helens effluents. *Opt. Eng.* **21**, 340–342.

McCormick, M. P., Ed. (1992). *Sixteenth International Laser Radar Conference*, NADA Conf. Publ. 3158. Massachusetts Institute of Technology, Cambridge.

McCormick, M. P., Swisser, T. J., Fuller, W. H., Hunt, W. H., and Osborn, M. T. (1984). Airborne and groundbased lidar measurements of the El Chichon stratospheric aerosol from 90 °N to 56 °S. *Geofis. Int.* **23–2**, 187.

Measures, R. M. (1984). *Laser Remote Sensing: Fundamentals and Applications*, Wiley, New York.

Measures, R. M., Ed. (1988). *Laser Remote Chemical Analysis.* Wiley (Interscience), New York.

Megie, G., Ed. (1984). *Twelfth International Laser Radar Conference*, Abst. Pap. Service d' Aern du C.N.R.S., Aix-en-Provence, France.

Megie, G., Ancellet, G., and Pelon, J. (1985). Lidar measurements of ozone vertical profiles. *Appl. Opt.* **24**, 3454–3463.

Menyuk, N., Killinger, D. K., and DeFeo, W. E. (1980). Remote sensing of NO using a differential absorption lidar. *Appl. Opt.* **19**, 3282–3286.

Menyuk, N., Killinger, D. K., and DeFeo, W. E. (1982). Laser remote sensing of hydrazine, MMH, and UDMH using a differential-absorption CO_2 lidar. *Appl. Opt.* **21**, 2275–2286.

Menzies, R. T. (1976). Laser heterodyne detection techniques. In *Laser Monitoring of the Atmosphere* (R. T. Hinkely, Ed.), Top. Appl. Phys., Vol. 14, pp. 297–353. Springer-Verlag, Heidelberg.

Menzies, R. T. (1991). Laser atmospheric wind velocity measurement. In *Optoelectronics for Environmental Science* (S. Martellucci and A. N. Chester, Eds.), pp. 103–116. Plenum, New York.

Menzies R. T., and Seals, R. K., Jr. (1977). Ozone monitoring with an infrared heterodyne radiometer. Science **197**, 1275–1277.

Milton, M. J. T., Bradsell, R. H., Jolliffe, B. W. Swann, N. R. W. and Woods, P.T. (1988). The design and development of a near-infrared DIAL system for the detection of hydrocarbons. *Int. Laser Radar Conf., 14th*, pp. 370–373.

Milton, M. J. T., Woods, P. T., Jolliffe, B. W., Swann, N. R. W., and McIlveen, T. J. (1992). Measurements of toluene and other aromatic hydrocarbons by differential-absorption LIDAR in the near-ultraviolet. *Appl. Phys.* **B55**, 41–45.

Mitra, S. (1986). *Mercury in the Ecosystem*, Technomic, Basel.

Molina, L. T., and Molina, M. J. (1986). Absolute absorption cross sections of ozone in the 185- to 350-nm region. *J. Geophys. Res.* **91D**, 14501–14508.

Ngriau, J. O., Ed. (1979). *The Biogeochemistry of Mercury in the Environment.* Elsevier/North-Holland, Amsterdam.

Nickolov, Zh., and Svanberg, S. (1986)."On the Possibilities of NO/NO_2 Simultaneous Detection by DIAL," Lund Rep. At. Phys. LRAP-71, LTH, Lund Institute of Technology, Lund, Sweden.

O'Neill, R. A., Buja-Bijunas, L., and Rayner, D. M. (1980). Field performance of a laser fluorosensor for the detection of oil spills. *Appl. Opt.* **19**, 863–870.

Osborn, M. T., Winker, D. M., Woods, D. C., and DeCoursey, R. J. (1992). Lidar observations of the Pinatubo volcanic cloud over Hampton, Virgina. *Nasa Conf. Publ.* **3158**, 91–94.

Reuter, R. (1991). Hydrographic applications of airborne laser spectroscopy. In *Optoelectronics for Environmental Science* (S. Martellucci and A.N. Chester, Eds.), pp. 149–160. Plenum, New York.

Revelle, R. (1982). Carbon dioxide and world climate. *Sci. Am.* **247**, 33–41.

Robertson, D. E., Crecelius, E. A., Fruchter, J. S., and Ludwick, J. D. (1977). Mercury emissions from geothermal power plants. *Science* **196**, 1094–1097.

Rothe, K. W., Brinkmann, U., and Walther, H. (1974a). Applications of tunable dye lasers to air pollution detection: Measurements of atmospheric NO_2 concentrations by differential absorption. *Appl. Phys.* **3**, 115–119.

Rothe, K. W., Brinkmann, U., and Walther, H. (1974b). Remote sensing of NO_2 emission from a chemical factory by the differential absorption technique. *Appl. Phys.* **4**, 181–184.

Rothman, L. S., Garmache, R. R., Goldman, A., Brown, L. R., Toth, R. A., Pickett, H. M., Poynter, R. L., Flaud, J. M., Camy-Peyret, C., Barbe, A., Husson, N., Rinsland, C. P., and Smith, M. A. H. (1987). The HITRAN Database: 1986 edition. *Appl. Opt.* **26**, 4058–4097.

Seinfeld, J. H. (1986). *Atmospheric Chemistry and Physics of Air Pollution*, Wiley, New York.

Shimizu, H., Sasano, Y., Nakane, H., Sugimoto, N., Matsui, I., and Takeuchi, N. (1985). Large scale laser radar for measuring aerosol distribution over a wide area. *Appl. Opt.* **24**, 617–626.

Shotland, R. M. (1966). Some observations of the vertical profile of water vapour by a laser optical radar. *Proc. Symp. Remote Sens. Environ., 4th*, University of Michigan, Ann Arbor, *1966*, pp. 273–283.

Shumate, M. S., Lundqvist, S., Persson, U., and Eng, S. T. (1982). Differential reflectance of natural and man-made materials at CO_2 laser wavelengths. *Appl. Opt.* **21**, 2386–2389.

Siegman, A. E. (1986). *Lasers.* University Science Books, Mill Valley, CA.

Slemr, F., Seiler, W., and Schuster, G. (1981). Latitudinal distribution of mercury over the Atlantic Ocean., *J. Geophys. Res.* **86D**, 1159–1166.

Sliney, D., and Wolbarsht, M. (1980). *Safety with Lasers and Other Optical Sources: A Comprehensive Handbook.* Plenum, New York.

Staer, W., Lahmann, W., Weitkamp, C., and Michaelis, W. (1984). Differential absorption lidar system for NO_2 and SO_2 monitoring. *Int. Laser Radar Conf., 12th*, pp. 281–284.

Stefanutti, L., Chairman (1988). *Fourteenth International Laser Radar Conference* Conf. Abstr. CNR-IROE/ILRC-CA, Innichen–San Candido, Italy.

Stefanutti, L., Castagnoli, F., Del Guasta, M., Morandi, M., Sacco, V. M., Zuccagnoli, L., Godin, S., Megie, G., and Porteneuve, J. (1992a). The Antarctic Ozone LIDAR system. *Appl. Phys.* **B55**, 3–12.

Stefanutti, L., Castagnoli, F., Del Guasta, M., Morandi, M., Sacco, V. M., Zuccagnoli., L., Kolenda, J., Kneipp, H., Rairoux, P., Stein, B., Weidauer, D., and Wolf, J. P. (1992b). A four-wavelength depolarization backscattering LIDAR for polar stratospheric cloud monitoring. *Appl. Phys.* **B55**, 13–17.

Stolarski, R. S. (1988). The Antarctic ozone hole. *Sci. Am.* **258** (1), 20–26

Svanberg, S. (1978). Fundamentals of atmospheric spectroscopy. In *Surveillance of Environmental Pollution and Resources by Electromagnetic Waves* (T. Lund, Ed.), pp. 37–66. Reidel, Dordrecht, The Netherlands.

Svanberg, S. (1980). Lasers as probes for air and sea. *Contemp. Phys.* **21**, 541–576.

Svanberg, S. (1985). Laser technology in atmospheric pollution monitoring. In *Applied Physics: Laser and Plasma Technology* (B. C. Tan, Ed.), pp. 528–548. World Science, Singapore.

Svanberg, S. (1990). Laser fluorescence spectroscopy in environmental monitoring. In *Optoelectronics for Environmental Science* (S. Martellucci and A.N. Chester, Eds.), pp. 15–27. Plenum, New York.

Svanberg, S. (1991). Environmental monitoring using optical techniques. In *Applied Laser Spectroscopy* (W. Demtröder and M. Inguscio, Eds.), pp. 417–434. Plenum, New York.

Svanberg, S. (1992). *Atomic and Molecular Spectroscopy: Basic Concepts and Practical Applications*, 2nd ed., Springer Ser. At. Plasmas, Vol. 6. Springer-Verlag, Heidelberg.

Svelto, O. (1989). *Principles of Lasers*, 3rd ed. Plenum, New York.

Thompson, B. A., Harteck, P., and Reeves, R. R. (1963). Ultraviolet absorption coefficients of CO_2, CO, O_2, H_2O, N_2O, NH_3, NO, SO_2, and CH_4 between 1850 and 4000 Å. *J. Geophys. Res.* **68**, 6431–6436.

Thompson, L. A., and Gardner, C. S. (1987). Laser guidestar experiment at Maona Kea Observatory for adaptive imaging in astronomy. *Nature* (*London*) **328**, 229–231.

Trush, B. A. (1988). The chemistry of the stratosphere. *Rep. Prog. Phys.* **51**, 1341–1371.

Uchino, O., Maeda, M., Yamamura, H., and Hirono, M. (1983). Observation of stratospheric vertical ozone distribution by a XeCl lidar. *Opt. Lett.* **8**, 347–351.

Udenfriend, S. (1962) *Fluorescence Assay in Biology and Medicine*, Vol. 1. Academic Press, New York.

Udenfriend, S. (1969). *Fluorescence Assay in Biology and Medicine*, Vol. 2. Acadamic Press, New York.

van de Hulst, H. C. (1957). *Light Scattering by Small Particles*. Wiley, New York.

Varekamp, J. C., and Buseck, P. R. (1981). Mercury emissions from Mount St Helens during September 1980. *Nature* (*London*) **293**, 555–556.

Varekamp, J. C. and Buseck, P. R. (1983). Hg anomalies in soils: A geochemical exploration method for geothermal areas. *Geothermics* **12**, 29–47.

Ward, T. V., and Zwick, H. H. (1975). Gas cell correlation spectrometer: GASPEC. *Appl. Opt.* **14**, 2896–2904.

Wayne, R. P. (1985). *Chemistry of Atmospheres*. Oxford Univ. Press (Clarendon), Oxford.

Wehry, E. L., Ed. (1976). *Modern Fluorescence Spectroscopy*, Vols. 1 and 2. Plenum, New York.

Weitkamp, C. (1981). The distribution of hydrogen chloride in the plume of incineration ships: Development of new measurement systems. *Int. Ocean Disposal Symp., 3rd*, Woods Hole, MA/USA, Oct 12–16, and Rep. GKSS 81/E/57, Geesthacht, Germany, 72 pp.

Werner, J., Rothe, K. W., and Walther, H. (1983). Monitoring of the stratospheric ozone layer by laser radar. *Appl. Phys.* **B32**, 113–118.

Wolf, J. P., Kölsch, H. J., Rairoux, P. and Wöste, L. (1990). Remote detection of atmospheric pollutants using differential absorption Lidar techniques. In *Applied Laser Spectroscopy* (W. Demtröder and M. Inguscio, Eds.), pp. 435–467. Plenum, New York.

Zanzottera, E. (1990). Differential absorption Lidar techniques in the determination of trace pollutants and physical parameters of the atmosphere. *CRC Crit. Rev. Anal. Chem.* **21**, 279–319.

Zuev, V. E., Chairman (1990). *Fifteenth International Laser Radar Conference*, Abstr. Pap., Parts I and II. Institute of Atmospheric Optics Publication, Tomsk, Russia.

CHAPTER

4

AIR MONITORING BY LASER PHOTOACOUSTIC SPECTROSCOPY

MARKUS W. SIGRIST

Infrared Physics Laboratory, Institute of Quantum Electronics, ETH Zurich, Switzerland

4.1. HISTORICAL PERSPECTIVES

Photoacoustics, also known as optoacoustics, was pioneered by Alexander Graham Bell, founder of the Bell Telephone Co., in the United States more than a century ago (Bell, 1880). The experiments of Bell with Tainter resulted in the construction of a "photophone," or an apparatus for the production of sound by light. The atmospheric transmission of voice by a sun beam between an emitter and receiver as shown in Fig. 4.1 was explored. The emitter consisted of a mirror activated by voice while the receiver section was made up of a hearing tube. Bell wrote to his father about this discovery: "I have heard articulate speech produced by sunlight, I have heard a ray of the sun laugh and cough and sing! I have been able to hear a shadow, and I have even perceived by ear the passage of a cloud across the sun's disk .... Can imagination picture what the future of this invention is to be .... "

It was soon found that the photoacoustic effect, i.e., the transformation of light to sound, occurs in all materials (solids, liquids, and gases), and Bell also soon recognized the spectroscopic potential of this technique. In the following years several renowned scientists studied this new phenomenon in detail (Rayleigh, 1881; Röntgen, 1881; Tyndall, 1881). During the following decades this important discovery was not pursued any further mainly owing to lack of sensitivity with respect to spectroscopic applications. In the late 1930s Viengerov (1938) presented the first gas analysis by using photoacoustic spectroscopy (PAS), and later Luft (1943) succeeded in improving the detection sensitivity considerably to a level that permitted the monitoring of gas

Air Monitoring by Spectroscopic Techniques, Edited by Markus W. Sigrist. Chemical Analysis Series, Vol. 127.
ISBN 0-471-55875-3

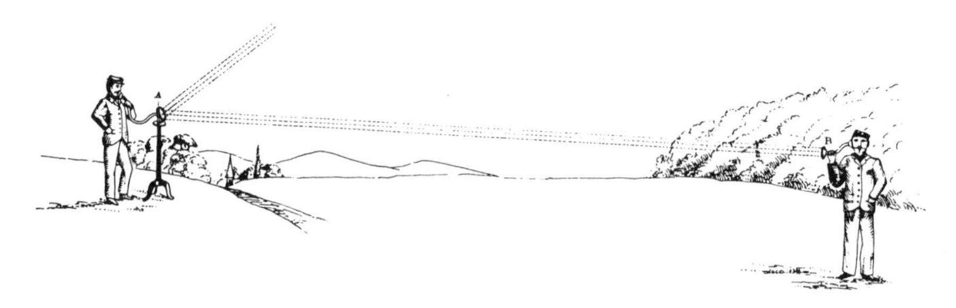

Figure 4.1. The "photophone" by Alexander Graham Bell (1880).

concentrations in the ppm, i.e., 10^{-6}, range. This was possible even with radiation sources of low spectral brightness such as lamps, because the photoacoustic technique directly measures the absorbed energy in the sample, in contrast to conventional detection methods that are based on the measurement of the transmitted radiation.

The real renaissance in photoacoustic research was connected with the progress made in the fields of laser development and electronics. While the former made radiation sources with high spectral brightness available, the latter provided sensitive microphones and lock-in detection and amplification. An important application with these new tools was trace gas analysis. Kerr and Atwood (1968) were the first to apply a laser to PAS. With a continuous wave (CW) CO_2 laser as radiation source they achieved a minimum detectable absorption coefficient α_{min} of 10^{-7} cm^{-1} for carbon dioxide (CO_2) buffered in nitrogen (N_2). The subsequent work by Kreuzer (1971), who reported on the sensitive detection of methane (CH_4) in N_2 with a limit of 10^{-8} (10 ppb) by the aid of a HeNe laser operating at 3.39 μm, initiated many activities in the field. With successful applications of PAS in condensed matter (Rosencwaig, 1978) photoacoustics experienced a renaissance which is evidenced today by periodic international conferences devoted to this field. The research themes of the photoacoustic and photothermal sciences span a host of physical, chemical, engineering, biological, and (bio)medical disciplines. Present activities include nondestructive evaluation of materials with the potential of depth-profiling, imaging, kinetic studies, environmental analysis, agricultural, biological and medical applications, phase transitions, heat and mass transfer, and many other areas (Hess, 1989a,b; Mandelis, 1992; Zharov and Letokhov, 1986).

The following discussion focuses on the application of photoacoustic and photothermal schemes to trace gas monitoring.

4.2. FUNDAMENTALS OF GAS PHASE PHOTOACOUSTICS

4.2.1. Photoacoustic Signal Generation

The theory of photoacoustic (PA) generation and detection in gaseous media has been mainly outlined by Kreuzer (1977), followed by numerous other authors (e.g., Hunter and Turtle, 1980; Tam, 1983). In general, two aspects have to be considered: first, the heat production in the gas by the absorption of radiation; secondly, the resulting generation of acoustic waves. The first aspect concerns the absorption processes encountered for a particular gas sample and the subsequent partial transfer of excitation energy to translational energy. The first step is thus the optical absorption resulting in the production of excited states. Assuming the simple case of an absorbing gas with molecule density N which can be modeled by a two-level system involving the ground state with population density $N - N'$ and the excited state with density N', the density N' can be calculated by the following rate equation:

$$\frac{dN'}{dt} = (N - N')\Phi\sigma - N'\left(\Phi\sigma + \frac{1}{\tau}\right) \tag{4.1}$$

Here Φ and σ represent the incident photon flux (in $cm^{-2} \cdot s^{-1}$) and the absorption cross section (in cm^2), respectively, whereas τ denotes the total lifetime of the excited state. The inverse time τ^{-1} can be expressed by the sum of reciprocal times τ_{nr} and τ_r of nonradiative and radiative relaxation, respectively:

$$\tau^{-1} = \tau_{nr}^{-1} + \tau_r^{-1} \tag{4.2}$$

It is only the nonradiative relaxation that contributes to the heat production and thus to the PA signal. In this sense the PA process is complimentary to fluorescence and additional nonthermal decay channels, as illustrated in Fig. 4.2 (Sigrist, 1989). In the case of air monitoring studies performed at atmospheric pressure the nonradiative decay rate due to collisions of the excited state exceeds the radiative decay rate by far. For infrared (IR) excitation, e.g., where the excitation of vibrational states is involved, the vibrational–translational (V–T) decay time τ_{nr} is on the order of 10^{-6}–10^{-9}s whereas τ_r is between 10^{-1} and 10^{-3}s (Hess, 1983). Consequently one obtains

$$\tau \approx \tau_{nr} \tag{4.3}$$

In addition, the photon flux Φ (and thus the excitation rate $\Phi\sigma$) is generally

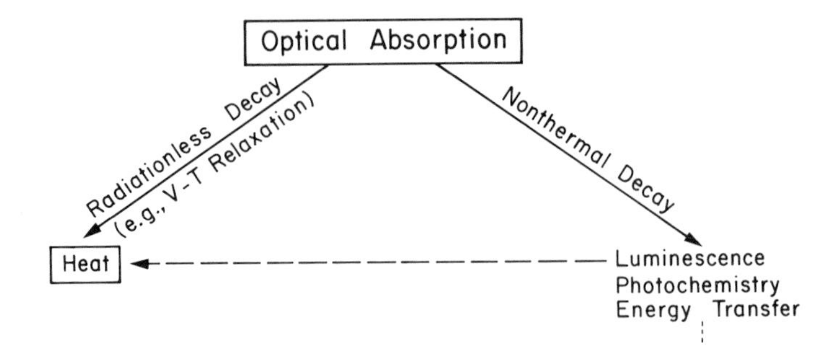

Figure 4.2. Principle of the photoacoustic effect.

kept small enough not to saturate the transition so that the population density N' in the excited state is only small ($N' \ll N$) and the stimulated emission from this state can be neglected. For typical experimental conditions for air monitoring with *low trace concentrations* and thus *small absorption,* Eq. (4.1) therefore reduces to

$$\frac{dN'}{dt} = N\Phi\sigma - \frac{N'}{\tau} \tag{4.4}$$

Two cases must now be distinguished, namely, modulated and pulsed excitation.

4.2.1.1. Modulated Excitation

Let us assume a sinusoidally modulated incident photon flux Φ whose dependence on position **r** and time t is given by

$$\Phi(\mathbf{r}, t) = \Phi_0(\mathbf{r})e^{i\omega t} \tag{4.5}$$

where only the real part has a physical meaning, with ω representing the circular modulation frequency. Introducing Eq. (4.5) into (4.4) one obtains for the density of the excited state:

$$N'(\mathbf{r}, t) = \frac{N\Phi_0(\mathbf{r})\sigma}{[1 + (\omega\tau)^2]^{1/2}} \cdot e^{i(\omega t - \theta)} \tag{4.6}$$

where

$$\theta = \arctan(\omega\tau) \tag{4.7}$$

indicates the phase lag of the modulation of the density N' with respect to the optical excitation.

The heat production rate $H(\mathbf{r}, t)$ is related to $N'(\mathbf{r}, t)$ by

$$H(\mathbf{r}, t) = N'(\mathbf{r}, t) \cdot E'/\tau_{nr} \tag{4.8}$$

where E' is the average thermal energy released owing to the nonradiative deexcitation of the excited state. Usually the collisional deexcitation process results in converting the excited state to the ground state. In this case, E' corresponds to the excitation energy, i.e., to the energy $h\nu$ of the absorbed photon, where h is the Planck constant and ν is the photon frequency.

Introducing Eqs. (4.3) and (4.6) and $E' = h\nu$, we can write Eq. (4.8) thus:

$$H(\mathbf{r}, t) = \frac{N\Phi_0(\mathbf{r})h\nu\sigma}{[1 + (\omega\tau)^2]^{1/2}} \cdot e^{i(\omega t - \theta)} = H_0 e^{i(\omega t - \theta)} \tag{4.9}$$

Since the intensity I_0 of the incident radiation is represented by

$$I_0(\mathbf{r}) = \Phi_0(\mathbf{r})h\nu \tag{4.10}$$

the amplitude H_0 of the heat production rate can be expressed as

$$H_0(\mathbf{r}) = \frac{N\sigma I_0(\mathbf{r})}{[1 + (\omega\tau)^2]^{1/2}} \tag{4.11}$$

For $\omega\tau \ll 1$, i.e., for modulation frequencies $\omega \ll 10^6\ \mathrm{s}^{-1}$, Eqs. (4.7) and (4.9)–(4.11) yield for the heat production rate $H(\mathbf{r}, t)$ the simple expression:

$$H(\mathbf{r}, t) = H_0(\mathbf{r})e^{i\omega t} \tag{4.12}$$

with

$$H_0(\mathbf{r}) = N\sigma I_0(\mathbf{r}) = \alpha I_0(\mathbf{r}) \tag{4.13}$$

where

$$\alpha = N\sigma \tag{4.14}$$

represents the usual absorption coefficient. The vanishing phase shift θ implies that the modulation of $H(\mathbf{r}, \mathrm{t})$ directly follows the modulation of the incident light without any phase lag. Equations (4.12) and (4.13) contain the essence of most PA trace gas detection studies since the conditions of slowly modulated light in the kilohertz range or below, i.e., $\omega \ll \tau^{-1}$, and the absence of optical saturation, i.e., $\Phi \cdot \sigma \ll \tau^{-1}$, are usually fulfilled.

The effects of saturation on photothermal measurements of IR excitation of gas molecules have been explicitly studied previously (Bialkowski and Long 1987). They need not be discussed any further here because saturation can easily be avoided simply by employing sufficiently low light levels. The range of appropriate modulation frequencies ω needs to be elucidated in some more detail. In addition to the upper limit for ω with $\omega \ll \tau^{-1}$, as assumed above for the derivation of Eqs. (4.12) and (4.13), there also exists a lower limit determined by the thermal diffusion from the heated gas zone to the walls of the employed PA gas cell. Thus, if the modulation frequency ω is too low, the heat diffuses away, resulting in a smoothing of the heat modulation. The corresponding diffusion time τ_{diff} is on the order of 0.1 s, which sets a lower limit for ω of a few s^{-1}. It should also be pointed out that care has to be taken with respect to the upper limit for ω, a requirement that is not always automatically fulfilled.The most important and often encountered exception concerns PA studies performed with a CO_2 laser on air samples containing CO_2 (Hammerich et al., 1992; Meyer and Sigrist, 1990). These molecules increase the lifetime of the excited states drastically, such that $\tau > 1/\omega$, which results in a decrease instead of an increase of the translational gas temperature. This phenomenon is known as kinetic cooling (Wood et al., 1971). In atmospheric air samples or any air sample of similar composition a fast resonant energy transfer occurs between the excited vibrational modes ν_3 of CO_2 and ν_1 of N_2. As has been outlined previously (Meyer and Sigrist, 1990) this results in an effective V–T deexcitation time τ_{eff} of ~ 0.8 ms for standard atmospheric conditions, also depending on the water vapor concentration. The important consequence is a possible phase reversal of the heat production rate H with respect to the modulation of the incident radiation. This phase shift of 180° or the resulting dip in the heat amplitude H_0 are clearly manifested in the calculated three-dimensional plots of Fig. 4.3a,b. Here the phase θ and amplitude H_0 are plotted vs. the concentrations c_{CO_2} and c_{H_2O} for an N_2–O_2–H_2O–CO_2 mixture. The plots shown are based on data for the 10 $R(20)$ $^{12}C^{16}O_2$ laser excitation at 10.25 μm with an intensity $I_0 = 20\ W/cm^2$, $\sigma_{CO_2} = 1.0 \times 10^{-22}\ cm^2$, $\sigma_{H_2O} = 3.5 \times 10^{-23}\ cm^2$, and a total pressure of 1 bar. As is evident from Fig. 4.3a, the phase reversal only occurs within rather narrow concentration ranges. Hence, although heat-rate phases θ different from 0° or 180° are expected rarely for low H_2O vapor and CO_2 concentrations, the actual phase θ should always be taken into account. This aspect is particularly important when one is considering actual atmospheric conditions with 350 ppm CO_2 and 1–2% absolute concentration of H_2O vapor. As Fig. 4.3a implies, the phase θ is subject to variations exactly around those concentrations.

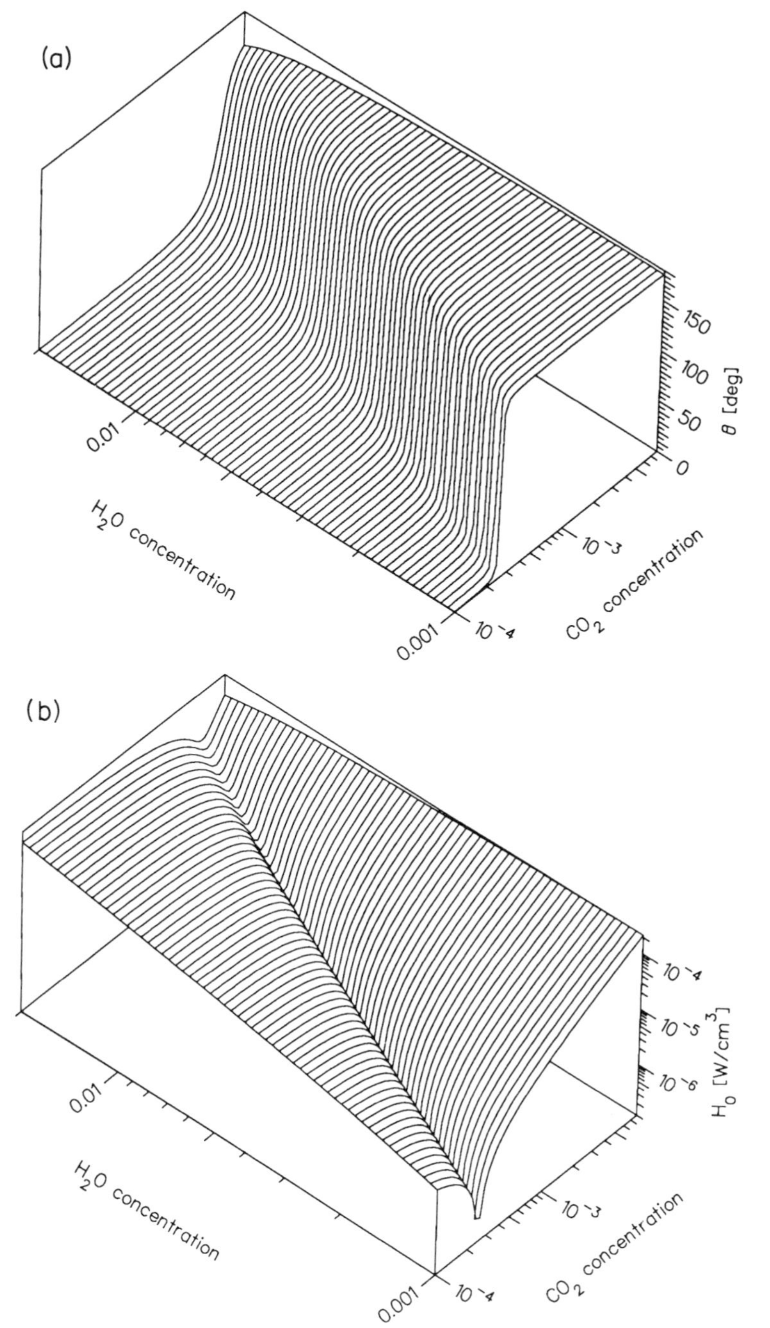

Figure 4.3. (a) Calculated phase θ and (b) amplitude H_0 of the heat production rate H for a CO_2–H_2O–N_2–O_2 mixture in relation to the concentrations c_{CO_2} and c_{H_2O} and for constant $C_{N_2} = 0.8$ and $C_{O_2} = 0.2$ (see text for details). From Meyer and Sigrist (1990).

In a second step the generation of the acoustic waves in the gas can now be calculated on the basis of the heat production rate $H(\mathbf{r}, t)$. Morse and Ingard (1986) have derived the inhomogeneous wave equation relating the acoustic pressure p and the heat source H:

$$\nabla^2 p - \frac{1}{c^2}\frac{\partial^2 p}{\partial t^2} = -\frac{\gamma - 1}{c^2}\frac{\partial H}{\partial t} \tag{4.15}$$

where c is the sound speed and γ is the ratio of specific heats of the gas for constant pressure and volume, respectively. It should be mentioned that all dissipative terms due to heat diffusion and dynamic viscosity have been neglected in Eq. (4.15). For many applications this simplified approach is sufficient, yet in some cases a more rigorous theory that takes relaxation effects into account is more appropriate (Hess, 1992). Acoustic wave generation has been intensively studied and reported in the literature for nonresonant and resonant PA cells (e.g., by Kreuzer, 1977), so-called spectrophones. The inhomogeneous wave equation (4.15) is usually solved for sinusoidal modulation of the incident radiation introduced in Eq. (4.5). In this case the Fourier transform of the pressure amplitude p can be expressed as a superposition of normal acoustic modes $p_k(\mathbf{r})$ of the spectrophone (Hess, 1983; Kreuzer, 1977; Tam 1983):

$$p(\mathbf{r}, \omega) = \sum_k A_k(\omega) p_k(\mathbf{r}) \tag{4.16}$$

where the normal mode $p_k(\mathbf{r})$ is a solution of the homogeneous wave equation

$$(\Delta^2 + \omega_k^2/c^2) p_k(\mathbf{r}) = 0 \tag{4.17}$$

At the cell wall the p_k must satisfy the boundary condition of vanishing gradient of p normal to the wall because the sound velocity which is proportional to the gradient of p vanishes at the wall. The general solution of Eq. (4.17) in the cylindrical geometry is given by the Bessel functions J_m of order m (Morse and Ingard, 1986):

$$p_k = J_m(k_r r) \cos(k_z z) \begin{cases} \sin(m\phi) \\ \cos(m\phi) \end{cases} \tag{4.18}$$

where (r, ϕ, z) are the cylindrical coordinates; R_0 is the radius of the gas cell; and k_r and k_z are wavenumbers with

$$k = \frac{\omega}{c} = (k_r^2 + k_z^2 + m^2/R_0^2)^{1/2} \tag{4.19}$$

In a PA experiment with the laser beam directed along the cylindrical axis, azimuthal modes are not excited and therefore $m = 0$. The corresponding eigenfrequencies f_{nmn_z} of the acoustic normal modes in a loss-free cylindrical cell of length L are given by

$$f_{nmn_z} = \frac{c}{2}\left[\left(\frac{\alpha_{mn}}{R_0}\right)^2 + \left(\frac{n_z}{L}\right)^2\right]^{1/2} \tag{4.20}$$

where α_{mn} is the nth root of the equation $dJ_m/dr = 0$ at $r = R_0$. The indices $n = 0, 1, 2, \ldots$; $m = 0, 1, 2, \ldots$; and $n_z = 0, 1, 2, \ldots$ refer to the radial, azimuthal, and longitudinal modes, respectively, as has been discussed in detail elsewhere (Hess, 1983). Figure 4.4 represents a geometrical illustration of these modes that occur in a cylindrical resonator.

After having delineated the basis $p_k(\mathbf{r})$ of Eq. (4.16) we now have to solve for the Fourier coefficients $A_k(\omega)$. They can be derived from the following equation, which uses the orthonormal conditions for the eigenfunctions p_k and also takes mode damping into account, thereby avoiding the physically unreasonable situation of $A_k \to \infty$ as $\omega \to \omega_k$:

$$A_k(\omega) = -\frac{i\omega(\gamma - 1)}{(\omega_k^2 - \omega^2 - i\omega\omega_k/Q_k)}\frac{1}{V_0}\int_V p_k^* H \, dV \tag{4.21}$$

Here V_0 is the cell volume; $Q_k = \omega_k/\Delta\omega_k$, the quality factor of the corresponding acoustic mode p_k; $\Delta\omega_k$, the width of the resonance at ω_k; and p_k^*, the complex conjugate of p_k. The integral is taken over the cell volume V and

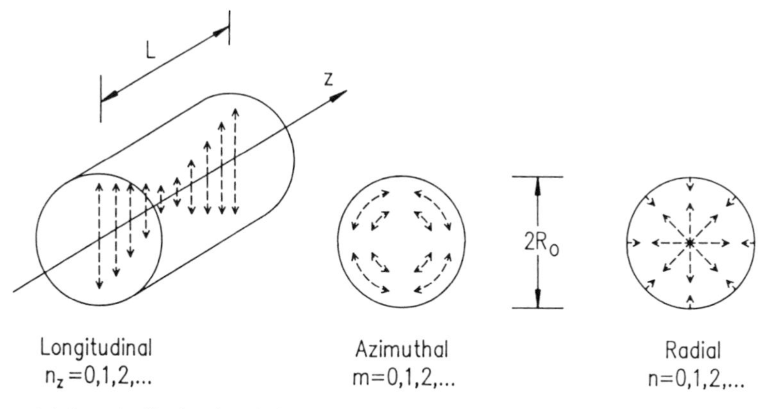

Figure 4.4. Longitudinal, azimuthal, and radial acoustic modes of a cylindrical resonator. From Hess (1992).

describes the geometrical coupling between laser radiation and the acoustical mode.

Remembering the simple relationship between the heat rate H and the intensity I according to Eq. (4.13), we consider two important cases in the following. In trace gas studies the absorption of the beam intensity along the cylindrical axis can be neglected. If we further assume a spatially constant beam intensity, the overlap integral of Eq. (4.21) vanishes for $k \neq 0$, that is,

$$\int_V p_k^* \, H \, dV = 0 \qquad \text{for} \quad k \neq 0 \tag{4.22}$$

The only nonzero mode is thus p_0 with a resonance frequency $\omega_0 = 0$. In contrast to resonant modes the zeroth mode corresponds to a spatially independent pressure change in the cell. The mode amplitude $A_0(\omega)$ is given by

$$A_0(\omega) = -\frac{i(\gamma - 1) N\sigma I}{\omega(1 - i/(\omega\tau_0))} \tag{4.23}$$

with $\omega_0 / Q_0 = 1/\tau_0$ representing the reciprocal damping time of p_0 resulting from heat conduction from the gas to the cell walls. The intensity I (assumed constant across the cell diameter) can be expressed by the power P, the length L, and the volume V_0 of the cylindrical cell as follows:

$$I = PL/V_0 \tag{4.24}$$

resulting in the amplitude

$$A_0(\omega) = -\frac{i(\gamma - 1)}{\omega(1 - i/(\omega\tau_0))} N\sigma P \cdot \frac{L}{V_0} \tag{4.25}$$

The corresponding pressure amplitude $p_0(\mathbf{r}, \omega) = p_0(\omega)$ is obtained by considering Eq. (4.16):

$$p_0(\omega) = A_0(\omega) \cdot p_0 \tag{4.26}$$

This result implies that the pressure signal is proportional to the absorbed power $N\sigma P$ yet decreases with ω^{-1} and V_0^{-1} in this nonresonant case. Low modulation frequencies and small cell volumes are thus preferable.

Similarly, if the first-order mode p_1 with resonance frequency ω_1 is excited, the amplitude $A_1(\omega)$ is given by

$$A_1(\omega) = -\frac{i\omega(\gamma - 1)\ N\sigma PL}{(\omega_1^2 - \omega^2 - i\omega\omega_1/Q_1)V_0} \tag{4.27}$$

resulting in a pressure amplitude

$$p_1(\omega) = A_1\ (\omega)\cdot p_1 \tag{4.28}$$

The ratio between the maximum amplitudes of the zeroth mode with $\omega_0 = 0$ and the first mode with $\omega = \omega_1$ is obtained from Eqs. (4.25) and (4.27):

$$\frac{A_1(\omega_1)}{A_0(0)} = \frac{Q_1}{\omega_1 \tau_0} \tag{4.29}$$

Equations (4.25) and (4.27) imply that the amplitude $A_0\ (\omega)$ exceeds $A_1(\omega)$ for low frequencies ω. At higher frequencies, $A_0(\omega)$ decreases with ω whereas $A_1(\omega)$ reaches a maximum at $\omega = \omega_1$.

The choice of the appropriate cell geometry is dependent on the application. If the measurements can be performed at low frequencies, nonresonant cells are preferable. In this case, the modulation frequency has to be chosen between a low frequency limit determined by the decrease of the microphone sensitivity and an upper limit where the $1/\omega$ dependence of $A_0\ (\omega)$ reduces the signal. Typical frequencies for this regime are in the region of some 10 Hz. Furthermore, the $1/V_0$ dependence of A_0 [Eq. (4.25)] implies the use of small-diameter cylinders to obtain maximum amplitudes. On the other hand, in situ studies are often performed in noisy environments. Since the noise amplitude decreases with frequency ω, the measurements are preferentially performed at frequencies ω in the kilohertz range. Hence, resonant cells offer several advantages in such cases. Appropriate cell designs are discussed in Section 4.3.3.

In practice, the laser power P has a radial dependence, e.g., Gaussian. Assuming weak absorption, i.e., negligible dependence of P on the position along the z axis of the cell, one finds for example for the pressure amplitude $p_1(r, \omega_1)$ at radius r in a cylindrical resonant cell operated at the first acoustic mode at ω_1 according to Eqs. (4.27) and (4.28):

$$p_1\ (r, \omega) = \frac{(\gamma - 1)\ N\sigma GPLQ_1}{\omega_1 V_0} p_1(r) \tag{4.30}$$

The geometrical factor G is on the order of one and takes the transverse beam profile into account. This pressure amplitude is usually detected by one or more microphones placed at appropriate positions $r = r_{mic}$.

4.2.1.2. *Pulsed Excitation*

If a single short radiation pulse, e.g., a laser pulse, is employed for the excitation of a gas, the pressure wave caused by collisional deexcitation is reflected back and forth within the PA cell until it vanishes owing to damping by dissipation processes. Various theoretical models for the derivation of the initial waveform as well as experimental studies have been presented, particularly for liquids (Akhmanov et al., 1989; Bozhkov et al., 1981; Diebold, 1989; Heritier, 1983; Lai and Young, 1982; Lyamshev and Sedov, 1981; Sigrist, 1986; Sigrist and Kneubühl, 1978; Sullivan and Tam, 1984). The calculations for liquids apply equally to gases (unless unusually low pressures are employed). The only difference is the contribution from electrostriction, which can generally be neglected for gases. Let us consider the pressure wave induced by pulsed excitation in a weakly absorbing gas corresponding to the case of an optically thin liquid. Let us further assume a cylindrical geometry with a Gaussian heating function of the form (Heritier, 1983):

$$H(r,t) = \frac{2N\sigma E_0}{\pi^{3/2}\, w_0^2 \tau_L} \exp\left(-\frac{2r^2}{w_0^2} - \frac{t^2}{\tau_L^2}\right) \tag{4.31}$$

Here, E_0 is the total energy of the pulse, and τ_L its width at the $1/e$ points. The corresponding beam radius is w_0 with the $1/e$ points at $w_0/\sqrt{2}$.

The wave equation (4.15) for the pressure $p(r, t)$ is solved by first writing the time-dependent quantities in terms of their Fourier transforms:

$$\begin{aligned} p(r,t) &= \int_{-\infty}^{\infty} p(r,\omega)\, e^{-i\omega t}\, d\omega \\ p(r,\omega) &= \frac{1}{2\pi}\int_{-\infty}^{\infty} p(r,t)\, e^{i\omega t}\, dt \end{aligned} \tag{4.32}$$

Substitution of the first of these as well as the corresponding Fourier transform of the heating function $H(r, \omega)$ into Eq. (4.15) yields:

$$\left(\nabla^2 + \frac{\omega^2}{c^2}\right) p(r,\omega) = \frac{i\omega(\gamma - 1)}{c^2} H(r,\omega) \tag{4.33}$$

This equation can be solved on the basis of the method using Green's function.

The Fourier transformation back into the time domain gives the final result for the pressure $p(r, t)$ far from the source, i.e., for $r \gg 2w_0$ (Diebold, 1989):

$$p(r, \hat{t}) = \frac{N\sigma E_0(\gamma - 1)}{8\pi w_0^2} \int_{-\infty}^{\infty} qH_0^{(1)}(q\hat{r}) \exp\left(-\frac{q^2}{8} - \frac{q^2\hat{\tau}_L^2}{4} - iq\hat{t}\right) dq \qquad (4.34)$$

Here $H_0^{(1)}$ is the Bessel function of the third kind (also called the Hankel function). The following dimensionless quantities have been defined:

$$q = \frac{\omega w_0}{c}, \hat{r} = \frac{r}{w_0}, \hat{\tau}_L = \frac{c\tau_L}{w_0}, \hat{t} = \frac{ct}{w_0}$$

The dimensionless distance $\hat{r}$ is related to the beam radius w_0, whereas the dimensionless times $\hat{\tau}_L$ and $\hat{t}$ correspond to propagation distances $c\tau_L$ and ct, respectively, of the acoustic wave with respect to w_0. Similar to the case of modulated radiation (Eq. 4.23), the generated pressure amplitude p is directly proportional to the absorbed energy $N\sigma E_0 = \alpha E_0$. A representative waveform calculated from Eq. (4.34) by using the fast Fourier transform method is shown in Fig. 4.5. The result refers to the parameters $\hat{r} = 10$, i.e., far away from the cylindrical source, and $\tau_L \ll 1$, i.e., for short pulse excitation where the propagation distance of the acoustic wave during the excitation pulse is much shorter

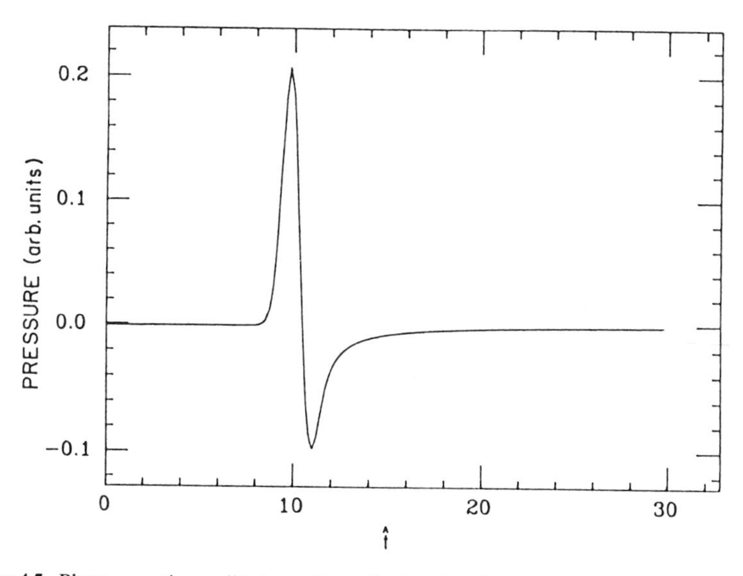

Figure 4.5. Photoacoustic amplitude vs. dimensionless time $\hat{t}$ for rapid energy release, computed from Eq. (4.34). From Diebold (1989).

than the beam radius w_0. With pulse widths τ_L assumed to be in the sub-microsecond-to-microsecond range and w_0 in the millimeter range, these conditions are generally fulfilled. The theoretical results have also been confirmed by experiments performed on liquids (Sullivan and Tam, 1984). A detailed analysis yields the following effective pulse width $\Delta\tau$ for the acoustical waveform (Diebold, 1989):

$$\Delta\tau = \left(\tau_L^2 + \frac{w_0^2}{2c^2} \right)^{1/2} \tag{4.35}$$

The acoustic pulse width $\Delta\tau$ is thus determined by the excitation pulse width τ_L and the acoustic transit time across the beam diameter. For nanosecond pulses the first term dominates, whereas for millisecond pulses the opposite is true.

A different approach is required for optically thick samples, i.e., for absorbing gases where the absorption along the beam path becomes appreciable and has to be included in the theoretical analysis according to the Beer–Lambert absorption law. However, in pollution studies with low concentrations of trace gases, this regime is generally not encountered and will thus not be discussed any further here.

Although pulsed radiation also yields a pressure signal proportional to the absorbed pulse energy, this regime is much less used in PA trace gas monitoring. One reason is that the straightforward, sensitive acoustic detection with microphones as used for modulated radiation is less advantageous owing to the limited frequency bandwidth of microphones. However, in addition to gas concentrations, a detailed analysis of the time dependence of the acoustic transients can also yield information on intramolecular energy transfer rates. An example is the study of multiphoton excitation of polyatomic molecules (Bailey et al., 1983; Beck and Gordon, 1988).

4.2.2. Photoacoustic Signal Analysis

The following discussion focuses on the regime of modulated excitation that is applied in the majority of studies and on the use of resonant PA cells. It is further differentiated between one- and multicomponent gas mixtures.

4.2.2.1. One-Component Samples

Referring to the case of a cell operated at the first mode at the frequency $\omega = \omega_1$ [see Eq. (4.30)], one obtains for the signal amplitude S recorded by a microphone placed at $r = r_{mic}$ the simple expression (omitting any phase term)

$$S = CPN\sigma \tag{4.36}$$

The cell constant C is given by (Eq. 4.30):

$$C = \frac{(\gamma - 1)GLQ_1}{\omega_1 V_0} R_{\text{mic}}\, p_1(r_{\text{mic}}) \tag{4.37}$$

where R_{mic} is the microphone sensitivity given in millivolts per pascal (usually 10–100 mV/Pa). The cell constant C in units of Vcm/W is generally not calculated on the basis of Eq. (4.37) but instead determined with the aid of Eq. (4.36) by calibration measurements under well-defined conditions, i.e., with a certified gas mixture of an absorbing species with known number density N (i.e., concentration) and absorption cross section σ diluted in a nonabsorbing buffer gas. As Eqs. (4.36) and (4.37) imply, the cell constant C and consequently the detection sensitivity depend on the cell geometry, the microphone sensitivity, and the nature of the acoustic mode. The factor

$$Q_1/\omega_1 = 1/\Delta\omega_1 \tag{4.38}$$

where $\Delta\omega_1$ denotes the resonance width, implies a high sensitivity at a narrow resonance. However, in view of the dependence of ω_1 on gas temperature, pressure, and composition, the accuracy of its control sets a practical lower limit to $\Delta\omega_1$ of a few hertz.

On the basis of Eqs. (4.36) and (4.14), one obtains for the minimum detectable absorption coefficient α_{min} the following equation:

$$\alpha_{\text{min}} = \frac{S_{\text{min}}}{CP} \tag{4.39}$$

where S_{min} is the noise-limited minimum microphone signal. Noise sources in PA gas sytems include acoustic noise from window heating, absorption of scattered radiation by the cell walls and by desorbing molecules, ambient noise, noise produced by gas flow, electronic noise, Brownian motion of the gas, and microphone noise (Rosencwaig, 1980). For example, our CO_2 laser PA system yields $S_{\text{min}}/P = 30$–50 nV/W, corresponding to the detected signal for a flow of nonabsorbing synthetic air (80% N_2/20% O_2) through the resonant cell (Meyer, 1988). With the mean cell constant $C \approx 3.5$ V·cm/W, this results in $\alpha_{\text{min}} \approx 10^{-8}$ cm^{-1}. This limit could be further improved by using one or more microphones with higher sensitivity R_{mic}, by intracavity absorption, multipath cells, etc. Referring to Eq. (4.36), we find that the minimum detectable concentration c_{min} of a species with absorption cross section σ mixed with a nonabsorbing gas at a total number density N_{tot} is given by

$$c_{\text{min}} = \frac{1}{N_{\text{tot}}} \cdot \frac{\alpha_{\text{min}}}{\sigma} = \frac{S_{\text{min}}}{N_{\text{tot}} CP\sigma} \tag{4.40}$$

With $\sigma \approx 10^{-18}\,\text{cm}^2$, $N_{\text{tot}} \approx 10^{19}\,\text{cm}^{-3}$, and $\alpha_{\min} \approx 10^{-8}\,\text{cm}^{-1}$ one easily gets $c_{\min} \approx 10^{-9}$, i.e., ppb concentrations corresponding to densities of micrograms per cubic meter.

4.2.2.2. *Multicomponent Samples*

In practice, one usually deals with multicomponent samples, e.g., ambient air, stack gas, or vehicle exhausts. It is therefore desirable to analyze the samples for a number of n different species, rather than just one. This can be accomplished by measuring the PA signal at a set of wavelengths λ_i $(i = 1, \ldots, m)$ chosen on the basis of the absorption spectra of the individual components. The microphone signal $S(\lambda_i)$ can thus be written as follows [see Eq. (4.36]:

$$S(\lambda_i) = S_i = CP_i N_{\text{tot}} \sum_{j=1}^{n} c_j \sigma_{ij} \tag{4.41}$$

with $i = 1, \ldots, m; j = 1, \ldots, n$; and $m \geqq n$.

Here $P_i = P(\lambda_i)$ denotes the laser power at λ_i; N_{tot}, the total number density; and c_j, the concentration of component j with absorption cross section σ_{ij} at λ_i. The sum is taken over the n components present in the sample. Again, a possible phase factor would have to be taken into account, as has been discussed in detail elsewhere (Meyer and Sigrist, 1990; Sigrist, 1988). Formally, the solution of Eq. (4.41) is

$$c_j = \frac{1}{CN_{\text{tot}}} \sum_{i=1}^{m} (\sigma_{ij})^{-1} \left(\frac{S_i}{P_i}\right) \tag{4.42}$$

where $(\sigma_{ij})^{-1}$ is the inverse of the matrix (σ_{ij}).

The effectiveness of this method in analyzing a multicomponent sample essentially depends on the nature of the matrix (σ_{ij}). The trivial case is represented by a diagonal matrix (σ_{ij}), i.e., when a set of wavelengths λ_i can be found with absorption of only one compound at each λ_i. However, this ideal case without interferences among the components can generally not be achieved. Instead, considerable effort is needed to discriminate among the various known components of the mixture, as well as to identify any unexpected components. In a previous study (Bernegger and Sigrist, 1990; Meyer and Sigrist, 1990) we have reported on a weighted iterative least squares algorithm on the basis of known absorption cross sections σ_{ij} that we applied to fit measured PA absorption spectra of polluted air from various sources. The iterative procedure was repeated until the concentrations c_j converged after a few iteration steps. Deviations between measured and calculated spectra at specific laser wavelengths λ_i or in certain spectral regions can be used to identify additional species that were originally not included in the fit.

A more robust multivariate calibration approach (e.g., see Haaland, 1990; Martens and Naes, 1989) is based on the use of principal component regression or partial least squares, where the predictive power of the selected laser wavelengths including the influence of outliers is considered. Its implementation for PA data analysis on various multicomponent air samples is presently being studied. Two further algorithms based on different criterions have recently been presented and successfully applied to some of our PA data on urban air (Kataev and Mitsel, 1991).

4.2.2.3. Determination of Sensitivity and Selectivity

The choice and number of laser wavelengths λ_i that are considered for the analysis of multicomponent mixtures strongly influence the accuracy of the concentrations determined (Frans and Harris, 1985). In order to quantify these effects the concepts of sensitivity and selectivity have been introduced. Referring to Eq. (4.40), we can again define the sensitivity as the minimum detectable concentration $c_{j,\min}$ of a substance j, yet now in the presence of other species. Therefore, $c_{j,\min}$ is often somewhat higher than in the one-component case owing to absorption interferences.

For multicomponent gas mixtures, the *selectivity* of detection may be even more important than the *sensitivity*. The selectivity describes the ability of the system to detect a small concentration of a selected species in the presence of other absorbing species possibly at even higher concentrations. The selectivity is essentially determined by interferences between the absorption spectra of the various substances and thus related to the orthogonality of the set of linear equations (4.41). Consequently, the selectivity can be influenced by a proper choice of wavelengths λ_i used for the measurement. On the basis of the $m \times n$ matrix (σ_{ij}) defined in Eq. (4.41), various parameters like partial sensitivity and selectivity as well as mutual cross sensitivies or rejection ratios have been introduced to quantify the interference effects (Bergmann et al., 1987; Junker and Bergmann, 1974; Kaiser, 1972; Kreuzer, 1974; Meyer and Sigrist, 1990; Meyer et al., 1988). Data for sensitivity and selectivity have been given previously for various gases and vapors measured with our CO and CO_2 laser PA systems (Bernegger and Sigrist, 1990; Meyer and Sigrist, 1990). As mentioned above, the selection of m different wavelengths λ_i for monitoring n different species (with $n \leqslant m$) is crucial for achieving high selectivity and sensitivity. For the case where the species and their average concentrations c_j are known, the evaluation of a set of λ_i is rather straightforward. The optimum λ_i are characterized by maximum contributions to the absorption of a given species j at λ_i relative to the total absorption α_i of all species at λ_i, i.e., for

$(\alpha_{ij}/\alpha_i) =$ maximum. In other cases, a full absorption spectrum of the sample taken in advance usually yields the necessary information. It is clear that a good selectivity is expected for compounds with well-structured spectra. A convincing example is the differentiation between isomers of dichlorobenzene in stack emission that was achieved with our CO_2 laser PA system (Meyer, 1988; Sigrist et at., 1989a). Further examples are discussed in Section 4.5.

4.3. EXPERIMENTAL ARRANGEMENTS FOR LASER PA TRACE GAS DETECTION

The typical setup for present PA air monitoring studies is schematically shown in Fig. 4.6. It has been discussed in various reviews (Sigrist, 1986; Tam, 1983, 1986; West, 1983; Zharov and Letokhov, 1986). The experimental arrangement consists of a tunable laser as radiation source whose beam is either modulated by some means, e.g., mechanically by a chopper, or directly pulsed. The radiation is directed through the PA cell that contains the gas sample under study. The generated acoustic waves are detected by an appropriate microphone or some other device whose signal is then further processed. In the example shown in Fig. 4.6 the microphone signal is fed to a lock-in amplifier locked to the modulation frequency. The average laser power P is recorded simultaneously by a power meter in order to normalize the microphone signal S according to Eq. (4.36). Numerous different versions and modifications of this general scheme have been discussed in the literature. The main features are outlined in the following paragraphs.

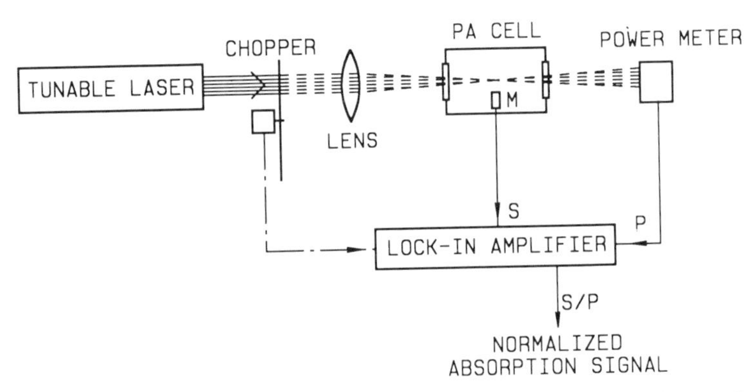

Figure 4.6. Schematic setup for gas-phase acoustic spectroscopy with tunable laser and microphone detection. *Key*: M, microphone; P, laser power; S, microphone signal.

4.3.1. Tunable Radiation Sources

Since the generated PA signal is proportional to the radiation power, as Eq. (4.36) implies, powerful radiation sources, particularly lasers offering high spectral power densities, are advantageous for achieving high detection sensitivity and selectivity. However, incoherent radiation sources such as lamps have also been employed in combination with filters. Examples are commercial PA gas monitors for the detection of single species (like those of Brüel & Kjær, Denmark, or Aritron, Switzerland). These devices use a small light bulb with either a chopper or direct current modulation as the modulated radiation source and appropriate filters to avoid absorption interferences with other species. At interference-free conditions detection limits in the sub-ppm range can be achieved for single compounds with such compact instruments.

Laser sources permit a considerable improvement in sensitivity and selectivity owing to their much higher spectral brightness. This is manifested in detection limits in the sub-ppb range that have been achieved for single species. Unfortunately, continuously tunable, intense lasers operating in the interesting mid-IR region are not readily available. Many studies have been performed with CO_2 lasers, including isotopes of CO_2, and some with CO lasers. These lasers offer high power, yet they are in general only step tunable. However, promising new laser developments will certainly extend their spectroscopic potential substantially. The CO laser has recently been operated on overtones, i.e., on $\Delta v = 2$ transitions, resulting in an additional tuning range between 2.86 and 4.07 μm, with stepwise tuning on 150 laser transitions (Gromoll-Bohle et al., 1989).

On the other hand, the progress in CO_2 laser technology has revolutionized the applications in two ways: first, sealed-off CO_2 lasers without need of flowing gas made field studies much easier; second, these lasers can easily be operated with isotopic CO_2 gas, which yields numerous additional laser transitions in the 9–12 μm wavelength range. Applications of isotopic CO_2 lasers are discussed below. An additional development concerns the compact waveguide CO_2 lasers. Since their gas pressure is higher than that of conventional CO_2 lasers, they offer the advantage of fine tuning around the pressure-brodened laser transitions. For example, Tang and Henningsen (1986) have built a pulsed waveguide CO_2 laser that is tunable over 450–500 MHz, single line and single mode at any of the ca. 80 strongest lines. In combination with photoacoustics at reduced gas pressure a powerful tool for spectroscopic investigations of molecules is obtained. Since the tuning range of 500 MHz corresponds to 5–10 Doppler widths, absorption line parameters (position, strength, pressure-broadened width) for lines of numerous molecules that fall within any of the tuning windows reached by the waveguide laser can be measured (Olafsson, 1990).

A further improvement of wavelength tunability is achieved with a high-pressure CO_2 laser. At a total pressure of the laser gas above ca. 10 bar the pressure-broadened laser transitions overlap and a continuous tunability within the 9 *R*, 9 *P*, 10 *P*, and 10 *R* laser branches can be realized (Carman and Dyer, 1979). In a very recent project we have built and operated such a CO_2 laser equipped with a near-grazing incidence grating resonator (Repond et al., 1992, 1993). A continuous tunability within the ranges indicated in Table 4.1 has been achieved at an emission bandwidth of ca. 0.5 GHz. This laser is presently being applied to PA trace gas studies for the first time.

Other lasers that have been applied to PAS on gases include a continuously tunable CO SFR (spin-flip Raman) laser, a DF (deuterium fluoride) laser, a lead salt diode laser ($PbS_{1-x}Se_x$), and a color center laser, all in the IR range, as well as an Ar^+ laser, dye and frequency-doubled dye lasers, and excimer lasers in the visible and UV spectral range. Tables 4.3 and 4.4 (in Sections 4.4.1 and 4.4.2, below) summarize PA studies performed with a variety of laser types.

Although the perfect laser system with continuous tunability throughout

Table 4.1. Continuous Tuning Ranges of High-Pressure CO_2 Laser

$^{12}C^{16}O_2$ Laser Transitions[a]	Wavelengths/Wavenumbers
9 *R*(32) to 9 *R*(8)	9.210–9.342 μm 1085.77–1070.46 cm^{-1}
Additional 9 *R*(6)	9.354 μm 1069.01 cm^{-1}
9 *P*(10) to 9 *P*(24)	9.473–9.586 μm 1055.63–1043.16 cm^{-1}
Additional 9 *P*(26), 9 *P*(28)	9.604, 9.261 μm 1041.28, 1039.37 cm^{-1}
10 *R*(30) to 10 *R*(6)	10.182–10.349 μm 982.10–966.25 cm^{-1}
10 *P*(10) to 10 *P*(30)	10.494–10.696 μm 952.88–934.89 cm^{-1}
Additional 10 *P*(8), 10 *P*(32)	10.477, 10.719 μm 954.55, 932.96 cm^{-1}

[a] The numbers 9 and 10 indicate the wavelength range (in μm) of the laser transition. The letters *P* and *R* refer to the branch the transition belongs to, with *P* corresponding to a jump in rotational quantum number of $\Delta J = +1$ and *R* corresponding to $\Delta J = -1$. Finally, the numbers in parentheses indicate the rotational quantum number *J* of the lower vibrational, i.e., of the final, state of the CO_2 laser transition.

the mid-IR range is not readily available, considerable progress has been made in recent times. Infrared generation based on nonlinear optical processes (Shen, 1976) appears most promising today mainly owing to the availability of new nonlinear optical materials such as $AgGaS_2$ and $AgGaSe_2$ and improved compact tunable solid state pump lasers. For example, a parametric oscillator with a Nd:YAG laser as a pump laser is now commercially available with a tuning range between 1.45 and 8 μm. On the other hand, CW difference frequency generation in a $AgGaS_2$ crystal pumped by two stabilized single-frequency CW dye/Ti:sapphire lasers and its spectroscopic application have recently been reported (Canarelli et al., 1992; Hielscher et al. 1992). A tuning range between 3 and 9 μm at an ultranarrow line width of < 0.5 MHz has been achieved, and an extension of the range to 18 μm by using $AgGaSe_2$ appears feasible (Simon et al., 1993). These new developments will certainly have a great impact on spectroscopic studies including PA investigations on trace gases.

4.3.2. Modulation Schemes

In PAS many different modulation methods are applied (Sigrist, 1986; Tam, 1983; West, 1983). One can differentiate between the modulation of the incident radiation and the modulation of the sample absorption itself. The first technique includes the widely used amplitude modulation of continuous radiation by mechanical choppers and electro-optic and acousto-optic modulators as well as modulation of the laser emission itself by pulsed excitation, Q-switching, or mode locking. In comparison to amplitude modulation, frequency or wavelength modulation of the radiation may improve the detection sensitivity by eliminating the continuum background caused by a wavelength-independent absorption, e.g., of the cell windows, known as window heating (Gandurin et al., 1987). For example, sensitivity enhancement by a factor of 10–50 for the detection of nitric oxide resulted by using frequency modulation of an SFR laser (Kreuzer and Patel, 1971). Obviously this type of modulation is most effective for absorbers with narrow line widths (such as atomic and diatomic species) and most easily performed with radiation sources whose wavelength can rapidly be tuned within a few wavenumbers, e.g., lead salt diode lasers with wavelength modulation directly by the current.

The second scheme mentioned above, i.e., a modulation of the absorption characteristics of the sample, is possible by applying modulated magnetic or electric fields to the samples; consequencly, the absorption wavelength of the sample is varied owing to the Zeeman or Stark effect, respectively. The result is again a suppression of the continuum background. For example, Kavaya et al. (1979) achieved a noise reduction by 500 times by using Stark modulation instead of the conventional chopper modulation. Of course, it should be remembered that Stark modulation in trace gas detection is restricted to molecules with a permanent electric dipole moment like ammonia (NH_3) or nitric oxide (NO). A simple calculation yields, for example, an absorption increase by a factor of 1.7 at the a$Q(5, 5)$ line of NH_3 measured at the 10 $R(20)$ $^{13}CO_2$ laser transition by applying an electric field strength of 12 kV/cm (W. Aeschbach, unpublished report, 1989). Di Lieto et al. (1982) combined the modulation of laser intensity and of Stark tuning for shifting molecular transitions into resonance with a fixed-frequency laser. The same authors also performed intermodulated PAS with counterpropagating beams chopped at different frequencies in combination with Stark tuning (Tonelli et al., 1983). A high sensitivity at low pressure was achieved and Doppler-free spectroscopy was obtained when the amplifier was locked to the sum frequency. More recently, intermodulated Stark spectroscopy with a chopped laser beam and a modulated Stark field at a different frequency was successfully applied to the sensitive monitoring of ammonia in the presence of absorbing water vapor and carbon monoxide (Sauren et al., 1991). The performance of a PA system operated at the sum and difference frequency sidebands of the modulation frequencies was found to be superior for ambient NH_3 monitoring compared to conventional schemes.

4.3.3. Design of PA Gas Cells

The PA cell serves as a container for the gas sample and for the microphone or some other device for the detection of the generated acoustic waves. An optimum design of the PA cell thus represents a crucial point for many applications. This is particularly true for trace gas monitoring, where the noise ultimately limits the detection sensitivity. Many cell configurations have been presented including acoustically resonant and nonresonant cells, single- and multipass cells, as well as cells placed intracavity (Dewey, 1977; Rosengren 1975; Tam, 1983). A selection of examples is given in Fig. 4.7. In nonresonant spectrophones operated in the $A_0(\omega)$ mode (Fig. 4.7a), a high sensitivity can only be attained for small volumes of 1–10 cm^3 and low modulation frequencies up to several tens of hertz, as outlined in Section 4.2. The main disadvantage is the relatively large continuum background signal caused by absorption at the windows (so-called window heating) and cell walls. An improvement in signal-to-noise ratio is obtained by the use of Brewster windows for reducing stray reflections (Patel and Kerl, 1977) or acoustic baffles as shown in Fig. 4.7b for minimizing the effect of window heating (Dewey, 1977). Resonant cells are usually considerably larger in volume, and it thus requires a fairly high Q-factor to overcome the problems of the greater cell volume and higher

Figure 4.7. Examples of PA cell designs for trace gas detection: (a) Simple nonresonant PA cell; (b) PA cell with window absorption effect minimized (From Dewey, 1977); (c) multipass, resonant PA cell (From Koch and Lahmann, 1978); (d) intracavity laser PA cell (From Bray and Berry, 1979).

modulation frequency. However, if properly designed, they offer the following advantages:

a. Reduction of surface-to-volume ratio minimizes wall adsorption effect.
b. Operation at higher frequencies reduces the background caused by environmental noise.
c. Operation with continuous gas flow is possible.

In the past, resonant cells operating on longitudinal, azimuthal, radial, or Helmholtz resonances have been developed (for reviews, see, e.g., Kamm, 1976; Tam, 1983; West, 1983; Zharov and Letokhov, 1986). Furthermore, the cell can be designed as a multipass cell, as demonstrated by Koch and Lahmann (1978) and shown in Fig. 4.7c. In their configuration the laser beam enters through a small uncoated orifice of a lens and is multiply reflected 65 times into the cell before exiting through the same orifice. With operation at an azimuthal resonance and maximizing of the overlap between the acoustic mode and the heating source H [see Eq. (4.21)], the aforementioned authors achieved a very high detection sensitivity of 0.1 ppb of SO_2 with only 1 mW of UV laser power. A further approach concerns the placement of the PA cell.

Barrett and Berry (1979), Bray and Berry (1979), and Smith and Gelfand (1980), for example, have discussed intracavity configurations like that shown in Fig. 4.7d. Such a scheme avoids the external multipass arrangement yet requires a careful alignment of the PA cell in the laser cavity. An interesting feature of intracavity PAS is the study of laser properties such as saturation intensities, small-signal gain, or optimum output coupling, as pointed out by Fiedler and Hess (1989).

In the following, some selected examples of PA cell designs are discussed in more detail. The construction of a "windowless" resonant PA cell by Gerlach and Amer (1980) or of an open PA chamber by Miklos and Lörincz (1989) or Angeli et al. (1991) represents a unique possibility of eliminating the window heating noise. Whereas the first authors operated their cell at the first radial resonance, the latter two groups used the (0,1,0) azimuthal mode and equipped the open cell with acoustic filters and an electronic resonance locking circuitry. In one of our own PA arrangements for trace gas monitoring (see Section 4.5) we have used a modified version of the "Gerlach and Amer" cell. In addition to the central cylinder we have added two buffer volumes (Meyer and Sigrist, 1990) such that both the laser beam and the gas flow enter and leave the central part of the cell at nodal positions of the operating first radial mode, thereby minimizing the noise caused by the flow. As a result we can operate the system continuously with gas flow rates of up to 1 L/min without a decrease of the signal-to-noise ratio.

In conjunction with a project on CO laser PAS we developed a novel cell operated on a longitudinal resonance (Bernegger and Sigrist, 1987). In this case the optimum cell design was found by numerical simulation of the acoustic properties of various cell geometries. For this purpose we developed a model using infinitesimal analog acoustic impedances (Morse and Ingard, 1986). Based on a matrix formalism for four terminals, a computer program was applied that permits the calculation of the frequency response of the PA amplitude at any position within a one-dimensional cell. The calculation yielded excellent agreement with experimental data. The resulting optimum design of the longitudinal resonant cell and its frequency response are presented in Fig. 4.8. Figure 4.8a shows the cell geometry, with the entrance and exit

Figure 4.8. (a) Top view of longitudinal resonant spectrophone. (b) Three-dimensional calculated plot of cell constant $F = C \cdot R_{mic}$ for air at 950 hPa and $T = 295$ K as a function of the modulation frequency $f = \omega/2\pi$ and the position x along the cell axis. The mode at 555 Hz represents the second longitudinal resonance, whereas that at 380 Hz corresponds to a Helmholtz resonance between the volumes of the cell and the pressure sensor. (c) Calculated frequency response at the cell center for absorption of equal amounts of laser radiation by the gas (solid line) and by the exit window (dashed line). The circles represent the frequency response measured with the microphone. From Bernegger and Sigrist (1990).

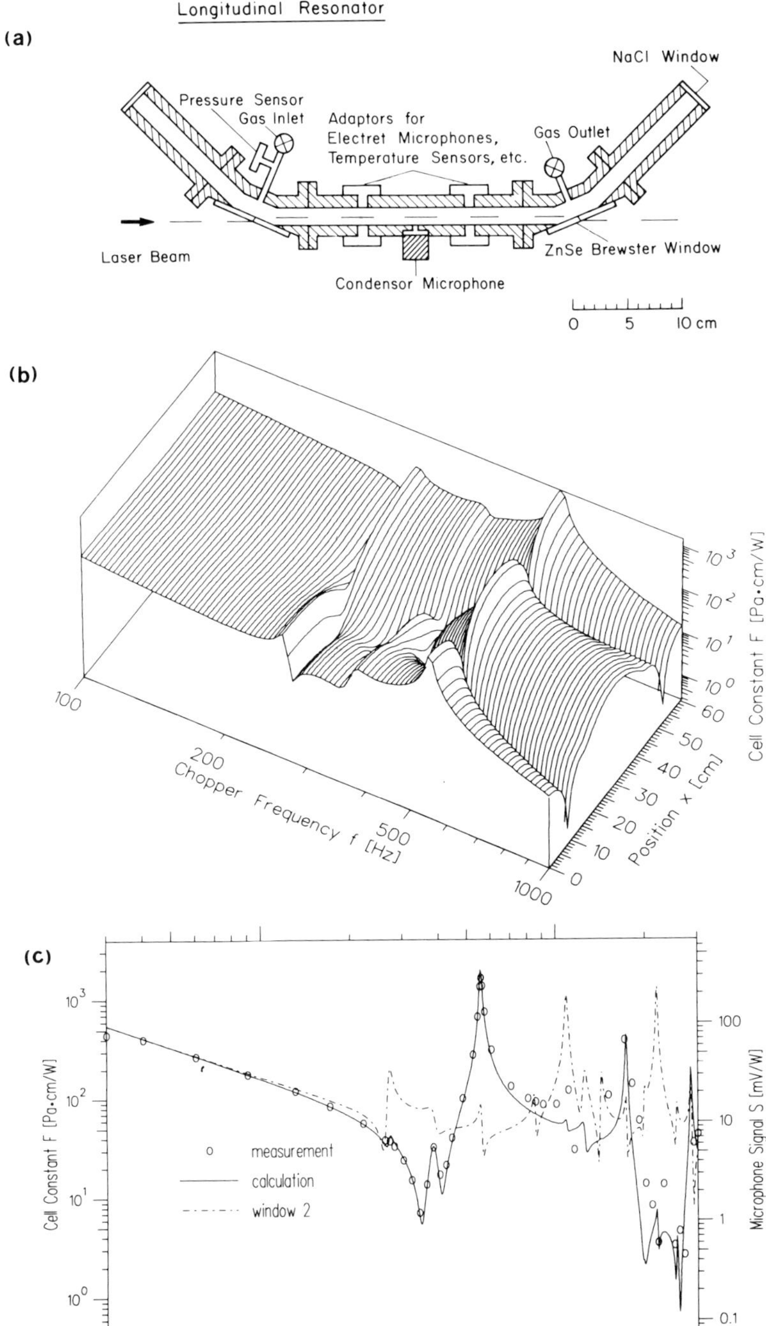
Longitudinal Resonator
(a)
NaCl Window
Pressure Sensor
Gas Inlet
Adaptors for
Electret Microphones,
Temperature Sensors, etc.
Gas Outlet
Laser Beam
ZnSe Brewster Window
Condensor Microphone
0 5 10 cm
(b)
Cell Constant F [Pa•cm/W]
Chopper Frequency f [Hz]
Position x [cm]
(c)
measurement
calculation
window 2
Cell Constant F [Pa•cm/W]
Microphone Signal S [mV/W]
Chopper Frequency f [Hz]

windows mounted at the Brewster angle. The NaCl window at one of the cell ends serves to transmit the residual part of the laser beam that is not totally polarized and is thus reflected off the exit Brewster window. Both the Brewster windows and the gas in- and outlets are located at the pressure nodes of the second longitudinal mode. Figure 4.8b represents a three-dimensional plot of the frequency response of the longitudinal resonator along the cell axis x. The acoustic cell constant $F = C \cdot R_{mic}$ in Pa·cm/W is calculated for air at a total pressure of 950 hPa and a temperature of 295 K. The main resonance at 555 Hz with a Q factor of 52 corresponds to the second longitudinal mode, whereas the small adjacent resonance at 380 Hz represents a Helmhotz resonance between the volumes of the cell and of the pressure sensor. The pressure nodes located at the window positions at $x = 15$ cm and $x = 45$ cm as well as the pressure maximum at the cell center at $x = 30$ cm are clearly seen. Finally, Fig. 4.8c shows a cross section of the plot of Fig. 4.8b for $x = 30$ cm, i.e., for the position of the condenser microphone. The solid and dashed lines represent the calculated frequency response due to the absorption by the gas and—assumed to be identical—by the exit window, respectively. This latter signal thus yields a relative measure of the noise caused by window heating. The maximum ratio between the solid and dashed lines at 555 Hz clearly demonstrates the advantage of operating the cell at the second longitudinal mode, where the pressure nodes coincide with the window positions. Furthermore, the agreement with experimental data, represented by circles, is manifested, as is the $1/f$ dependence ($f = \omega/2\pi$) for the nonresonant case at the lower frequencies, as predicted by Eq. (4.25).

The application of the Stark or Zeeman effect discussed in the previous subsection requires a special cell design. For example, Sauren et al. (1990) presented a resonant cell where an electric field strength of max. 5 kV/cm could be applied at a modulation frequency of 0.1 Hz and a gas flow rate of 0.5 L/min. The authors achieved an ultimate detection limit for ammonia of 0.4 ppb. Most recently, Thöny (1993) reported on a special design of a PA Stark cell which allows the application of electric field strengths of up to 17 kV/cm.

Furthermore, for some applications such as for studies on liquid samples with low vapor pressures a heatable cell is needed. For example, Jalink and Bicanic (1989) have proposed a resonant PA heat pipe cell whose design is presented in Fig. 4.9. The central region of the stainless steel tube (the evaporator) is externally heated by a resistive heater, thereby causing vaporization of liquid and hence generation of a driving force required for the transport of a sample to the water-cooled sections. There it condenses before being returned to the evaporator by the action of capillary forces. The contamination of the windows and the microphone is prevented by a sharp boundary that is formed between the vapor section and the adjacent volumes filled with a buffer gas. The operational temperature range of this cell is limited

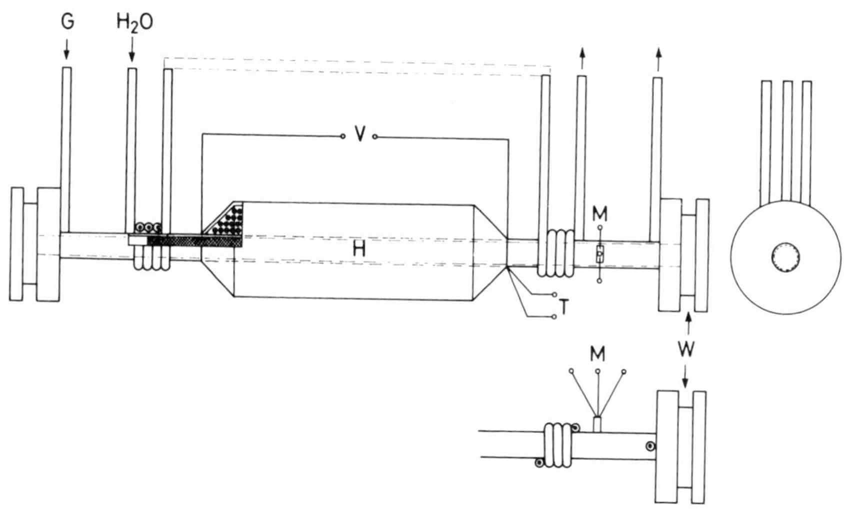

Figure 4.9. Design of a PA heat-pipe cell with windows (W), gas inlet (G), water cooling (H_2O), resistive heater (H), thermocouple (T) with voltage, and microphone (M). From Jalink and Bicanic (1989).

only by the choice of working fluid and cell material. The feasibility of the cell has successfully been demonstrated by taking PA spectra of geraniol ($C_{10}H_{18}O$), an essential oil fragrant, at 403 K (Bicanic et al., 1989). An improved version of the heat pipe cell is claimed to permit measurements even at much higher temperatures, e.g., 900 K for CO_2 (Bicanic et al., 1992).

Sample cells used in pulsed PAS must be of different design from those used in CW PAS. The reason is that extraneous signals generated by the large photon fluxes provided by pulsed laser sources need to be minimized. Since the repetition rate of pulsed lasers is usually low, there is no benefit from using acoustic cell resonances and the nonresonant mode of operation thus predominates. Another point to be considered is the need to optimally monitor the ballistic acoustic wave first generated by sample absorption. The window heating signal caused by absorption at the windows should thus be of the highest possible optical quality. The microphone or any other acoustic detector should be placed close to the focal region of the laser radiation in order to maximize the collection efficiency of the acoustic signal. An interesting solution has recently been presented by Carrer et al. (1992). The cell geometry is designed as a toroidal or parabolic reflector that focuses the microphone signal, thereby improving the detection sensitivity by an order of magnitude. Also in this regime, a multipass arrangement might be implemented to more effectively use the source radiation (Herriot et al., 1964; Owyoung et al., 1978; Siebert et al., 1980). In one of the schemes, the microphone signal scaled with the number of radiation passes up to a total of 17 passes (Siebert et al., 1980).

Finally, attention has to be paid to the condition and quality of the cell surfaces. They influence the background level due to light scattering and their tendency to adsorb molecules. Various cell materials and surface treatments as well as the application of surface coatings have been studied for this purpose (Beck, 1985; Hinderling et al., 1987)

4.3.4. Detection Schemes

The thermal and acoustic disturbances generated in the gas sample can be detected either by pressure sensors or by refractive index sensors (Bailey and Cruickshank, 1989; Sigrist, 1986; Tam, 1986; Tran, 1986; West, 1983). The appropriate choice depends on sensitivity, ease of operation, and ruggedness as well as on a potential requirement for noncontact detection. The most common schemes are illustrated in Fig. 4.10 and briefly outlined in the following paragraphs.

The conventional and most widely used PA scheme (Fig. 4.10a) uses commercial microphones as pressure sensors. These include miniature electret

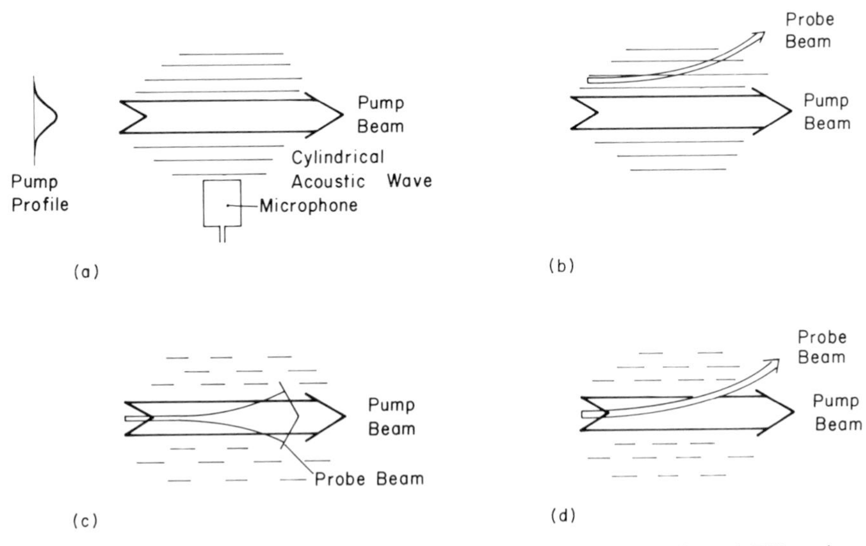

Figure 4.10. Geometrical configurations for photoacoustic (PA) and photothermal (PT) sensing schemes. (a) Conventional PA method with microphone detection. (b) Deflection of a displaced, collinear probe beam by the pressure-induced change of the refractive index. (c) Detection of the generated thermal lens by a collinear probe beam. (d) Deflection of a collinear probe beam by the thermally induced change of the refractive index.

microphones such as Knowles or Sennheiser models with typical responsivities R_{mic} of 10 m V/Pa as well as condenser microphones, e.g., Brüel & Kjær models, with typical R_{mic} of 100 m V/Pa. These devices are easy to use and usually also sensitive enough for trace gas studies with very low absorptions. Most often, the detection threshold is neither determined by R_{mic} nor by the electrical noise of the microphone but rather by other noise sources (absorption by desorbing molecules from the cell walls, window heating, external noise, etc.). The responsivity R_{mic} only weakly depends on frequency. The electret microphones produced for hearing aids exhibit a rather flat frequency response between, say, 20 Hz and 20 kHz. In special applications, a temperature dependence of R_{mic} may have to be taken into account. Since these acoustic detectors are in contact with the gas sample, special precautions are necessary for studies on aggressive media. In these cases, noncontact schemes based on refractive index sensors are preferable.

Figure 4.10b represents the scheme of PA deflection where the acoustic disturbance is detected by monitoring the deflection of a collinear or transverse probe beam, generally a He–Ne laser beam in combination with a position sensor. This setup is applied for pulsed laser excitation and offers an improved temporal resolution and, in the case of the transverse probe beam, also spatial resolution. However, the detection sensitivity is lower than that achieved with conventional PAS.

Photothermal (PT) techniques, notably thermal lensing and PTdeflection, have also been proved to be suitable methods for low-absorbance measurements (Bailey and Cruickshank, 1989; Bailey et al., 1989; Gupta, 1992; Higashi et al., 1983; Jackson et al., 1981; Tran, 1986). However, they have preferentially been applied to studies on solids and liquids. The thermal lens technique shown in Fig. 4.10c is based on the formation of a lens, usually of a divergent lens, that is caused by the modification of the refractive index by the generated temperature rise. The lens formation is again monitored by a weak probe beam either collinear or transverse to the pump beam (Fang and Swofford, 1983). Kawasaki et al. (1987) reported on the detection of ammonia by thermal lensing. However, their detection limit of 6% was rather poor at that time and limited by the small power of the pump beam. Tran and Franko (1989) introduced a novel dual-wavelength thermal lens spectrometer. The sample is excited by two wavelengths alternately and the corresponding thermal lens signals are monitored by a He–Ne probe laser. A detection limit of 300 ppb for NO_2 was estimated by using 30 mW excitation beams modulated at 4.6 Hz. In another example, Takahashi et al. (1991) applied the thermal lens method to investigate combination bands of acetylene vapor with a pulsed tunable IR laser. Both studies imply that in comparison to acoustic detection the thermal lens techique is less sensitive, by about 100 times in the latter example. The same conclusion was drawn by Patel and Kerl (1977). Thermal lensing appears thus less suitable for trace gas monitoring than acoustic detection unless a noncontact scheme is mandatory.

A different approach of noncontact detection is based on PT deflection spectroscopy as illustrated in Fig. 4.10d (Gupta, 1992). This scheme again uses a pump and probe beam arrangement. The pump beam is usually a pulsed laser for excitation of the gas sample, while the collinear or transverse probe beam is a He–Ne laser beam whose transient deflection induced by the absorption-related transient change of the refractive index is monitored by a photodiode in combination with a razor blade or by a position sensor. Fournier et al. (1980) have presented a simple PT arrangement that is feasible for in situ studies. They achieved a detection threshold of 5 ppb for ethene diluted in nitrogen. This corresponds to a minimum detectable absorption coefficient $\alpha_{\mathrm{min}} = 10^{-7}\ \mathrm{cm}^{-1}$. Rose et al. (1982) reported on the application of PT deflection to combustion diagnostics. They claim to have obtained sensitivities below 1 ppm for monitoring OH molecules with good spatial resolution (transverse probe beam) and good temporal resolution (single shot). Bialkowski and He (1988) obtained a detection limit of 0.7 ppm for Freon 12 (CCl_2F_2) in argon at a pulse energy of the pump laser of only 1 mJ. An intracavity dye laser PT deflection technique has been demonstrated to yield a similar sensitivity, with $\alpha_{min} = 10^{-9}\ \mathrm{cm}^{-1}$ as the intracavity PA method (Reddy et al., 1982). However, Sell (1985), who successfully applied this technique to take absorption spectra of a flowing gas stream, pointed out the difficulty of deriving accurate absorption coefficients by this scheme in comparison to the conventional PA technique. This might be the main reason why PT deflection spectroscopy has not found wide application in trace gas monitoring. However, various successful studies on multiphoton and saturation type nonlinear absorption behavior by the PT deflection method have been reported, e.g., for analyzing mixtures at two halocarbon species with analyte levels below 10 ppm in argon (Bialkowski and Long, 1987).

Photothermal spectroscopy (Rosengren, 1973) is based on a direct measurement of the induced temperature variation by a thermal sensor, e.g., a pyroelectric detector. Hartung et al. (1981) built a photo- or optothermal (PT or OT) detector with a pyroelectric poly (vinyl difluoride) (PVF_2) foil. The authors claim to have reached a detection sensitivity corresponding to a minimum absorption coefficient $\alpha_{\mathrm{min}} = 2 \times 10^{-8}\ \mathrm{cm}^{-1}$ at a laser power of 1 W and a detection bandwidth of 1 Hz (Hartung and Jurgeit, 1982). This limit is comparable to routine PA measurements, particularly at low gas pressures. Another OT detector for gases has been introduced by Guorong et al. (1984). They placed a platinum wire parallel to the optical axis of the cell close to the laser beam and measured the induced change of resistance by a Wheatstone bridge. The authors used this simple detector for the study of collisional multiphoton energy deposition in SF_6 by a pulsed TEA (transversely excited atmospheric pressure) CO_2 laser.

Another noncontact scheme based on fiber optics has been introduced by Leslie et al. (1981). In this arrangement the induced pressure fluctuations are detected by a fiber-optic coil in one arm of an all-fiber Mach–Zehnder interferometer. The pressure modulation induces a phase modulation of the light propagating in the fiber sensor coil. However, in spite of later improvements (Leslie et al., 1983), the minimum detectable absorption was 1.6×10^{-7} $cm^{-1} \cdot W/Hz^{1/2}$ and thus still above that of a common microphone-based device. Further noncontact techniques such as a laser Schlieren microphone (Choi and Diebold, 1982) or, alternatively, protected transducers (Marinero and Stuke, 1979a) have also been applied.

An interesting technique for the *remote* detection of gases by PA detection has been proposed by Brassington (1982). With reference to the lidar principle (see Chapter 3), it has been named padar for *p*hoto*a*coustic *d*etection *a*nd *r*anging. The technique involves sending out a laser pulse tuned to an absorption line of the gas. When the laser pulse hits a localized concentration of the gas an acoustic pulse is generated that can be detected by a parabolic microphone sited close to the laser. The range is determined from the time delay between laser pulse and microphone signal. An excellent range resolution of < 1 cm and a maximum range of 100 m are estimated. No estimates on detection sensitivity are given, yet the method could prove useful for remote leak detection.

Table 4.2 summarizes the main features of PA and PT schemes for trace gas detection. It should be emphasized that most studies are performed with conventional microphone detection while the noncontact methods discussed are applied to special cases.

4.4. LABORATORY STUDIES ON AIR SAMPLES

This section summarizes the state of the art of trace gas detection by laser PA techniques in the laboratory by means of numerous examples. However, since the field has grown tremendously since the pioneering studies, no attempt is made to include all important references. Since most investigations are performed by conventional PAS as discussed above, related techniques like Raman (Barrett and Berry, 1979; Rotger et al., 1992; Siebert et al., 1980; West et al., 1980), Doppler-free (Marinero and Stuke, 1979b), saturation (Di Lieto et al., 1979), and harmonic saturated PAS (Klimcak and Gelbwachs, 1985) as well as PAS combined with gas chromatography (Nickolaisen and Bialkowski, 1986; Zharov, 1985) are not considered any further in this review. In the following

Table 4.2. Summary of Spectroscopic Schemes Based on Photoacoustic (PA) and Photothermal (PT) Effects

Technique	Sensitivity α_{min} (cm^{-1})	Experimental Setup	Special Features
PA (microphone)	10^{-9}-10^{-10}	Simple	Quantitative; mechanical contact; sensitivity to mechanical and acoustical noise
PA deflection (collinear)	$\sim 10^{-7}$	Simple to align	Noncontact; temporal resolution when pulsed; sensitivity to pointing noise
PA deflection (transverse)	$< 10^{-7}$	Simple to align	Noncontact; spatial resolution; temporal resolution when pulsed; sensitivity to pointing noise
Thermal lensing	10^{-8}	Difficult to align	Noncontact; temporal resolution when pulsed; sensitivity to pointing and intensity noise
PT deflection (collinear)	$\sim 10^{-8}$	Difficult to align	Noncontact; temporal resolution when pulsed; sensitivity to pointing noise
PT deflection (transverse)	$< 10^{-8}$	Simple to align	Noncontact; spatial resolution; temporal resolution when pulsed; sensitivity to pointing noise

two subsections we differentiate between PAS studies on certified gases and on multicomponent gas mixtures.

4.4.1. Certified Gases

The first PA studies that demonstrated the high sensitivity of laser PAS on gases and thus paved the way for numerous applications of the PA effect were all performed on one-component trace gases diluted in a nonabsorbing buffer gas, usually nitrogen (N_2) at atmospheric pressure and room temperature. In Table 4.3 numerous examples of such studies are given. The table lists the laser source with the corresponding wavelength range, the species studied, as well as the minimum detectable concentration, which lies in the ppb to sub-ppb range. It is obvious that many investigations have been performed with CO_2 lasers including isotopic lasers owing to the favorable wavelength range for spectroscopy and ease of laser operation. It should be pointed out that the studies on certified gases are valuable not only for demonstrating the feasibility of the technique but primarily for establishing a library of absorption spectra. This forms the basis for analyzing measured spectra of multicomponent mixtures according to the procedure outlined in Section 4.2.

Two examples of calibration spectra are shown in Figs. 4.11 and 4.12. The absorption spectrum of ethene (ethylene, C_2H_4) at CO_2 laser wavelengths is recorded in Fig. 4.11. It was taken with a certified mixture of C_2H_4 in N_2, usually 10–100 ppm, at a total pressure of 950 hPa and a temperature of 295 K (Meyer, 1988). The absolute absorption cross sections were derived after calibration on eight selected laser transitions and literature data (Boscher et al., 1979), taking pressure and temperature dependence into account (Persson et al., 1980). The C_2H_4 spectrum exhibits a characteristic absorption peak at the 10 $P(14)$ laser transition at 949.49 cm^{-1}. This peak is caused by the proximity of the 10 $P(14)$ laser transition to the Q-branch of the ν_7-vibration band of C_2H_4 centered at 948.7715 cm^{-1} (Herlemont et al., 1979). The series of the other sharp peaks in the spectrum originates from the rotational structure of this band. The 10 $P(14)$ laser wavelength exhibits a five fold higher absorption than the other wavelengths and thus represents the optimum choice for continuous C_2H_4 monitoring with respect to both sensitivity and selectivity. The maximum absorption cross section amounts to 1.3×10^{18} cm^2, which according to Eq. (4.40) yields a minimum detectable C_2H_4 concentration of ca. 0.7 ppb in N_2 or pure synthetic air (80% N_2/20% O_2) with our system.

The second example concerns the isomer m-xylene (C_8H_{10}), an aromatic hydrocarbon. Its absorption spectrum at CO laser wavelengths taken with N_2 as buffer gas at a total pressure of 950 hPa at a temperature of 295 K is plotted in Fig. 4.12 (Bernegger and Sigrist, 1990). In this case the calibration with an

Table 4.3. **Examples of Laboratory Photoacoustic Studies on One-Component Gas Mixtures (Trace Species Diluted in Nonabsorbing N_2)**

Laser	Spectral Region	Species	c_{min} (ppbv)[a]	References
HeNe	3.39 μm	CH_4	200	Kreuzer (1971)
CO	5.2–6 μm	10 different trace gases	$\leqslant 1$	Kreuzer et al. (1972)
CO_2	9.2–10.8 μm			
HeNe	3.39 μm	CH_4, *n*-butane	10^3	Dewey et al. (1973); Goldan and Goto (1974)
CO_2	9.2–10.8 μm	NH_3	<3	Max and Rosengren (1974)
CW dye	0.57–0.62 μm	NO_2	10	Angus et al. (1975)
DF	3.8 μm	CH_4, N_2O	0.5–5	Deaton et al. (1975)
HF	2.7–2.9 μm	HF, NO, CO_2	10^3	Gomenyuk et al. (1975)
CO_2	9.2–10.8 μm	C_2H_4, NH_3, Freons, etc.	$\leqslant 4$	Schnell and Fischer (1975)
CO SFR	~5.3 μm	NO	$\leqslant 0.1$	Patel and Kerl (1977)
CO	4.75 μm	CO	150	Gerlach and Amer (1978)
CO_2	9.2–10.8 μm	C_2H_4	0.3	Kritchman et al. (1978)
Frequency-doubled CW dye	290–310 nm	SO_2	0.12	Koch and Lahmann (1978)
CO_2	9.2–10.8 μm	Explosives	0.2–25	Crane (1978)
CO_2	9.2–10.8 μm	SF_6	0.01	Nodov (1978)
CO_2	9.2–10.8 μm	Various	ppb to ppm	Konjevic and Jovicevic (1979)
CO_2	9.2–10.8 μm	Hydrazines	<30	Loper et al. (1980)
$PbS_{1-x}Se_x$	4.8 μm	CO	5×10^4	Vansteenkiste et al. (1981)
DF	3.6–4 μm	CH_4 (in Ar)	$\alpha_{min} = 7 \times 10^{-8}\ cm^{-1}$	Leslie and Trusty (1981)
CO_2	9.2–10.8 μm	C_2H_4	—	Brewer et al. (1982)

CO_2	9.2–10.8 μm	C_2H_4, NH_3	~50	Chen et al. (1982)
CO_2 incl. isotopes	9–12 μm	22 species out of 250	—	Hubert et al. (1983)
CW dye	~600 nm	I_2, Br_2	3 ng/cm^3, 48 ng/cm^3	Keller et al. (1983)
XeCl (pulsed)	308 nm	Acetaldehyde	25	Leugers and Atkinson (1984)
CO_2	9.2–10.8 μm	Various	Few	Antipov et al. (1984)
CO_2	9.2–10.8 μm	SF_6, C_2H_4	3	Fung and Lin (1986)
CO_2	9.2–10.8 μm	Hydrazines, toxic vapors	≤10	Loper et al. (1986)
CO, CO_2	5–6, 9–11 μm	Freons	Few	Zelinger (1986)
CO/CO_2/HeNe	IR	NO, NO_2, NH_3, C_2H_4, saturated hydrocarbons	0.1–10^3	Gandurin et al. (1987)
$PbS_{1-x}Se_x$	4.6 μm	CO	44	Hurdelbrink (1986)
CO_2	9.2–10.8 μm	Various	0.2–4.6	Harren et al. (1987)
CO_2	9.2–10.8 μm	19 gases and vapors	<ppb to ppm	Meyer (1988)
CO	5.15–6.35 μm	18 gases and vapors	<ppb to ppm	Bernegger (1988)
$PbS_{1-x}Se_x$	5.94 μm	H_2O vapor	10^4	Johnson et al. (1988)
CW dye	~600 nm	H_2O vapor	$\alpha_{min} = 4 \times 10^{-9}$ cm^{-1}	Bondarev et al. (1988)
Frequency-doubled pulsed dye	~220 nm	HCOOH, CH_3COOH	140 120	Vujkovic Cvijin et al. (1988)
CO_2	9.2–10.8 μm	N_2O	2×10^4	Bicanic et al. (1988)
Diode	~10.53 μm	C_2H_4	200	Harren et al. (1989)
Nd:YAG (pulsed)	532 nm	NO_2	5	Kato and Sato (1991)
CO	4.8–8.4 μm	$COCl_2$	Few	Luo et al. (1991)
CO_2	9.2–10.8 μm	CO_2, NH_3, C_2H_4	7.4×10^3, 0.4, 1.3	Angeli et al. (1992)
CO_2	9.2–10.8 μm	CH_3F, CH_3Cl, SO_2	$\alpha_{min} = 10^{-7}$ cm^{-1}	Radak et al. (1993)

[a] ppbv: parts per billion per volume.

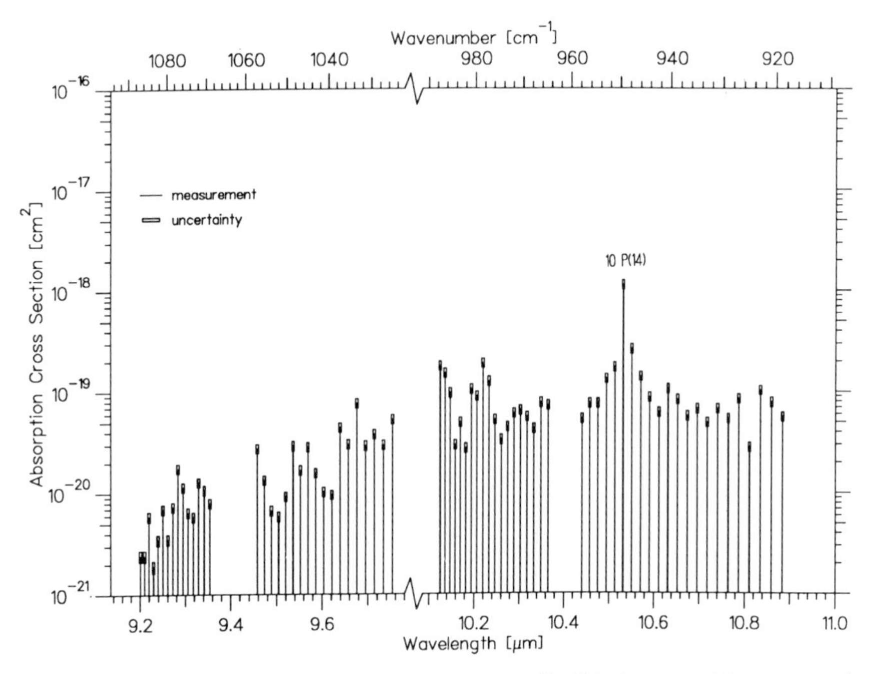

Figure 4.11. Absorption cross sections of ethene (C_2H_4) at $^{12}C^{16}O_2$ laser transitions measured with our PA setup with the carrier gas N_2 at a total pressure of 950 hPa and a temperature of 295 K. From Bernegger and Sigrist (1990).

estimated relative error of 10% is based on the measurements performed with our CO_2 laser PA system (Meyer, 1988). Xylene, like other aromatic hydrocarbons, exhibits absorption throughout the entire CO laser emission range. The maximum absorption cross section amounts to 10^{-19} cm^2, i.e., an order of magnitude lower than the corresponding data for ethene at CO_2 laser wavelengths. Hence, the detection limit for *m*-xylene under interference-free conditions is higher, namely, ca. 19 ppb. It should be noted that each molecule shows specific features in its absorption spectrum, which enables one to distinguish even among the three isomers of xylene (Sigrist et al., 1989b).

4.4.2. Multicomponent Gas Mixtures

Most gas samples such as stack gas or ambient air contain more than one absorbing component. Therefore, the problem of absorption interferences introduced in Section 4.2.2 has to be considered. Some examples of PA investigations on multicomponent samples are listed in Table 4.4. Most of them have

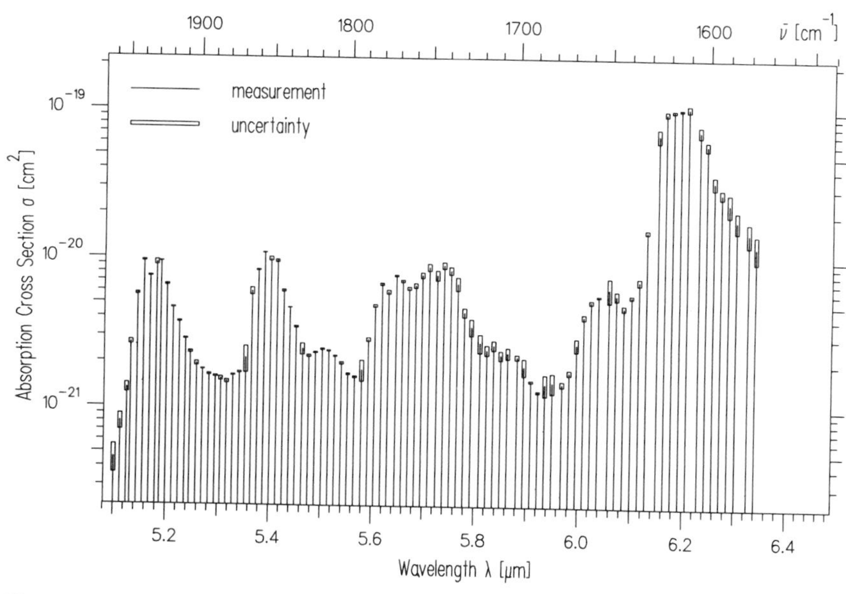

Figure 4.12. Absorption cross sections of *m*-xylene (C_8H_{10}) at $^{12}C^{16}O$ laser lines measured with our PA setup with the carrier gas N_2 at a total pressure of 950 hPa and a temperature of 295 K. From Bernegger and Sigrist (1990).

been aimed at the detection of one or a few specific components rather than at performing a chemical analysis, which would have been far more sophisticated. In some cases the sample was pretreated (filtered, purified), resulting in extremely low detection thresholds. The pretreatment of samples can thus be advantageous in the effort to gain in sensitivity. Some form of sample pretreatment is also used by other methods such as the preconcentration applied in gas chromatography. However, in these cases it must be ensured that the concentration of a species to be determined is not influenced or altered by such processes.

Probably the first PA study on multicomponent gas samples was devoted to the detection of NO in samples taken from room air, car exhausts, and ambient air collected near a busy road (Kreuzer and Patel, 1971). These measurements were performed with a CO spin-flip Raman (SFR) laser tuned to absorption frequencies of NO. The derived NO concentrations ranged from ca. 2 to 100 ppm. Another investigation was concerned with the detection of ethene (C_2H_4) in air samples taken from a main road and from fruit storage chambers (Perlmutter et al., 1979). In this case a discretely tunable CO_2 laser was applied and the investigators found a minimum detectable and identifi-

Table 4.4. Examples of Laboratory Photoacoustic Studies on Multicomponent Gas Mixtures

Laser	Spectral Region	Species	c_{min} (ppbv)	References
CO SFR	~ 5.5. μm	NO (in exhausts)	10	Kreuzer and Patel (1971)
$CO/^{13}CO_2$	~ 6 μm/~ 11 μm	Explosives (in air)	1–10	Claspy et al. (1976)
CO_2/high-pressure pulsed	9.2–10.8 μm	HDS (in H_2S)	100	Zharov (1977)
Pulsed dye	480–625 nm	NO_2 (in air)	4	Claspy et al. (1977)
CO_2	9.2–10.8 μm	C_2H_4 (in air)	5	Shtrikman and Slatkine (1977)
Ar^+	~ 488 nm (multiline)	NO_2 (in air)	5	Ioli et al. (1979)
CO_2	9.2–10.8 μm	SF_6, NH_3, O_3, C_2H_4, C_6H_6, SO_2 in air	ppbv to ppmv	Adamowicz and Koo (1979)
CO_2	9.2–10.8 μm	C_2H_4 (in air)	$\geqslant 5$	Perlmutter et al. (1979)
CO_2	9.2–10.8 μm	30 species in dry air	50	Wang et al. (1982)
Kr^+	406.7531 nm	NO_2 (in air)	2	Poizat and Atkinson (1982)

DF	3.64 μm	HCl (in air)	50	Fried and Berg (1983)
Frequency-doubled pulsed dye	300.05 nm	SO_2 in purified air	0.2	Vujkovic Cvijin et al. (1987)
CO_2	9.2–10.8 μm	Freons (in air)	20	Zelinger et al. (1988)
Frequency-doubled pulsed dye	303.59 nm	H_2CO in selected gas mixtures	51	Boutonnat et al. (1988)
CO_2	9.2–10.8 μm	NH_3 (in filtered air)	—	Bicanic et al. (1989)
Frequency-doubled pulsed dye	300–310 nm	SO_2/NO_2 in H_2O/N_2	1–2 nL/L	Gilmore et al. (1989)
Nd:YAG harmonics and Raman	UV–vis	SO_2, NO_2 in stack gases	$\alpha_{min} \leq 10^{-7}$ cm^{-1}	Carrer et al. (1990)
CO_2	9.2–10.8 μm	C_2H_4, H_2S, O_3 etc. (in air)	$\alpha_{min} = 7 \times 10^{-10}$ cm^{-1}	Harren et al. (1990)
CO	5–6 μm	Car exhausts	$\geqslant 10^3$	Bernegger and Sigrist (1990)
CO_2	9.2–10.8 μm	H_2S, D_2S, HDS, HTS	250×10^3 (HTS in H_2S)	Petkowska et al. (1991)
CO_2	9.2–10.8 μm	NH_3 (in air)	3	Sauren et al. (1991)
$^{13}C^{16}O_2$	10.8 μm	NH_3 (in air)	1	Trushin (1992)

able concentration of C_2H_4 of about 5 ppb in most cases. However, this limit could be increased up to 50 ppb primarily owing to inaccuracy in a priori knowledge of IR spectral signatures of interfering gases. Thus, the detection sensitivity was limited by interfering absorptions rather than by background noise.

Agricultural applications of PA trace gas detection have been put forward by several researchers in the Netherlands. For example, Bicanic et al. (1989) have investigated the formation of ammonia during the spoilage of inoculated beef at room temperature. Further experiments concentrated on the sensitive detection of ethene, which is known as a plant growth regulator (Woltering, 1987; Yang and Hoffman, 1984). With the aid of a PA intracavity setup and a NaOH scrubber for removing CO_2, a detection limit of 20 ppt of C_2H_4 in air was obtained (Harren et al., 1990.) This high sensitivity permitted the detection of C_2H_4 production from a single orchid flower as a function of time. Of particular interest was the increase of C_2H_4 production following emasculation and during the wilting process.

The problem of interfering absorptions with respect to the analysis of multicomponent gas mixtures originating from motor vehicle exhausts has been addressed in our laboratory by Bernegger and Sigrist (1990). For this purpose two identical PA cells of the special design presented above in Fig. 4.8 were implemented as sample and reference cell in a dual-beam setup. A CW CO laser (Edinburgh Instruments, Type PL 3) tunable on ca. 90 transitions between 5.15 μm ($\approx 1940\,cm^{-1}$) and 6.35 μm ($\approx 1575\,cm^{-1}$) with output powers between a few milliwatts and ca. 1 W was used as radiation source. The computer-controlled arrangement as shown in Fig. 4.13 has been described in detail previously (Bernegger and Sigrist, 1990; Sigrist, 1992). Since H_2O vapor is present in any sample (even after the introduction of a dry ice cooling trap) its strong absorption was always recorded simultaneously, i.e., at identical spectral composition of the CO laser beam both in the sample and reference cell. Since the H_2O vapor content in the reference cell was exactly known, the H_2O vapor content of the sample could be derived mathematically.

This PA system was applied to the analysis of exhaust gases from cars, trucks, and motorcyles. The exhausts samples were obtained from vehicles at the Swiss Federal Laboratory for Testing Materials and Research (whose acronym here is EMPA*) in Tedlar sampling bags and brought to our laboratory in special sampling chambers. The analysis of the dried and diluted samples was performed as described in Section 4.2.2 on the basis of a weighted iterative least squares fit of the measured PA spectrum, with numerous spectra of components expected to be present such as residual H_2O vapor, nitric

* EMPA, Eidgenössische Material Prüfungs Anstalt.

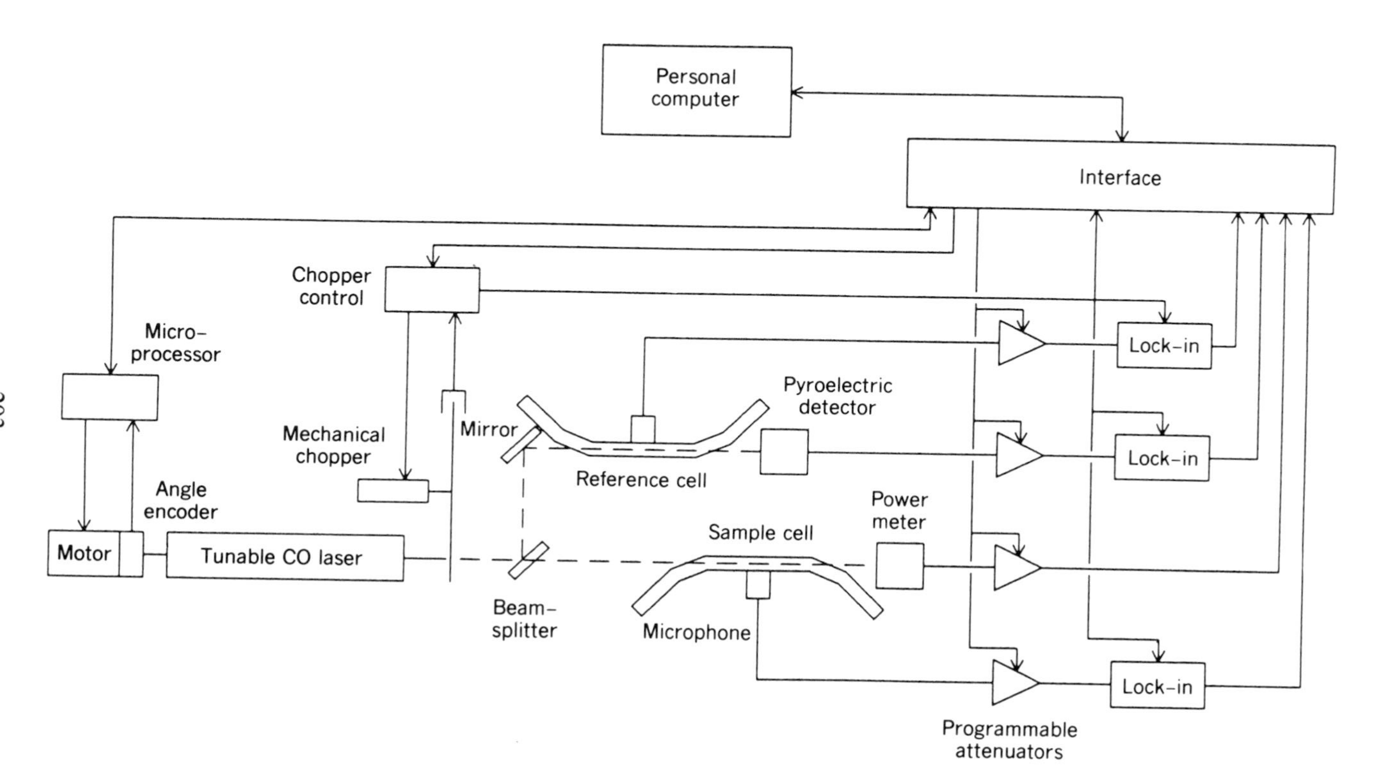

Figure 4.13. Experimental setup for automated CO laser PA spectroscopy with sample and reference cell.

oxides, and various volatile organic compounds (VOCs). Thereby the observed linear increase of the H_2O vapor concentration during the measurement was also taken into account. As a consequence of the necessary drying process the concentration of H_2O vapor but also that of NO were reduced in comparison to undried samples.

Figure 4.14 shows a comparison between a measured spectrum (solid lines) and a fit (open bars) for an exhaust sample of a Jeep after subtraction of the residual H_2O vapor contribution. The iteration converges after a few steps, yielding good agreement between the measured and the calculated spectrum (Sigrist et al., 1989b). The small deviations, particularly around 5.9 μm, are presumably due to additional exhaust constituents not taken into account for the analysis. Some typical results of VOC analyses derived from samples collected at idling operation of the engines of a car (operated without a catalytic converter), a diesel truck, and a two-stroke motorcycle are presented in

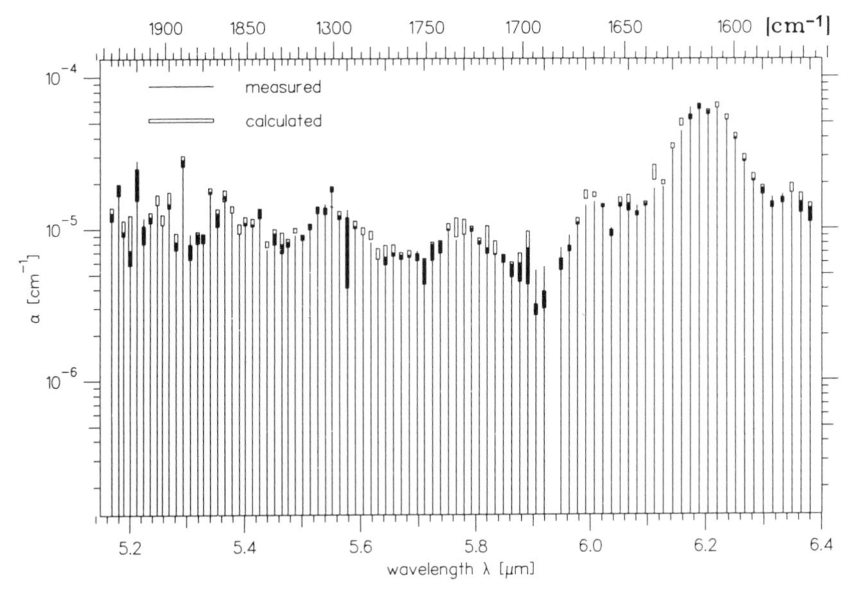

Figure 4.14. Absorption spectrum of sampled, dried, and diluted car exhausts from a Jeep, without catalytic converter, after subtraction of residual contribution by H_2O vapor absorption. The PA measurements (solid lines) are fitted with a calculated spectrum (open bars) according to the iterative algorithm described in the text. The bar lengths indicate the uncertainity of the calculated data, resulting from uncertainties of calibration spectra of individual components. Good agreement between theory and experiment is obtained.

Table 4.5. Analysis of Vehicle Exhausts with CO Laser Photoacoustic System

	Gas Concentrations (ppmv)		
	Car (Without Catalytic Converter)	Truck (Diesel)	Two-Stroke Motorcycle
Ethene (C_2H_4)	206	23	234
Propene (C_3H_6)	107	23	284
Formaldehyde (CH_2O)	< 1	3	12.6
Acetaldehyde (C_2H_4O)	4.1	8.5	5.2
Acrolein (C_3H_4O)	2.4	6.9	6.5
Benzene (C_6H_6)	78	2	217
Toluene (C_7H_8)	196	20	405
o-Xylene (C_8H_{10})	103	11	148
m-Xylene (C_8H_{10})	188	6	95
p-Xylene (C_8H_{10})	67	3	20

Table 4.5. The concentrations are given in ppm and refer to the pure exhausts, i.e., the dilution factor is taken into account. The errors generally vary between 5% (for the higher concentrations) and 10% (for the lower concentrations). The concentrations of the 10 most abundant VOCs could be determined separately, including even the different isomers of xylene. To our knowledge it is the first time that this high selectivity and sensitivity has been achieved with pure PA techniques. Gas chromatography as a more conventional method in this field can also distinguish among the xylene isomers yet often lacks the ability to confirm the identity of a peak since many compounds can elute at the same retention time. It is also interesting to note that the analysis time could be substantially reduced by considering only 14 carefully selected laser lines (instead of the entire spectrum of 90 lines). In this case the derived concentrations are in good agreement with those of a complete analysis, yet the errors are somewhat larger.

As Table 4.5 implies, the comparison between different vehicles reveals large differences in the exhaust composition in terms of relative abundance and absolute concentrations. Both depend on the type of vehicle (gas- and diesel-fueled cars, trucks, motorcycles) and the condition of the engine (cold or hot, fuel injection, etc.). The application of a catalytic converter under optimum conditions was found to reduce the NO and VOC concentrations by a factor of approximately 10.

4.5. AIR MONITORING IN THE FIELD

4.5.1. A Mobile Laser PA System

Hitherto, most PA studies on trace gases have been devoted to investigations on collected samples of different origins. However, a profound knowledge of tropospheric and stratospheric chemistry as well as of emission processes and industrial process control requires temporally and spatially resolved information on the distribution of numerous compounds. Therefore, in situ or remote measurements are necessary in addition to detailed analyses in the laboratory. Unlike lidar or long-path absorption measurements in the open atmosphere, PA schemes are not suited for remote studies except possibly for short ranges as discussed by Brassington (1982; see Section 4.3.4, above). However, PAS can be applied to in situ measurements even in the stratosphere, as was already demonstrated in the mid-1970s (Patel et al., 1974). Although those measurements were successful and yielded valuable data, it took several years before new activities began to develop PAS into a versatile tool for field studies.

At our laboratory we have been developing a mobile station for ambient air measurements for the last few years. Our present computerized CO_2 laser PA system is schematically shown in Fig. 4.15. It has been described in detail previously (Meyer and Sigrist, 1990; Sigrist, 1992). Briefly, a sealed-off, low-pressure CO_2 laser is used that can be operated with different CO_2 isotopes. The tuning range comprises ca. 70 transitions between 9.2 and 10.8 μm for a $^{12}C^{16}O_2$ laser tube or ca. 65 transitions between 9.6 and 11.4 μm for a $^{13}C^{16}O_2$ laser tube. The computer control ensures a proper wavelength selection with a long-term frequency stability of 10^{-3} cm^{-1}. The laser beam is modulated with a chopper and directed through a resonant PA cell operated at the first radial resonance. Its frequency of ca. 2.7 kHz enables measurements also to be carried out in noisy environments, e.g., close to a busy road. In a modified version of the original setup, the cell is now equipped with two miniature electret microphones located opposite to each other on the cylindrical axis of the cell to further enhance the detection sensitivity. So far, calibration spectra of more than 20 compounds buffered in synthetic air have been taken with this system using both $^{12}C^{16}O_2$ and $^{13}C^{16}O_2$ lasers. As the situation requires, more spectra are added to our "library," which forms the basis for the multicomponent analysis discussed in Section 4.2.2.

For in situ measurements the air to be sampled is pumped continuously through the cell at atmospheric pressure and with a flow rate of typically 0.5–1 L/min. The air is not pretreated by any means except for measurements in dusty environments where a filter is inserted at the air inlet. Our PA system with an extracavity cell offers a great dynamic range, with a detection limit corresponding to a minimum absorption coefficient α_{min} of approximately

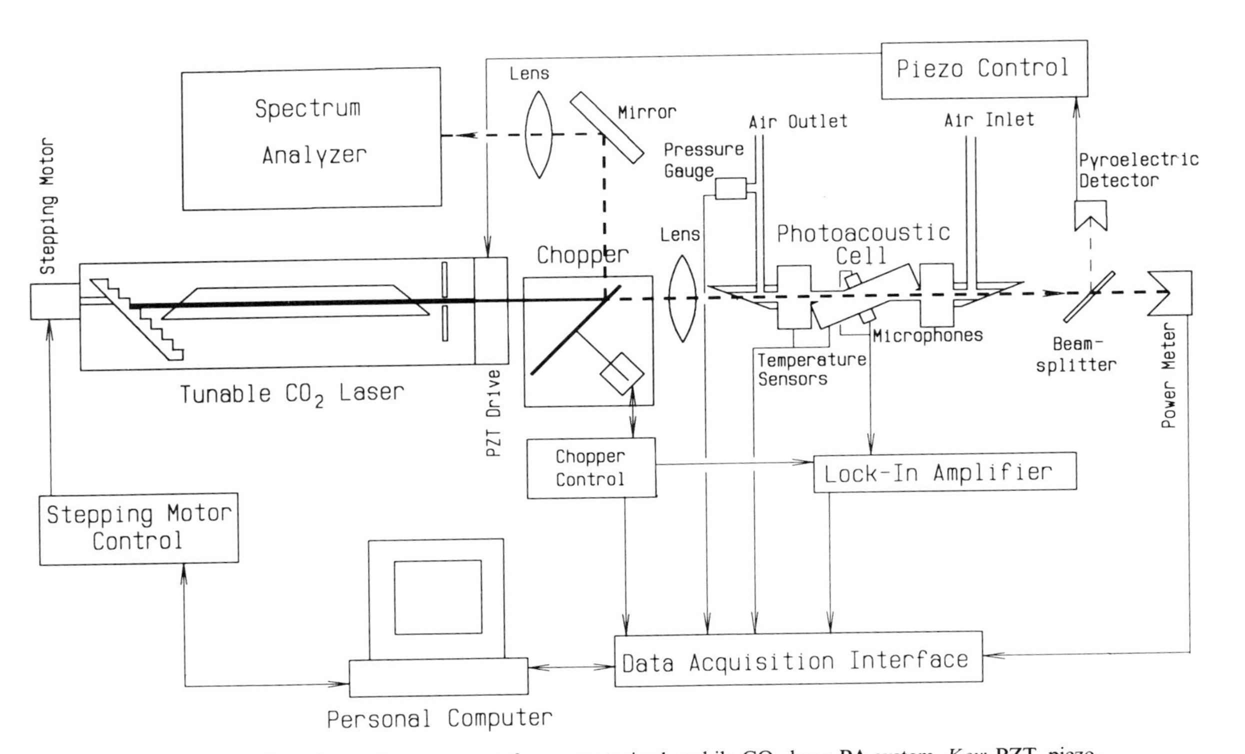

Figure 4.15. Experimental arrangement for computerized mobile CO_2 laser PA system. *Key*: PZT, piezoelectric transducer.

10^{-8} cm^{-1}. The dynamic range comprises more than 5 orders of magnitude in trace gas concentrations, which is important for field studies where concentrations of compounds often exhibit appreciable variations both spatially and temporally. Our system is installed in a mobile air-conditioned trailer (shown below in Fig. 4.20; see Section 4.5.3). The trailer is equipped with an extensible mast where devices for meteorological parameters such as wind direction and speed, air temperature, relative humidity and pressure can be mounted. Hitherto, the trailer has been operated at various locations in Switzerland without degradation of the system performance by transportation.

In the following subsections, some representative in situ studies performed by us and various other research groups are discussed that demonstrate the potential of PAS in different areas.

4.5.2. Stack Gas Emissions

Power plants burning fossil fuels emit nitric oxides (NO_x) that play a key role in acid precipitation and photochemical smog formation. A commonly used reduction scheme is based on the injection of ammonia (NH_3) into the combustion process. This technique requires a detection system for fast and selective monitoring of NH_3 down to the ppm level in order to ensure both a proper NO_x reduction and compliance with NH_3 emission limits.

Olafsson et al. (1989) have constructed and implemented a fully computerized system on the basis of CO_2 laser PAS. A nonresonant PA cell was used that could be heated to 125 °C in order to conform to the temperature of the power plant exhaust at the sampling point. In order to minimize absorption interferences the PA cell was operated at a reduced pressure of 12 mbar at a flow rate of 50 mL·atm/min. The NH_3 absorption line width is thereby reduced to 140 MHz fwhm (full width at half-maximum). The waveguide CO_2 laser was operated at 160 mbar and had a single mode tunability of ± 230 MHz by varying the length of the resonator with a piezoelectric element. This permitted a scan beyond the line center of the s*R*(5,0) NH_3 absorption line, which is centered 190 MHz below the 9 *R*(30) CO_2 laser transition center. Potential interference problems associated with line center absorption by the presence of 10–15% CO_2 were solved by exploiting the kinetic cooling effect on the phase of the PA signal (see Section 4.2.1). The PA system was successfully applied to in situ monitoring of CO_2 and NH_3 stack gas concentrations at a Danish power plant. A typical result is shown in Fig. 4.16 for a 6-h run. CO_2 and NH_3 could be monitored simultaneously by appropriate tuning of the laser to the corresponding line centers and additional recording of the PA phases. A reliable detection of NH_3 under the rough measurement conditions was achieved down to the 1 ppm level.

A different approach to PA stack gas monitoring has been reported by

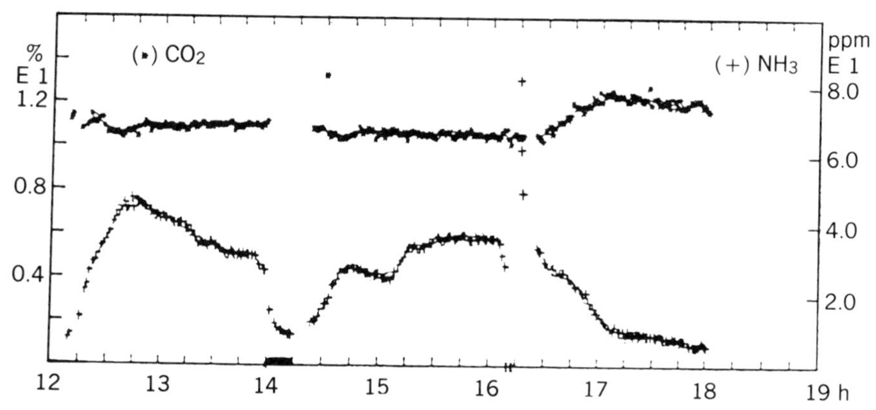

Figure 4.16. Temporal concentration profiles of CO_2 (upper trace—left scale) and ammonia (NH_3, lower trace—right scale) derived from PA field data taken at the stack emission of a Danish power plant. From Olafsson et al. (1989).

Carrer et al. (1990). A pulsed Nd:YAG laser equipped with second and fourth harmonic generators and a Raman cell filled with H_2 yielding a 4156 cm^{-1} shift was applied. Special PA cells with improved sensitivity for pulsed PA detection were constructed (see Section 4.3.3). Since the fuel gas mixture to be monitored consisted of varying amounts of several compounds including CO_2, CO, H_2O, NH_3, SO_2, NO_x, and O_3, preliminary laboratory tests were devoted to detailed analyses of interferences.

A further area with increasing relevance in view of federal regulations for emission control concerns *industrial* stack gases. These emissions consist of numerous components [mainly volatile organic compounds (VOCs)] with strongly varying concentrations. The continuous monitoring of the main constituents is thus of great interest. Present monitoring devices for this purpose often comprise gas chromatographic (GC) systems that offer high sensitivity and reasonable selectivity yet often lack the desired time resolution.

We investigated two emission sources with our mobile CO_2 laser PA system at a pharmaceutical production site of a chemical company in Basel, Switzerland (Meyer, 1988; Sigrist, 1992; Sigrist et al., 1989a,b). The measurements were performed by pumping the stack gas exhausts from the top of the building via a Teflon tube through the PA cell in the trailer. However, the first studies at any emission source with only vaguely known composition usually consist of taking a full PA spectrum of a sealed-off sample in the closed cell. As in the case of other multicomponent mixtures discussed above (see Section 4.4.2), the major absorbing components and their concentrations can generally be derived on the basis of previously taken calibration spectra of components presumed to be present and their implementation into the algorithm described in Section 4.2.2. Typical results of analyses for two different emission sources (A and B) are given in Table 4.6.

Apart from H_2O vapor and CO_2, various VOCs with concentrations in the ppm range could be identified. The absolute concentration of H_2O vapor is in reasonable agreement with data derived from the independent measurements of the relative humidity and the temperature. For source A a comparison was possible with independent GC data. The agreement is within the measurement error of both methods, especially if one takes into account the fact that the GC and PA sampling did not occur exactly at the same location in the exhaust stream and that the GC monitor was placed much closer to the sampling location than our trailer. The concentrations of some compounds (*o*- and *p*-xylene and ethylbenzene for source A; methanol and CO_2 for source B) are obviously below the detection threshold valid for those species in these particular gas mixtures. These rather high thresholds result from strong absorption interferences with the other compounds in the mixture. For example, methanol vapor could easily be detected at ppb concentrations under interference-free conditions. The high selectivity that can be achieved is demonstrated by the differentiation between the isomers of *m*- and *o*-dichlorobenzene. This separation can usually not be achieved with conventional GC unless GC is combined with mass spectrometry (GC–MS). It is worth noting that the (unexpected) presence of *m*-dichlorobenzene in the exhausts was only found by our PA analysis, namely, on the basis of the enhanced absorption at the 9 $R(28)$ laser transition that could not be explained by the absorption of other components, notably not by *o*-dichlorobenzene or chlorobenzene. This example demonstrates that PAS is not only suited for the detection of individual compounds in a multicomponent mixture but can also be useful for a detailed analysis and for the identification of additional components.

At emission source A we evaluated the continuous recording of single constituents. For this purpose a small part of the exhaust stream was pumped through our PA cell at a flow rate of ca. 0.5 L/min. Instead of taking full spectra the laser was repetitively tuned to transitions with high relative absorptions of the corresponding compounds, as discussed in Section 4.2.2. For the case of methanol and ethanol vapor in the mixture of source A the maximum relative absorptions occur for the $^{12}C^{16}O_2$ laser transitions 9 $P(34)$ and 9 $R(26)$, respectively, whereas for the same compounds but in the mixture of source B the corresponding transitions are 9 $P(34)$ and 9 $R(12)$, respectively. This example shows that the selection of appropriate laser wavelengths is determined by both the species to be monitored and the absorption features of other compounds present in the mixture. An example of simultaneous monitoring of methanol and ethanol vapor for a 6-h run is shown in Fig. 4.17a,b. Both PA

Table 4.6. Photoacoustic Analyses of Stack Gas Emissions

Source A			Source B		
Compound		Concentration	Compound		Concentration
Methanol	(CH_3OH)	15.8 ± 0.8 ppm[a]	Ethene	(C_2H_4)	1.3 ± 0.4 ppm
Ethanol	(C_2H_5OH)	4.7 ± 0.6 ppm[b]	Methanol	(CH_3OH)	0.2 ± 0.4 ppm
Toluene	(C_7H_8)	12.9 ± 3 ppm[c]	Ethanol	(C_2H_5OH)	6.4 ± 1.8 ppm
o-Xylene	(C_8H_{10})	0 ± 2 ppm	Toluene	(C_7H_8)	47.5 ± 13.6 ppm
m-Xylene	(C_8H_{10})	32.6 ± 5 ppm	Chlorobenzene	(C_6H_5Cl)	27.8 ± 3.1 ppm
p-Xylene	(C_8H_{10})	0 ± 5 ppm	*o*-Dichlorobenzene	$(C_6H_4Cl_2)$	4.4 ± 1.2 ppm
Ethylbenzene	(C_8H_{10})	0 ± 4 ppm	*m*-Dichlorobenzene	$(C_6H_4Cl_2)$	80.8 ± 3.2 ppm
Water Vapor	(H_2O)	$0.4 \pm 0.06\%$ abs.[d]	Carbon dioxide	(CO_2)	466.5 ± 355.0 ppm
			Water vapor	(H_2O)	$1.7 \pm 0.3\%$ abs.[e]

Comparison with independent GC measurement: [a]1.02 ppm; [b]5.55 ppm; [c]8.2 ppm.
Comparison with capacitive humidity meter: [d]0.52 % abs.; [e]1.5 % abs.

concentration profiles recorded with a temporal resolution of approximately 5 min are compared to independent data deduced from GC measurements, indicated by the × symbols in Fig. 4.17. In general, good agreement is obtained except for times of strong concentration fluctuations for which the time resolution of 20 min of the GC system is inadequate. This excellent time resolution of PAS, which has previously been clearly demonstrated (e.g., by Sigrist, 1992; Sigrist et al., 1989a,b), can be of great importance for various applications such as for the continuous monitoring of hazardous gases and vapors or generally for chemical process control.

4.5.3. Ambient Air Monitoring

In comparison to stack gas emissions, the concentrations of trace gases in ambient air are usually considerably lower and the variety of species present may be much larger. Up to now only a few studies have addressed the challenging problem of PA in situ monitoring of atmospheric trace species with mobile systems.

Emphasis has been put on the monitoring of ammonia (NH_3), which plays a significant role in the atmosphere because it is the only highly soluble common base present. In contrast to the conventional indophenol blue colorimetric method for the detection of gaseous NH_3, PAS offers several advantages because it permits real-time measurements without the need for sample preparation (Solyom et al., 1992).

Antipov et al. (1987) have developed a mobile NH_3 monitor based on a CO_2 laser PA system installed in an automobile. The residual noncompensated background originating from interfering absorptions by H_2O vapor and CO_2 corresponded to a NH_3 concentration of 7.7 ppb. Since this is a systematic background it could be compensated by an electric signal yielding a real detection limit of ca. 0.5 ppb, determined by the variation of the concentrations of the background components. The system is applied to regional NH_3 measurements in the atmosphere and to study the rate at which NH_3 is emitted into the air by ammonium-fertilized soil.

Recently, Rooth et al. (1990) have applied a mobile PA system to the in situ monitoring of ambient NH_3, H_2O vapor, and CO_2. The computer-controlled apparatus used a line-tunable waveguide CO_2 laser as radiation source. The system was successfully operated at a heath in the central Netherlands almost continuously for several months. During the measuring campaign in the field, the H_2O vapor and CO_2 concentrations were obtained once every 40 min while the NH_3 concentration was determined at a rate of 10 times per hour. An example of the data is shown in Fig. 4.18a–c, where the hourly averaged concentration profiles during one month of H_2O vapor (Fig. 4.18a) and NH_3 (Fig. 4.18b) are plotted. An enlarged view of NH_3 data taken at 6-min intervals

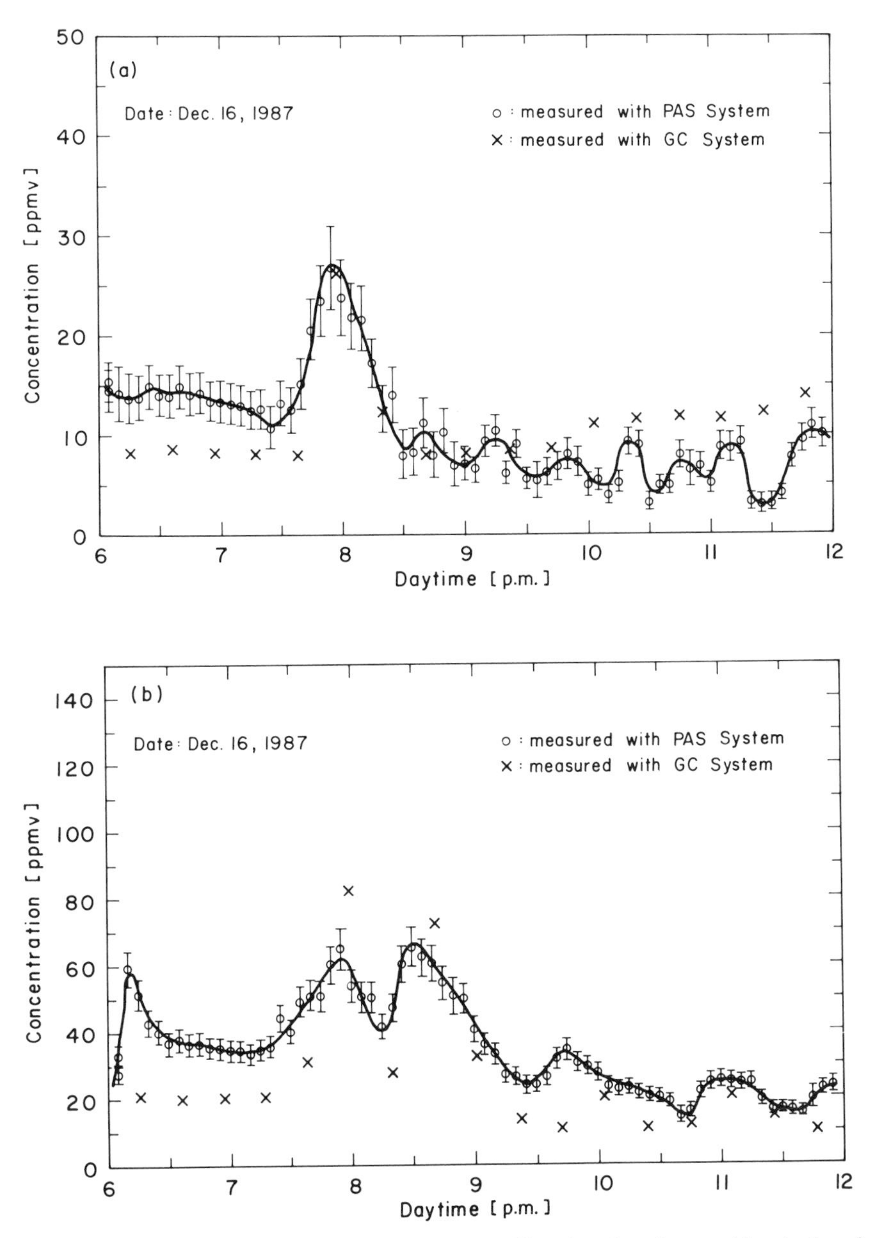

Figure 4.17. Simultaneous temporal concentration profiles of methanol vapor (a) and ethanol vapor (b) in the exhaust stream of a pharmaceutical production plant in Basel, Switzerland. Data derived from our own PA measurements are compared to GC data.

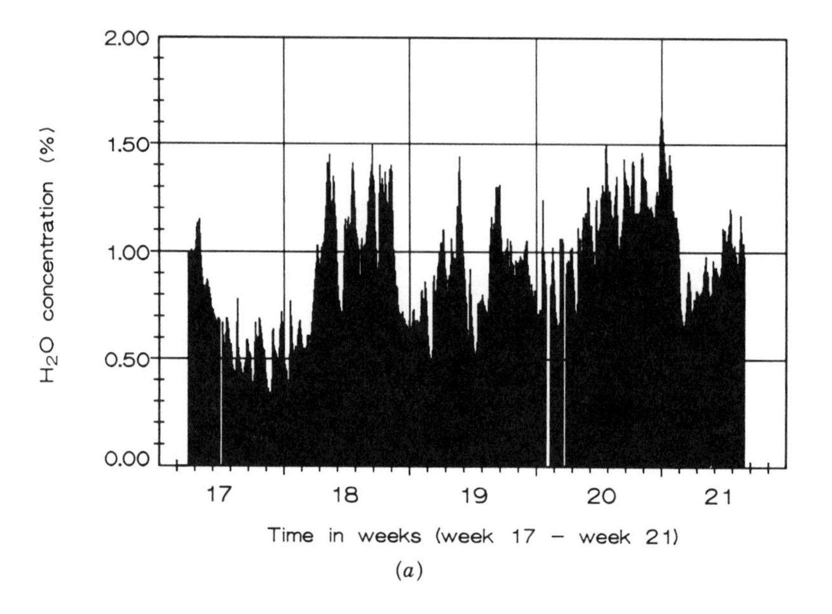
2.00
1.50
1.00
0.50
0.00
H2O concentration (%)
17
18
19
20
21
Time in weeks (week 17 – week 21)

(*a*)

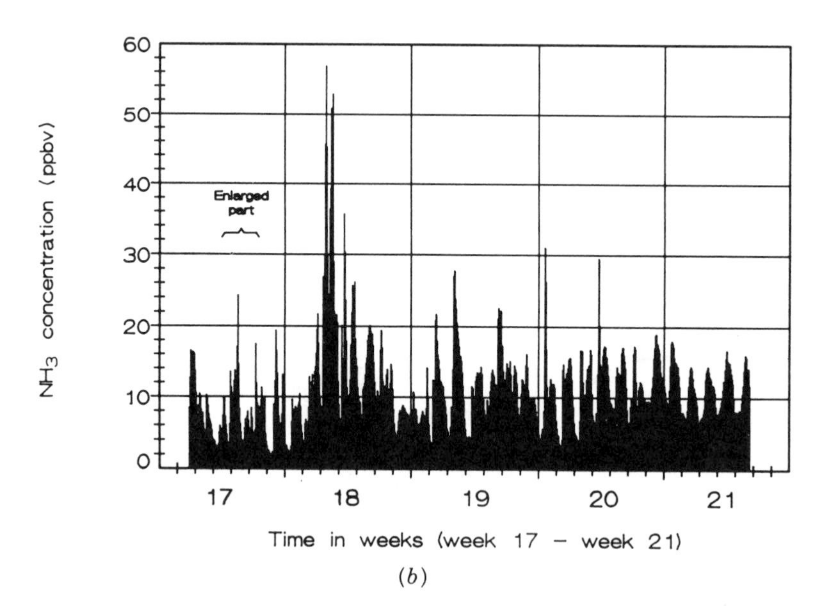
60
50
40
30
20
10
0
NH3 concentration (ppbv)
Enlarged part
17
18
19
20
21
Time in weeks (week 17 – week 21)

(*b*)

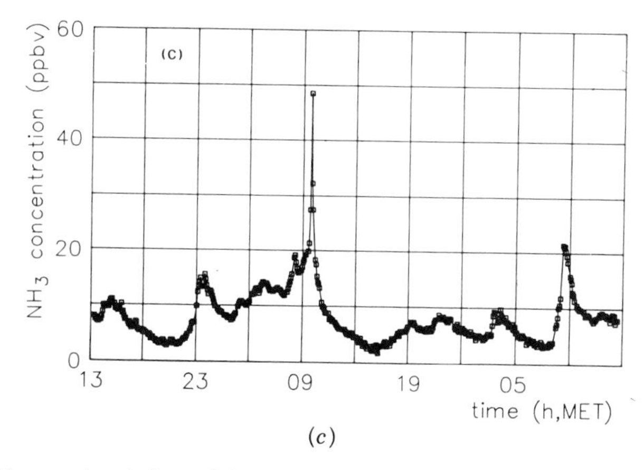

(c)

Figure 4.18. Temporal variations of the hourly averaged concentrations of H_2O vapor (a) and NH_3 (b) measured with a CO_2 laser PA system on a heath in Elspeet, Netherlands, for 4 weeks. The enlarged section (c) shows the initial NH_3 data taken at 6-min intervals on the third and fourth day of the measurements (April 26, 13 h, through April 28, 15 h, 1989). From Rooth et al. (1990).

is presented in Fig. 4.18c. The NH_3 data showed good agreement with denuder measurements taken simultaneously, yet the performance of the PA NH_3 monitor with respect to reliability and time response proved to be superior to that of the denuder system although capital costs were higher.

Further NH_3 measurements in outside air have been reported by Sauren et al. (1993). However, in this case the outside air was drawn through Teflon tubing before being admitted into the laboratory CO_2 laser PA system. The authors developed a PA Stark cell and intermodulated PA Stark spectroscopy (IMPASS) for their measurements. In this case the laser was chopped at a frequency f_{chop} whereas the electric field applied to the PA Stark cell was modulated at a higher frequency f_{Stark} such that $f_{Stark} - f_{chop} = f_{res} = f_{det}$, where f_{res} indicates a resonance frequency of the cell that corresponds to the detection frequency f_{det}. Potential absorption interferences by other trace components could be overcome by Stark shifting the NH_3 absorption line into resonance with the chosen 10 *R*(6) and 10 *R*(8) CO_2 laser transitions. A detection limit of 2–3 ppb NH_3 in air was achieved, which is well below the typical background concentration of 10–20 ppb.

Besides NH_3, methane (CH_4) production is also strongly influenced by biological activities. The basic process is the release of CH_4 by microbiological metabolism. Fiedler et al. (1992) have recently developed a portable PA monitor for CH_4 detection at the source. The device uses a nonresonant PA cell and a He–Ne laser operated at the 3.39 μm line, which is well absorbed by

CH_4. The PA sensor was installed in a cowshed for continuous measurement of the CH_4 concentration. The integration time of 1 min enabled resolution of the large fluctuations of the CH_4 concentration between 40 and 200 ppm. Comparison with PA and GC analysis of air samples performed at the laboratory yielded good agreement with the on-line measurements.

Our mobile CO_2 laser PA system described in Section 4.5.1 has also been employed in field measurements on ambient air in both urban and rural environments. Figures 4.19a,b show an example of urban air monitoring performed in a park in the small city of Biel, Switzerland, on two consecutive sunny summer days. The diurnal concentration profiles of H_2O vapor, CO_2, and C_2H_4 were deduced from repetitive PA measurements taken at the 10 $R(20)$, 10 $P(20)$, and 10 $P(14)$ laser transitions at which these gases dominate the absorption. As a result of the rather small cross sensitivities, possible short-term fluctuations of the CO_2 and H_2O concentrations of, e.g., 10 ppm and 0.1%, respectively, cause uncertainties of the derived C_2H_4 concentration of only 1.6 and 2.2 ppb, respectively. The estimated uncertainty of the C_2H_4 concentration is thus approximately ± 2 ppb. The top of Fig. 4.19a,b illustrates the good agreement between our PA data on the absolute H_2O vapor concentration and independent data derived from measurements of the relative humidity and temperature. The CO_2 concentration fluctuates only slightly around an average value of 350 ppm. On the other hand, C_2H_4 as a major exhaust component from motor vehicles (see Table 4.5) exhibits pronounced daily concentration peaks related to traffic. This coincidence was observed although our trailer was installed behind some trees at a distance of ca. 10 m from a main road. The C_2H_4 concentration varied between the detection threshold of ca. 2 and 35 ppb, corresponding to 44 $\mu g/m^3$. At other downtown locations peak concentrations exceeding 50 ppb were observed (Meyer and Sigrist, 1990). The first C_2H_4 peak in Fig. 4.19a occurring around 1:30 a.m. on July 2, 1986, was presumably caused by local truck movements in a nearby parking lot. Our PA data on C_2H_4 in Fig. 4.19a,b and in Table 4.5 demonstrate the large dynamic range that can easily be covered by PAS and thus renders this technique suitable for both emission and ambient air studies. The ozone (O_3) data in Fig. 4.19a,b originate from measurements taken by the Environmental Agency at the canton of Berne with a conventional O_3 sensor. The wind data stem from measurements by the Geographical Institute of the University of Berne, Switzerland. While the low wind speeds on July 2 hardly affected the C_2H_4 concentration, stronger winds in the afternoon of the next day (Fig. 4.19b) resulted in a pronounced yet transient dilution of the pollutants (Schüpbach et al., 1991). As seen in Fig. 4.19a the photochemical smog production, represented by the increasing O_3 concentration in the afternoon, clearly reduced the morning and noon peak concentrations of the photosmog precursor component C_2H_4. As Fig. 4.19b implies the photosmog production

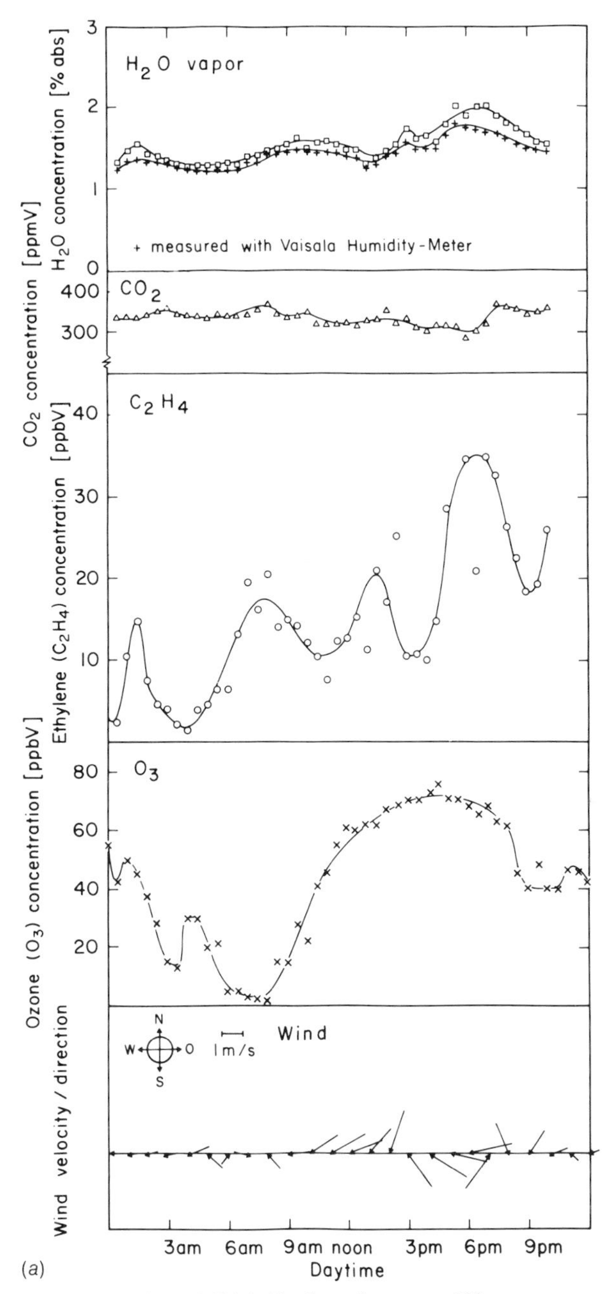

Figure 4.19(a). For legend, see page 219.

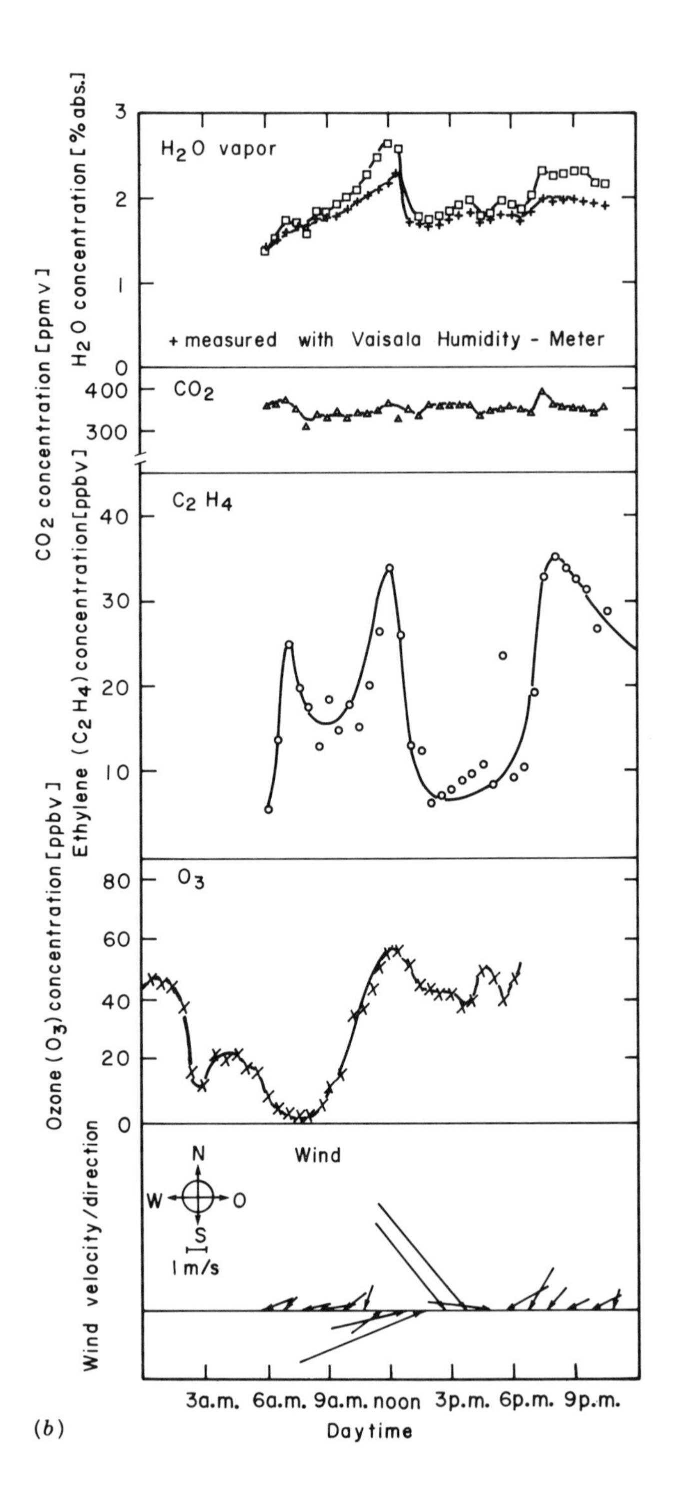
H2O vapor
+ measured with Vaisala Humidity - Meter
H2O concentration [% abs.]
3
2
1
0
CO2
400
300
CO2 concentration [ppmv]
C2 H4
40
30
20
10
Ethylene (C2 H4) concentration [ppbv]
O3
80
60
40
20
0
Ozone (O3) concentration [ppbv]
Wind
N
W
O
S
1 m/s
Wind velocity/direction
3a.m. 6a.m. 9a.m. noon 3p.m. 6p.m. 9p.m.
Daytime

(*b*)

Figure 4.20. Photograph of our mobile CO_2 laser PA system for air monitoring. The air inlet is located next to the ladder below the air condition. The picture shows the measurement site at a farm-house near Morschach, in central Switzerland. In the background the receiver of a DOAS system can be seen.

was reduced on July 3 owing to cloudy skies, a light shower around noontime, and strong winds.

Further measurements with our mobile system concerned ambient air monitoring in a rural area. In the frame of a nationwide interdisciplinary research program named POLLUMET (*Pollu*tion and *Met*eorology) aimed at a three-dimensional monitoring and modeling of the air polution over the complex topography of Switzerland we performed measurements near Morschach (central Switzerland) at an altitude of 825 m above sea level. Figure 4.20 shows a photograph of our trailer located close to a farmhouse during that measurement campaign.

Measurements were performed during an extended stable weather period with high air pressure favoring photosmog production. In addition to H_2O vapor, CO_2 and C_2H_4 we succeeded in monitoring NH_3 and O_3 with our system. For this purpose sequential measurements were performed at specific CO_2 laser transitions that were selected on the basis of characteristic absorption features of the five species, minimum absorption interferences, and good laser performance. As a result, the following nine laser transitions were chosen:

Figure 4.19. Temporal concentration profiles of H_2O vapor, CO_2, and C_2H_4 in urban air derived from measurements taken with our mobile CO_2 laser PA system. The O_3 concentration profile and wind data were obtained by commercial equipment. The absolute H_2O vapor concentration is compared to independent measurements of the relative humidity and temperature. Location: a city park in Biel, Switzerland. Date: (a) July 2, 1986; (b) July 3, 1986.

10 $R(20)$, which is characteristic for H_2O vapor absorption; 10 $P(20)$ for CO_2; 10 $P(34)$ for NH_3; 9 $P(14)$, 9 $P(24)$, and 9 $P(26)$ for O_3; 10 $P(14)$ for C_2H_4, and, in addition, 9 $P(34)$ and 10 $R(34)$ where none of these molecules exhibits an appreciable absorption. The analysis of the PA data was again performed on the basis of calibration spectra and the algorithm described in Section 4.2.2. The phase of the PA signal was also taken into account because of the influence by atmospheric CO_2. Figure 4.21 shows the result of such an analysis for a sunny summer day with considerable photosmog production. The

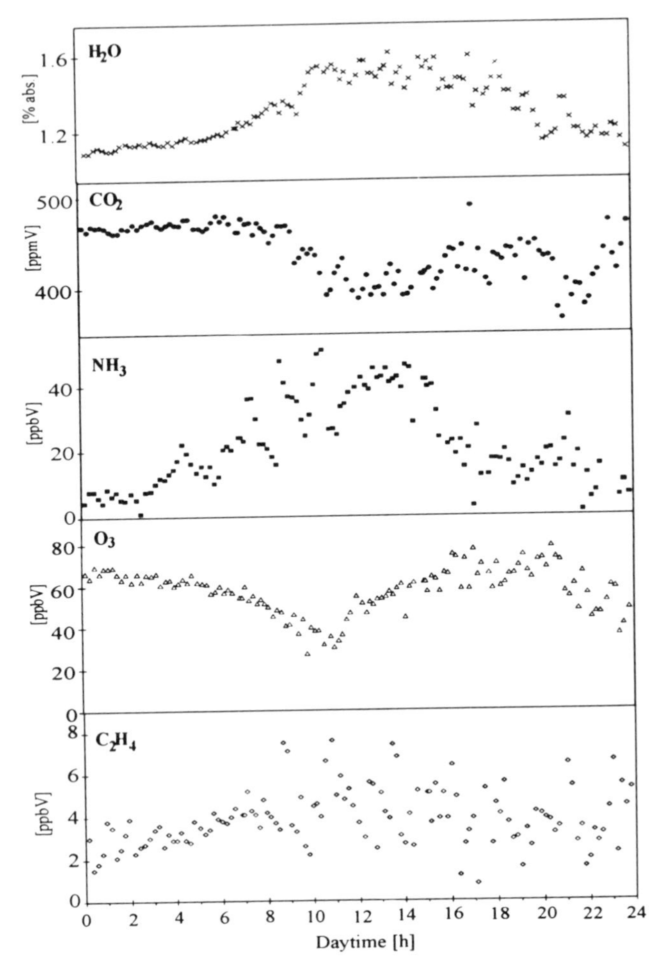

Figure 4.21. Temporal concentration profiles of H_2O vapor, CO_2, NH_3, O_3, and C_2H_4 in rural air, all derived from data obtained with our mobile CO_2 laser PA system. Location: Morschach (central Switzerland), 825 m above sea level. Data: July 26, 1990. From Thöny (1993).

concentration profiles of H_2O vapor, CO_2, NH_3, O_3,and C_2H_4 are plotted vs. daytime hours. In agreement with independent measurements of relative humidity and temperature, the absolute concentration of H_2O vapor varied between 1.0% and 1.6% following the daytime profile of the temperature. The CO_2 concentration varied around the unexpectedly high value of ca. 450 ppm. It is not yet clear whether this enhanced CO_2 concentration could be caused by local sources, e.g., by the nearby cowshed, because no independent CO_2 monitor was available for comparison at that time. It should be mentioned that a preliminary analysis of the PA data, which was based on inaccurate calibration data, yielded not only lower CO_2 but also unexpectedly low O_3 concentrations, yet did not influence the derived H_2O vapor and NH_3 concentrations noticeably (Thöny et al., 1992). The NH_3 concentration usually shows a variation during the day with typical concentrations between a few ppb and 50 ppb. However, local agricultural activities can strongly influence these data, as occurred two days later when the NH_3 concentration quite suddenly reached 135 ppb (Sigrist and Thöny, 1993; Thöny et al., 1992). This concentration jump occurred around 8 p.m. when the evening downhill wind transported air masses enriched with NH_3 from a freshly manured field to our trailer. As seen in Fig. 4.21 the O_3 concentration shows the typical behavior for a hot summer day, namely, a decrease during the night and a broad maximum in the afternoon. The derived O_3 concentrations are typical for a summer smog period and our data are supported by independent measurements with a conventional O_3 monitor (G. Pagani, unpublished report, 1992). It is also known that O_3 levels are often higher in rural locations than in urban environments. The lack of local emission sources explains the fact that the O_3 decrease during the night is much less pronounced than in urban areas, where the NO_x emissions by morning traffic destroy the remaining O_3 (see Fig. 4.19). The absence of emission sources of organic species is reflected in the low C_2H_4 concentrations that fluctuate around 5 ppb, i.e., well below urban concentrations.

The use of nine characteristic laser transitions to detect five species resulted in a temporal resolution of 10 min for the measurements. The present system, having faster tuning capability, would improve the time resolution to ca. 3 min. A yet further improvement on the cost of concentration accuracy could of course be achieved by using fewer laser lines. On the other hand, the accuracy of the results could be enhanced if time resolution were not of primary concern.

To our knowledge it is the first time that five different compounds, organic and inorganic, have been recorded at such low concentrations by PAS. At present, we are pursuing the monitoring with our system of further species, particularly of VOCs, in ambient air. For this purpose we are also applying a $^{13}C^{16}O_2$ laser and possibly other isotopes. In addition to the extension of wavelengths, the use of isotopic CO_2 lasers helps to reduce absorption interferences. In particular, the interference by natural CO_2 and thus the kinetic cooling effect mentioned in Section 4.2.1 is drastically reduced since atmospheric CO_2 contains ca. 99% $^{12}C^{16}O_2$ and, e.g., only ca. 1% $^{13}C^{16}O_2$. For example, the PA absorption spectra of natural CO_2 measured with the $^{12}C^{16}O_2$ and $^{13}C^{16}O_2$ laser are plotted in Figs. 4.22 and 4.23, respectively. The three peaks in Fig. 4.23 are caused by close coincidences of a $^{13}CO_2$ laser transition with a pressure-broadened $^{12}CO_2$ absorption line (Sigrist and Thöny, 1993). The separation between the 9 *R* (32) $^{13}CO_2$ laser transition and the 9 *P* (28) absorption line of $^{12}CO_2$ amounts, for example, to only 1.25×10^{-2} cm^{-1}, which is considerably less than the fwhm of typically 0.05–0.1 cm^{-1} at atmospheric pressure. As a result, the use of a $^{13}CO_2$ laser still permits the sensitive detection of atmospheric CO_2 yet essentially avoids interferences with CO_2 on most laser transitions. However, interferences with other species present in polluted air may still cause difficulties for selective detection. This problem can be addressed by improving the coincidences

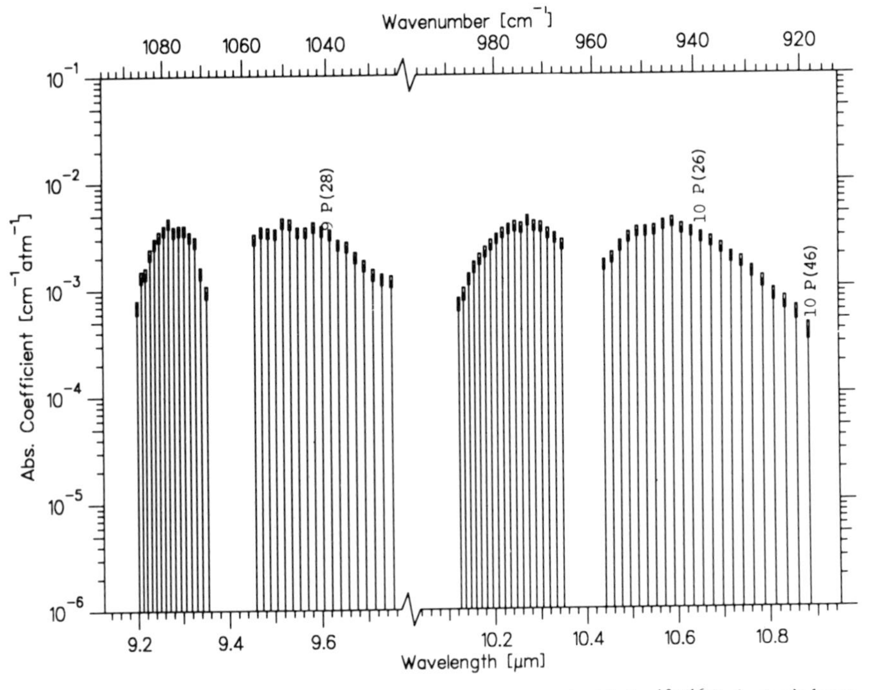

Figure 4.22. PA absorption spectrum of natural CO_2 measured with the $^{12}C^{16}O_2$ isotopic laser. Measurement conditions: 31 Pa CO_2 buffered in synthetic air (80% N_2; 20% O_2) at a total pressure of 950 hPa and a temperature of 295 K.

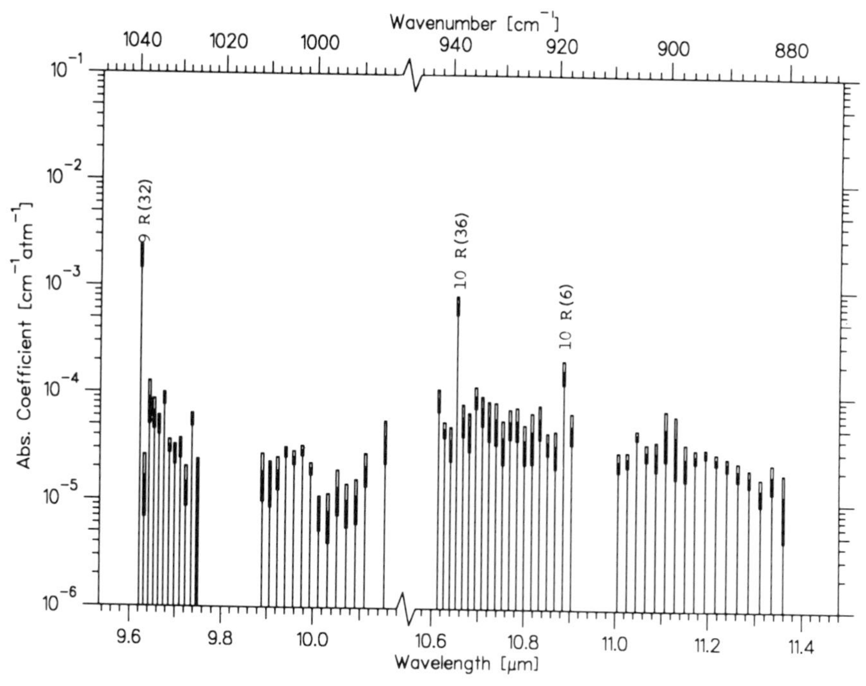

Figure 4.23. PA absorption spectrum of natural CO_2 measured with the $^{13}C^{16}O_2$ isotopic laser. Measurement conditions: 955 Pa CO_2 buffered in synthetic air (80% N_2; 20% O_2) at a total pressure of 955 hPa and a temperature of 298 K.

between molecular absorption lines and laser lines. That can be achieved either by the tuning of absorption lines, e.g., by applying a Stark field, or by using a continuously tunable laser like a high-pressure CO_2 laser (Repond et al., 1992, 1993) instead of the common line-tunable CO_2 laser (see Table 4.1) presently used in our mobile system.

4.5.4. Stratospheric Studies

The only PA studies performed in the stratosphere so far have been concerned with NO and H_2O monitoring (Burkhardt et al., 1975; Patel, 1976; Patel et al., 1974). Measurements of the temporal variation of stratospheric NO are important in validating its postulated role in catalytic destruction of stratospheric O_3. The authors carried out in situ measurements at an altitude of 28 km with a balloon-borne laser PA instrument. A InSb SFR laser with a CO pump laser and a variable magnetic field was used as tunable radiation source in the 5.3 μm region. At these wavelengths the NO absorption is caused by the

$v = 0$ to $v = 1$ vibrational–rotational band while the corresponding ν_2 band of H_2O is responsible for the H_2O absorption. For the NO and H_2O measurements the entire setup consisting of the SFR laser, PA cell, minicomputer, and electronic apparatus was packaged into a pressurized vessel to be sent aloft into the stratosphere suspended from a balloon. Sampling of the stratospheric air was done by continuously drawing it into the PA cell. Two balloon flights were performed with the objective of obtaining NO and H_2O concentrations as a function of time during a 24-h period at an altitude of ca. 28 km. The expected H_2O line width of 250–300 MHz due to Doppler and collisional broadening was in agreement with the width of 4–5 G of the magnetic field of the SFR laser. The measured H_2O concentration was 7.5×10^{11} molecules/cm^3, corresponding to a volume mixing ratio of 1.5 ppm at 28 km altitude. This value indicates a very dry stratosphere since the saturated H_2O concentration at the ambient temperature of 225 K is $\sim 1 \times 10^{15}$ molecules/cm^3.

The proper identification of stratospheric NO was based both on line positions as well as on relative line widths in comparison to laboratory experiments. The detection threshold was $\sim 1.5 \times 10^8$ NO molecules/cm^3 corresponding to a volumetric mixing ratio of ~ 0.2 ppb at a height of 28 km. The results from measurements of two different balloon flights are summarized in Fig. 4.24. A distinct increase of the NO concentration is observed from $< 1.5 \times 10^8$ molecules/cm^3 to a value around 2×10^9 molecules/cm^3 after sunrise. In the late afternoon, the NO concentration drops again within a period of ca. 90 min from the daytime value to concentrations $< 3 \times 10^8$ molecules/cm^3 at sunset. These in situ measurements represented the first accurate determination of the temporal variation of stratospheric NO. The results strongly supported the theorized role of NO as a catalyst in the destruction of O_3 and its importance in the stratospheric O_3 balance.

4.6. CONCLUSIONS AND OUTLOOK

The feasibility and versatility of laser PAS for air monitoring has been demonstrated by numerous examples. Minimum detectable concentrations of gaseous pollutants are often in the ppb or sub-ppb range, depending on the molecular absorption cross section and on possible absorption interferences. This latter aspect is crucial for analyzing multicomponent gas mixtures, and a new iterative formalism has been developed for this purpose. In some cases, an excellent selectivity can be achieved that is demonstrated, for example, by the individual detection of 10 different compounds including the isomers of xylene in motor vehicle exhausts or of the isomers of dichlorobenzene in stack emissions.

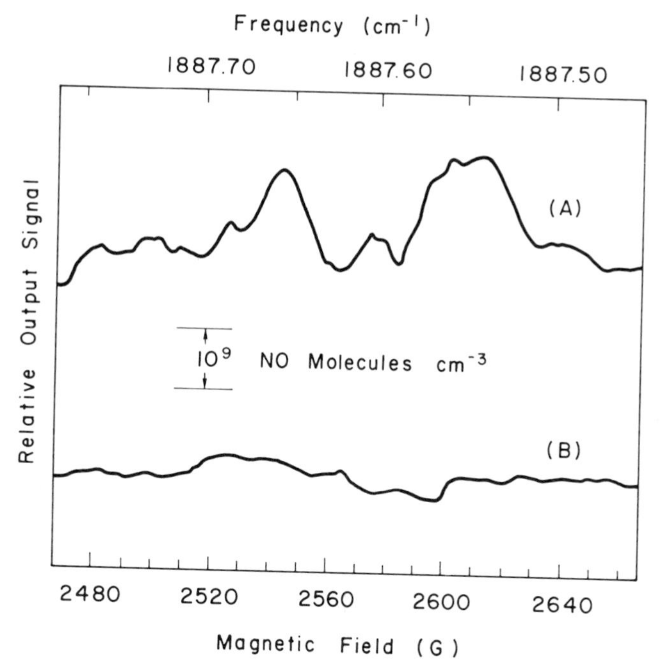

Figure 4.24. Stratospheric PA measurement of nitric oxide (NO). Relative output signals from the PA cells are plotted as a function of the magnetic field that determines the SFR laser frequency. The NO absorption signals are expected to occur at 2545 and 2605 G, corresponding to 1887.63 and 1887.55 cm^{-1}, respectively. The concentration scale for NO in molecules cm^{-3} is as shown: (A) NO run 1 taken between 11:00 and 11:28 M.S.T. (17:00–17:28 U.T.) on May 22, 1974, at a balloon altitude of 28 km and a local temperature of 225 K. (B) NO run 13 taken between 20:13 and 20:40 M.S.T. (02:13–02:40 U.T.) well past the expected UV sunset at a balloon altitude of 21.3 km. The disappearance of NO at sunset has thus been ascertained. From Burkhardt et al. (1975).

Another important parameter is the temporal resolution to be achieved. An almost simultaneous and continuous monitoring of several components of interest can be obtained by sequential measurements at appropriate laser wavelengths. The latter are selected to yield maximum sensitivity at minimum interference for the selective detection of the individual components. An example is the continuous monitoring of five compounds (CO_2, H_2O vapor, ethene, ozone, and ammonia) in rural air at time intervals of a few minutes with our mobile CO_2 laser PA system. The excellent time resolution can be of importance for surveillance of toxic gases, process control in chemical industry, etc.

A further aspect concerns the large dynamic range that permits the monitoring of trace gas concentrations exceeding a range of 5 orders of magnitude. Low concentrations of pollutants in rural air can thus be detected with the same instrument as that detecting high concentrations at emission sources.

It should be emphasized that PA monitoring is not restricted to laboratory measurements but can be performed in situ with mobile systems, as has been demonstrated by various examples concerning stack emission, ambient air, or even stratospheric air. However, unlike long-path or lidar measurements performed in the open atmosphere, PA studies are restricted to point monitoring and remote sensing is excluded. On the other hand, PAS investigations can be carried out in the entire IR wavelength range, notably also outside atmospheric windows, since a high sensitivity is already achieved with short absorption path lengths. This can represent advantages with respect to both accessible wavelength ranges and required sample volumes.

Although the great advantage of present PA systems is that a large variety of organic and inorganic species can be detected, this also can present the main problem, which is related to absorption interferences in multicomponent mixtures. Detection thresholds for a specific substance in a given mixture may therefore be considerably higher than for interference-free conditions. This phenomenon can often aggravate a quantitative analysis of unknown gas mixtures with numerous species of similar absorptivities. Hence, PA studies have so far concentrated on the monitoring of specific species in gas or air samples with at least roughly known composition, PAS being used to quantify them. However, as examples have demonstrated, real analyses of multicomponent samples and the identification of unknown species is feasible on the basis of spectral fits. In this respect, the potential of a PA system strongly depends on the availability of tunable IR sources. Since the PA signal is proportional to the radiation power, high detection sensitivity can only be achieved with lasers with CW powers in the milliwatt-to-watt range.

The future of laser PAS is thus expected to address the issue of laser development with respect to wide continuous tunability. Present efforts include high-pressure CO_2 lasers, difference frequency generation in nonlinear optical materials as well as optical parametric oscillators. For practical aspects the compactness of the laser source and of the entire system as well as easy operation, automation, and portability of the apparatus are important. One can imagine the development of small laser PA instruments with restricted applications, e.g., with lower sensitivity for emission control.

Based on the great potential of laser PAS, we expect an increasing number of successful applications of in situ trace gas monitoring for air pollution surveillance and many other areas of current and future interest.

ACKNOWLEDGMENTS

The author acknowledges fruitful cooperation with numerous present and previous co-workers and graduate students in the field of laser photoacoustic spectroscopy applied to trace gas monitoring, in particular S. Bernegger, R. Kästle, P.L. Meyer, P. Repond, E. Schmid, and A. Thöny. He is also indebted to several researchers who have contributed material to this chapter as well as for many stimulating discussions with them. Furthermore, he would like to thank Mrs. B. Blättler for typing the manuscript and Mrs. I. Wiederkehr and Mrs. H. Hediger for the preparation of the figures. This work was supported by the Swiss National Science Foundation and ETH Zurich.

REFERENCES

Adamowicz, R. F., and Koo, K. P. (1979). Characteristics of a photoacoustic air pollution detector at CO_2 laser frequencies. *Appl. Opt.* **18**, 2938–2946.

Akhmanov, S. A., Gusev, V. E., and Karabutov, A. A. (1989). Pulsed laser optoacoustics: Achievements and perspective. *Infrared Phys.* **29**, 815–838.

Angeli, G. Z., Bozoki, Z., Miklos, A., Lörincz, A., Thöny, A., and Sigrist, M. W. (1991). Design and characterization of a windowless resonant photoacoustic chamber equipped with resonance locking circuitry. *Rev. Sci. Instrum.* **62**, 810–813.

Angeli, G. Z., Solyom, A. M., Miklos, A., and Bicanic, D. D. (1992). Calibration of a windowless photoacoustic cell for detection of trace gases. *Anal Chem.* **64**, 155–158.

Angus, A. M., Marinero, E. E., and Colles, J. J. (1975). Opto-acoustic spectroscopy with a visible CW dye laser. *Opt. Commun.* **14**, 223–225.

Antipov, A. B., Kapitalov, B. A., Ponomarev, Y. N., and Kaposlekova, B. A. (1984). *Opto-acoustic Laser Spectroscopy of Molecular Gases* (in Russian). Nauka, Novosibirsk.

Antipov, A. B., Artyomov, B. M., Artyomov, E. M., Birjulin, V. B., Zharov, V. P., Kolesnikov, S. A., and Ponomarev, Y. N. (1987). Opto-acoustic technique for gas analysis. In *Dig. Int. Top. Meet. Photoacoust. Photothermal Phenom. 5th*, Univ. Heidelberg/FRG, pp. 186–187.

Bailey, R. T., and Cruickshank, F. R. (1989). Thermal lensing. In *Photoacoustic, Photothermal and Photochemical Processes in Gases* (P. Hess, Ed.), Top. Curr. Phys., Vol. 46, Chapter 3. Springer-Verlag, Berlin.

Bailey, R. T., Cruickshank, F. R., Guthrie, R., Pugh, D. and Weir, I. J. M. (1983). Short time-scale effects in the pulsed source thermal lens. *Mol. Phys.* **48**, 81–95.

Bailey, R. T., Cruickshank, F. R. and Pugh, D. (1989). Optothermal and thermal lens techniques in chemistry. *Chem. Br.*, January, pp. 39–46.

Barrett, J. J., and Berry, M. J. (1979). Photoacoustic Raman spectroscopy (PARS) using CW laser sources. *Appl. Phys. Lett.* **34**, 144–146.

Beck, K. M., and Gordon, R. J. (1988). Theory and application of time-resolved optoacoustics in gases. *J. Chem. Phys.* **89**, 5560–5567.

Beck, S. M. (1985). Cell coatings to minimize sample (NH_3 and N_2H_4) adsorption for low-level photoacoustic detection. *Appl. Opt.* **24**, 1761–1763.

Bell, A. G. (1880). On the production and reproduction of sound. *Am. J. Sci.* **20**, 305–324.

Bergmann, G., von Oepen, B., and Zinn, P. (1987). Improvement in the definitions of sensitivity and selectivity. *Anal. Chem.* **59**, 2522–2526.

Bernegger, S. (1988). CO-laser photoacoustic spectroscopy of gases and vapors for trace gas analysis, Ph.D. Thesis, ETH Zurich, Switzerland.

Bernegger S., and Sigrist, M. W. (1987). Longitudinal resonant spectrophone for CO-laser photoacoustic spectroscopy. *Appl. Phys.* **B44**, 125–132.

Bernegger, S., and Sigrist, M. W. (1990). CO-laser photoacoustic spectroscopy of gases and vapours for trace gas analysis. *Infrared Phys.* **30**, 375–429.

Bialkowski, S. E., and He, Z. F. (1988). Ultrasensitive photothermal deflection spectrometry using an analyzer etalon. *Anal. Chem.* **60**, 2674–2679.

Bialkowski, S. E., and Long, G. R. (1987). Quantitative discrimination of gas-phase species based on single-wavelength nonlinear intensity dependent pulsed infrared laser excited photothermal deflection signals. *Anal. Chem.* **59**, 873–879.

Bicanic, D. D., Bizzarri, A., Zuidberg, B. F. J., Bot, G. P. A., de Jong, T., and Wegh, H. C. P. (1988). CO_2 laser based N_2O monitor. *Springer Ser. Opt. Sci.* **58**, 143–144.

Bicanic, D. D., Harren, F., Reuss, J., Woltering, E., Snel, J., Voesenek, L. A. C. J., Zuidberg, B., Jalink, H., Bijnen, F., Blom, C. W. P. M., Sauren, H., Kooijman, M., van Hove, L., and Tonk, W. (1989). Trace detection in agriculture and biology. In *Photoacoustic, Photothermal and Photochemical Processes in Gases* (P. Hess, Ed.), Top. Curr. Phys., Vol. 46, Chapter 8, Springer-Verlag, Berlin.

Bicanic, D., Jalink, H., Chirtoc, M., Sauren, H., Lubbers, M., Quist, J., Gerkema, E., van Asselt, K., Miklos, A., Solyom, A., Angeli, G. Z., Helander, P., and Vargas, H. (1992). Interfacing photoacoustic and photothermal techniques for new hyphenated methodologies and instrumentation suitable for agricultural, environmental and medical applications. *Springer Ser. Opt. Sci.* **69**, 20–27.

Bondarev, B. V., Kapitanov, V. A., Kabtsev, S. M., and Ponomarev, Y. N. (1988). A high-sensitive laser optoacoustic spectrometer with a narrow band dye laser (in Russian). *Opt. Atmos.* **1**, 18–24.

Boscher, J., Schäfer, G., and Wiesemann, W. (1979). "Gasfernanalyse mit CO_2-Laser," Rep. BF-R-63.616.4. Battelle Institut, Frankfurt.

Boutonnat, M., Gilmore, D. A., Keilbach, K. A., Oliphant, N., and Atkinson, G. H. (1988). Photoacoustic detection of formaldehyde as a minority component in gas mixtures. *Appl. Spectrosc.* **2**, 1520–1524.

Bozhkov, A. I., Bunkin, F. V., Kolomenskii, A. A., and Mikhalevich, V. G. (1981). Thermo-optical methods of sound excitation in a liquid. *Sov. Sci. Rev. Sect. A* **3**, 459–551.

Brassington, D. J. (1982). Photo-acoustic detection and ranging—A new technique for the remote detection of gases. *J. Phys. D* **15**, 219–228.

Bray, R. G., and Berry, M. J. (1979). Intramolecular rate processes in highly vibrationally excited benzene. *J. Chem. Phys.* **71**, 4909–4922.

Brewer, R. J., Bruce, C. W., and Mater, J. L. K. (1982). Optoacoustic spectroscopy of C_2H_4 at the 9- and 10-μm $C^{12}O_2^{16}$ laser wavelengths. *Appl. Opt.* **21**, 4092–4100.

Burkhardt, E. G., Lambert, C. A., and Patel, C. K. N. (1975). Stratospheric nitric oxide: Measurements during daytime and sunset. *Science* **188**, 1111–1113.

Canarelli, P., Benko, Z., Curl, R., and Tittel, F. K. (1992). Continuous-wave infrared laser spectrometer based on difference frequency generation in $AgGaS_2$ for high-resolution spectroscopy. *J. Opt. Soc. Am.* **B9**, 197–202.

Carman, T. W., and Dyer, P. E. (1979). Continuous tuning characteristics of a small high pressure UV preionised CO_2 laser. *Opt. Commun.* **29**, 218–222.

Carrer, I., Monguzzi, M., Zanzottera, E., Civitano, L., and Sani, E. (1990). Stack gas monitoring by pulsed laser optoacoustic spectroscopy. *Springer Ser. Opt. Sci.* **62**, 366–368.

Carrer, I., Fiorina, L., and Zanzottera, E. (1992). High-sensitivity cell for pulsed photoacoustic spectroscopy in gases and liquids. *Springer Ser. Opt. Sci.* **69**, 568–571.

Chen, C., Liu, Y., Ming, C., Wang, C., Wang, W., Xu, J., Li, Z., and Wang, L. (1982). A study of an opto-acoustic CO_2 laser spectrometer. *Chin. Phys.* **2**, 451–452.

Choi, J. G., and Diebold, G. J. (1982). Laser Schlieren microphone for optoacoustic spectroscopy. *Appl. Opt.* **21**, 4087–4091.

Claspy, P. C., Pao, Y.-H., Kwong, S., and Nodov, E. (1976). Laser optoacoustic detection of explosive vapors. *Appl. Opt.* **15**, 1506–1509.

Claspy, P. C., Ha, C., and Pao, Y.-H. (1977). Optoacoustic detection of NO_2 using a pulsed dye laser. *Appl. Opt.* **16**, 2972–2973.

Crane, R. A. (1978). Laser optoacoustic absorption spectra for various explosive vapors. *Appl. Opt.* **17**, 2097–2102.

Deaton, T. F., Depatie, D. A., and Walker, T. W. (1975). Absorption coefficient measurements of nitrous oxide and methane at DF laser wavelengths. *Appl. Phys. Lett.* **26**, 300–303.

Dewey, C. F., Jr. (1977). Design of optoacoustic systems. In *Optoacoustic Spectroscopy and Detection* (Y. H. Pao, Ed.), Chapter 3. Academic Press, New York.

Dewey, C. F., Jr., Kamm, R. D., and Hackett, C. E. (1973). Acoustic amplifier for detection of atmospheric pollutants. *Appl. Phys. Lett.* **23**, 633–635.

Diebold, G. J. (1989). Applications of the photoacoustic effect to studies of gas phase chemical kinetics. In *Photoacoustic, Photothermal and Photochemical Processes in Gases* (P. Hess, Ed.), Top. Curr. Phys., Vol. 46, Chapter 6. Springer-Verlag, Berlin.

Di Lieto, A., Minguzzi, P., and Tonelli, M. (1979). Sub-Doppler optoacoustic spectroscopy. *Opt. Commun.* **31**, 25–27.

Di Lieto, A., Minguzzi, P., and Tonnelli, M. (1982). Design of an optoacoustic cell for laser-Stark spectroscopy, *Appl. Phys.* **B27**, 1–3.

Fang, H. L., and Swofford, R. L. (1983). The thermal lens in absorption spectroscopy. In *Ultrasensitive Laser Spectroscopy* (D. S. Kliger, Ed.), Chapter 3. Academic Press, New York.

Fiedler, M., and Hess, P. (1989). Laser excitation of acoustic modes in cylindrical and spherical resonators: Theory and applications. In *Photoacoustic, Photothermal and Photochemical Process in Gases* (P. Hess, Ed.), Top. Curr. Phys., Vol. 46, Chapter 5. Springer-Verlag, Berlin.

Fiedler, M., Gölz, C., and Platt, U. (1992). Photoacoustic field measurements of methane. *Springer Ser. Opt. Sci.* **69**, 9–11.

Fournier, D., Boccara, A. C., Amer, N. M., and Gerlach, R. (1980). Sensitive in situ trace gas detection by photothermal deflection spectroscopy. *Appl. Phys. Lett.* **37**, 519–521.

Frans, S. D., and Harris, J. M. (1985). Selection of analytical wavelengths for multicomponent spectrophotometric determinations. *Anal. Chem.* **57**, 2680–2684.

Fried, A., and Berg, W. W. (1983). Photoacoustic detection of HCl. *Opt. Lett.* **8**, 160–162.

Fung, K. H., and Lin, H.-B. (1986). Trace gas detection by laser intracavity photothermal spectroscopy. *Appl. Opt.* **25**, 749–752.

Gandurin, A. L., Gerasimov, S. B., Zheltukhin, A. A., Konovalov, I. P., Kornilov, S. T., Melnik, G. F., Mikhalevich, Y. Y., Ogurok, D. D., Petrishchev, V. A., and Chirikov, S. N. (1987). Optoacoustic gas analyzer for NO, NO_2, NH_3, C_2H_4, and saturated hydrocarbon pollutants. *Appl. Spectrosc.* **45**, 886–890.

Gerlach, R., and Amer, N. M. (1978). Sensitive optoacoustic detection of carbon monoxide by resonance absorption. *Appl. Phys. Lett.* **32**, 228–231.

Gerlach, R., and Amer, N. M. (1980). Brewster window and windowless resonant spectrophones for intracavity operation. *Appl. Phys.* **23**, 319–326.

Gilmore, D. A., Oliphant, N., Boutonnat, M., and Atkinson, G.H. (1989). Ultraviolet laser photoacoustic spectrometric determination of sulfur dioxide in mixtures containing larger amounts of nitrogen dioxide. *Anal. Chim. Acta* **218**, 101–110.

Goldan, P. D., and Goto, K. (1974). An acoustically resonant system for detection of low-level infrared absorption in atmospheric pollutants. *J. Appl. Phys.* **45**, 4350–4355.

Gomenyuk, A. S., Zharov, V. P., Oguruk, D. D., Ryabov, E. A., Tumanov, O. A., and Shaidurov, V. O. (1975). Optoacoustic detection of low concentrations of hydrogen fluoride, nitric oxide, and carbon dioxide in gases using radiation of pulsed hydrogen fluoride laser. *Sov. J. Quantum Electron.* (*Engl. Transl.*) **4**, 1001–1004.

Gromoll-Bohle, M., Bohle, W., and Urban, W. (1989). Broadband CO laser emission on overtone transitions $\Delta v = 2$. *Opt. Commun.* **69**, 409–413.

Guorong, M., Yihan, B., and Changlin, L. (1984). Measurement of collisional multiphoton energy deposition by a new optothermal detector. *Opt. Commun.* **51**, 25–28.

Gupta, R. (1992). Principles of photothermal spectroscopy in fluids. In *Principles and Perspectives of Photothermal and Photoacoustic Phenomena* (A. Mandelis, Ed.), Chapter 3. Elsevier, New York.

Haaland, D. M. (1990). Multivariate calibration methods applied to quantitative FT-IR analyses. In *Practical Fourier Transform Infrared Spectroscopy* (J. R. Ferraro and K. Krishnan, Eds.), Chapter 8. Academic Press, San Diego, CA.

Hammerich, M., Olafsson, A., and Henningsen, J. (1992). Photoacoustic study of kinetic cooling. *Chem. Phys.* **163**, 173–178.

Harren, F. J. M., Sikkens, C., and Bicanic, D. (1987). Photoacoustics in agriculture. In *European Conference on Optics, Optical Systems and Applications* (S. Sottini and S. Trigari, Eds.), SPIE **701**, 525–531.

Harren, F. J. M., Bijinen, F., Lindenbaum, C., Reuss, J., Boesenek, L. A. C. J., and Blom, C. W. P. M. (1989). Sensitive photoacoustic trace detection of ethylene. In *Monitoring of Gaseous Pollutants by Tunable Diode Lasers* (R. Grisar, G. Schmidtke, M. Tacke, and G. Restelli, Eds.), pp. 289–293. Kluwer Academic, Dordrecht, The Netherlands.

Harren, F. J. M., Reuss, J., Woltering, E. J., and Bicanic, D. D. (1990). Photoacoustic measurements of agriculturally iinteresting gases and detection of C_2H_4 below the ppb level. *Appl. Spectrosc.* **44**, 1360–1368.

Hartung, C., and Jurgeit, R. (1982). Optimization of optothermal detector for absorption spectroscopy in the low pressure range. *Appl. Phys.* **B27**, 39–42.

Hartung, C., Jurgeit, R., Danz, R., and Elling, B. (1981). Laser spectroscopy by an optothermal detector (OT). *Ferroelectrics* **34**, 21–26.

Heritier, J. M. (1983). Electrostrictive limit and focusing effects in pulsed photoacoustic detection. *Opt. Commun.* **44**, 267–272.

Herlemont, F., Lyszik, M., and Lemaire, J. (1979). Laser spectroscopy of ethylene with waveguide CO_2 and N_2O lasers. *J. Mol. Spectrosc.* **74**, 400–408.

Herriot, D., Kogelnik, H., and Kompfner, R. (1964). Off-axis paths in spherical mirror interferometers. *Appl. Opt.* **3**, 523-526.

Hess, P. (1983). Resonant photoacoustic spectroscopy. *Top. Curr. Chem.* **111**, 1–32.

Hess, P., Ed. (1989a). *Photoacoustic, Photothermal and Photochemical Processes in Gases*, Top. Curr. Phys., Vol. 46, Springer-Verlag, Berlin.

Hess, P., Ed. (1989b). *Photoacoustic, Photothermal and Photochemical Processes at Surfaces and in Thin Films*, Top. Curr. Phys., Vol. 47, Springer-Verlag, Berlin.

Hess, P. (1992). Principles of photoacoustic and photothermal detection in gases. In *Principles and Perspectives of Photothermal and Photoacoustic Phenomena* (A. Mandelis, Ed.), Chapter 4. Elsevier, New York.

Hielscher, A. H., Miller, C. E., Bayard, D. C., Simon, U., Smolka, K. P., Curl, R. F., and Tittel, F. K. (1992). Optimization of a midinfrared high-resolution difference-frequency laser spectrometer. *J. Opt. Soc. Am.* **B9**, 1962–1967.

Higashi, T., Imasaka, T., and Ishibashi, N. (1983). Thermal lens spectrophotometry with argon laser excitation source for nitrogen dioxide determination. *Anal. Chem.* **55**, 1907–1910.

Hinderling, J., Sigrist, M. W., and Kneubühl, F. K. (1987). Laser-photoacoustic spectroscopy of water-vapor continuum and line absorption in the 8 to 14 μm atmospheric window. *Infrared Phys.* **27**, 63–120.

Hubert, M. H., Ryan, J. S., and Crane, R. A. (1983). "Laser Optoacoustic Detector Measurement of Signatures of a Selection of Environmental Contaminants," Rep. No. 83–715–1. Ultra Lasertech Inc., Mississauga, Canada.

Hunter, T. F., and Turtle, P. C. (1980). Optoacoustic and thermo-optic detection in spectroscopy and in the study of relaxation processes. In *Advances in Infrared and Raman Spectroscopy* (R. J. H. Clark, Ed.), Vol. 7, Chapter 5. Heyden, London.

Hurdelbrink, W. (1986). Die Photoakustische Infrarot-Laserspektroskopie zur Schadgasanalyse. Ph.D. Thesis, Technical University, Stuttgart.

Ioli, N., Violino, P., and Meucci, M. (1979). A simple transversely excited spectrophone. *J. Phys. E* **12**, 168–170.

Jackson, W. B., Amer, N. M., Boccara, A. C., and Fournier, D. (1981). Photothermal deflection spectroscopy and detection. *Appl. Opt.* **20**, 1333–1344.

Jalink, H., and Bicanic, D. (1989). Concept, design, and use of the photoacoustic heat pipe cell. *Appl. Phys. Lett.* **55**, 1507–1509.

Johnson, S. A., Bone, S. A., Davies, P. B., Cummins, P. G., and Staples, E. J. (1988). Infrared lasers for optoacoustic detection of vapours. *Springer Ser. Opt. Sci.* **58**, 133–136.

Junker, A., and Bergmann, G. (1974). Auswahl, Vergleich and Bewertung optimaler Arbeitsbedingungen für die quantitative Mehrkomponenten-Analyse. *Fresenius' Z. Anal. Chem.* **272**, 267–275.

Kaiser, H. (1972). Zur Definition von Selektivität, Spezifität und Empfindlichkeit von Analysenverfahren. *Fresenius' Z. Anal. Chem.* **260**, 252–260.

Kamm, R. D. (1976). Detection of weakly absorbing gases using a resonant optoacoustic method. *J. Appl. Phys.* **47**, 3550–3558.

Kataev, M. Yu., and Mitsel, A. A. (1991). Detection of gases with the help of an optoacoustic gas analyzer. *Atmos. Opt.* **4**, 705–712.

Kato, K., and Sato, K. (1991). Trace gas analysis by pulsed laser induced photoacoustic spectroscopy using resonant cell (in Japanese). *Bunseki Kagaku* **40**, 283–288.

Kavaya, M. J., Margolis, J. S., and Shumate, M. S. (1979). Optoacoustic detection using Stark modulation. *Appl. Opt.* **18**, 2602–2606.

Kawasaki, S., Imasaka, T., and Ishibashi, N. (1987). Thermal lens spectrophotometry using a tunable infrared laser generated by a simulated Raman effect. *Anal. Chem.* **59**, 523–525.

Keller, R. A., Nogar, N. S., and Bomse, D. S. (1983). Photoacoustic detection of intracavity absorption. *Appl. Opt.* **22,** 3331–3334.

Kerr, E. L., and Atwood J. G. (1968). The laser illuminated absorptivity spectrophone: A method for measurement of weak absorptivity in gases at laser wavelengths. *Appl. Opt.* **7**, 915–921.

Klimcak, C. M., and Gelbwachs, J. A. (1985). Harmonic saturated spectroscopy applied to molecular photoacoustic detection. *Appl. Opt.* **24**, 247–253.

Koch, K. P., and Lahmann, W. (1978). Optoacoustic detection of sulphur dioxide below the parts per billion level. *Appl. Phys. Lett.* **32**, 289–291.

Konjevic, N., and Jovicevic, S. (1979). Spectrophone measurements of air pollutants absorption coefficients at CO_2 laser wavelengths. *Spectrosc. Lett.* **12**, 259–274.

Kreuzer, L. B. (1971). Ultralow gas concentration infrared absorption spectroscopy *J. Appl. Phys.* **42**, 2934–2943.

Kreuzer, L. B. (1974). Laser optoacoustic spectroscopy—A new technique of gas analysis. *Anal. Chem.* **46**, 235A–244A.

Kreuzer, L. B. (1977). The physics of signal generation and detection. In *Optoacoustic Spectroscopy and Detection* (Y.-H. Pao, Ed.), Chapter 1. Academic Press, New York.

Kreuzer, L. B., and Patel, C. K. N. (1971). Nitric oxide air pollution: Detection by optoacoustic spectroscopy. *Science* **173**, 45–47.

Kreuzer, L. B., Kenyon, N. P., and Patel, C. K. N. (1972). Air pollution: Sensitive detection of ten pollutant gases by carbon monoxide and carbon dioxide lasers. *Science* **177**, 347–349.

Kritchman, E., Shtrikman, S., and Slatkine M. (1978). Resonant optoacoustic cells for trace gas analysis. *J. Opt. Soc. Am.* **68**, 1257–1271.

Lai, H. M., and Young K. (1982). Theory of the pulsed optoacoustic technique. *J. Acoust. Soc. Am.* **72**, 2000–2007.

Leslie, D. H., and Trusty, G. L. (1981). Measurements of DF laser absorption by methane using an intracavity spectrophone. *Appl. Opt.* **20**, 1941–1947.

Leslie, D. H., Trusty, G. L., Dandridge, A., and Giallorenzi, T. G. (1981). Fibre-optic spectrophone. *Electron. Lett.* **17**, 581–582.

Leslie, D. H., Miles, R. O., and Dandridge, A. (1983). Laser-gas spectrophone measurements with a fiber-optic interferometer. *J. Phys. (Paris), Colloq.* **C6**, 537–540.

Leugers, M. A., and Atkinson, G. H. (1984). Quantitative determination of acetaldehyde by pulsed laser photoacoustic spectroscopy. *Anal. Chem.* **56**, 925–929.

Loper, G. L., Calloway, A. R., Stamps, M. A., and Gelbwachs, J. A. (1980). Carbon dioxide laser absorption spectra and low ppb photoacoustic detection of hydrazine fuels. *Appl. Opt.* **19**, 2726–2734.

Loper, G. L., Gelbwachs, J. A., and Beck, S. M. (1986). CO_2-laser photoacoustic spectroscopy applied to low-level toxic-vapor monitoring. *Can. J. Phys.* **64**, 1124–1131.

Luft, K. F. (1943). Ueber eine neue Methode der registrierenden Gasanalyse mit Hilfe der Absorption ultraroter Strahlen ohne spektrale Zerlegung. *Z. Tech. Phys.* **5**, 97–104.

Luo, X., Shi, F. Y., Lin, J. X. (1991). CO-laser photoacoustic detection of phosgene $COCl_2$. *Int J. Infrared Millimeter Waves* **12**, 141–147.

Lyamshev, L. M., and Sedov, L. V. (1981). Optical generation of sound in a liquid: Thermal mechanism (review). *Sov. Phys.–Acoust.* (*Engl. Transl.*) **27**, 4–18.

Mandelis, A., Ed. (1992). *Principles and Perspectives of Photothermal and Photoacoustic Phenomena.* Elsevier, New York.

Marinero, E. E., and Stuke, M. (1979a). Quartz optoacoustic apparatus for highly corrosive gases. *Rev. Sci. Instrum.* **50**, 241–244.

Marinero, E. E., and Stuke, M. (1979b). Doppler-free optoacoustic spectroscopy. *Opt. Commun.* **30**, 349–350.

Martens, H., and Naes, T. (1989). *Multivariate Calibration.* Wiley, Chichester.

Max, E., and Rosengren, L.-G. (1974). Characteristics of a resonant opto-acoustic gas concentration detector. *Opt. Commun.* **11**, 422–426.

Meyer, P. L. (1988). Air pollution monitoring with a mobile CO_2-laser photoacoustic system. Ph.D. Thesis, ETH Zurich, Switzerland.

Meyer, P. L., and Sigrist, M. W. (1990). Atmospheric pollution monitoring using CO_2-laser photoacoustic spectroscopy and other techniques. *Rev. Sci. Instrum.* **61**, 1779–1807.

Meyer, P. L., Bernegger, S., and Sigrist, M. W. (1988). On-line monitoring of air pollutants with a mobile computer-controlled CO_2-laser photoacoustic system. *Springer Ser. Opt. Sci.* **58**, 127–130.

Miklos, A., and Lörincz, A. (1989). Windowless resonant acoustic chamber for laser-photoacoustic applications. *Appl. Phys.* **B48**, 213–218.

Morse, P. M., and Ingard, K. U. (1986). *Theoretical Acoustics.* Princeton University Press, Princeton, N.J.

Nickolaisen, S. L., and Bialkowski, S. E. (1986). Species-selective detection in gas chromatography by photothermal deflection spectroscopy. *J. Chromatogr.* **366**, 127–133.

Nodov, E. (1978). Optimization of resonant cell design for optoacoustic gas spectroscopy (H-type). *Appl. Opt.* **17**, 1110–1119.

Olafsson, A. (1990). Photoacoustic molecular spectroscopy with tunable waveguide CO_2 lasers. Ph.D. Thesis, University of Copenhagen, Denmark.

Olafsson, A., Hammerich, M., Bülow, J., and Henningsen, J. (1989). Photoacoustic detection of NH_3 in power plant emission with a CO_2 laser. *Appl. Phys.* **B49**, 91–97.

Owyoung, A., Patterson, C. W., and McDowell, R. S. (1978). CW stimulated Raman gain spectroscopy of the ν_1 fundamental of methane. *Chem. Phys. Lett.* **59**, 156–162.

Patel, C. K. N. (1976). Spectroscopic measurements of the stratosphere using tunable infrared lasers. *Opt. Quantum Electron.* **8**, 145–154.

Patel, C. K. N., and Kerl, R. J. (1977). A new optoacoustic cell with improved performance. *Appl. Phys. Lett.* **30**, 578–579.

Patel, C. K. N., Burkhardt, E. G., and Lambert, C. A. (1974). Spectroscopic measurements of stratospheric nitric oxide and water vapor. *Science* **184**, 1173–1176.

Perlmutter, P., Shtrikman, S., and Slatkine, M. (1979). Optoacoustic detection of ethylene in the presence of interfering gases. *Appl. Opt.* **18**, 2267–2274.

Persson, U., Marthinsson, B., Johansson, J., and Eng, S. T. (1980). Temperature and pressure dependence of NH_3 and C_2H_4 absorption cross sections at CO_2 laser wavelengths. *Appl. Opt.* **19**, 1711–1728.

Petkowska, L. T., Radak, B., and Miljanic, S. S. (1991). CO_2-laser photoacoustic spectroscopy of deuterated and tritiated forms of hydrogen sulphide. *Infrared Phys.* **31**, 303–309.

Poizat, O., and Atkinson, G. H. (1982). Determination of nitrogen dioxide by visible photoacoustic spectroscopy. *Anal. Chem.* **54**, 1485–1489.

Radak, B. B., Petkovska, L. T., and Miljanic, S. S. (1993). CO_2-laser coincidence with the absorption of CH_3F, CH_3Cl and SO_2, studied at variable pressure by photoacoustic spectroscopy. *Infrared Phys.* (to be published).

Rayleigh, J. W. (Lord) (1881). The photophone. *Nature* (*London*) **23**, 274–275.

Reddy, K. V., Heller, D. F., and Berry, M. J. (1982). Highly vibrationally excited benzene: Overtone spectroscopy and intramolecular dynamics of C_6H_6, C_6D_6, and partially deuterated or substituted benzenes. *J. Chem. Phys.* **76**, 2814–2837.

Repond, P., Marty, T., and Sigrist, M. W. (1992). A continuously tunable CO_2 laser for photoacoustic detection of trace gases. *Helv. Phys. Acta* **65**, 828–829.

Repond, P., Thöny, A., Jaquet, L., and Sigrist, M. W. (1993). Air monitoring by photoacoustic spectroscopy. In *Proc. LASERS* '92 (C. P. Wang, Ed.), STS Press, McLean, VA, pp. 982–988.

Röntgen, W. C. (1881). Ueber Toene, welche durch intermittierende Bestrahlung eines Gases entstehen. *Ann. Phys.* (Leipzig) [3] **12**, 155–159.

Rooth, R. A., Verhage, A. J. L., and Wouters, L. W. (1990). Photoacoustic measurement of ammonia in the atmosphere: Influence of water vapor and carbon dioxide. *Appl. Opt.* **29**, 3643–3653.

Rose, A., Pyrum, J. D., Muzny, C., Salamo, G. J., and Gupta, R. (1982). Application of the photothermal deflection technique to combustion diagnostics. *Appl. Opt.* **21**, 2663–2665.

Rosencwaig, A. (1978). Photoacoustic spectroscopy. *Adv. Electron. Electron Phys.* **46**, 207–311.

Rosencwaig, A. (1980). Photoacoustics and Photoacoustic Spectroscopy, Chem. Anal. Vol. 57, Chapter 6. Wiley (Interscience), New York.

Rosengren, L.-G. (1973). An optothermal gas concentration detector. *Infrared Phys.* **13**, 173–182.

Rosengren, L.-G. (1975). Optimal optoacoustic detector design. *Appl. Opt.* **14**, 1960–1975.

Rotger, M., Lavorel, B., and Chaux, R. (1992). High-resolution photoacoustic Raman spectroscopy of gases. *J. Raman Spectrosc.* **23**, 303–309.

Sauren, H., Bicanic, D., Hillen, W., Jalink, H., van Asselt, C., Quist, J., and Reuss, J. (1990). Resonant Stark spectrophone as an enhanced trace level ammonia concentration detector: Design and performance at CO_2 laser frequencies. *Appl. Opt.* **29**, 2679–2682.

Sauren, H., Bicanic, D., and van Asselt, K. (1991). Photoacoustic detection of ammonia at the sum and difference sidebands of the modulating laser and Stark electric fields. *Infrared Phys.* **31**, 475–484.

Sauren, H., Gerkema, E., Bicanic, D., and Jalink, H. (1993). Real time and in situ determination of ammonia concentrations in the atmosphere by means of intermodulated Stark resonant CO_2 laser photoacoustic spectroscopy. *Atmos. Environ.* **27A**, 109–112.

Schnell, W., and Fischer, G. (1975). Carbon dioxide laser absorption coefficients of various air pollutants. *Appl. Opt.* **14**, 2058–2059.

Schüpbach, E., Jeannet, P., and Sigrist, M. W. (1991). Untersuchungen zum Sommersmog. In *Luftschadstoffe und Lufthaushalt in der Schweiz.* (E. Schüpbach and H. Wanner, Eds.), Chapter 6. Verlag der Fachvereine, Zürich.

Sell, J. A. (1985). Photoacoustic and photothermal deflection spectroscopy of propane at CO_2 laser wavelengths. *Appl. Opt.* **24**, 152–153.

Shen, Y. R., Ed. (1976). *Nonlinear Infrared Generation.* Top. Appl. Phys., Vol. **16**. Springer-Verlag, Berlin.

Shtrikman, S., and Slatkine, M. (1977). Trace-gas analysis with a resonant optoacoustic cell operating inside the cavity of a CO_2 laser. *Appl. Phys. Lett.* **31**, 830–831.

Siebert, D. R., West, G. A., and Barrett, J. J. (1980). Gaseous trace analysis using pulsed photoacoustic Raman spectroscopy. *Appl. Opt.* **19**, 53–60.

Sigrist, M. W. (1986). Laser generation of acoustic waves in liquids and gases *J. Appl. Phys.* **60**, R83–R121.

Sigrist, M. W. (1988). Atmospheric trace gas monitoring by laser photoacoustic spectroscopy. *Springer Ser. Opt. Sci.* **58**, 114–121.

Sigrist, M. W. (1989). Laser photoacoustic spectroscopy. *Europhys. News* **20**, 167–170.

Sigrist, M. W. (1992). Environmental and chemical trace gas analysis by photoacoustic methods. In *Principles and Perspectives of Photothermal and Photoacoustic Phenomena* (A. Mandelis, Ed.), Chapter 7. Elsevier, New York.

Sigrist, M. W., and Kneubühl, F. K. (1978). Laser-generated stress waves in liquids. *J. Acoust. Soc. Am.* **64**, 1652–1663.

Sigrist, M. W., and Thöny, A. (1993). Atmospheric trace gas monitoring by CO_2 laser photoacoustic spectroscopy, In *Optical Methods in Atmospheric Chemistry* (H. I. Schiff and U. Platt, Eds.), SPIE **1715**, 174–184.

Sigrist, M. W., Bernegger, S., and Meyer, P. L. (1989a). Atmospheric and exhaust air monitoring by laser photoacoustic spectroscopy. In *Photoacoustic, Photothermal and Photochemical Processes in Gases* (P. Hess, Ed.), Top. Curr. Phys., Vol. 46, Chapter 7. Springer-Verlag, Berlin.

Sigrist M. W., Bernegger, S., and Meyer, P. L. (1989b). Infrared-laser photoacoustic spectroscopy. *Infrared Phys.* **29**, 805–814.

Simon, U., Benko, Z., Curl, R. F., Tittel, F. K., and Sigrist, M. W. (1993). Design considerations of an infrared spectrometer based on difference frequency generation in $AgGaSe_2$. *Appl. Opt.* **32** (in press).

Smith, W. H., and Gelfand, J. (1980). Quantitative laser photoacoustic detector spectroscopy: HD 5–0 transitions. *J. Quant. Spectrosc. Radiat. Transfer* **24**, 15–17.

Solyom, A., Angeli, G. Z., Bicanic, D. D., and Lubbers, M. (1992). Determination of ammonia using carbon dioxide laser photoacoustic spectroscopy compared with conventional spectrophotometry. *Analyst* **117**, 379–382.

Sullivan, B., and Tam, A. C. (1984). Profile of laser-produced acoustic pulse in a liquid. *J. Acoust. Soc. Am.* **75**, 437–441.

Takahashi, M., Okuzawa, Y., Fujii, M., and Ito, M. (1991). Thermal lensing and photoacoustic spectra of gaseous acetylene by pulsed tunable infrared laser. *Can. J. Chem.* **69**, 1656–1658.

Tam, A. C. (1983). Photoacoustics: Spectroscopy and other applications. In *Ultrasensitive Laser Spectroscopy* (D. S. Kliger, Ed.), Chapter 1. Academic Press, New York.

Tam, A. C. (1986). Applications of photoacoustic sensing techniques. *Rev. Mod. Phys.* **58**, 381–431.

Tang, F., and Henningsen, J. O. (1986). A 500 MHz tunable CO_2 waveguide laser for optical pumping. *IEEE J. Quantum Electron.* **QE-22**, 2084–2087.

Thöny, A. (1993). New developments in CO_2-laser photoacoustic monitoring of trace gases. Ph.D. Thesis, ETH Zurich, Switzerland.

Thöny, A., Brand, B., Ferro-Luzzi, M., and Sigrist, M. W. (1992). Photoacoustic in situ monitoring of trace gases in a rural environment. *Springer Ser. Opt. Sci.* **69**, 28–30.

Tonelli, M., Minguzzi, P., and Di Lieto, A. (1983). Intermodulated optoacoustic spectroscopy. *J. Phys. (Paris), Colloq.* **C6**, 553–557.

Tran, C. D. (1986). Intracavity He-Ne laser photothermal deflection as a sensitive method for trace gas analysis *Appl. Spectrosc.* **40**, 1108–1110.

Tran, C. D., and Franko, M. (1989). Dual-wavelength thermal-lens spectrometry as a sensitive and selective method for trace gas analysis. *J. Phys. E* **22**, 586–589.

Trushin, S. A. (1992). Photoacoustic air pollution monitoring with an isotopic CO_2 laser. *Ber. Bunsenges. Phys. Chem.* **96**, 319–322.

Tyndall, J. (1881). Action of an intermittent beam of radiant heat upon gaseous matter. *Proc. R. Soc. London* **31**, 307–317.

Vansteenkiste, T. H., Faxvog, F. R., and Roessler, D. M. (1981). Photoacoustic measurement of carbon monoxide using a semiconductor diode laser. *Appl. Spectrosc.* **35**, 194–196.

Viengerov, M. L. (1938). *Dokl. Akad. Nauk SSSR* **19**, 687.

Vujkovic Cvijin, P., Gilmore, D. A., Leugers, M. A., and Atkinson, G. H. (1987). Determination of sulfur dioxide by pulsed ultraviolet laser photoacoustic spectroscopy. *Anal. Chem.* **59**, 300–304.

Vujkovic Cvijin, P., Gilmore, D. A., and Atkinson, G. H. (1988). Determination of gaseous formic acid and acetic acid by pulsed ultraviolet photoacoustic spectroscopy. *Appl. Spectrosc.* **42**, 770–774.

Wang, W., Chen, C., and Liu, Y. (1982). Laser photoacoustic spectra of gaseous molecules. *Kexue Tongbao (Engl. Transl.)* **27**, 400–404.

West, G. A. (1983). Photoacoustic spectroscopy. *Rev. Sci. Instrum.* **54**, 797–817.

West, G. A., Siebert, D. R., and Barrett, J. J. (1980). Gas phase photoacoustic Raman spectroscopy using pulsed laser excitation. *J. Appl. Phys.* **51**, 2823–2828.

Woltering, E. J. (1987). Effects of ethylene on ornamental pot plants: A classification. *Sci. Hort. (Amsterdam)* **31**, 283–294.

Wood, A. D., Camac, M., and Gerry, E. T. (1971). Effects of 10.6-μ laser induced air chemistry on the atmospheric refractive index. *Appl. Opt.* **10**, 1877–1884.

Yang, S. F., and Hoffman, N. E. (1984). Ethylene biosynthesis and its regulation in higher plants. *Annu. Rev. Plant. Physiol.* **35**, 155–189.

Zelinger, Z. (1986). Opto-acoustic detection of laser infrared signal in gases (in Czech). *Chem. Listy* **80**, 673–690.

Zelinger, Z., Papouskova, Z., Jakoubkova, M., and Engst, P. (1988). Determination of trace quantities of freon by laser optoacoustic detection and classical infrared spectroscopy. *Collect. Czech. Chem. Commun.* **53**, 749–755.

Zharov, V. P. (1977). Development and research of laser optoacoustic instruments for analysis of gas media. Ph.D. Thesis (in Russian) (MVTU, Moscow).

Zharov, V. P. (1985). Laser opto-acoustic spectroscopy in chromatography. In *Laser Analytical Spectrochemistry* (V. S. Letokhov, Ed.), Chapter 5. Adam Hilger, Bristol.

Zharov, V. P., and Letokhov, V. S. (1986). *Laser Optoacoustic Spectroscopy*. Springer-Verlag, Berlin.

CHAPTER

5

THE USE OF TUNABLE DIODE LASER ABSORPTION SPECTROSCOPY FOR ATMOSPHERIC MEASUREMENTS

H. I. SCHIFF, G. I. MACKAY, AND J. BECHARA

Unisearch Associates Inc.,
Concord, Ontario, Canada

5.1. INTRODUCTION

Tunable diode laser absorption spectroscopy (TDLAS), operating in the middle infrared region, is gaining increased acceptance as the method of choice for trace gas measurements where sensitivity, specificity, and fast response are required.

The middle infrared (MIR), between 2 and 15 μm, is a rich spectral region for atmospheric measurements where almost all species of interest have strong fundamental absorptions while the major constituents, nitrogen and oxygen, do not. In fact, the absorption coefficients of most molecules are so strong in this region that very high spectral resolution is required to avoid interferences between species, particularly from the ubiquitous water and carbon dioxide. Spectrometers capable of resolving individual rotational–vibrational features, with line widths typically 2×10^{-3} cm^{-1} ($\sim$ 60 MHz), are required to provide unequivocal identification of constituents of polluted air samples or even of clean air samples where high sensitivities are required.

Spectrometers with broad-band light sources such as Fourier transform infrared (FTIR) and nondispersive infrared (NDIR) are generally incapable of such high resolution. Laser sources such as the CO and CO_2 lasers used in photoacoustic spectroscopy have narrow line widths but generally rely on accidental resonance between the emission lines of the laser and the absorption lines of the target molecule. Tunable diode lasers have line widths less

Air Monitoring by Spectroscopic Techniques, Edited by Markus W. Sigrist. Chemical Analysis Series, Vol. 127.
ISBN 0-471-55875-3

than Doppler line widths and can be continuously tuned over a desired wavelength region.

The first atmospheric measurements by TDLAS used a long-path retroreflector to achieve the desired sensitivity for CO measurements in cities (Hinkley and Kelley, 1971; Ku et al., 1975). Higher sensitivity and resolution were demonstrated by the use of multiple-pass cells through which air was continuously sampled at reduced pressure (Reid et al., 1978, 1980). The resulting decrease in pressure broadening of the lines not only increased the sensitivity but also lessened the likelihood of interferences. Further improvements in sensitivities were subsequently achieved by the use of wavelength modulation and line-locking techniques. These and other developments will be discussed in some detail in the following sections of this chapter.

TDLAS systems have been used extensively for a variety of atmospheric measurements including those that require fast response, such as aircraft or gas flux measurements, and those that require high sensitivities, such as measuring trace species in remote areas or trace constituents in complex gas mixtures. They have been used both to monitor air pollutants and as research tools to study atmospheric processes. Its high specificity and freedom from interferences has led to the use of TDLAS as a standard against which other, less definitive techniques can be compared and calibrated. A wide variety of gaseous species have been measured on the ground from both stationary and mobile platforms, as well as aloft from aircraft, balloons, and rockets, and from ships at sea.

Section 5.2 contains a general discussion of the absorption relationships on which measurements in the IR region are based. The main features and present status of tunable diode laser spectrometry are then reviewed in Section 5.3, and the components of a system suitable for atmospheric measurements are described in Sections 5.4 and 5.5. Noise sources and detection limits are considered in Section 5.6. A typical operating system is then described in some detail in Section 5.7, followed by a review of some of the more recent measurements made by such systems in Section 5.8. Finally, in Section 5.9 some new developments actively being pursued to improve and extend the applications of the method for atmospheric measurements will be described.

For details of the spectroscopic and industrial applications of tunable diode lasers, the reader is referred to earlier reviews (Hinkley et al., 1976; Eng et al., 1980; Eng and Ku, 1982; Hirota, 1985; Webster et al., 1988). Two of these reviews (Hinkley et al., 1976; Webster et al., 1988) also cover atmospheric pollution measurements. The proceedings of three conferences on monitoring of atmospheric pollutants by tunable diode laser spectroscopy (Grisar et al., 1987, 1989, 1992) provide an excellent overview of aspects of tunable diode laser technology, instrumentation, and applications. The proceedings of the SPIE (Society of Photooptical Instrumentation Engineers; see Schiff, 1991) present an overview of several spectroscopic techniques for the measurement of atmospheric gases including TDLAS, FTIR, DOAS (differential optical absorption spectroscopy), and LIF (laser–induced fluorescence).

5.2. ABSORPTION RELATIONS

The fundamental, vibrational absorption bands in the MIR contain a number of discrete rotational lines, the width and shape of which depend on temperature and pressure. For some gases the lines are well separated under normal atmospheric pressures and temperatures. For other molecules pressure broadening results in overlapping lines, and the pressure in the sampled air can be reduced to minimize this effect. Larger molecules often possess so many closely spaced lines that thermal or Doppler broadening is sufficient to cause line overlapping. Reducing the temperature is not a practical solution to this problem, although a system for doing so has been described (Bauerecker et al., 1992). Even though tunable diode lasers (TDLs) have spectral resolutions smaller than Doppler line widths, advantage can be taken of this property only if the Doppler line width of the molecule is smaller than the separation between adjacent lines.

5.2.1. The Beer–Lambert Law

The fraction of light intensity transmitted through a gas is given by the familiar Beer–Lambert law:

$$P/P_0 = \exp(-\sigma(\bar{v})NL) \tag{5.1}$$

where P and P_0 are the transmitted and incident powers respectively; L is the absorption path length in centimeters; N, the concentration of the absorbing molecules in molecules per cubic centimeter; and $\sigma(\bar{v})$, the wavenumber-dependent absorption cross section in square centimeters per molecule.

Under the relatively low concentrations of the gases that are measured in the atmosphere, the expression reduces to

$$a = P_{abs}/P_0 = \sigma_0 NL \tag{5.2}$$

where a is the absorbance; σ_0 is the absorption coefficient at the line center ($\bar{v}_0$); and $P_{abs} = P_0 - P$ is the power absorbed by the gas at that wavelength.

For a given path length, the concentration is proportional to the absorbance. The best detection sensitivity is achieved when both the path length, L and cross section σ_0 are as large as possible. Methods of realizing long path lengths are discussed in Section 5.4.4.

Of more importance than the absorption coefficient at the line center is the line strength S, which is the integral of the absorption cross section $\sigma(\bar{\nu})$ over the wavenumber range of the line:

$$S = \int_{-\infty}^{\infty} \sigma(\bar{\nu})\, d\nu \tag{5.3}$$

Other characteristics that are important for atmospheric measurements are the line shape and width. The value of S for any given line depends on the temperature but not on the pressure of the gas. The absorption cross section and line shape and width, however, depend both on temperature and pressure. The dependence of the line shape and width on these parameters is usually classified into three pressure regimes.

5.2.2. Doppler Broadening

At pressures below about 10 Torr, the line shape and width are essentially defined by Doppler broadening arising from random thermal motion of the gas molecules.

The line shape of a Doppler-broadened line is a Gaussian function that depends on the line width, $\Delta\bar{\nu}_D$, defined as the full width at half the maximum height (fwhm) of the line. For a Doppler-broadened line it is given by

$$\Delta\bar{\nu}_D = 7.16 \times 10^{-7} \bar{\nu}_0 (T/M)^{1/2} \tag{5.4}$$

where $\bar{\nu}_0$ is the frequency (in wavenumbers) of the center of the line; T, the temperature in kelvins; and M, the molecular mass. For example, methane ($M = 16$) at a temperature of 295 K has a Doppler-broadened line width (fwhm) for the line at 3100 cm^{-1} of 9.53×10^{-3} cm^{-1}. The much heavier molecule CF_2Cl_2 ($M = 121$) has a Doppler width for a line at 930 cm^{-1} of 1.04×10^{-3} cm^{-1}. Nevertheless the line width of a Pb-salt TDL is considerably narrower than the Doppler line widths of any molecule.

At line center, the peak absorption cross section σ_0 of a Doppler-broadened line reduces to the simple relationship

$$\sigma_0 = 0.94 \times S/\Delta\bar{\nu}_D \tag{5.5}$$

from which the absorption cross sections for gases of atmospheric interest can be calculated to lie in the range 10^{-18}–10^{-16} $cm^2 \cdot molecule^{-1}$. Such cross sections are sufficiently large to permit sensitive measurements of gaseous concentrations.

5.2.3. Pressure Broadening

For pressures above 100 torr, the line shapes and widths are essentially determined by collision-induced broadening. In this case, the line shape is defined by a Lorentzian function (Hinkley et al., 1976).

The pressure-broadened line width depends, in general, on the gas partial pressure (self-broadening), the total pressure (foreign gas broadening), and the temperature. Since the concentrations of the gases that interest us in the atmosphere are relatively low, we need not concern ourselves with self-broadening. For a given temperature the line width increases linearly with pressure:

$$\Delta\bar{\nu}_{\mathrm{L}} = 2b_c p \tag{5.6}$$

where $\Delta\bar{\nu}_{\mathrm{L}}$ is the Lorentzian line width (fwhm); p is the gas pressure; and b_c is known as the pressure broadening coefficient, which is usually determined experimentally.

For pressure-broadened lines, the absorption cross section at line center ν_0 is given by

$$\sigma_0 = 2S/(\pi\Delta\bar{\nu}_{\mathrm{L}}) = S/(\pi b_c p) \tag{5.7}$$

Under conditions where pressure broadening is dominant, the peak absorption cross section decreases linearly with increasing total pressure. Pressure-broadened line widths are typically on the order of 0.1 cm^{-1} at atmospheric pressure, that is, 1–2 orders of magnitude larger than purely Doppler line widths. This excessive broadening results in a much smaller peak cross section and in progressive overlapping and blending of the molecular absorption lines of many larger molecules.

For a number of heavy polyatomic molecules, partial overlap is already observed at pressures below 10 torr, because of the density of spectral lines (several hundred lines per cm^{-1}) resulting in an unresolved absorption background underlying the line spectrum (Jennings, 1978; Bechara et al., 1989; Becker et al., 1989; Kosichkin et al., 1990). The contrast between the line absorption and the background decreases with increasing pressure, leading to a decrease in specificity.

But, since the TDLAS method essentially operates on a differential absorption principle (see Section 5.4.2), the broad background does not contribute to the measured absorption signal. Consequently, the effective absorption cross section, σ_{eff}, which contributes to the measured signal at a particular total pressure, is smaller than σ_0 given by Eq. (5.7).

In addition, absorption of atmospheric pressure-broadened lines of some atmospheric constituents, e.g., H_2O and CO_2, which are present in large concentrations, renders some regions of the MIR completely opaque except for so-called windows extending approximately in the ranges 7–14, 4.5–5, and 3–4 μm. Consequently, measurements at atmospheric pressures in the lower troposphere with TDLs in the MIR are confined to some diatomic molecules (e.g., NO or CO) or light polyatomic molecules (e.g., CH_4 or NH_3) with absorptions in the atmospheric windows.

5.2.4. The Voigt Regime

To overcome these limitations, measurements are made at pressures between 10 and 100 torr in multireflection cells or at altitudes between 10 and 40 km. In this pressure range, known as the Voigt regime, the line shape and width are described by a convolution of Doppler and pressure broadening.

In general, the Voigt line width, $\Delta\bar{\nu}_V$, is computed numerically and tables can be consulted (Abramovitz and Stegun, 1970). However, a quick estimate for $\Delta\bar{\nu}_V$ as a function of $\Delta\bar{\nu}_L$ and $\Delta\bar{\nu}_D$ can be obtained with the following approximation (Hinkley et al., 1976):

$$\Delta\bar{\nu}_V = (\Delta\bar{\nu}_L{}^2 + \Delta\bar{\nu}_D{}^2)^{1/2} \tag{5.8}$$

A more accurate approximation has been derived by Olivero and Longbothum (1977), with a claimed accuracy of 2 parts per 10,000:

$$\Delta\bar{\nu}_V = C_1\,\Delta\bar{\nu}_L + (C_2\,\Delta\bar{\nu}_L^2 + \Delta\bar{\nu}_D^2)^{1/2} \tag{5.9}$$

where $C_1 = 0.5346$ and $C_2 = 0.2166$.

Typical line widths for atmospheric species in the Voigt pressure regime fall in the range 5×10^{-3} to 2×10^{-2} cm^{-1}, smaller than the spacing of the rotational lines of many molecules of interest. Operation in this regime therefore increases sensitivity and specificity and allows measurements in the spectral regions where H_2O and CO_2 have strong absorptions. However, this limits measurements to point sampling or to higher altitudes.

5.3. TUNABLE DIODE LASERS

Tunable diode lasers (TDLs) operating in the MIR are semiconductors consisting of lead salts (PbS, PbSe, PbTe) and their mixed alloys with themselves, with tin salts (e.g., SnS or SnSe), and other materials (e.g., Cd, Eu, or Yb). These semiconductors form *p–n* junctions with small energy bandgaps that can result in emissions in the 3–30 μm spectral region, depending on the composition of the alloy. The application of a forward bias current (typically 0.1–1.0 A) to the p–n junction produces population inversion between the nearly empty conduction band and the nearly full valence band, providing the gain mechanism through which stimulated emission occurs via electron-hole recombination.

Early forms of diode lasers consisted of a crystal chip approximately 400 μm in length and 200 μm in width and height, with end faces cleaved along natural crystal planes, the p–n junction being formed by diffusion. Electrical contacts are evaporated onto the crystal, which is then mounted on a rigid block. This type of diode is called a homojunction laser since both the p and n regions are made of the same material.

Reflections from the end faces of the crystal provide the feedback necessary to excite continuous wave (CW) laser action once the current exceeds a threshold value. The threshold current density for a simple homojunction-type diode is too high at room temperature to permit CW operation, and these diodes are generally operated at temperatures below 50 K. The output powers are typically 0.01–0.5 mW, and the output radiation is linearly polarized in the p–n junction plane. The laser beam profile is highly elliptical because of the rectangular shape of the emitting surface. Owing to the small dimensions of the active region, the beam divergence angle can be 20° or more.

The wavelength of the output radiation is given, to a first approximation, by the bandgap of the semiconductor, which in turn is determined by the composition of the crystal. Diode lasers can therefore be manufactured to operate over the chosen wavelength by choice of the alloy material and the composition. The bandgap energy E_g defines the peak wavelength of the gain envelope within which laser emission occurs.

In general, the Fabry–Pérot resonator formed by the crystal end faces defines the resonance condition for laser emission, which occurs in a series of longitudinal modes separated in wavenumber by $\Delta\bar{\nu} = (1/(2nl))$ where l is the cavity length (defined by the separation of the end faces, typically 0.4 mm) and n is the temperature-dependent refractive index of the material, with values in the range 4.5–7, for commonly used Pb-salt materials. The separation between the modes is therefore typically 1–2 cm^{-1}. In practice, multimode emission with variable intermodal separation occurs, especially in diffused homojunction diodes.

Early work carried out with the diodes held in liquid helium Dewar systems (Butler et al., 1963) was therefore limited by availability and cost of the refrigerant. These limitations were overcome by the introduction of reliable closed-cycle cryostats capable of achieving temperatures as low as 12 K.

The line width of a TDL is extremely narrow. If a Dewar system is used to cool the diode, the laser half-width can be less than 3×10^{-6} cm^{-1} (Hinkley and Freed, 1969). However, temperature and current instabilities, mechanical vibrations due to piston motion in closed-cycle cryostats, and some optical effects can increase the line width to 10^{-4} cm^{-1} (Eng and Ku, 1982; Harward and Hoell, 1979). But since Doppler half-widths are on the order of 10^{-3} cm^{-1}, TDL systems can obtain absorption spectra that are limited by Doppler or pressure broadening.

One of the most useful features of tunable diode lasers is their wavelength tunability. Tunability can be accomplished by changing the temperature of the diode or by subjecting it either to a variable magnetic field (Calawa et al., 1969) or to a variable hydrostatic pressure (Besson et al., 1965). Since these devices must be operated at a stable reduced temperature, varying this parameter is the most convenient method. Small temperature changes suitable for fine tuning can be made by varying the current through the diode. The first development of this technique by Hinkley (1970) opened the application of these devices to high-resolution absorption spectroscopy.

Temperature tuning of a given diode permits coarse wavelength variations over a range of about 200 cm^{-1}. However, the tuning rate of the gain curve with temperature is usually different from the tuning rate of the cavity mode, which is determined by the temperature-dependent refractive index n as well as by the bandgap. Consequently, temperature or current tuning of the diode results in a series of continuous portions 1–2 cm^{-1} wide, separated by gaps where no emission occurs, a phenomenon known as mode hopping. This can be a serious limitation for spectroscopic applications but, fortunately, generally not for atmospheric measurements requiring only coincidence of the laser output with a single well-chosen rotational line. Mode hopping is circumvented in new diode designs such as distributed feedback (DFB) or distributed Bragg reflectors (DBR) (Hsieh and Fonstad, 1980; Shani et al., 1989; Fach et al., 1992), or cleaved-coupled cavity (Lo, 1984; Linden and Reeder, 1984) or external cavity configurations (Muertz et al., 1992).

The tunning characteristics of a particular laser are shown in Fig. 5.1 (Werle et al., 1992), which shows a mode map of one of the best available liquid-phase epitaxy lasers (see below) manufactured by Fujitsu, Inc. The measured single-mode power at 6.3 μm was about 0.4 mW. Even for this predominantly single-mode laser, small ($<1\%$) spurious side modes are evident. Figure 5.2 shows the output laser power measured at the same time.

The frequency response of these diodes exceeds 10 GHz, which permits construction of fast response instruments and the use of rapid modulation techniques to achieve high sensitivities. Both aspects are discussed in Section 5.4.2.

At present homojunction laser diodes are manufactured to cover the spectral range from 6.25 to 27 μm. The first versions of these diodes did not have well-defined active regions, which broadened the spatial distribution of the emitted radiation and increased the number of modes. Major efforts were therefore made to decrease and define the surface geometry.

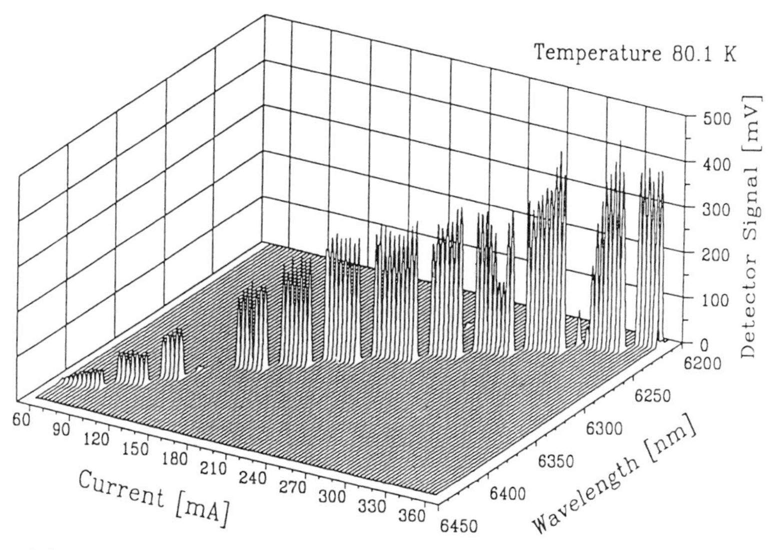

Figure 5.1. Mode map of a single-mode Pb-salt diode laser. Reproduced from Werle et al. (1992).

Mesa-stripe structures (Butler et al., 1983) in which an area on the order of 10 μm in depth and as low as 20 μm in width is formed by etching out two grooves in the surface. The chip surfaces surrounding these grooves have insulating layers deposited on the outlying surfaces to prevent the current from flowing through these regions. This leads to charge carrier confinement, which results in lower threshold and higher operating temperatures. The main disadvantage of the diffusion growth technique is difficulty in controlling the quality of the crystals. Obtaining good diodes by diffusion growth still remains, if not in the realm of black magic, at least as a rather hit-and-miss process.

More recently, molecular beam epitaxy (MBE) techniques have been used that overcome some of these difficulties. General Motors laboratories (Partin and Lo, 1981; Partin et al., 1984) first developed heterostructure crystals by deposition in a molecular beam system. A typical device consists of PbEuSeTe quaternary alloy on a PbTe substrate. Wall (1991) has described the development of these diodes in which stripe widths as small as 2 μm have been fabricated. The advantages claimed for these devices include (1) low threshold currents, (2) higher probability of single mode operation, and (3) higher temperature operation. Lasers have been constructed that are reported to tune over 800 cm^{-1} and to operate in single mode to 30 times the threshold current at 100 K. CW operation at temperatures as high as 203 K has been achieved

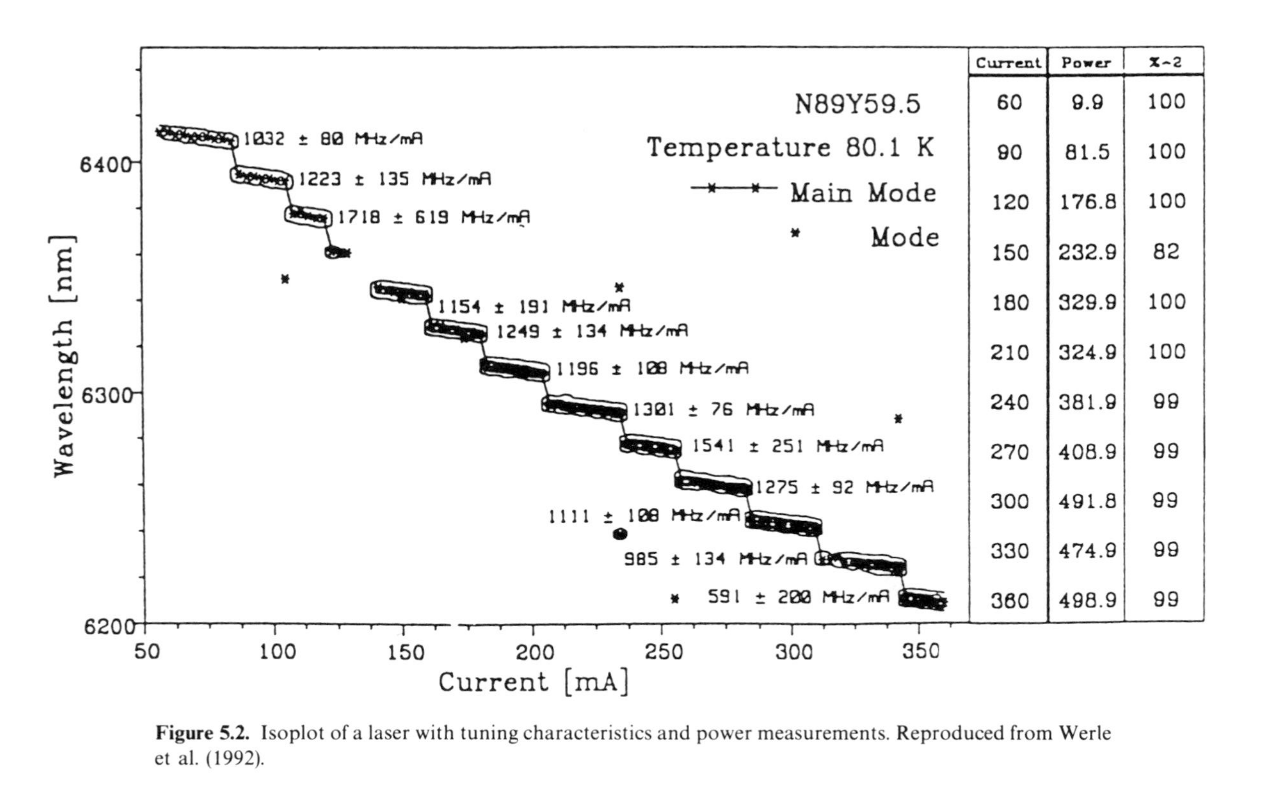

Current	Power	X-2
60	9.9	100
90	81.5	100
120	176.8	100
150	232.9	82
180	329.9	100
210	324.9	100
240	381.9	99
270	408.9	99
300	491.8	99
330	474.9	99
360	498.9	99

Figure 5.2. Isoplot of a laser with tuning characteristics and power measurements. Reproduced from Werle et al. (1992).

(Feit et al., 1991). The maximum wavelength that can be reached with presently available devices is 6.5 μm. Extension of the spectral coverage of these lasers to longer wavelengths has been reported (Feit et al., 1990).

A new type of TDL is the quantum well laser (Partin, 1988; Feit et al., 1992) based on the principle that an active laser with a thickness of "quantum dimensions" (0.1 μm or less) exhibits discrete energy levels in the valence and conduction bands. The bandgap energy depends on the alloy composition and on the thickness of the active layer. The potential advantages of these diodes include dominant single-mode operation and higher efficiency, improved optical confinement, lower threshold currents, higher output power, and higher operating temperatures—all highly desirable properties.

Unfortunately, the superior diodes manufactured by Fujitsu in Japan (Shinohara, 1989) are no longer available. At present Pb-salt diodes suitable for CW operation in the MIR are manufactured by Laser Photonics Inc. (Orlando, Florida, USA) and by the Fraunhofer Institute for Measurement Techniques at Freiburg, Germany. Both manufacturers are actively developing the new types of diodes that will find welcome reception in the TDLAS community. For more detailed information on the development and technology of Pb-salt diode lasers the reader is referred to the literature (Linden, 1985; Lo and Partin; 1984; Preier 1979; Tacke, 1989; Preier et al., 1989; Wall, 1991).

In addition to operating in the CW mode, Pb-salt diodes have also been used in a pulsed mode (Preier and Riedel, 1974; Preier et al., 1977; Klingerberg and Winckler, 1987; Lach et al., 1989; Riedel et al., 1991, 1992; Wolf et al., 1989). Russian scientists have been particularly active in developing the pulsed technique (Britov et al., 1977; Zasavitskii et al., 1984; Kuritsyn, 1986; Nadezhdinskii and Prokhorov, 1992; Nadezhdinskii and Stepanov, 1991; Kuznetzov et al., 1992). The difficulty of controlling the mode tuning during the pulse has limited its use in atmospheric measurements. On the other hand, the recent availability of Pb-salt diode lasers operating in the pulsed mode at or near room temperature (Spanger et al., 1988; Feit et al., 1990) permits the use of thermoelectric devices that can simplify the cooling requirements. Although the laser diodes manufactured by the aforementioned Russian groups are based apparently on the same materials and technologies as those of other suppliers, they have been developed for operation primarily in the pulsed mode (see Section 5.4.2).

The technology for another class of tunable diode lasers, based on the compounds between elements of group III (Ga, Al, and In) and group V(P, As, and Sb), is much further advanced. That is because they have widespread commercial application for optical fiber communications and for consumer electronics such as CD players and laser printers.

These diodes emit laser radiation in the near-infrared (NIR) region, between 0.59–2.5 μm. Relatively few molecules have absorptions in this region, and these are due to overtone and combination bands that are 2–3 orders of magnitude weaker than fundamental band absorptions. For that reason, these diodes have not previously seen much application in atmospheric measurements. However, they do possess a number of characteristics that have recently aroused increased interest (Mohebati and King, 1988; Cassidy, 1988; Cooper et al., 1991; Johnson et al., 1991; Uehara and Tai, 1992; Stanton and Hovde, 1992; Cooper and Martinelli, 1992).

NIR diodes have power outputs in the milliwatt range and operate at or near room temperature, which permits the use of small inexpensive thermoelectric devices for temperature control and tuning. For the wavelength regions 760–900, 1280–1350, and 1480–1560 nm, used in telecommunication and consumer products, these diodes are relatively inexpensive and readily available. Molecules that absorb in these regions include NO_2 (790 nm), HF (1330 nm), NH_3 (1515 nm), C_2H_2 (1530 nm), and HI (1540 nm). Single-mode lasers (DFB or external cavity) can also be obtained in these ranges, generally at a higher price. Diodes with wavelengths suitable for measuring other molecules, CO and CO_2 (1570 nm), H_2S (1580 nm), and CH_4 (1650 nm), are more difficult to obtain at present, particularly with single-mode characteristics, because most major manufacturers are reluctant to select lasers for a specified wavelength. However, lasers suitable for CH_4 at 1650 nm have recently become available on order owing to the increased interest in environmental monitoring. Prices for lasers in this category, however, are even higher than Pb-salt diodes.

The fact that relatively few molecules absorb in this spectral region may actually be an advantage for measurement of those that do, since there is much less chance for mutual interferences. Moreover, the ratio of pressure to Doppler broadening in this region, although not negligible for some gases, is frequently not as large as in the MIR, which facilitates long-path measurements at atmospheric pressure. Some of the loss of sensitivity due to the weak absorption in this region can be regained by the use of high-frequency modulation (FM) techniques, which are particularly applicable to this class of diodes for reasons discussed in Section 5.4.2.3. It is very likely that they will see increased application for measurements of a select number of species under environmental conditions where their concentrations exceed 10 ppbv (parts per billion by volume).

Further information on the physics of semiconductors in general can be found in standard books on solid state physics, e.g., *Introduction to Solid State Physics* (Kittel, 1971). Numerous books deal with the basics of laser operation in general and of semiconductor lasers in particular, e.g., *Quantum Electronics* by Yariv (1981), *Principles of Lasers* by Svelto (1989), *Physics of Semiconductor Laser Devices* by Thompson (1980).

5.4. COMPONENTS OF TDL SPECTROMETERS

A number of essential components are required to construct a TDLAS system for atmospheric measurements. In addition to selecting tunable diode lasers that emit radiation in the desired wavelength region, a system also requires: (1) cooling and temperature control of the diodes; (2) TDL current control and modulation systems to increase sensitivity; (3) an optical system to direct the radiation through the sampled air; (4) an optical path of sufficient length to attain the desired sensitivity; (5) a detector and associated optics to detect the transmitted radiation; (6) gas handling systems for sampling, calibration and background measurements; and (7) computer hardware and software to control the system and for collection, manipulation, display, and storage of data.

Figure 5.3 shows a schematic of a simple laboratory TDLAS system that includes these components. The diode laser is held in a source assembly maintained to ± 0.001 K. The laser radiation, which is generally emitted as a multilobe pattern within a $f/1$ cone, is collected and focused by a $f/1$ lens, L1. The plane fold mirrors, PM1 and PM2, enable the beam to be maneuvered for accurate entry into the White cell. The exiting beam is directed by a plane mirror, PM3, through a sample cell and another $f/1$ lens, L2, onto a liquid-nitrogen-cooled mercury cadmium telluride (HgCdTe, MCT) detector the output of which is fed into an oscilloscope for visual representation, then to a lock-in amplifier for processing. The spectrometer can be operated in either an amplitude modulation (AM) or a wavelength modulation (WM) mode.

We shall now discuss some of the components of TDLAS systems in greater detail.

5.4.1. Laser Temperature Control

As mentioned in Section 5.3, commercially available Pb-salt diode lasers operate at temperatures below 120 K. The temperature tuning rate of most diodes is greater than 1 $cm^{-1} \cdot K^{-1}$, requiring a temperature stability in the order of 0.001 K. Up to four diodes can be held on a cold-finger in a cryocooler. Temperature control and tuning is provided by a heating element used in combination with a semiconductor temperature sensor in a stabilized feedback loop.

Both closed-cycle refrigerators and Dewar systems containing liquid cryogens, have been used to achieve the required operating temperatures. Liquid helium can be used in Dewars for temperatures below 77 K and liquid nitrogen above 77 K. Because liquid helium is expensive and not readily available, it is not used in modern atmospheric measurement systems. Liquid nitrogen, although relatively inexpensive, is presently restricted to diodes that operate in the spectral region between 3.3 and 6.5 μm. New developments with MBE-

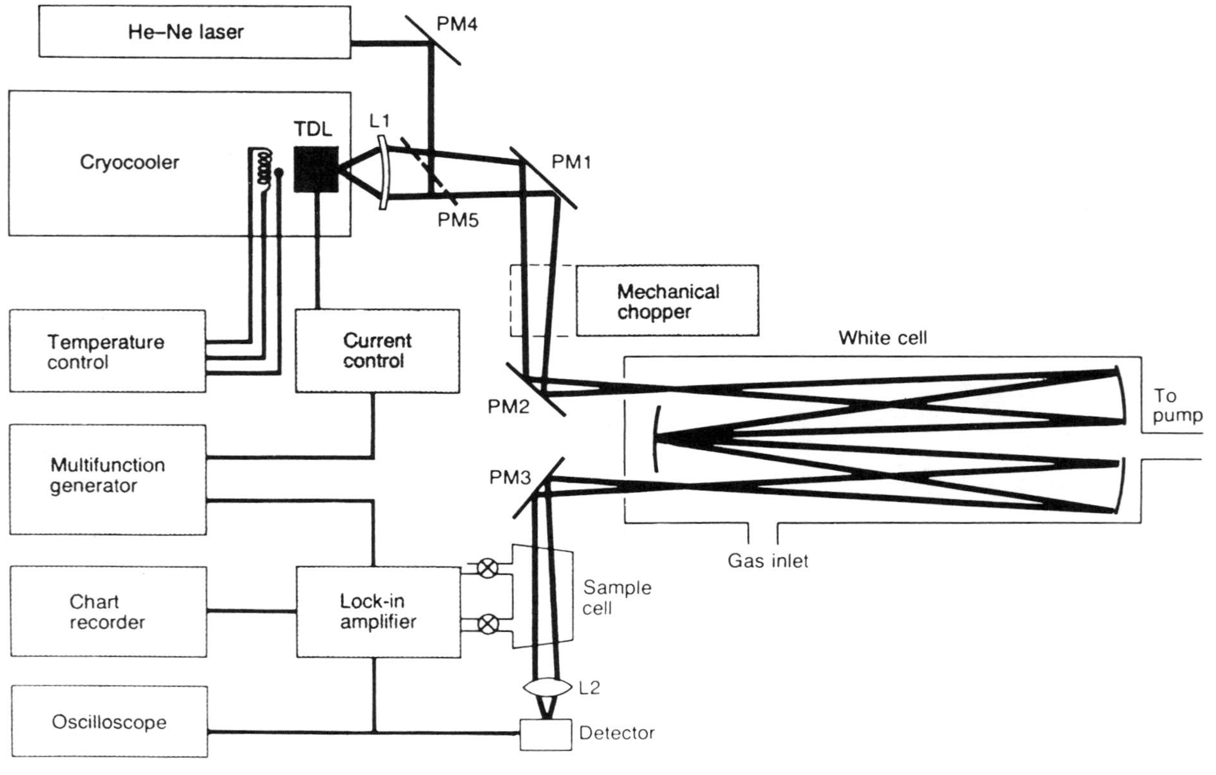

Figure 5.3. Schematic of a laboratory tunable diode laser system: L1 and L2 are the lenses used to focus the beam onto the White cell entrance and the detector, respectively, and PM1–PM5 are gold-coated mirrors used for beam alignment.

grown double-heterostructure lasers should extend the operating range above 77 K to longer wavelengths and permit the use of liquid nitrogen Dewars. Dewars must, however, be filled at regular intervals, usually every day. Automatic feed systems are available that can extend the time between operator attention to some 3–4 days but are somewhat bulky.

The development in the late 1970s of reliable closed-cycle helium coolers capable of achieving temperatures as low as 12 K permitted the development of mobile TDLAS systems for field measurements, free from the need for periodic supply of liquid cryogens. Three types of cryocoolers are available commercially. All closed-cycle cooler configurations use a thermodynamic compression-expansion cycle to cool an evacuated chamber (cold-head) that contains up to four diode lasers. The cooler consists of a helium compressor, a heat exchanger to remove the heat of compression, an expander with a reciprocating displacer element, and a regenerative heat exchanger to alternately remove and restore heat to the gas.

Early versions of such cooling systems were expensive, cumbersome, weighed a total of close to 100 kg and required ~ 2 kW of power. They did, however, provide about 10,000 h of operation before requiring maintenance. Mechanical vibrations due to the compression–expansion cycle, however, resulted in low-frequency excess noise above the intrinsic laser noise, even when the diodes are mounted on vibration-isolated cold-fingers (Eng et al., 1979). Liquid nitrogen Dewars were preferred over the early versions of closed-cycle coolers for operation on relatively small aircraft and balloons, where power and weight are major considerations.

Recently, a miniature integral Stirling cooler, originally built to military specifications, has become available. The cooler is capable of removing 1.0 W of heat at 80 K with a power consumption of only 75 W. A cold-head accommodating two diode lasers has been specially developed at Unisearch Associates Inc. (Nadler et al., 1993) for use in conjunction with the cooler shown in Figure 5.4, in a compact system for monitoring trace gases in the 3–5 μm region (see Section 5.5). The cold-head is coupled to the cryocooler by means of shock absorbers. The cooler has an operating life greater than 5000 h before requiring maintenance.

Figure 5.5 shows the schematic of the cold-head design. The laser platform is hard mounted to the vacuum housing through the use of thermally isolating machinable ceramic materials. The laser mount is coupled to the cold-finger of the cryocooler by means of a flexible metal strap that also acts to reduce the transfer of piston vibrations to the lasers. The dimensions of the complete cold-head assembly are $H \times L \times W = 26.6 \times 24.13 \times 20.32$ (cm) and the weight is 7 kg (comparable to a fully loaded Dewar). Foil heaters attached to the laser mount provide the required heating for set-temperature selection and stabilization. The cool-down period of the system is about 30 min from room tempera-

Figure 5.4. Photograph of the UniSERACH model CRU77-2 self-contained laser cold-head assembly capable of holding two diode lasers in the temperature range 75–110 K.

ture to 65 K. The temperature operation range is 75–110 K, with a stability better than 10^{-3} K. A small turbo- or ion pump ensures continuous purging for periods as long as the lifetime of the cryocooler pump (> 5000 h).

The increased availability of high-temperature diode lasers operating at or near room temperature should eventually allow the use of thermoelectric coolers. Six-stage Peltier coolers can be used to achieve temperatures as low as 200 K, but their application to CW operation must await future development of these Pb-salt diodes. Peltier coolers are currently used for the tuning and temperature control of NIR GaAs lasers and for Pb-salt lasers used in a pulsed mode (Preier et al., 1977).

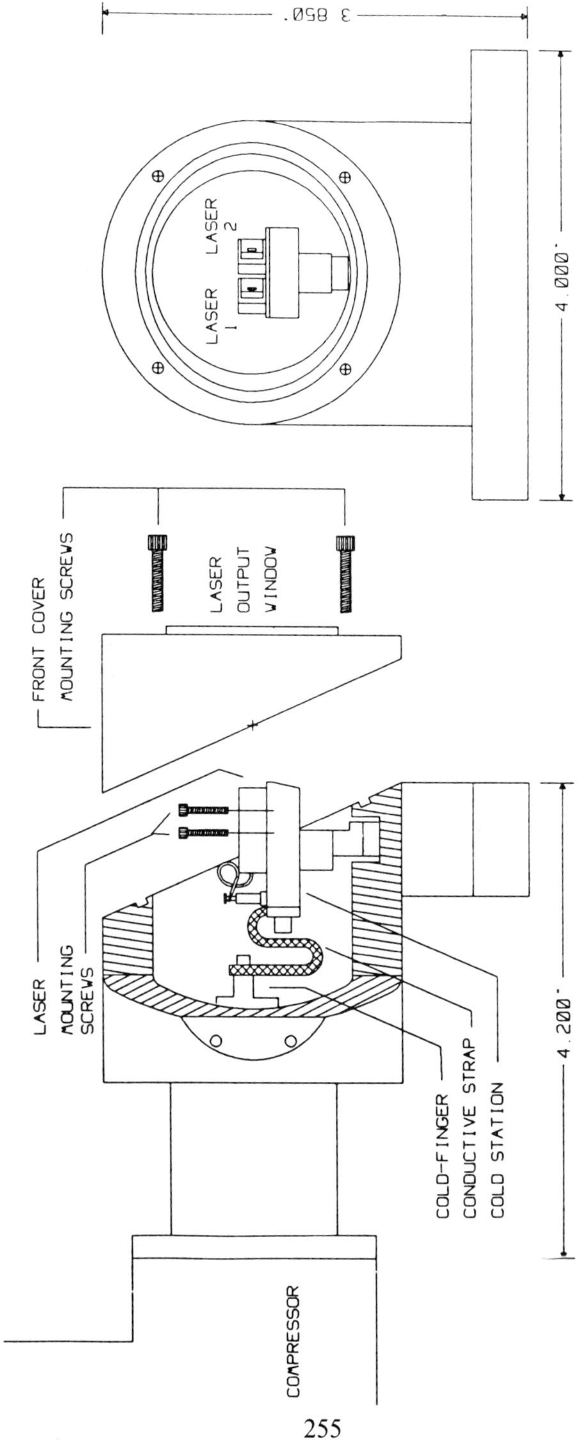

Figure 5.5. Schematic of the UniSEARCH model CRU77-2 laser cold-head assembly.

5.4.2. Current Control and Modulation Techniques

Once temperature stabilization is achieved the laser is tuned by altering the laser current. TDLs typically tune at a rate of about $5 \times 10^{-3}\,\mathrm{cm}^{-1}\cdot\mathrm{mA}^{-1}$, which requires current stability better than 0.05 mA. A sawtooth or other electrical current waveform of sufficient amplitude to tune the laser across the absorption feature is applied to the laser current and is repeated at frequencies of about 10 kHz. The absorption signal is accumulated in a suitable data acquisition system.

To discriminate against background signals, modulation techniques are employed. Absorption measurements with TDLAS can be made either by AM, WM, or FM techniques or by fast scan integration.

5.4.2.1. Amplitude Modulation

In the amplitude modulation (AM) mode, the beam is chopped mechanically at frequencies below 1 kHz while the laser diode current is slowly changed to scan across the spectral feature. The signal is detected at the chopper frequency with a lock-in amplifier that gives a direct measure of the power incident on the detector. The signal is displayed as intensity vs. wavenumber.

An example of an AM absorption spectrum is shown in Figure 5.6 for HNO_3 in the 1722 cm^{-1} region. The concentration of the absorbing species can be calculated directly from the spectrum, provided the line shape, width, and strength are available for determining the absorption cross section. The sloping background in this spectrum is due to the increasing power of the TDL with increasing current.

This method is, however, limited to peak absorbances greater than 0.1% by uncertainties in subtracting two large numbers (P and P_0). The sloping background and the $1/f$ noise of the diode itself (see Section 5.6.1) at the relatively low operating frequency contribute to the detectable signal limitation. The relatively low sensitivity is insufficient for atmospheric trace gas measurements. It is, however, occasionally useful in providing absolute calibration compared with the other modulation techniques, which can provide only relative absorption values.

5.4.2.2. Wavelength Modulation

The limitations of the AM mode can be overcome by wavelength modulation (WM) (Reid et al., 1978, 1980; Reid and Labrie, 1981; Hastie et al., 1983; Slemr et al., 1986; Mackay et al., 1991). A 1–25 kHz sine or triangular electrical wave is superimposed on the laser current to produce a frequency-(wavelength)-modulated laser output. Measurement of the detector output at this frequency

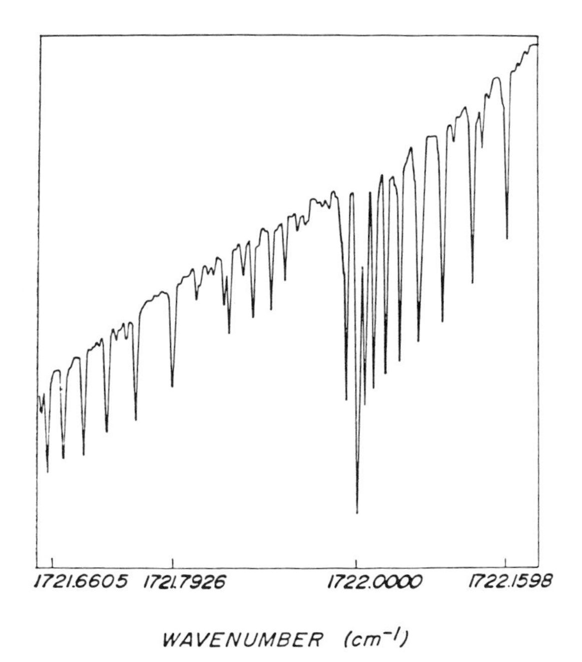

Figure 5.6. The absorption spectrum of HNO_3 taken in the region of 1722 cm^{-1} by an 11-cm cell. Total pressure = 1.0 torr; HNO_3 partial pressure = 0.01 torr; and laser power is ~ 1 μW.

gives a signal related to the variation in power with wavelength rather than to the power incident on the detector itself. The signal obtained by scanning the modulated radiation over a single absorption line has a shape that resembles the derivative of the absorption feature. The signal is a true derivative only if the modulation amplitude is much smaller than the absorption line width, but under these conditions the signal is relatively small. Much larger signals are obtained with larger modulation amplitudes and reach a maximum when the modulation amplitude is 2.2 times the line half-width (Reid and Labrie, 1981), which is less than 1% of the total laser current. Under these conditions the shape of the signal has a severely broadened derivative appearance, which we therefore prefer to call the "$1f$" signal.

The modulated $1f$ signal often carries a substantial zero offset due to the variation in laser power with applied current. To reduce this offset and maximize the signal at the line center, the detector output is analyzed at twice the modulation frequency, $2f$. Figure 5.7 shows a comparison between the AM and $2f$ WM detection methods.

In addition to eliminating the need to determine small differences between large signals, the use of the $2f$ WM method also reduces the $1/f$ laser noise and

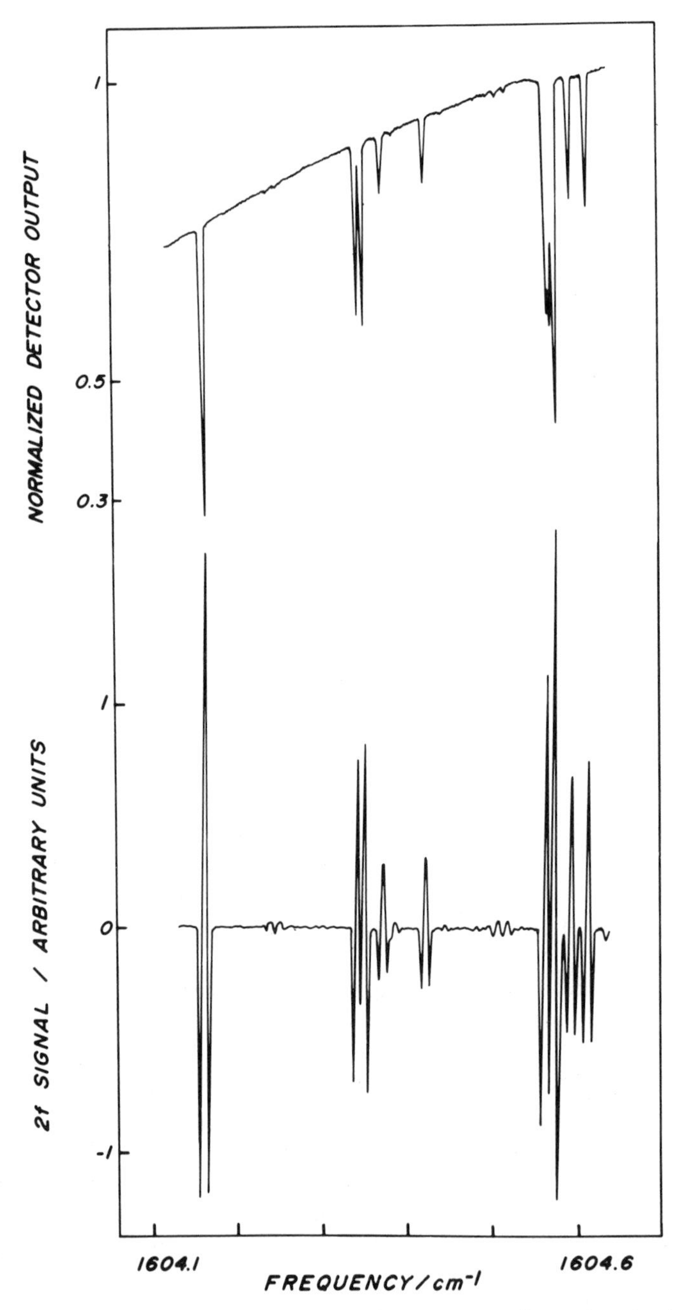

Figure 5.7. The NO_2 spectrum in the region 1604.13–1604.65 cm^{-1}. AM detection (top) and $2f$ detection (bottom) schemes are shown for a 1% mixture of NO_2 in air in an 11-cm cell. Total pressure = 10 torr.

the sensitivity to thermal fluctuations or other signals that do not show a wavelength dependence. Along with the absence of sloping backgrounds and zero offsets these advantages permit routine absorption measurements of better than 10^{-5}.

The price that has to be paid for these advantages is the loss of information about the unabsorbed laser power, P_0. The use of a dual-beam system in which the unattenuated laser power is measured separately and simultaneously would be one solution to this problem. More commonly, the system is calibrated with a known concentration of the target gas (see Section 5.5.1).

5.4.2.3. *High-Frequency Modulation*

Extending the modulation to still higher frequencies—in the 100 MHz to 10 GHz range—has gained considerable attention recently. The advantage is that the decrease in the intensity noise of the laser with frequency should approach shot or quantum noise at these frequencies, representing an improvement in detection limits of several orders of magnitude. Modulation at these radio frequencies is referred to as *high-frequency modulation*, or simply as FM.

Bjorklund (1980) was the first to apply the FM technique to a tunable dye laser using an external electro-optic modulator. GaAlAs diode lasers were the first TDLs to be modulated (Lenth, 1983, 1984) at gigahertz frequencies. Application of the FM technique to Pb-salt diode lasers was first demonstrated by Lenth and co-workers (Lenth et al., 1985; Gehrtz et al., 1986) and by Cooper and Watjen (1986).

When the diode laser is modulated at these frequencies, the emitted beam consists of a strong carrier at the natural emission frequency (wavelength) and weaker sidebands, offset above and below the carrier by the modulation frequency. These sidebands have signals that are equal in magnitude and opposite in sign. Cancellation therefore occurs when the wavelength is scanned with no absorber present. If the sidebands are scanned across an absorption feature, a signal appears at the absorption frequency with the appearance of a first derivative. The modulation frequency must be larger than the half-width of the absorption line so that the absorption feature is probed by only one sideband at a time.

The major drawback is that relatively high-speed detectors are required that are expensive and not readily available. The alternative is the two-tone FM technique (Janik et al., 1986; Cooper and Watjen, 1986). The laser current is simultaneously modulated with two radio frequencies (rf) in the gigahertz range separated from one another by a few megahertz. With this technique a beat signal proportional to the signal absorption is detected at the difference frequency, permitting the use of relatively inexpensive detectors.

In theory, the application of FM techniques should lead to shot-noise-limited sensitivities below 10^{-7} (Lenth, 1984; Gehrtz et al., 1985; Cooper and Warren, 1987; Werle et al., 1989a). In practice however, little success has been achieved in approaching these levels (Wang et al., 1988). The main problems are the ill-defined mode structures of current generation Pb-salts diodes. High-frequency modulation of these diodes also produces simultaneous amplitude modulation of the laser output (Cooper and Warren, 1987), which contributes to the noise seen at the detector. Sensitivities around 10^{-6} are the best that have been achieved except for a few cases where specially selected diodes have been used (Carlisle et al., 1989).

The use of multipass cells also limits the advantage of the FM technique because of attenuation of the laser power on the detector (Werle and Slemr, 1991). The signal-to-noise ratio reaches a maximum at a lower number of passes compared to the WM technique. Sensitivities are also limited by water vapor absorptions in the optical path outside the cell and by etalons.

These limitations have led to relatively few applications of FM techniques to atmospheric measurements (Werle et al., 1991; Cooper et al., 1991), and future improvements will have to await improvements in Pb-salt diode fabrication. But even though FM techniques do not lead to much greater sensitivities than those of WM techniques, they do offer several other advantages. Because of the large detection bandwidth they allow fast detection, which is important for certain applications such as “eddy correlation” flux measurements. The use of small rf electronic components instead of lock-in amplifiers also enable systems with the same sensitivities to be built that are smaller, lighter, and less expensive than are WM systems. Finally, the use of FM techniques with the much better fabricated NIR diodes should result in improvements in sensitivities.

Stanton and co-workers (see Bomse et al., 1992; Silver, 1992) have critically compared various modulation techniques for Pb-salt diodes, including wavelength modulation at low (kilohertz) and intermediate (megahertz) frequencies and single and two-tone high-frequency modulation. Detection with modulation at frequencies below 100 kHz was limited by laser excess ($1/f$) noise. The theoretically superior detection from high-frequency one-tone or two-tone modulation was not achieved owing mainly to inefficient coupling of the high-frequency modulation waveform to the laser current. The best detection limits were attained with high-frequency wavelength modulation in the 10 MHz range. With this technique they detected absorbances below 10^{-6}, with 1 Hz detection bandwidth, limited by detector thermal noise. They also concluded that systems containing multipass cells will be limited to 10^{-7} by etalon fringes (see Section 5.6.1).

5.4.2.4. *Sweep Integration*

In the sweep integration method developed by Jennings (1980), a sawtooth or other waveform is applied to the laser current at frequencies of a few tens to a few hundred hertz and phase locked to a signal averager. The resulting spectrum is integrated (hence the term *sweep integration*) a sufficient number of times to reduce noise. Cassidy and Reid (1982c) were able to demonstrate absorption measurements of 10^{-5} in 50 s using a 200-m path length in a low-pressure White cell. The method provides a direct absorption measurement that, at first sight, appears to obviate the need for calibration procedures. However, because of the change in laser power with current, rapid sweeping of the current also produces amplitude modulation that can amount to 20% of the signal. In addition, the laser must operate in single mode to obtain an accurate measure of I_0.

Background signal subtraction is still a requirement that has been met (Silver and Stanton, 1987) by simultaneously chopping the laser output to provide a measure of the total power. For atmospheric measurements the sensitivity achievable by the sweep integration method was inferior to that from the WM and FM techniques. The sensitivity was limited by the speed with which the absorption feature could be swept. More recently, however, fast digital processing components have become available that permit integration of high-frequency sweeps, which is expected to lead to higher sensitivity. Sweep frequencies up to 500 kHz (Hayward et al., 1989) have been achieved by superimposing a sinusoidal current modulation upon the TDL current. An automated analog subtraction circuit and/or digital filtering were/was used to remove the undesired AM background and to achieve the desired sensitivity. The system sensitivity was reported to be 1.5×10^{-5} in 10-ms integration time and was limited by detector noise.

Spaink and co-workers (1990) have described a tunable diode laser spectrometer using sweep integration at 1 kHz. Numerical data processing methods were used for background subtraction and base line correction. Comparisons between sweep integration and the $2f$ method for measurements of NO were found to give similiar sensitivities, although the authors anticipate further improvements in the sweep integration method.

Since sweep integration obviates the need for lock-in amplifiers, the overall size of a TDLAS system can be reduced. But fast DSP (digital signal processor) circuit boards are expensive at present, and the overall cost of the system is unlikely to be any lower. New software will also be required, but this is not expected to be a major problem. Improvement in the signal-to-noise ratio over that of the $2f$ technique has yet to be demonstrated.

5.4.2.5. *Pulsed Lasers*

Russian scientists have been particularly active in using diode lasers in the pulsed mode (Britov et al., 1977; Zasavitskii et al., 1984; Kuritsyn, 1986; Kosichkin et al., 1990; Nadezhdinskii and Prokhorov, 1992; Nadezhdinskii and Stepanov, 1991; Kuznetzov et al., 1992), although other groups have also used pulsed laser techniques (Preier and Riedel, 1974; Preier et al., 1977; Klingerberg and Winckler, 1987; Lach et al., 1989; Wolf et al., 1989; Riedel et al., 1991). The combination of pulsed tunable diode lasers with fast-scan signal detection has not yet found wide application in atmospheric monitoring, mainly owing to the difficulty in controlling the mode tuning during the pulse and to the lack of appropriate instrumentation for the analysis of the fast signal time evolution. However, the appeal of lasers operating in a pulsed mode at or near room temperature (Spanger et al., 1988; Feit et al., 1990) could result in increased interest.

In the fast-scan pulsed-mode operation described by Kuritsyn (1986), the laser is supplied with a repetitive rectangular pulse, milli- to microseconds wide, repeated at a few hundred hertz. The pulses are detected with a fast HgCdTe detector, the signal of which is amplified with a broad band amplifier and fed to a sampling voltage converter. The converter measures the detector signal amplitude in a fixed time aperture, e.g., 4 ns, and converts it to a dc signal that is then digitized by a 10-bit analog-to-digital converter.

The data acquisition system acts as a dual-channel digital signal averager. The spectrum is obtained by a point-by-point recording of the pulse shape distortion due to the absorbing gas. The time to measure a single spectrum is a few seconds and depends on the resolution required (number of data points) and the sensitivity (number of averages per point). The data acquisition time can be decreased to a fraction of a second by simultaneous sampling at several points in the pulse duration and by using shorter scan times.

For atmospheric monitoring applications, two data points are usually sufficient to measure the concentration with a two-channel boxcar having an appropriate time delay. The first channel records the signal amplitude at the peak line absorption, and the second records the background signal. An absorption detection limit of 10^{-5} could be achieved in a 6-min scan time, which is comparable to that obtained with CW harmonic detection.

In addition to the Russian work, Preier et al. (1977) reported the operation of a double-heterostructure $PbS_{1-x}Se_x$ diode in the pulsed mode at temperatures near 200 K, permitting thermoelectric cooling and temperature control. Four- and seven- stage Peltier coolers were required to allow operation down to 195 and 173 K, respectively. The emission of the laser could be tuned between 5.5 and 6.1 μm and was used to detect NO absorption. These authors reported that the shape of the laser emission did not change from pulse to pulse and that there were no detectable variations in operating conditions and output after 500 h of operation with 1-μs pulses at 1 kHz. The sensitivity of the system was quite low, however.

Pulsed-mode operation has been used to study automobile exhaust (Klingerberg and Winckler, 1987; Lach et al., 1989). The beams from six diode lasers were combined, with each diode operating in the pulsed mode. The shape of the current pulses applied to each diode produced linear wavelength tuning over the range of 0.5–1.0 cm^{-1}. The pulse width was 20–30 μs, and the repetition rate 1 kHz. A single path length of 1 m provided a detection limit of a few ppmv.

Riedel and co-workers (see Wolf et al., 1989; Riedel et al., 1991) developed a fast, pulse-operated exhaust gas analysis system. A rise time of less than 3 ms and detections limits of 100 ppmv for NO and 20 ppmv for CO were reported. The laser is driven by current pulses 100 ms wide at a repetition rate of 5 kHz.

Improved sensitivities should be possible by combining pulsed TDL systems with high-frequency modulation (FM) techniques. This combination has been reported (Lenth et al., 1981; Gallagher et al., 1982) with pulsed dye lasers, but no work has yet been reported on FM with pulsed TDLs.

5.4.2.6. *Heterodyne Detection*

A brief mention must be made of the use of Pb-salt diodes in heterodyne systems for remote sensing of atmospheric gases. Heterodyne detection consists of combining the radiation from a probe laser or a broad-band source, e.g., sunlight, with that of a laser used as a local oscillator on the detector or photomixer. The heterodyne signal is detected at the difference frequency which usually lies in the radio-frequency (rf) range. The use of TDLs in heterodyne detection remains limited by the low power of Pb-salt TDLs. The sensitivity is determined by the excess laser noise and by the relatively poor wavelength stability of TDLs compared to that of fixed frequency gas lasers.

A heterodyne spectrometer using a tunable diode laser as the local oscillator has been described (Glenar et al., 1982) for remote measurements near 8 μm with the sun as the broad-band source. A tunable diode laser heterodyne spectrometer has been described (McElroy et al., 1990) to record high-resolution (3×10^{-4} cm^{-1}) solar spectra in the 9.6 μm band of ozone. A signal-to-noise ratio of 120:1 was reported for a 1/8-s integration time and 1.3×10^{-3} cm^{-1} resolution, equivalent to 30% of the shot noise-limited value.

A general review of the theory and applications of IR heterodyne spectroscopy is presented in Kingston (1978). Applications to atmospheric monitoring are reviewed by Menzies (1976) and Webster et al. (1988).

5.4.3. Collimating Optics

Because of the short cavity length of diode lasers, the radiation is generally emitted as a multilobe pattern within an $\sim f/1$ cone. Lenses and/or mirrors must therefore be used to collect and focus the radiation into absorption cells and onto the detectors. Early spectrometers used lenses, but more recent systems tend to employ off-axis parabolas (OAP) or off-axis ellipsoids (OAE) to collimate the laser emission so that it may be more easily maneuvered about the optical table without restriction of distance from the optical element to the final focal point. Additional OAPs or OAEs are employed to focus the beam into absorption cells and onto detectors.

5.4.4. Long-Path Techniques

The Beer–Lambert law, Eq.(5.1), shows that measurement sensitivity increases with absorption path length as well as with the concentration and absorption coefficient of the target gas. The gases of interest in clean and polluted atmospheres span a wide concentration range from parts per million to fractional parts per billion, which therefore require corresponding optical path lengths from several meters to hundreds of meters.

Two fundamental methods are available to achieve these long path lengths. One is an open, double-pass system in which the TDL beam is directed to a retroreflector placed at a suitable distance and is then returned to the detector system. One advantage of this arrangement is the absence of walls and, consequently, absence of sampling problems. All remote-sensing instruments encounter the problem of atmospheric turbulence that occurs with frequencies between 0.1 and 10 Hz. Measurements must therefore be taken at a rate faster than 10 Hz. Fortunately tunable diodes can be tuned at rates much faster than this, and the necessary signal-to-noise ratio can be achieved by accumulating a sufficient number of these rapid measurements. A more serious problem is atmospheric pressure broadening, which not only decreases sensitivity but limits the species that can be measured, as discussed in Section 5.2. The method has been used successfully at atmospheric pressure (Partridge and Curtis, 1989) for molecules with relatively small pressure-broadening coefficients and/or wide line spacing such as CO, CH_4 and HCl. It has also been used very successfully (Menzies et al., 1983; Webster and May, 1987, 1992; May and Webster, 1989; Webster et al., 1990a) from a stratospheric balloon at altitudes of about 30 km, where pressure-broadening is no longer a serious problem. Another characteristic of long-path remote measurements is that they provide average concentration over the entire path length. This may be construed as either an advantage or a disadvantage, depending on the objectives of the measurements.

The desired long path can also be achieved by using multipass cells, which generally provide point sampling. Reduced pressures are used to decrease pressure broadening. However, the questions of sampling artifacts must then be addressed. There are two types of multipass cells currently in use, both named after their inventors—the White cell (see White, 1942, 1976; Horn and Pimentel, 1971) and the Herriott cell (Herriott et al., 1964; Herriott and Schulte, 1965). [The White cell and its modifications are discussed in detail in Chapter 6, and only one example will be described in Section 5.7 of this chapter.]

The Herriott multipass cell was originally developed for the study of optical delay and reflectivity (Herriott et al., 1964; Herriott and Schulte, 1965)]. The cell consists of two spherical concave mirrors having identical focal lengths, f, placed coaxially at a distance, d, such that $d < 2f$. A collimated beam is injected through an off-axis hole in one of the mirrors. The injected beam bounces back and forth, experiencing periodic focusing and defocusing until, for certain values of the mirrors, separation d, it exits through the entrance hole. The entrance angle of the beam governs the geometrical distribution of the consecutive spots, usually the pattern formed is elliptical. In cases when $d \ll f$, the distribution of the reflected spots may continue for many cycles (orbits) before finally exiting the system. The individual orbits may be separated by means of precession induced when one (or both) of the mirrors has astigmatic properties. When both mirrors are astigmatic and, in addition, have different focal lengths the reflective spots may be packed in a "designer's choice" Lissajous pattern to optimally suit a given system.

For use in absorption spectroscopy a somewhat simpler approach is used that satisfies two basic criteria. The input and output beam must be clearly separated to facilitate external manipulation, and the output beam must be a product of a single reflection, i.e., without overlap components from neighboring spots that cause etalon fringes. Since complex spot patterns are very difficult to align and maintain in the IR, Herriott cells are usually designed to consist of only one orbit by mirror separations for which $f < d < 2f$ (Altmann et al., 1981). The light beam exits through the entrance hole only for certain values of d, defined by

$$d/f = 2(1 - \cos(\pi/n)) \tag{5.10}$$

where n is the number of the beam round trips (equal to the number of spots on each mirror) inside the system.

For $d = f$ the total number of spots is 6 with 3 on each mirror. As d is increased, the number of spots increases, tending to infinity as d tends to $2f$. For $d = 2f$ the number of spots collapses to 4. Hence, the optical path length can be changed by altering the mirror separation with values of d/f satisfying

Eq. (5.10). The beam exits the cell at an angle different from that of the input beam, which allows the exit beam to reach the detector.

Weitkamp and co-workers (see Altmann et al., 1981) found that the direction of the exit beam is independent of the path length, that the alignment of the mirrors with respect to angle and distance was not critical, and that the path length was not very sensitive to changes in entrance beam slope or mirror distance.

A low-volume Herriott cell designed for flux measurement has recently been described by McManus and Kebabian (1990) in which an astigmatic mirror arrangement was used to increase the path length-to-volume ratio. This configuration distributes the spots in the orbit to produce a larger distance between the spot corresponding to the output beam (N) and the ($N - 1$) spot.

The Herriott cell is finding increased application in atmospheric trace gas monitoring with tunable diode lasers owing to its stability and low etalon fringe amplitude (see Section 5.6) (Webster et al., 1990b; Anderson and Zahniser, 1991).

However, there is little advantage over the modified White cell (Horn and Pimentel, 1971; White, 1976; Nadler et al., 1993) in terms of the path length-to-volume ratio, and the Herriott cell suffers from the need for precise alignment of the input beam, which is less critical for the modified White cell. In addition, it is easier to vary the path length of the White cell than of the Herriott cell, giving the White cell an advantage for applications where flexibility is required. For mobile systems where the stability of the absorption cell optics is critical, the Herriott cell may be preferred.

5.4.5. Detectors

Semiconductor photodetectors in which incident light is directly converted to conducting electrons are suitable for TDLAS systems. Photodetectors are sensitive only to light at wavelengths below a certain cutoff wavelength, λ_c. Below λ_c the spectral response, defined as the ratio of detector output to power input, decreases in a quasi-monotonic fashion, whereas above λ_c it falls rather abruptly, giving the spectral response curve the appearance of a somewhat rounded sawtooth.

Photodetectors are of two types, photoconductive and photovoltaic, depending on how the generated photoelectrons are measured. For a photoconductive (PC) detector, an external reverse bias voltage is applied and the change in internal resistance is measured, either as a change in current or as a voltage across a load resistor in series with the detector. A photovoltaic (PV) detector is used without bias, and the generated photovoltage or current is measured. PV operation requires the presence of an internal potential barrier such as p–n junctions in photodiodes to separate the photogenerated charge carriers in the absence of external bias.

Most commonly used materials for photodetectors in the MIR are InSb and PbSe for the spectral range below 5 μm, and $Pb_{1-x}Sn_xTe$ and $Hg_{1-x}Cd_xTe$ for wavelengths greater than 5 μm. All these detectors operate at or above liquid nitrogen temperature. PbSe can be used at room temperature for wavelengths shorter than 4 μm. The HgCdTe alloys can also be used at thermoelectrically cooled temperature below 5 μm.

The $Hg_{1-x}Cd_x$ detectors are particularly useful for applications that employ wide-band detection in the MIR or high-frequency modulation because of their fast response (> 1.5 GHz). In addition, their cutoff wavelength can be varied by changing the alloy composition, x. These detectors, sometimes called HCT or MCT photodetectors, with different composition can cover the MIR range between 2 and 24 μm.

In the NIR region, InAs and InGaAs photodiodes operating at or near room temperature in the range 1–3.5 μm, PbS (PC) detectors operating between 77 and 295 K in the range 1–4 μm, Ge (0.8–1.8 μm) and Si (< 1 μm) photodiodes operating at room temperature are commonly used.

The performance of a photodetector is usually characterized by the spectral detectivity $D^*(\lambda)$, which is defined as

$$D^*(\lambda) = (S/N)(AB)^{1/2}/P \tag{5.11}$$

where S/N is the rms (root mean square) current or voltage signal-to-noise ratio; A, the detector area; B, the detection bandwidth; and P, the incident power of a monochromatic source.

The minimum incident power detectable per unit bandwidth, termed noise equivalent power (NEP), is then given by

$$\text{NEP} = (A)^{1/2}/D^*(\lambda) \tag{5.12}$$

The detectivity of a particular photodetector decreases with increasing operating temperature. It is also a function of the wavelength and of the modulation frequency and is usually quoted at the wavelength of peak response and for a particular modulation frequency (typically, 1 kHz); D^* values for photodetectors used in the MIR lie in the range 10^9–10^{12} cm·Hz$^{1/2}$·W^{-1}. Given the detector area, NEP values can be derived via Eq. (5.12). Detectors with NEP values in the range 10^{-11}–10^{-13} W/Hz$^{1/2}$ are currently available.

The properties of some commonly available photodetectors are summarized in Table 5.1.

Table 5.1. Properties of Some Commonly Available Photodetectors for TDLAS[a]

Material/(Mode)	Range (μm)	T Operation (K)	D^* (cm · Hz$^{1/2}$/W)
Ge (PV, PC)	0.8–1.8	295	6×10^{11}
InAS (PV)	1 –3.5	77,195	2×10^{11}
		295	6×10^{9}
InGaAs (PV)	0.9–1.8	295	1×10^{10}
PbSe (PC)	1 –6	77,195	2×10^{10}
	1 –5	295	2×10^{9}
InSb (PC, PV)	1 –5.5	77	1×10^{11}
HgCdTe (PC, PV)	3 –7	77–295	$10^{9}–5 \times 10^{11}$
	7 –14	77	5×10^{10}
	14 –24	77	2×10^{10}
PbSnTe (PV)	4 –13	77	2×10^{10}
	6 –18	4	1×10^{11}

[a] From manufacturers' data sheets, e.g., Barnes, EG&G Judson, Optikon.

5.5. REQUIREMENTS FOR AN ATMOSPHERIC TDLAS SYSTEM

For use in atmospheric measurements, TDLAS systems should be capable of operating continuously, with a minimum of operator attention, in rigorous environments. They must be rugged enough to be transported to the measurement site and to operate on moving platforms when required. Measurements made with systems that meet these criteria are discussed in Section 5.7. But before these systems can be used in the field a number of requirements must be met.

First, a suitable tunable diode must be selected. This is usually a compromise between the line strengths of the molecule and the power and mode structure of the diode at the wavelengths of the absorption features. The selected absorption line must also be free of interferences from other gases in the measured air. A sampling system must be designed to minimize wall effects, and any such effects for either the sampling system or the absorption cell must be characterized. This is particularly important for such "sticky" gases as HNO_3 and NH_3. The use of modulation techniques requires the development of calibration procedures for the gases of interest. The accuracy of the measurement is dictated by the accuracy with which the concentration of the calibration standards are known. Calibration procedure must also take into account any surface effects in the sampling lines or on the walls of the absorption cell.

Other considerations include subsystems for keeping the instrument tuned on the selected line, for measuring more than one gas simultaneously, for automatically calibrating and providing background spectra, and for acquiring and storing the data.

In addition to discussing some of these requirements, the following subsections will also address the factors that determine the noise and detection limits of a TDLAS system.

5.5.1. Calibration Systems

In principle AM techniques permit the concentration of the absorbing species to be calculated directly from the Beer–Lambert law if the total path length and absorption coefficients are known. But, as discussed in Section 5.4.2, the AM technique does not provide the sensitivities required for most atmospheric measurements. Sweep integration techniques should also be able to provide direct concentration measurements, but care must be taken to avoid interference from residual AM modulation. The calibration procedures described below avoid this danger and those from sampling artifacts.

Calibration is required when frequency or wavelength modulation is used since information on the unabsorbed laser power is lost. Several methods are available for calibrating a TDLAS system. The first is to measure the signal, I_{ref}, when the beam is passed through a short cell of known length containing a known concentration, C_{ref}, of the gas to be measured (Fried et al., 1984; Weitkamp, 1984; Anlauf et al., 1988). The ambient concentration of the gas, C_{amb}, can then be determined from the signal, I_{amb}, when the beam is passed through a length, L_{amb}, of the ambient air:

$$C_{amb} = (C_{ref}\, I_{amb}/I_{ref})(L_{ref}/L_{amb}) \tag{5.13}$$

This method is suitable for use with open, long-path measurements.

An ingenious but rather limited method has been described (Restelli and Cappellani, 1985) that takes advantage of the relatively constant mixing ratio of some atmospheric trace gases such as N_2O. The spectral features of the target molecules lying near a N_2O line (within the same laser mode) are probed and the concentration is derived from the relative strengths of the two lines, taking into account the difference in absorption cross sections and other relevant molecular parameters such as line shape and width which influence the $2f$ signal. The limitation, of course, is the requirement that the line pairs must lie close enough together that changes in laser intensity with wavelength can be neglected.

Neither of the foregoing methods is appropriate when the air is drawn through a multipass sampling cell because of possible sampling artifacts. To avoid the possibility of the sampling line or the cell itself absorbing some of the target gas from the sampled air, the only reliable approach is to add a known amount of the calibration gas to the sampled air stream at the entrance to the sampling line. When possible, the concentration of the "spike" should be chosen to provide an increase in measured mixing ratio comparable to the mixing ratio of the gas in the ambient air. In this way, any surface effects that may occur will be the same for the sampled and spiked air.

Suitable calibration mixtures of the more stable gases such as NO, CH_4, or CO can be purchased commercially. The accuracy of the known concentration is usually provided by the manufacturer and traced to some standard such as those of the NIST (National Institute of Standards and Technology; formerly the U.S. National Bureau of Standards). For other gases whose integrity in steel containers is not reliable such as NO_2, HNO_3, H_2O_2, and HCHO, permeation devices are frequently used. It is advisable to check the permeation rates of these devices on a regular basis by gravimetric or other methods. For some constituents, permeation devices can be purchased commercially. For other species, permeation devices can be constructed by the user (Slemr et al., 1986; Harris et al., 1987, 1989b; Mackay et al., 1988, 1991; Schiff and Mackay, 1989).

5.5.2. Line Locking

The fast response time of diode lasers permit the selected rotational line to be scanned with frequencies greater than 1 kHz. Signal-to-noise ratios can be improved by accumulating the signals over periods from $\sim$0.1 s to many minutes depending upon the time response or sensitivity desired. The improvement in sensitivity, which from a purely statistical point of view should increase as the square root of the integration time, requires that the laser frequency does not drift over the scanned spectral range during the integration period. One reason for such frequency drifts is changes in ambient temperature of the electronic control circuitry.

Stabilization of the laser emission can be achieved by locking the frequency relative to an etalon (Reich et al., 1986) or to a reference cell containing a high concentration of the target gas (Schiff et al., 1987; Mackay et al., 1991). Better temperature compensation can be attained with the reference cell option, which makes it the preferred method under field operating conditions.

Either analog or digital line locking can be used. With analog line locking part of the laser beam is passed through the reference cell and detected at the modulation frequency ($1f$). The signal over a rotational line resembles a first derivative passing through zero at the line center, which provides the control point. The output from the lock-in amplifier is used in a feedback circuit to control either the laser current or the laser temperature. Since the response of the laser is faster to changes in current than to temperature, this is the preferred control method whenever short integration times are used.

As long as the laser frequency remains at line center, the output of the $1f$ reference is zero and no change in the laser current occurs. If the frequency drifts from line center, either a positive or negative voltage is superimposed on the laser drive current to shift the frequency of the laser emission back to line center.

For digital line locking the computer determines the line center either by detecting the zero crossing of the $1f$ signal or the maximum in the $2f$ signal from the reference detector. This information is passed to a digital-to-analog (D/A) converter that supplies the appropriate voltage to the laser current or temperature control circuits.

Laser temperature control offers advantages under conditions where the measurements are made near the detection limit when longer integration times are desired. Temperature control directly corrects for the drifts in the electronics of the temperature stabilization circuitry, which otherwise would shift the laser temperature. These instabilities in the temperature control electronics are the major cause of laser emission frequency variations. Temperature control is, therefore, the best way to ensure that the output power of the laser remains constant and that the ultimate sensitivity is maximized over long periods of time.

For laser temperature control a feedback circuit supplies an offset voltage to the laser temperature setpoint control to maintain the absorption line center at the midpoint of the scan.

5.5.3. Multiplexing

To understand atmospheric processes or to identify related species in complex mixtures such as automobile exhaust, simultaneous measurements of several species are often desired. The tuning ranges of currently available diode lasers are of the order of 200 cm^{-1}, consisting of 1–2 cm^{-1} continuous segments, separated by similar gaps with no laser emission. Occasionally strong absorption lines for several species of interest can be found that lie within one of the continuous segments. For example, we have been able to measure CH_4, C_2H_6, HCHO, and C_6H_6 with a diode operating in the 3050 cm^{-1} range. In principle, it is possible to temperature tune a given diode to access other modes, each with its own 200 cm^{-1} tunable range. In practice, however, it is often easier to use separate diodes for each desired species along with some method of signal multiplexing.

Three multiplexing methods have been described. One, which can be called time multiplexing, sequentially measures the absorption of the beams from a series of diodes. It is the most straightforward method but suffers the limitation of time resolutions of about a second, which is good enough for most applications but not for some aircraft studies or for eddy correlation flux measurements that require time responses of 0.1 s or better. For the latter applications, however, multispecies measurements are generally not required. The other two methods, called frequency and wavelength multiplexing, provide truly simultaneous measurements and are therefore capable of high time resolution but are more technically difficult.

A time multiplexing system in which the signals from two diodes are measured has been described (Schiff et al., 1987). Two cryostats are used, each containing a laser source assembly housing four laser diodes. One TDL from each assembly can be selected, and the laser beam from each is directed to a selection mirror that flips back and forth to permit each beams to enter the absorption cell in turn. A single detector is used for both beams with a dwell time on each beam of typically 3 s. Another pair of diodes can be chosen with this arrangement, requiring 15 min for the operating temperatures of the new pair to become stabilized.

A modification of this method to permit the sequential switching between four diodes has been described (Harris et al., 1989a). A balloon-borne system (Menzies et al., 1983) utilizes four diodes held in separate cold-head/liquid He Dewars the beams from which are picked up in turn by a rotating periscope.

Wavelength multiplexing has been achieved by Sachse et al. (1991) with two diodes mounted in separate cold stations in a single LN_2 Dewar. The beams are combined directly through the absorption cell and detected separately by InSb and Hg/Cd/Te detectors, which respond to different spectral regions. This approach restricts the measurements to species that lie above and below the responses of the two detectors—in this case, about 5 μm. This restriction could be lifted if beams from different lasers were to enter and leave the multipass cell at different positions and were then focused on separate detectors. To our knowledge this option has not yet been tried. But both options require separate detector circuits for each species and, when compared with using separate TDLAS systems for each species, offer only the relatively small advantage of sharing a common absorption cell.

In the frequency multiplexing approach (Hastie and Miller, 1985; Riedel et al., 1992), a number of lasers are each modulated at different frequencies. Their beams are then combined to enter the absorption cell simultaneously and are focused on a single detector. Demodulation of the detector signal permits separation of the contributions from individual species. Although the instrumental complexity is not much greater than that for time multiplexing, the method does encounter additional noise due to cross-talk between the noise of the individual diodes. Muecke et al. (1991) investigated this effect and found that the noise levels of each diode add according to the square root law. For a two-diode system they found that the detection limits were 40% worse than those for each species measured separately. The detection limits for a weak absorbing species will deteriorate even further when the other species is a strongly absorbing one. They also investigated the time multiplexing system and found that the detection limit was decreased by an amount equaling the duty cycle of the scanning device, which in their case was 30%.

5.5.4. Operating Control and Data Acquisition Systems

The operating control and data acquisition systems used in the Unisearch TDLAS instruments are described in some detail in Section 5.7. Other investigators use similar systems with relatively minor variations. Some of these variations that are useful under special circumstances are worth noting. For example, a method has been developed (Fried et al., 1991a,b) for measuring, almost simultaneously, two signals that differ in amplitude by as much as a factor of 26. Two lock-in amplifiers with different gain settings are used to scan the two features. A variation of the rapid current jump method described in Section 5.6 has also been used that permits two absorption features separated by as much as 0.424 cm^{-1} to be scanned rapidly. A third beam channel has been used (Sachse and Hill, 1987) in front of the White cell to correct for changes in amplitude of the laser output caused by the current sweep or by any absorption outside cell and to help minimize the $1/f$ noise of the laser.

The great improvement in computer speeds in recent years has led to the development of rapid scanning techniques that employ fast DSP boards to acquire and accumulate the data (Hayward et al., 1989; Spaink et al., 1990). As in the system that we shall describe in Section 5.7, the measured absorption is compared (fitted) to a reference. In cases where the mode structure of the laser is well defined, line strength information can be employed to provide absolute concentration measurements.

5.6. SOURCES OF NOISE AND DETECTION LIMITS

The performance of any analytical system is judged by a number of factors, including sensitivity, detection limits, freedom from interferences of other gases, and the precision and accuracy of the measurements. Sensitivity is determined by the minimum detectable signal that is distinguishable from noise or other unwanted signals appearing at the output of the detection system. Detection limits can also be defined in terms of the signal-to-noise ratio (S/N) but, in addition, will also depend on interferences, if any, from other gases in the sampled air. In this section we will discuss noise sources in TDLAS systems as well as distinguishing between precision and accuracy.

Sensitivities are frequently defined in terms of the minimum detectable absorbed power, corresponding to $S/N = 1$. The signal S is the average electrical signal corresponding to the absorbed power P_{abs} and is given by

$$S = Cge\eta P_{abs}/h\nu_0 \tag{5.14}$$

where g and η are the detector gain and quantum efficiency of the detector respectively; e is the electronic charge; and C, a multiplication factor equal to 1 for AM techniques and to less than 1 for WM techniques. Usually, modulation amplitudes and other conditions can be chosen to produce C values close to unity.

5.6.1. Noise Sources

Noise is a general term used to designate all unwanted signals observed at the output of a detection system and masking the information (true signal) of interest. Several sources of noise are present in an absorption measurement with TDLAS systems. These include (1) detector noise, (2) laser excess noise resulting from fluctuations in laser power and frequency, and (3) interference fringes.

5.6.1.1. Detector Noise

A full treatment of the origin and types of noise in semiconductor photodetectors can be found elsewhere (Kruse, 1980; Kingston, 1978; Spears, 1983; Bachor and Manson, 1990). We will confine our discussion to the three principal components (Kruse, 1980) of photodetector noise: Johnson (or thermal) noise, shot noise, and $1/f$ noise.

Thermal noise is due to random fluctuations of the electronic charges and is present in all conducting components of the detection system including the preamplifier. The thermal noise current of the detector system, i_T, is given by

$$i_T = (4kTB/R)^{1/2} \tag{5. 15}$$

where k is Boltzmann's constant; T, the temperature; B, the detection bandwidth; and R, the detector system resistance. As the name implies, thermal noise depends on the temperature. It also depends on the detection bandwidth but not on the frequency or the light intensity. Thermal noise can be reduced by cooling the detector. It can also be reduced by increasing the resistance, R, but at the price of an increase in response time.

Shot noise is due to the random generation of electrons in photodiodes. The rms shot noise current i_S, is given by

$$i_S = (2e^2\eta BP/h\nu)^{1/2} \tag{5.16}$$

where P is the transmitted power and $h\nu$ the photon energy. Shot noise is independent of modulation frequency but is proportional to the square root of the laser power and of the detection bandwidth.

Detector $1/f$ noise is thought to be caused by potential barriers in semiconductors and semiconductor contacts, although rigorous theoretical analysis of this noise source is not currently available. The general expression for $1/f$ noise current takes the form:

$$i_{1/f} = (KB/f^{b})^{1/2}\,(e\eta P^{a}/h\nu) \tag{5.17}$$

where K is a proportionality factor, and b and a are constants with values close to unity. The $1/f$ noise depends greatly on manufacturing procedures, particularly with respect to contacts and surfaces. It dominates the detector noise spectrum at low frequency (< 1 kHz) and drops below the thermal and/or shot noise levels at higher frequencies. The rms $1/f$ noise current shows an approximate linear dependence on the photocurrent and therefore on the light intensity.

Since all three detector noise sources depend linearly on B, reduction of the detection bandwidth by low pass filtering or signal integration results in noise reduction. Again, the price to be paid is an increase in time response.

The foregoing expressions show that $1/f$ noise can be made smaller than shot noise by operating at higher frequencies (> 1 kHz). Since thermal noise is independent of laser power, Eqs. (5.15) and (5.16) can be used to determine the laser power required to achieve shot-noise-limited detection. For example, with a laser emitting at 10 μm and with a detector having a quantum efficiency of 0.8, shot-noise-limited conditions should be attained at laser powers greater than 25 μW, a value that is attainable with current generation diode lasers.

In practice, however, shot-noise-limited detection is difficult to achieve owing to laser noise and interference fringes.

5.6.1.2. *Laser Excess Noise*

The sensitivity of TDLAS measurements is often limited by fluctuations in laser power, particularly when relatively high laser power and low modulation frequencies are employed. These power fluctuations, usually termed laser excess noise, can be traced either to the inherent behavior of the laser diode (intrinsic noise) or to external effects such as injection current noise, temperature instabilities, on mechanical vibrations from, for example, closed-cycle cryocooling and optical feedback. Excess laser noise due to the first three sources can be minimized by using battery-driven or highly stabilized current sources together with liquid cryogen Dewars and by line locking the laser to an absorption feature or to a Fabry–Pérot etalon. Noise due to uncontrolled optical feedback is more difficult to reduce. It is produced by spurious reflections and scattering in the external optical system that feed back into the laser cavity, resulting in the deterioration of the laser spectral and noise characteristics (Nillson and Buus, 1990; Harward and Hoell, 1979; Spilker et al., 1992). In general, the use of reflecting optics, proper alignment (or misalignment) of optical components, and the use of optical isolators and apertures can be useful, although reduction in laser power can result (Harward and Hoell, 1979; Spilker et al., 1992; Carlisle and Cooper, 1989a).

Intrinsic laser noise is produced by photon and carrier density fluctuations, index of refraction fluctuations, and partition noise resulting from mode competition in multimode lasers (Armstrong and Smith, 1965; Welford and Mooradian, 1982; Henry, 1982; Li and Abraham, 1989; Kikuchi, 1989). It depends greatly on the manufacturing process and design of the laser and on the operating current and temperature.

Investigations of the amplitude noise characteristics of some Pb-salt diode lasers have been reported in the literature (Eng et al., 1979; Werle et al., 1989b; Fischer et al., 1991; Fischer and Tacke, 1991; Spilker et al., 1992). All the lasers investigated exhibited a $1/f$-type frequency-dependent noise with a cutoff frequency ranging from 10 MHz to well above 100 MHz, except for the DBR lasers (Fischer and Tacke, 1991), which showed a nearly white (frequency-independent) spectrum. In general, the relative intensity noise (RIN) was found to decrease with increasing injection current. In addition, noise spikes due to mode hops, partition noise due to mode competition, and feedback effects due to spurious reflections in the optical system were observed. Mode competition noise was particularly serious with some diodes, increasing the noise by 2 orders of magnitude.

The rms current i_{ex} at the detector, caused by laser excess noise, can be expressed as

$$i_{ex} = (e\eta P_{ex}/h\nu)\,(B/f^{b})^{1/2} \tag{5.18}$$

where B defines the frequency dependence of the laser excess noise and ranges between 0.8 and 1.5, and P_{ex} defines the magnitude of the laser power fluctuations at 1 Hz in a 1-Hz detection bandwidth; P_{ex} is approximately proportional to laser power and depends on the intrinsic noise of the diode laser and on the external effects of the particular measuring system. Values of P_{ex} can vary widely from system to system. Careful system design and diode laser selection can limit P_{ex} to values of 10^{-4}–10^{-5} of the laser power, P.

Shot-noise-limited detection is determined when the excess noise index, P_{ex}/P, satisfies the condition:

$$P_{ex}/P < (f^{b}h\nu/P)^{1/2} \tag{5.19}$$

For example, for a $10\,\mu m$ laser width, $P = 100\,\mu W$, and $b = 1$ ($1/f$ noise dependence), shot-noise-limited detection is achieved at 10 MHz for an excess noise index 5×10^{-5}.

As in the case of detector noise, laser excess noise is detection bandwidth dependent and can be reduced with appropriate techniques. At high laser power and/or at modulation frequencies below 100 kHz, laser excess noise determines the detection limit.

5.6.1.3. Residual Amplitude Modulation

One of the great hopes of those developing the high-frequency modulation technique (see Section 5.4.2.3) was to achieve shot-noise-limited noise. Unfortunately this goal has not been attained with present-generation Pb-salt diodes, largely because of the relatively poor spatial and mode tuning characteristics of these diodes. High-frequency modulation of these diode lasers results in simultaneous amplitude modulation of the laser injection current and therefore in the wavelength, giving rise to excess low-frequency or residual amplitude modulation (RAM), which then becomes the main noise source.

In principle, RAM can be reduced by dual-beam subtraction techniques (Gehrtz et al., 1985; Carlisle and Cooper, 1989b), but with an increase in complexity prohibiting its use in the field. Employing two-tone techniques, Carlisle et al. (1989) found that in most cases lasers could be operated in a saturated region of the power vs. current curve to yield low values of RAM. Theoretical and experimental studies (Silver, 1992; Bomse et al., 1992; Whittaker et al., 1988) showed that higher harmonics could be used to obtain shot-noise limited sensitivities.

5.6.1.4. Interference Fringes

Another source of noise in TDLAS systems is interference fringes from Fabry–Pérot etalons between reflecting or scattering surfaces such as mirrors, detector and laser head windows, semiconductor surfaces, and components of multipass cells. These fringes have spacing between maxima at wavelengths in the range 10^{-3}–$10^{-2}\,cm^{-1}$, which is in the same wavelength range as Doppler and pressure-broadened molecular line widths. The fringe spacing F corresponds to Fabry–Pérot optical cavities arising from reflecting or scattering surfaces separated by a distance L and given by

$$F = 1/(2Ln) \tag{5.20}$$

where n is the refractive index of air, which is close to unity. Measurement of the line spacing frequently permits identification of the optical elements responsible for the etalon.

Elimination of this noise source cannot be made by simple subtraction because thermal and mechanical instabilities of the optical system causes the etalon spacing to change with time. The effect of etalons can be minimized by using reflective rather than refractive optics where possible, avoiding parallel surfaces (e.g., by using Brewster angle windows), avoiding sharp focus, and selection of the diode laser mode (Hastie et al., 1983) and of the modulation waveform and amplitude (Iguchi, 1986; Koga et al., 1981).

A jitter (Reid et al.,1980) or a two-tone modulation method (Cassidy and Reid, 1982b) can be used to reduce fringe amplitude. A second modulation wave at frequencies not harmonically related to the basic modulation frequency is applied to average the fringes. However, this averaging also decreases the absorption signal.

The waveform of the modulation signal has been found (Koga et al., 1981; Iguchi, 1986) to affect the relative magnitude of the absorption signal and etalon fringes, and alternate waveforms have been suggested.

Both the jitter and waveform selection are efficient in suppressing etalons with etalon spacing either much larger or smaller than the molecular line width. But for fringes comparable to the line width they are inefficient and tend to wash out the signal.

A mechanical method can also be used to average out the fringes by oscillating the path length with an external device. For example, an oscillating Brewster angle plate can be placed in the beam path at right angles to the plane of incidence (Webster, 1985). The resulting periodic change of the path length averages and reduces the amplitude of the fringes by an order of magnitude. An oscillating mirror driven by a piezoelectric transducer asynchronously to the modulation frequency has also been used (Silver and Stanton, 1988).The oscillation amplitude is adjusted to obtain a change in path length greater than half the fringe separation, resulting in effective fringe averaging. A mirror mounted on a speaker driven at audio frequencies has also been used (Chou et al., 1989) and resulted in a 20-fold reduction in fringe amplitude. Pressure modulation was also used (Fried et al., 1990) to reduce fringe amplitude in a small-base White cell, which could not be reduced by other methods. However, the pressure could not be modulated very fast, precluding the use of this technique for applications requiring a response time faster than 1 s.

Postdetection Fourier transform digital filtering can also be used in some cases where the signal and interference spectra are distinguishable in the frequency domain (Cappellani et al., 1987). In systems using high-frequency modulation, postdetection electronic or digital filtering reduces the fringe amplitude. A high-pass filter is usually placed between the detector and a double-balanced mixer to reject low-frequency noise, fluctuations, and interference fringes.

As mentioned in Section 5.4.2, etalon fringes can be the limiting noise in high-frequency modulation (FM) systems. Complete elimination of the interference fringes is usually not attainable. In most systems etalon noise can be reduced to levels below other limiting noise components.

Other sources of noise include differences in temperature or pressure of the gas in the reference cell and the sampled air with resultant differences in line shapes. The air in the optical path of point sampling systems can produce broad-band interferences because of the larger pressure broadening relative to that in the multipass cell. This is generally not a problem when FM techniques are used but may, under special conditions, require flushing of the optical path with some noninterfering gas.

Interferences between atmospheric gases is less serious than for most other spectroscopic methods because of the high spectral resolution of TDLAS systems. An accidental resonance between the selected spectral feature and that from another gas is usually readily detected by a distortion of the line shape. In that event another absorption feature can be selected. A stringent test for the absence of interferences is provided by measuring the mixing ratio of the gas at two different absorption lines. The probability of two gases having two absorptions lines that overlap exactly and that simultaneously have the same relative intensities is vanishingly small. Although interferences do not normally present a problem with TDLAS systems it must be pointed out that for certain cases, such as measurements of very small concentrations (in the few pptv region), it may be difficult to select an absorption line that is free from interferences from weak H_2O and CO_2 absorption features.

5.6.2. Detection Limits

The sensitivity of the TDLAS system can be defined in terms of the minimum detectable absorption ($a = P_{abs}/P_0$) for a signal-to-noise ratio, S/N, equal to 1. The general expression of the total rms noise current i_N is obtained by taking the square root of the quadrature summation of all the noise contributions discussed above and assuming that interference fringes have been reduced to a negligible level.

Under shot-noise-limited conditions, the minimum detectable absorbance a_{min} is given by

$$a_{min} = (2Bh\nu/\eta P_0)^{1/2} \tag{5.21}$$

For typical conditions, with $P_0 = 100\ \mu W$ at $10\ \mu m$, $\eta = 0.8$, and $B = 1$ Hz, the detection limit is calculated to be $a_{min} = 2.2 \times 10^{-8}$. This is, of course, a theoretical limit that has been approached to date in only a few cases with laboratory systems using high-frequency modulation techniques.

Most reported atmospheric measurements have been made with the $2f$ technique at frequencies below 100 kHz. For these frequencies $1/f$ noise generally, not shot noise, determines the detection limit. For this case the minium detectable absorption can be approximated as

$$a_{\min} = (P_{ex}/P_0)(B/f)^{1/2} \tag{5.22}$$

where the terms are defined as in Eq. (5.18).

A typical modulation frequency used in the $2f$ technique is 10 kHz, and a value of $P_{ex}/P_0 = 10^{-4}$ is typical for most Pb-salt lasers. For these values the minimum detectable absorption calculated from Eq. (5.22) is 10^{-6} for a 1-Hz detection bandwidth. The sensitivities reported in the literature for TDLAS systems using the $2f$ technique with multireflection cells are between 1×10^{-5} and 3×10^{-6} (Reid et al., 1978, 1980; Hastie et al., 1983; Sachse et al., 1987; Sachse and Hill 1987,1991; Fried et al., 1984, 1991a; Schiff et al., 1987; Schiff and Mackay, 1989; Mackay et al., 1991). The modulation frequencies ranged between a few kilohertz to a few tens of kilohertz with laser powers at the detectors in the tens of microwatts range.

Mechanical vibrations, temperature and current instabilities, optical feedback, and interference fringes present in the systems can account for the lower sensitivities obtained in practice as compared with theoretically expected values. Systems using long-path retroreflector configurations are even less sensitive (10^{-3}–10^{-4}) owing to atmospheric turbulence (Cassidy and Reid, 1982a; Reid et al., 1985; Partridge and Curtis, 1989).

The detection limit for a particular atmospheric species is usually quoted as the minimum mixing ratio ($M_{\min}$) detectable in a given detection bandwidth (usually 1 Hz) and is related to the absorption sensitivity by

$$M_{\min}\,(ppbv) = a_{\min} \times 10^9/(\sigma_0 L N_t) \tag{5.23}$$

where σ_0 is the absorption cross section at the center of the monitored line; L, the path length; and N_t the total molecular density of the gas. Detection limits in the ppbv and sub-ppbv range are achievable with standard TDLAS systems using multipass cells operated at low pressures.

Becker et al. (1991) assumed a value of $a_{\min} = 5 \times 10^{-6}$ for a 1-s integration time to estimate detection limits of a TDLAS system with a multipass cell ($L = 132$ m; $p = 10$ torr). An integration time of 1 s per data point means that the response time or total integration time is about 1 min (64 points per spectrum) when the full line shape is recorded. Column 5 of Table 5.2 gives ranges of atmospheric mixing ratios for some important atmospheric trace gases; column 6 gives the estimated detection limits for the same species.

Table 5.2. Detection Limit M_{min} (ppbv) of TDLAS for Some Atmospheric Trace Gases

Species	Wave-number ν_0 (cm^{-1})	Line Strength S (10^{-20} cm)	Doppler Width $\Delta\nu D$ (10^{-3} cm^{-1})	Typical Atmospheric Concentrations (ppbv)	M_{min} Calculated (ppbv)	M_{min} Measured (ppbv)
CH_4	1301.4	2.6	4.02	1700	0.25	0.25[a]
C_2H_2	743.3	60	1.8	< 30	0.01	—
C_2H_4	950.0	11.4	2.2	~ 10	0.04	—
C_2H_6	822.3	0.33	1.86	< 50	1.5	10[a]
C_6H_6	686.4	0.1	0.96	1–10	2.0	—
H_2CO	2781	3.4	6.3	0.5–75	0.25	0.05[b]
CH_3OH	1030	0.7	2.25	~ 1	0.7	1.0[b]
CH_3OOH		1320.9	—	—	< 1	2.5
HCOOH	1107	2.6	2.06	< 1	0.2	1.0[b]
COS	2050.9	101	3.6	< 0.1	0.005	0.005[c]
CO	2111.5	36.6	4.9	50–1000	0.02	0.1[d]
CO_2	2339.4	320	4.6	3×10^5	0.002	—
N_2O	1250.6	3.1	2.3	310	0.2	0.05[b]
NO	1890.9	5.2	4.3	0.01–100	0.1	0.04[b]
NO_2	1600	18	2.9	0.02–1000	0.03	0.025[b]
HNO_3	1720	12	2.7	0.1–50	0.05	0.1[b]
H_2O_2	1285.7	2.7	2.7	< 1	0.2	0.1[b]
SO_2	1360.7	3.4	1.7	1–100	0.1	0.5[b]
O_3	1050	2.5	1.9	10–500	0.2	—
NH_3	1065	20	1.7	~ 1	0.03	0.025[b]
HCl	2900	48	5.9	—	0.02	0.2[e]

[a] Unisearch Associates Inc., EMS series TDLAS system (unpublished data). Path length: 12 m; total integration time: 1s.

[b] Unisearch Associates Inc. TAMS 150 TDLAS system (Schiff et al., 1987; Mackay et al., 1991). Path length: 150 m; total integration time: 3–5 min.

[c] Data from Fried et al., (1991a). Path length: 120 m; total integration time:2 min.

[d] Data from Fried et al. (1984). Path length: 40 m; total integration time not reported.

[e] Data from Sachse and Hill (1987) and Sachse et al. (1987). Path length: 17 m; total integration time 13 s; precision: 1 ppbv.

Minimum detectable absorbances can be improved by increasing the integration time of the measurements, i.e., detection limits can be traded off for time resolution up to a point. In principle one should be able to gain by the square root of the integration time. In practice a limit is reached because of drifts of laser and detector power and noise. With presently available lasers it is difficult to gain S/N by integrating for more than about 15 min.

Mackay and co-workers (see Slemr et al., 1986; Schiff et al., 1987; Mackay et al., 1991) have defined an operational method for determining the minimum detection limit (MDL). This is done by subtracting two background spectra and using the fitting procedure to compare the difference to the reference spectrum. The last column of Table 5.2 gives MDL values obtained this way during actual field measurement studies of a number of species using the Unisearch models TAMS-150 and EMS-050. Detection limits reported by other groups are also included for certain molecules. For gases with relatively high concentrations such as CH_4, CO, CO_2, and N_2O, the precision of the instrument is frequently of more interest than the MDL in defining the performance of the system.

5.6.3. Precision and Accuracy

Few pairs of scientific terms are confused more frequently than precision and accuracy. *Precision* is a measure of the range of values obtained from a series of repeated measurements of samples that are believed to have identical concentrations. *Accuracy* is a measure of how close the measurement value is to the "true" value. To follow the trend in concentration of an atmospheric constituent with time at one location using the same instrument, precision is the important consideration. It is also more important than accuracy for measuring gaseous fluxes by the "eddy correlation" technique. On the other hand, accuracy is the important criterion for determining whether certain emissions meet regulatory standards or when one is comparing measurements made at different locations or with different instruments.

Precision depends on how closely the instrument is operating to its detection limits and whether the background signal remains constant during the signal integration period. Precision is also affected by the dynamic range of the measurement. For example, if one attempts to measure ratios of $^{13}C/^{12}C$ in CH_4 with an accuracy of 0.1 part per thousand, a precision of better than 1 part in 10^6 would be necessary—a nontrivial task. Similarly eddy correlation flux measurements require measurements of changes in concentrations of a few tenths of a ppbv in ambient concentrations that may be in the ppm range for CO_2 and CH_4.

In addition to the factors that govern precision, overall accuracy is also determined by the accuracy with which sampled air and calibration flow rates are measured and especially by the accuracy with which the concentrations of the calibration gases are known. Bottled gas sources are available that are certified to be accurate to better than 2%, which permits total accuracies of about 15% to be achieved under field measuring conditions.

5.7. A TYPICAL TDLAS SYSTEM FOR ATMOSPHERIC MEASUREMENTS

As an example of a complete TDLAS atmospheric measurement system, we now describe the compact Unisearch model EMS-012 system (Mackay et al., 1991; Schiff et al., 1990, 1991; Nadler et al., 1993) designed to operate from an aircraft, on a meteorological tower, or from a mobile laboratory. It is a miniaturized version of the company's TAMS-150 system reported in earlier publications (Schiff et al., 1987; Harris et al., 1989b; Mackay et al., 1988, 1990). A photograph of the EMS-012 system is shown in Figure 5.8. The entire system can be contained in a single 19-in. rack with a total weight of ~ 200 kg. For aircraft applications the modular sections can readily be rearranged to best fit the available space. Power consumption can be under 1000 W at 50 or 60 Hz for a typical configuration consisting of 100 W for the cryocooler, 200–300 W for the electronics, and 500–1000 W for the vacuum pumps depending on the response time requirements. The entire system is ruggedized for field operation. Optical elements are motorized for automated alignment, and the operating and data acquisition systems are under computer control. This permits automatic operation once the initial conditions are selected from the software menu.

Figure 5.9 is a schematic diagram of the optical system showing a single laser beam. Two lasers can be housed in the miniature, integral Stirling cryocooler and laser cold-head assembly. The temperature of the lasers is maintained in the 65–120 K range with a stability of 10^{-3} K by means of the closed-cycle helium cryocooler, a heater, and a servo-temperature controller. One diode may be operated at a time, and multispecies monitoring capability is provided through the application of the "Scan-and-Fit" software described below. The lasers operate in CW mode, and the current is modulated at 50 kHz to permit phase sensitive detection techniques to be employed as described in Section 5.4.2.

The laser radiation from the selected diode is collected and collimated by an off-axis parabolic mirror, OAP1. The plane-fold mirror, M1, directs the beam onto a second off-axis parabolic mirror, OAP2, which focuses the beam into the absorption cell and finally onto the reference detector. Prior to striking OAP2, the beam is chopped mechanically to provide total laser power measurements. The beam splitter, BS, permits half the beam to be directed by plane mirrors M2 and M3 into the absorption cell. The other half of the beam is diverted through a reference cell containing a high concentration of the

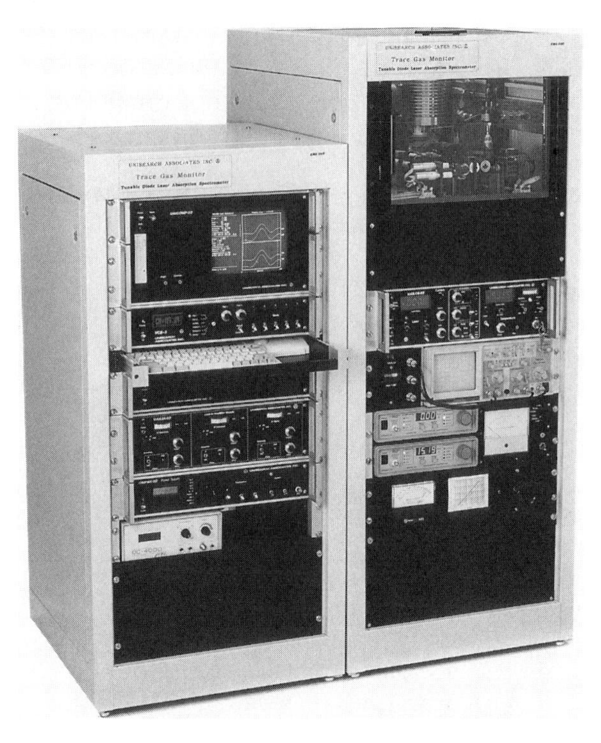

Figure 5.8. Photograph of the EMS-012 tunable diode laser absorption spectrometer designed for trace gas monitoring, assembled in the dual rack configuration.

target gas(es) and onto the reference detector. The signal from this detector is used for long-term frequency stabilization through temperature line locking (see Section 5.5.2).

The stages on which OAP1 and M1 are mounted are motorized in the x direction to select the desired laser and in the y direction to collimate the beam. The horizontal and azimuthal directions of the steering mirrors, M2 and M3, are also motorized to automatically select and optimize the optical path. Primary alignment is accomplished by manually rotating OAP1 180° to intercept the emission from a visible diode laser (VDL) that is maximized on the two detectors by manually adjusting the collimating stage and mirrors M1–M4. OAP1 is then rotated back to the beam from the selected laser. The

Figure 5.9. Schematic diagram of the optical layout and block diagram of the control and electronics system of the EMS-012 tunable diode laser absorption spectrometer system designed for trace gas monitoring. Key: OAP1 & OAP2, off-axis parabolic mirrors; M1–M4, beam directing mirrors; LCS, LSA, laser assembly; *Reference Cell*, cell used for line locking; *Chopper*, mechanical chopper; VDL, visible diode laser for optical alignment; *, diode laser; BS, beam-splitter; DAS, digital acquisition system.

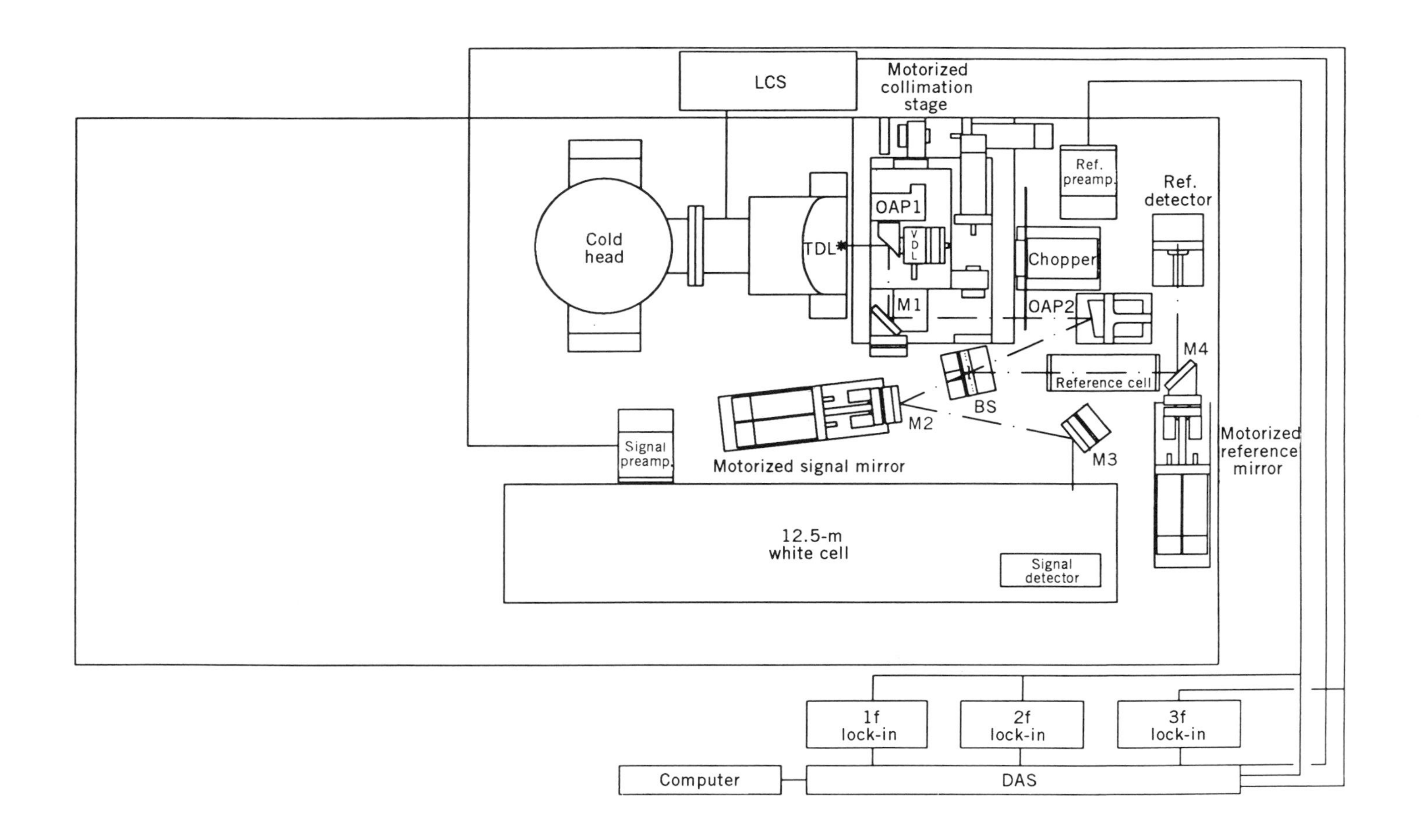
LCS
Motorized collimation stage
Cold head
TDL
OAP1
VDL
M1
Ref. preamp.
Ref. detector
Chopper
OAP2
M4
Reference cell
BS
M2
M3
Motorized signal mirror
Motorized reference mirror
Signal preamp.
12.5-m white cell
Signal detector
1f lock-in
2f lock-in
3f lock-in
Computer
DAS

power incident on the reference Hg/Cd/Te detector is maximized by computer-controlled adjustment of the collimation stage and mirror M4 in a sequential manner using a second degree simplex routine. A similar routine applied to the motorized stage M2 maximizes the laser power at the signal detector after the beam has passed through the White cell. The detectors may be cooled either thermoelectrically or by liquid N_2 depending on the required detection limits and on the spectral region employed.

The time response of any TDLAS system is usually limited by the residence time of the sampled air in the multipass cell. To minimize this time, the system is provided with two low-volume cell options. For applications requiring moderate sensitivity a White cell with a 25-cm base path, a 12-m total path length, and a volume of 0.6 L is available. This system can provide measurements of most atmospheric species with detection limits in the 1–10 ppbv range. For measurements requiring more sensitivity a modified Horn–Pimentel design cell with a 65-cm base path, a 50-m total path length and a volume of 1.4 L is available. This cell has a double corner-cube reflector to provide four times the number of reflections in the same volume as a standard White cell. A total path length of 50 m is obtained with this cell, about a third of the path length of the Unisearch TAMS-150 cell but with less than a twentieth of the volume. Detection limits are in the 0.2–2 ppbv range. Other configurations are available as special orders.

Sample air is drawn through 6.35 mm OD, 1.2 mm wall, PFA-type Teflon tubing. Particles are removed from the air by a 2 μm pore size Teflon filter located at the tubing entrance. A Teflon needle valve immediately downstream of the particle filter maintains a constant flow. With a moderate size rotary vacuum pump (e.g.350 L/min) the residence time of the gas in the cell can be reduced below 200 ms. The air exiting the absorption cell is monitored with a calibrated mass flowmeter. A motorized valve, which is referenced to an MKS baratron pressure gauge, automatically controls the pressure in the absorption cell at the desired pressure (usually in the range of 10–100 torr).

The operation of the system and the data acquisition system is under computer control. The laser current is scanned over a few tens of milliamperes by a 4096 step ramp generated by the PC (personal computer). This corresponds to a spectral scan of $< 0.5\ cm^{-1}$ in a time of ~ 20 ms. The amplitude of the ramp is chosen so that any window of 128 channels will contain a preselected absorption feature. The laser current is adjusted so that the first window is situated in the center of the ramp. Additional windows can then be selected at either shorter or longer wavelengths from the first. In this way, a single laser can be used to monitor several species simultaneously, as long as the absorption features lie within the $0.5\ cm^{-1}$ tunability range of currently available diodes. This procedure is a variation of the jump scan technique developed by Fried et al. (1991a,b).

The laser is modulated at 50 kHz, and the $2f$ signal measured at 100 kHz. The output from the signal and reference detectors are monitored simultaneously. The pressure in the reference cell is adjusted to be close to that in the absorption cell so that the line shape of the two signals will be similar. The chopper is operated at ~ 80 Hz so that the power on both the reference and signal detectors can be measured and normalized. The signals are corrected for the power incident on each detector and the ambient signal, measured in each 128-channel window, is least square fitted, point by point, to the corresponding reference signal. This provides a mixing ratio value, as determined by the chopper frequency, every 20 ms. The operator can choose to average these 20 ms spectra over a time period from 0.1 to 1000 s. The average of the measured mixing ratio can be presented to an analog output for external acquisition (e.g., chart recorder) and stored, along with the spectral trace of the average signal, in digital form on an internal hard drive for future manipulation.

The absorption feature(s) and the operating parameters such as calibration mixing ratios, calibration frequency, and averaging times can be selected from the computer menu. The system can then operate automatically under computer control. The computer commands valves to introduce a known mixing ratio of the target gas at the inlet to the sampling system. Frequency of calibration is determined by the operator depending on factors such as ambient signal variation and stablity of the optical system. Periodically, when the computer determines that the alignment of the system has degraded beyond prescribed limits, the simplex routine is automatically invoked to realign the optics. A new calibration is then performed to ensure optimum accuracy of the measurements.

5.8. MEASUREMENTS WITH TDLAS SYSTEMS

The unique characteristics of TDLAS have been exploited to measure many gases in the atmosphere over a variety of environmental conditions. Its high specificity has permitted positive identification and unequivocal measurements of components in complex gas mixtures as well as providing an ideal reference standard against which other, less definitive techniques can be compared. Its high resolution makes real-time isotope measurements possible. Its high sensitivity permits measurements of species in pristine environments or identification of trace constituents in more polluted air. Advantage has been taken of the rapid time response of these systems to make eddy correlation flux measurements and to get spatial resolution from fast-moving aircraft. TDLAS measurements have been made in the very clean troposphere and stratosphere, in polluted urban and rural air, in smog chambers, and at landfills, industrial sites, and indoor workplaces, as well as in studies of automobile exhausts.

Systems have been deployed on mobile vans, ships, aircraft, balloons, and even rockets. The following sections describe some examples of these applications.

5.8.1. Ground-Based Measurements of Ambient Air Concentrations

In this subsection we describe some of the measurements made during field campaigns. Some of these campaigns were designed to measure the composition of atmospheric species, others were undertaken to make comparisons between measurement techniques, while still others combined both objectives. Species that have been measured by TDLAS include O_3, N_2O, NO, NO_2, HNO_3, CO, CO_2, H_2O_2, HCHO, NH_3, CH_4, C_2H_4, C_2H_6, HCl, HF, SiF_6, CF_2Cl_2, H_2S, OCS, SO_2, CS_2, and benzene with detection limits as low as 10 pptv within 5-min integration times.

One of the earliest instrument comparisons was made (Walega et al., 1984) between a TDLAS system and a NO chemiluminescence (CL) detector equipped with convertors for NO_x and HNO_3. Good agreement was found between the two instruments for measurements of NO in ambient, captive, and synthetic air to which a known amount of NO was added. The CL instrument also agreed with TDLAS measurements for synthetic mixtures of NO_2 but gave higher readings for ambient and captive air, which indicated that the convertor in the CL instrument was also converting other substances, possibly peroxy acetyl nitrate (PAN), to NO. Reasonable agreement between the techniques was obtained for HNO_3 measurements in captive air but not for Los Angeles ambient air for which the CL method gave erratic results.

A comparison of measurement techniques for NO_2 was conducted by the Atmospheric Environment Service of Canada in rural Ontario (Mackay et al., 1991). Several CL systems were intercompared with the TDLAS measurements. Figures 5.10 and 5.11 show ambient levels of NO_2 measured by the TDLAS system and the TECAN chemiluminescence analyzer at this rural site when it was impacted by relatively clean air from northern Ontario when the airflow came from the more polluted southwest quadrant. Figures 5.12 and 5.13 respectively show the correlation plots of the TDLAS vs. the TECAN and Monitor Labs (ML) CL analyzers. Conversion of NO_2 into NO is accomplished photolytically in the TECAN instrument and by a hot molybdenum convertor in the ML instrument. Basically, both instruments measure NO from its chemiluminescent reaction with O_3. The measured NO signal without converter is subtracted from that with converter to determine NO_2. The results shown in Figs. 5.10 and 5.11 and the small amount of scatter in the correlation plot (Fig. 5.12) show that the agreement between the TDLAS and the TECAN is very good during high- and low-pollution episodes corresponding to a range in NO_2 concentrations. The ML instrument shows poorer agreement, with considerable scatter in the correlation plot (Fig. 5.13) at all

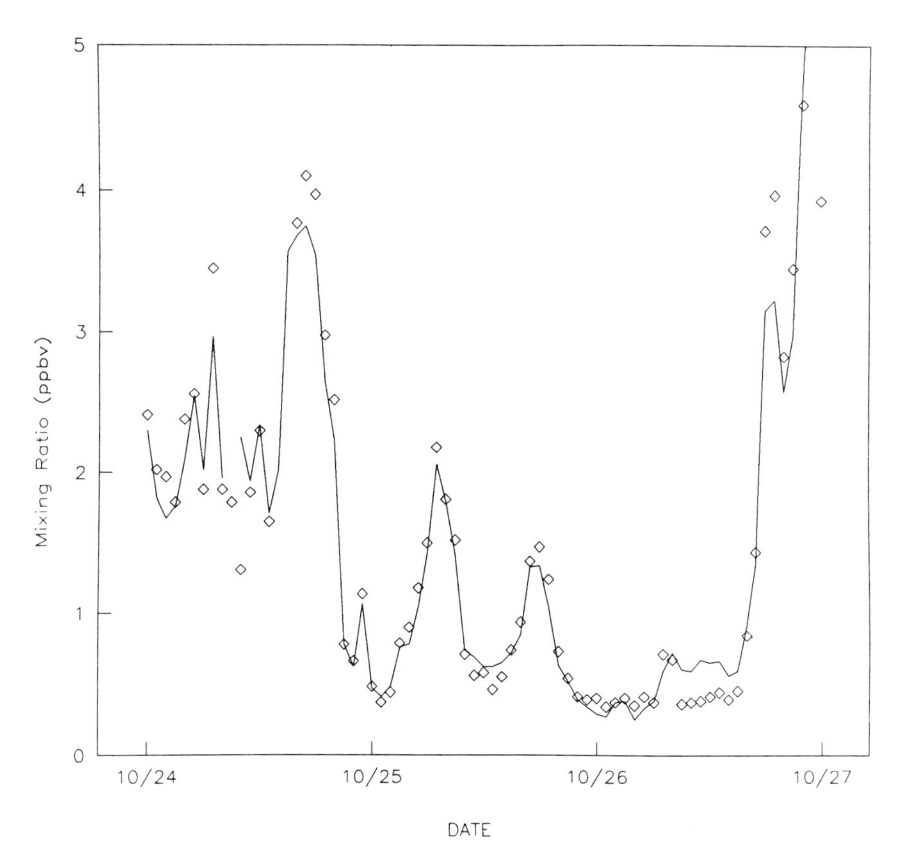

Figure 5.10. NO_2 mixing ratios as measured by the TDLAS (solid line) and TECAN chemiluminescence analyzer (◇) at the Atmospheric Environment Service of Environment Canada, Centre for Atmospheric Research facility at Egbert, Ontario, during the period Oct. 24–26, 1990. The TECAN data were kindly provided by Dr. K. Anlauf, Atmospheric Environment Service.

NO_2 levels. This is probably due to interference from other nitrogen compounds that are also converted to NO by the molybdenum converter.

In 1985, the State of California Air Resources Board sponsored a "Nitrogen Species Methods Comparison Study" at Claremont, in the southern part of that state. TDLAS, DOAS, filter techniques, and a vareity of CL instruments including a Luminox® NO_2 analyzer participated in the intercomparison (Mackay et al., 1988). Figure 5.14 shows good agreement between the DOAS and TDLAS systems. The comparison between TDLAS and the Luminox instrument (shown in Fig. 5.15) is also quite good, except for two nighttime periods in which the Luminox values were much larger than the TDLAS

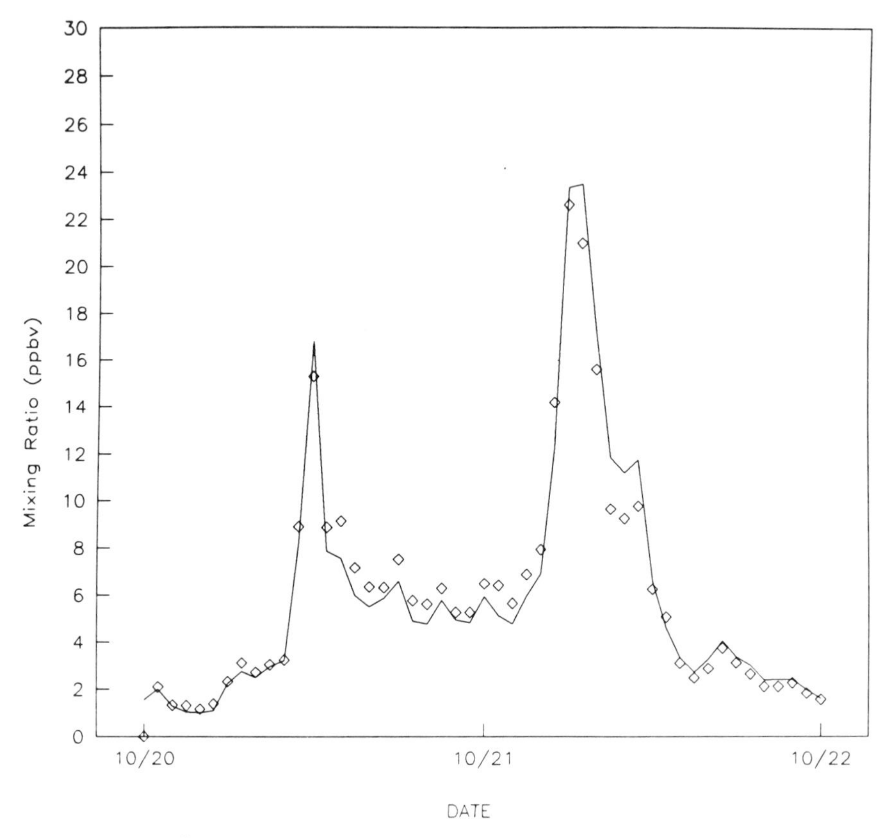

Figure 5.11. NO_2 mixing ratios as measured by the TDLAS (solid line) and TECAN chemiluminescence analyzer (◇) at the Atmospheric Environment Service of Environment Canada, Centre for Atmospheric Research facility at Egbert, Ontario, during the period Oct. 20–21, 1990. The TECAN data were kindly provided by Dr. K. Anlauf, Atmospheric Environment Service.

values. There were considerable discrepancies between some of the other CL analyzers and the TDLAS system.

A number of comparisons have been made of HNO_3 measuring techniques. Three different denuder techniques were compared with a TDLAS system (Fox et al., 1988). Gaseous nitric acid was measured in the effluent stream of a continuous stirred tank reactor in which HNO_3 was generated photochemically in a HC/NO_x mixture. This study showed that a short nylon denuder configuration agreed well with the TDLAS, whereas lower values were obtained with the other denuders.

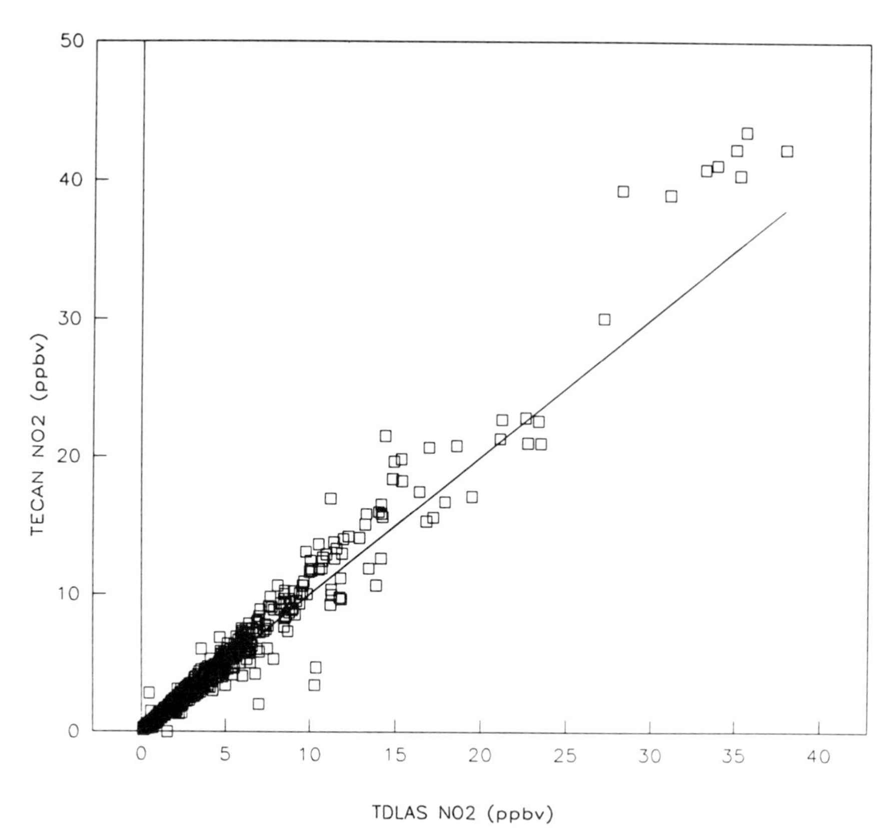

Figure 5.12 Correlations between ambient NO_2 mixing ratio data sets measured by the TDLAS and TECAN Chemiluminescence Analyzer at the Atmospheric Environment Service of Environment Canada, Centre for Atmospheric Research facility at Egbert, Ontario, during the period Oct. 3–Nov. 3, 1990. The solid line represents the regression line; slop = 1.12; intercept = −0.23; R = 0.97. The TECAN data were kindly provided by Dr. K. Anlauf, Atmospheric Environment Service.

A major HNO_3 intercomparison was carried out between 18 instruments as part of the aforementioned California study at Claremont (Hering et al., 1988). Filter packs and a variety of denuder type instruments were compared with TDLAS and FTIR spectroscopic methods. Two co-located TDLAS systems were operated during the study and agreed with one another within 10% at all times. Except during the midday period, the HNO_3 levels were below the FTIR detection limit. For the entire 8-day study period, nitric acid concentrations reported by the different instruments varied by as much as a factor of 4. During the periods when the FTIR was able to measure HNO_3 the

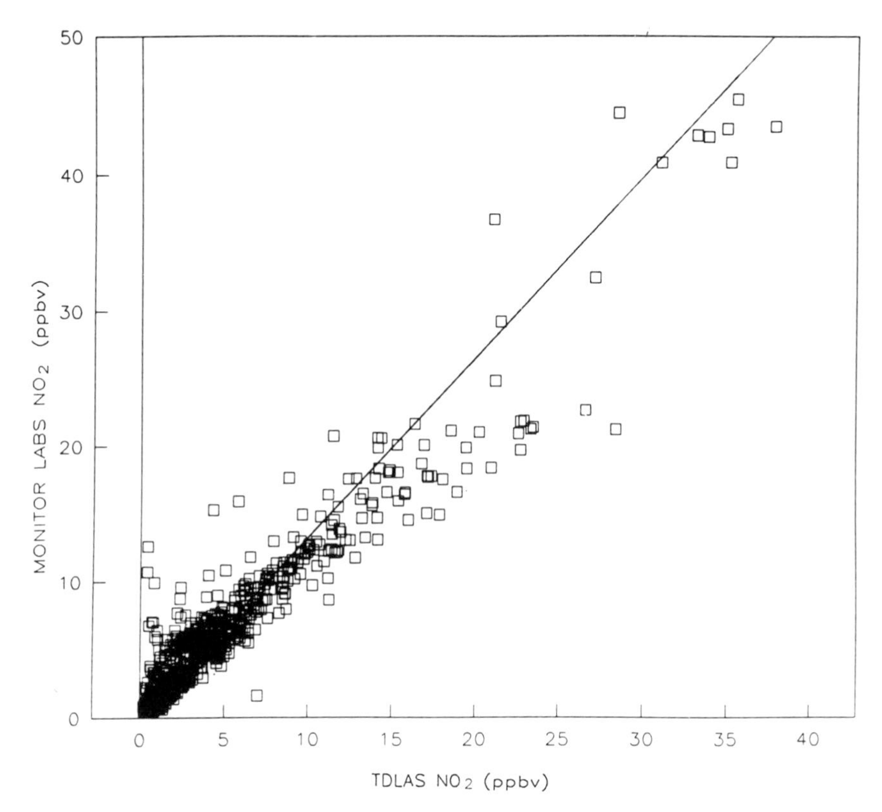

Figure 5.13. Correlations between ambient NO_2 mixing ratios data sets measured by the TDLAS and Monitor Labs chemiluminescence analyzer at the Atmospheric Environment Service of Environment Canada, Centre for Atmospheric Research facility at Egbert, Ontario, during the period Oct. 3–Nov. 3, 1990. The solid line represents the regression line; slope = 1.12; intercept = 0.68; R = 0.91. The Monitor Labs data were kindly provided by Dr. K. Anlauf, Atmospheric Environment Service.

values correlated well but were, on average, 10% higher than the TDLAS values. For example, Fig. 5.16 shows the results obtained by the two techniques on Sept. 14, 1985 (Mackay et al., 1988). Analysis of the comparison of the TDLAS and the filter pack measurements have been published separately (Anlauf et al., 1988).

Formaldehyde (HCHO) has been measured at a wide variety of tropospheric sites by TDLAS. Measurements were reported (Harris et al., 1989b) at four sites in North America with identification of two distinct types of diurnal

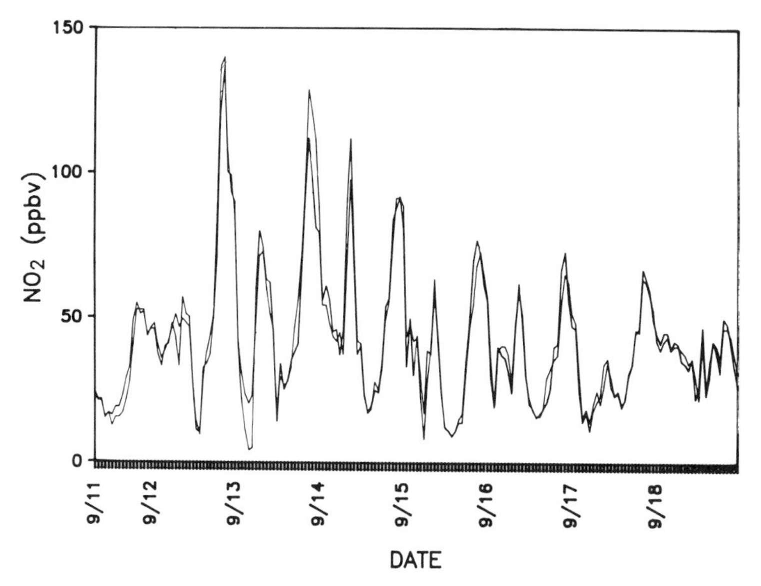

Figure 5.14. Comparison between 1 h average NO_2 mixing ratio measurements made simultaneously with the TDLAS (dark line) and DOAS (light line) instruments at Claremont, California, September 1985. The DOAS data was kindly provided by Dr. A. Winer, State-wide Air Pollution Research Center, University of California, Riverside.

behavior. In the absence of local pollution sources, the HCHO diurnal variation is weak and HCHO is not at night. The authors concluded that the lifetime of HCHO with respect to dry deposition is greater than 50 h at the least-polluted site (the Atlantic Ocean coastal region). At sites downwind of pollution sources, the HCHO maximizes near solar maximum and declines in the afternoon as a result of photochemical formation from primary pollutants. Measurements of HCHO made in southern California at Claremont, during an unseasonable wet and cloudy 8-day period in August 1985 (Mackay et al., 1988) gave HCHO mixing ratios between 1–3 ppbv with only a weak diurnal variation.

A more extensive evaluation of techniques for measuring ambient formaldehyde was carried out in the Los Angeles area during the "Carbonaceous Species Methods Comparison Study" funded by the State of California Air Resources Board (Lawson et al., 1990). Measurement methods included 2, 4-dinitrophenyl hydrazine (DNPH)-impregnated cartridges, an enzymatic

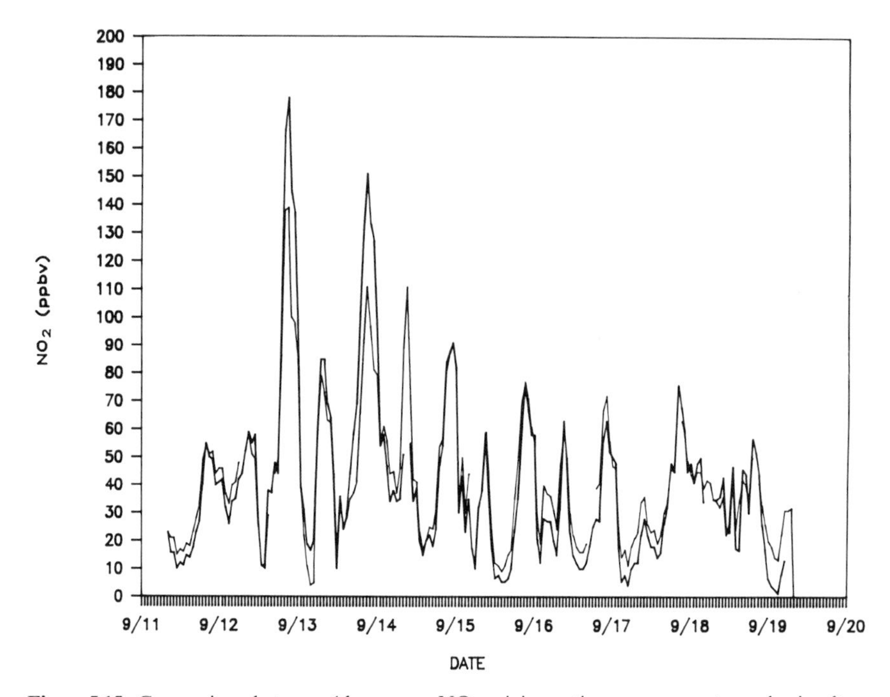

Figure 5.15. Comparison between 1 h average NO_2 mixing ratio measurements made simultaneously with the TDLAS (light line) and Luminox (dark line) instruments at Claremont, CA, during September 1985. From Schiff et al. (1987).

technique, a diffusion scrubber, FTIR, differential optical absorption spectroscopy (DOAS), and TDLAS.

The enzymatic technique and diffusion scrubber reported concentrations some 25% higher and 25% lower than the spectroscopic mean, respectively, for the entire study period. The DNPH cartridges, the only routine monitoring method in the study, yielded values 15–20% lower than the spectroscopic mean, with somewhat lower values over longer sampling periods.

Figure 5.17 shows results obtained by the three spectroscopic methods over the entire 8-day measurement period. The agreement is good (within 15% of the mean of the three measurements), with the exception of the DOAS measurements at night when the concentration was at or below the detection limits of that system.

An interesting feature of these measurements is the occurrence of two maxima that were evident each day. The peak in the early afternoon is readily understood as correlating with photochemical HCHO formation. The early

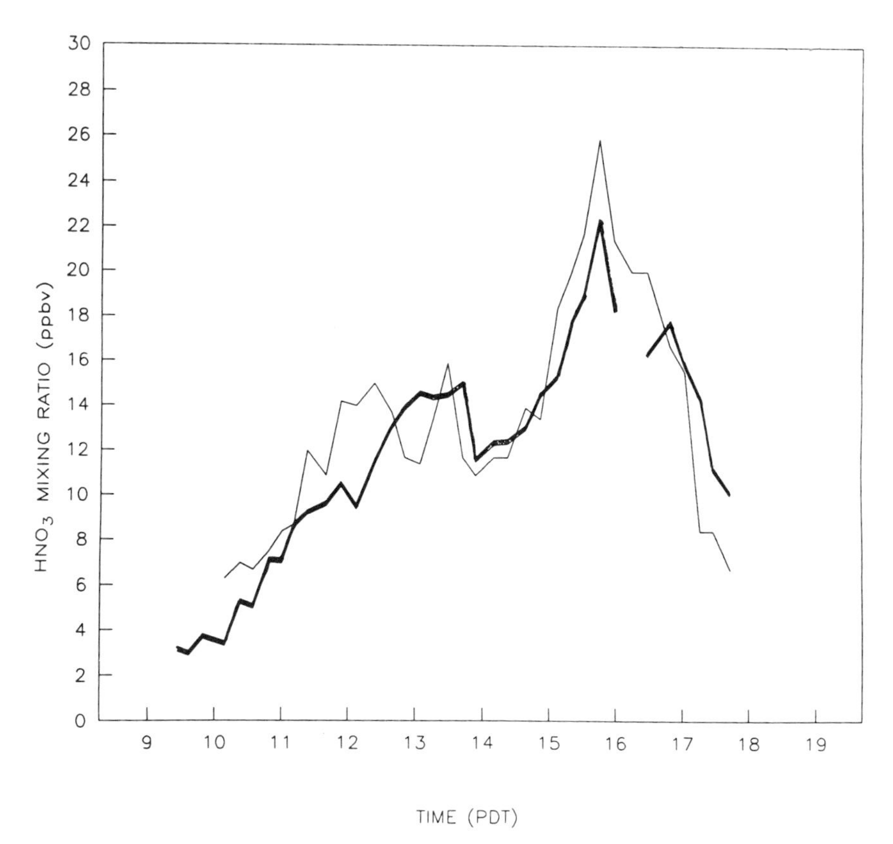

Figure 5.16. Comparison of the 5 min. average HNO_3 mixing ratios measured by the TDLAS system (heavy line) and the corresponding value obtained by the FTIR (light line) on Sep. 14, 1985, at Claremont, California (PDT = Pacific Daylight Time). The TDLAS data points have been corrected to account for a 15% error in the calibration technique. From Mackay et al. (1988).

morning peak correlates with maxima in particle carbon and provides convincing evidence for the direct emission of HCHO from automobile exhaust. Furthermore, these data suggest that primary formaldehyde may be a greater contributor to total formaldehyde levels in California's South Coast Air Basin than previously believed (Lawson et al., 1990).

A laboratory intercomparison of formaldehyde measurement techniques was carried out at ambient concentrations between a TDLAS system and four techniques based on derivatization followed by fluorometric analysis or by high-pressure liquid chromatography (HPLC) with detection by ultraviolet

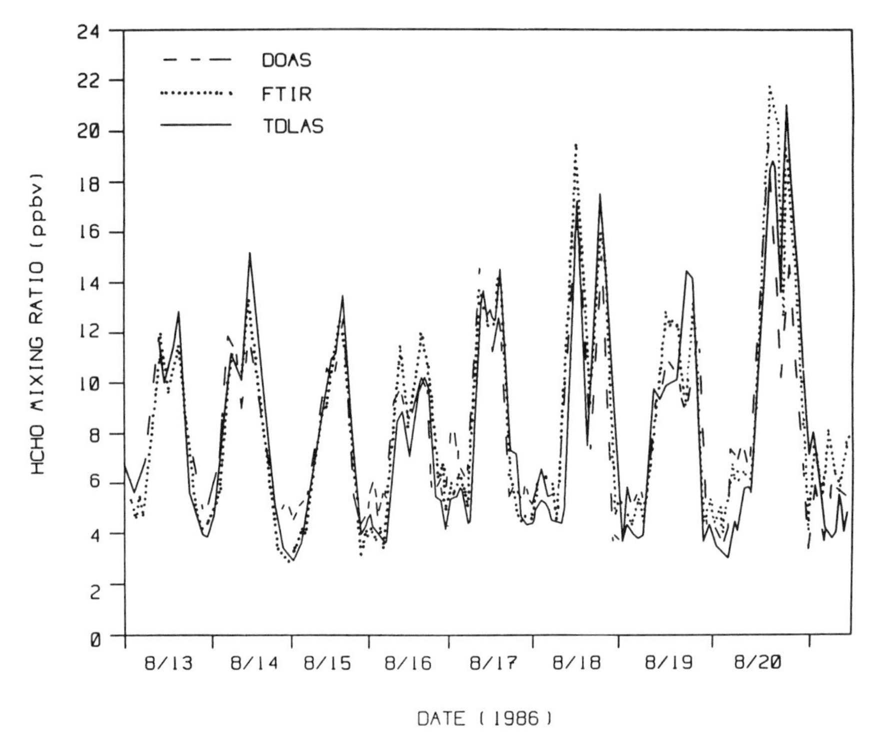

Figure 5.17. Time series plot for formaldehyde concentrations by differential optical absorption spectroscopy (DOAS), Fourier transform infrared spectroscopy (FTIR), and tunable diode laser absorption spectroscopy (TDLAS) at Glendora, California, during August 1986. From Lawson et al. (1990).

(UV) absorption (Kleindienst et al., 1988a). HCHO was generated by bubbling zero air through a solution of aqueous HCHO. The techniques agreed within 15% and 30%. None of the techniques showed interferences from O_3, NO_2, SO_2, or H_2O_2. Subsequently a comparison was made of ambient rural North Carolina air measurements between the TDLAS and a continuous scrubbing enzymatic–fluorometric detection technique, the latter yielding values some 15% higher than the TDLAS values. The results of 36-h measurement period are shown in Fig. 5.18.

HCHO measurements were also performed in a smog chamber of the University of North Carolina. Agreement between the enzymatic and TDLAS techniques was good for photochemically generated HCHO. However, when direct injection of HCHO (paraformaldehyde heated to vaporize) was em-

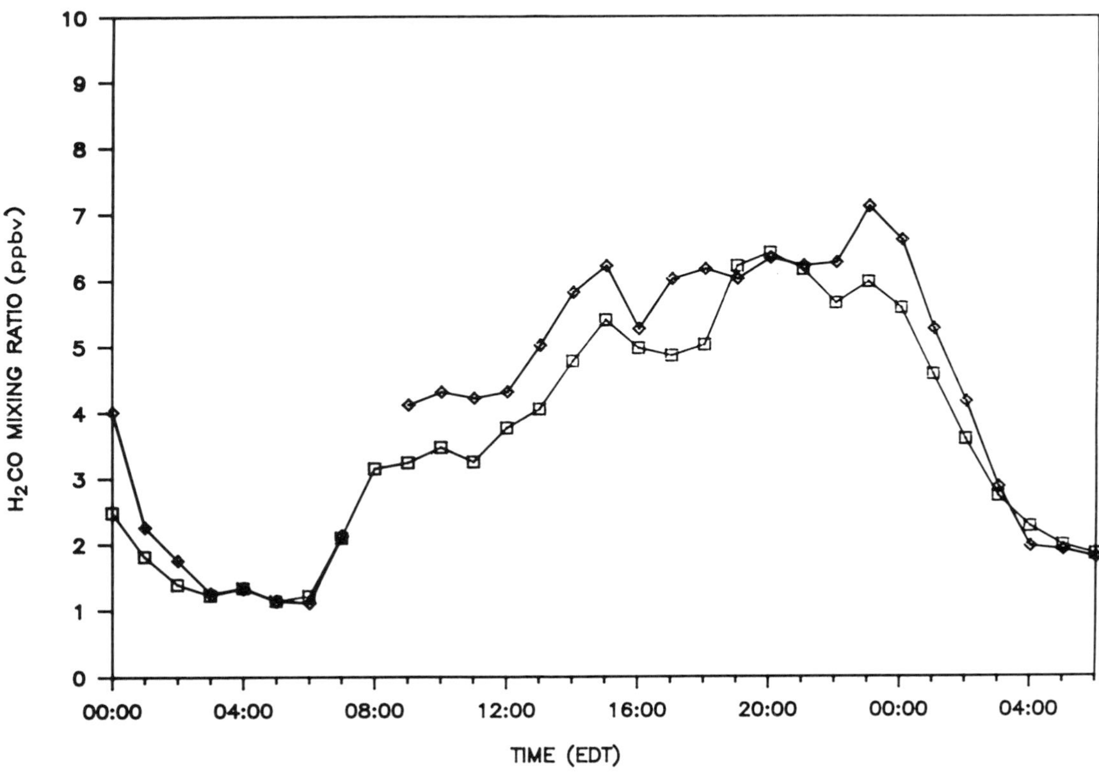

Figure 5.18. Ambient HCHO concentrations measured by TDLAS (□) and the continuous scrubbing fluorometric detection technique (◇) during the period June 24–26, 1986, at Research Triangle Park North Carolina (EDT = Eastern Daylight Time). From Kleindienst et al. (1988a).

ployed to generate mixtures in the same range as those provided by the photochemistry, the enzymatic technique always reported significantly higher values than those of TDLAS, by as much as a factor of 2. A possible reason for the discrepancy is the detection of undissociated paraformaldehyde by the enzymatic technique whereas the TDLAS will only measure the monomer. This is a good example of the specificity of the TDLAS technique and its use as a comparison standard.

Furthermore, the added HCHO was found to disappear rather rapidly after injection, which indicated that it was depositing rapidly on the walls of the chamber. This could have considerable significance in the efforts to model the chemistry occurring in the chamber.

Measurement of hydrogen peroxide (H_2O_2) by TDLAS was first reported (Slemr et al., 1986) for rural air. Five-minute average mixing ratios up to 2.9 ppbv were measured during the summer months of 1984 and 1985, with 1-h averages reaching as high as 1 ppbv. During other periods mixing ratios were below the detection limits of 300 pptv.

Improved sensitivity of 150 pptv for 3-min averages was achieved in a California study in the summer of 1986 (Mackay et al., 1990). Daytime values generally ranged from 250 to 500 pptv with the maxima occurring at midafternoon and falling to below detection limits during the night. During the last 4 days of the mission midafternoon values rose to 1.8 ppbv.

During the winter of 1990 a study (Schiff et al., 1992) was undertaken at the Grand Canyon, Arizona, to determine the effect on visibility of the emissions coming from a coal-fired generation station located some 150 km to the southwest of the canyon. The TDLAS system was chosen to provide high-quality, interference-free H_2O_2 data with good time resolution and at levels expected to be in the 100 pptv range.

Figure 5.19 shows the average diurnal behavior for H_2O_2 determined during the period Jan. 10–Mar. 31, 1990, measured on the rim of the Grand Canyon. The hourly averages were in the 30–600 pptv range for most of the study period. The system detection limit was typically 25 pptv, so that the peroxide levels were usually above the detection limits. The diurnal behavior was opposite to what had been observed at other sites with the maxima occurring during the night and the minima during the mornings. This has been interpreted as the canyon acting as a large lung. Subsidence during the night of air into the canyon brought relatively high-concentration air past the instrument located at the top of the canyon. In the morning the reverse occurred, with air that had been depleted of H_2O_2 by dry deposition during the night being convected up from the canyon past the sampling inlet.

An H_2O_2 instrument intercomparison has been reported (Kleindienst et al., 1988b) between TDLAS and techniques involving fluorescence from an enzymatically produced complex, and from the chemiluminescent reaction with

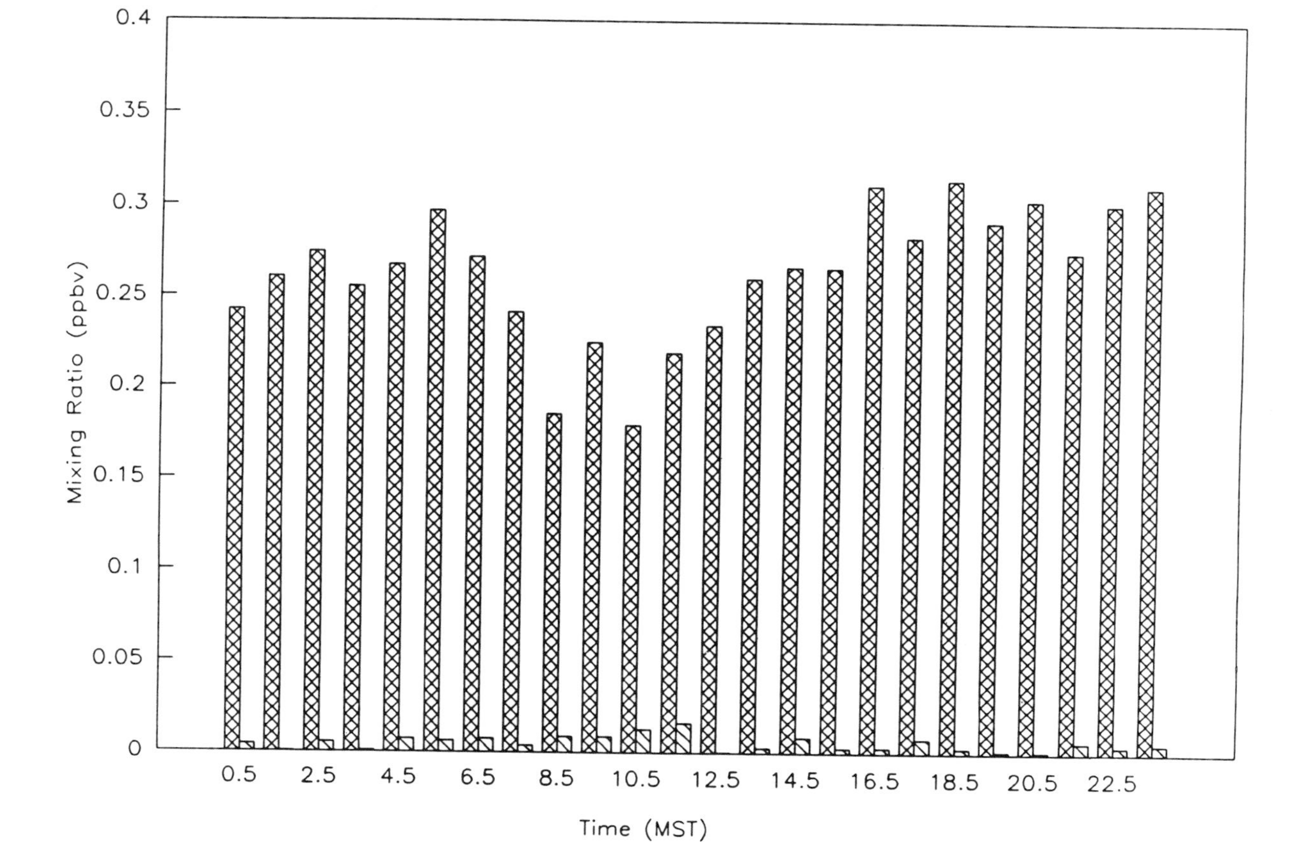

Figure 5.19. Average diurnal behavior for H_2O_2 determined by TDLAS during the period Jan. 10–Mar. 31, 1990, at Desert View, Arizona (MST = Mountain Standard Time). From Schiff et al. (1992).

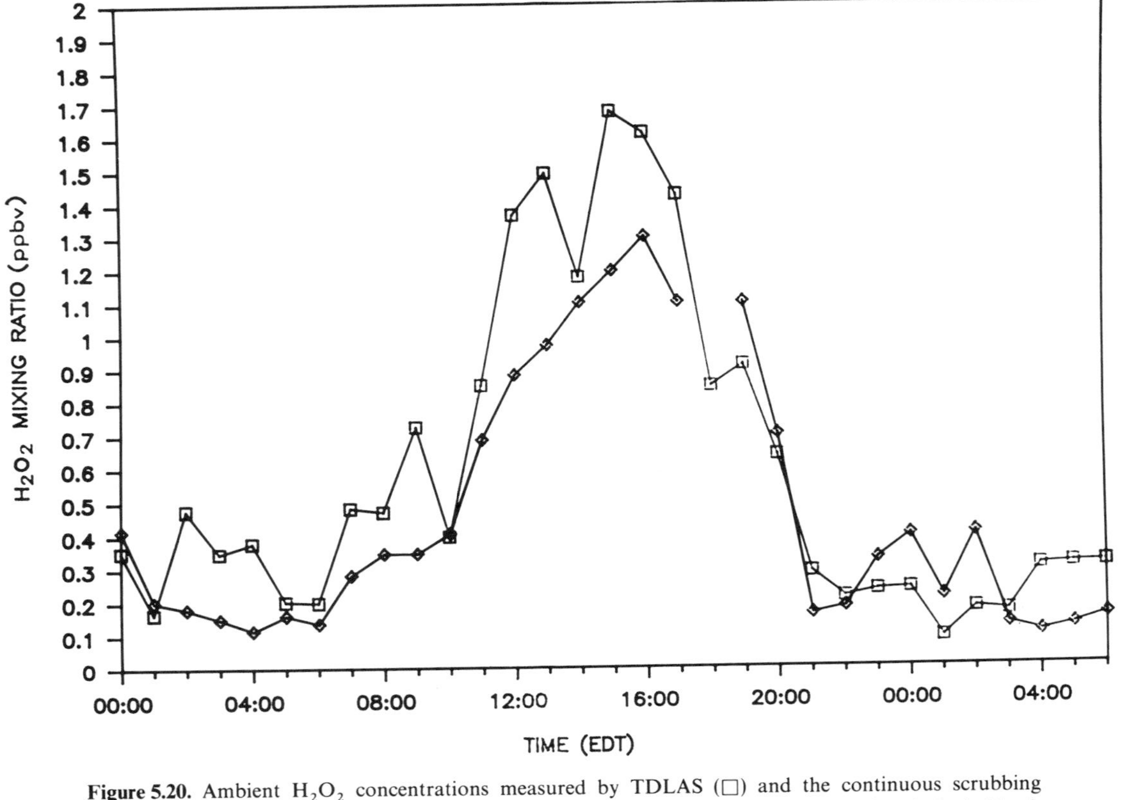

Figure 5.20. Ambient H_2O_2 concentrations measured by TDLAS (□) and the continuous scrubbing fluorometric detection technique (◇) during the period June 24–26, 1986, at Research Triangle Park, North Carolina. From Kleindienst et al. (1988b).

luminol. For pure samples in zero air, in the presence of common interferences, agreement between the three techniques of 14–23% was achieved when compared to standard values. The CL method suffered from negative interference due to SO_2. Agreement was much worse in irradiated HC/NO_x mixtures, where the enzymatic–fluorometric method suffered from interference from organic peroxides. For measurements of rural North Carolina air, agreement within 30% was obtained between the enzymatic–fluorescence technique and the TDLAS, with the enzymatic results generally lower than those of the TDLAS (see Fig. 5.20). The study concluded that TDLAS can be used as an interference-free method to identify interfering species and that care must be taken when one is interpreting the results obtained from the other techniques.

Simultaneous TDLAS measurements of H_2O_2 and HCHO were made during the Mauna Loa Observatory Photochemistry Experiment II (MLOPEX II) fall and winter intensives (Mackay et al., 1993a). The measurements were made 3.4 km above sea level on the side of Mt. Mauna Loa, Hawaii, a site characterized by strong diurnal upslope and downslope winds. Moreover, air reaching this site generally has not been in contact with continental land masses for at least 5 days. Thus, during downslope conditions, air at the site is representative of the free troposphere at altitudes between 3 and 4 km over much of the northern Pacific. Measurements under these conditions therefore provide a unique opportunity for developing and testing models of the remote free troposphere.

The half-hourly average mixing ratios varied between 1300 pptv and the instrument detection limit of 50–100 pptv for H_2O_2 and between 400 pptv and the instrument detection limit ($\sim$ 25 pptv) for HCHO. Strong diurnal behavior shown in Fig. 5.21 was observed for HCHO with maxima in the 12:00–17:00 time period, whereas Fig. 5.22 shows essentially no diurnal variation for H_2O_2. Clean tropospheric air sampled during downslope conditions between 20:00 and 06:00 had HCHO mixing ratios averaging 80 pptv in the fall and 160 pptv in the winter. These values are considerably lower than model predictions which are shown in Table 5.3, strongly suggesting inadequacies in the model.

Figure 5.22 shows a comparison of the average diurnal measurements of H_2O_2 obtained with the TDLAS system and those made with the enzyme fluorescence technique of NCAR (National Center for Atmospheric Research, Boulder, CO), (Heikes et al., 1987). The general behavior measured by the two methods correlated very well. However, differences in the magnitude of the measured ambient H_2O_2 mixing ratio on the order of 20% were observed with the enzyme technique higher than the TDLAS. These discrepancies were entirely attributed to difference in calibration standards.

During this study, comparisons were made between HCHO measurements made with the TDLAS and three other methods—an immobilized enzyme

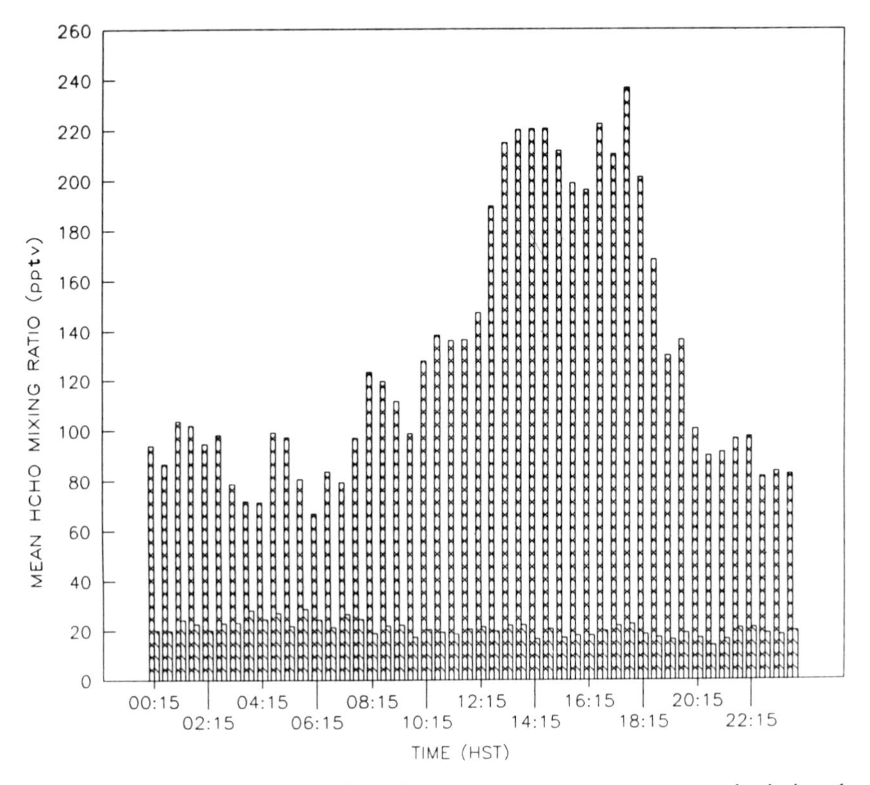

Figure 5.21. Tunable diode laser absorption spectrometer measurements made during the Mauna Loa Observatory Photochemical Experiment II Mt. Mauna Loa, Hawaii: $\frac{1}{2}$-h average HCHO mixing ratios for the period 15:00 Oct. 5 through 07:30 Oct. 23, 1991 (HST = Hawaiian Standard Time). The lower bar graph indicates the 95% confidence limits of the measurements. From Mackay et al. (1993a).

system, a flowing enzyme system, and a continuous DNPH method. Preliminary results from this intercomparison indicated a deviation by a factor of 8 between the highest and lowest with the TDLAS values lying in the lower middle range.

A system using a combination of sweep integration, $2f$ modulation, and a jump scanning technique has been described (Fried et al., 1991a, b) that measures OCS (carbonyl sulfide) with precisions in the range of 0.3–1% for a detection bandwidths of 1.7 Hz. Preliminary measurements of ambient air with this system were in the 0.5 ppbv range. The system was also used (Fried et al., 1991c) in an intercomparison study with a gas filter correlation (GFC) method for CO in the 100–1500 ppbv concentration range. Precisions were 4% for the TDLAS and 10% for the GFC. Measurements by the two systems

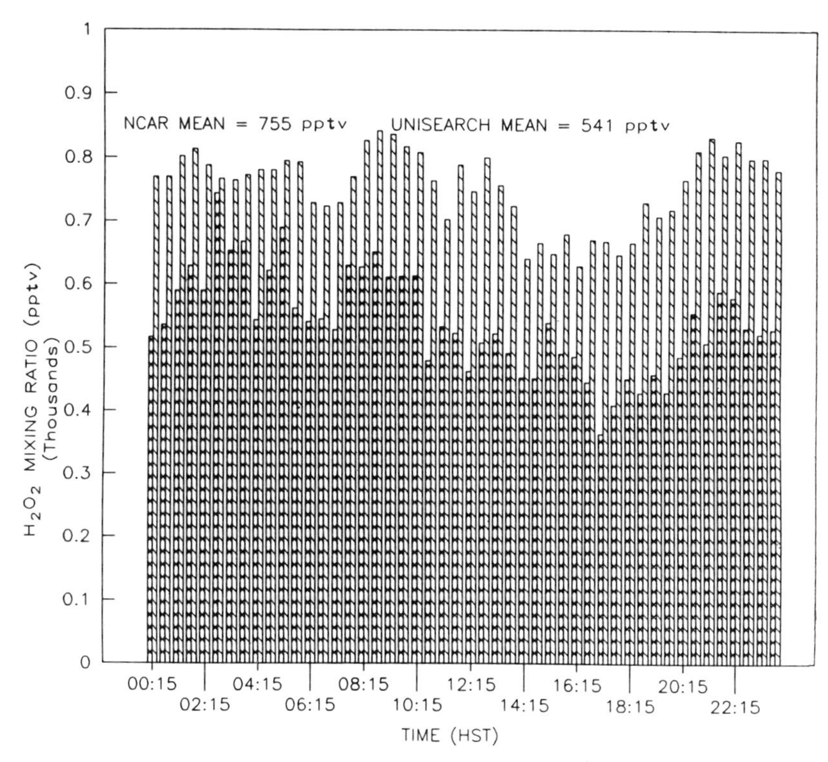

Figure 5.22. Diurnal behavior of H_2O_2 measured by tunable diode laser absorption spectrometer measurements (lower) and a continuous scrubbing fluorometric detection technique (upper) made during the Mauna Loa Observatory Photochemical Experiment II, Mt. Mauna Loa, Hawaii: $\frac{1}{2}$-h average H_2O_2 mixing ratios for the period 15.00 Oct. 5 through 07.30 Oct. 23, 1991. From Mackay et al. (1993a).

Table 5.3. Mixing Ratios Calculated from Photochemical Equilibrium Conditions vs. Observed Values

	Calculated[a]		Observed (Standard Deviation)		
	Without Heterogeneous Loss	With Heterogeneous Loss	May 1988[a]	Oct. 1991	Feb. 1992
HCHO	300	255	105 (42)	80 (62)	160 (75)
			UP SLOPE[b]	192 (99)	228 (117)
H_2O_2	2200	1400	1053 (265)	526 (294)	570 (242)
			UP SLOPE[b]	484 (270)	469 (170)

[a] Data from Liu et al. (1992).

[b] Data referring to periods with up slope wind direction, i.e., wind ascending from the valley.

correlated well, with the TDLAS giving values some 6% lower. The GFC system was found to suffer from interferences from O_3.

5.8.2. Ground-Based Flux Measurements

Understanding the strengths of the sources of emissions and sinks of natural and anthropogenic gases requires accurate flux measurements. The preferred method for such measurements is the so-called eddy correlation technique (Hicks and McMillen, 1988; Hicks et al., 1989; Luebken et al., 1991; Verma et al., 1992). It is based on the correlation of fluctuations in the concentrations of the measured species with fluctuations in the vertical wind speed. The time scale for these fluctuations ranges from a tenth of a second to tens of minutes, requiring instruments with response times approaching 10 Hz.

The TDLAS system described in Section 5.7 has been used to measure CH_4 concentrations with a precision of better than 0.5% and a response time better than 0.1 s. It has been used successfully to measure CH_4 fluxes of 120–270 $mg \cdot m^{-2} \cdot day^{-1}$ from a meteorological tower located in a bog area in northern Minnesota (Verma et al., 1992). Based on this study, noise sources were identified and eliminated, resulting in improved performance in a 1992 study.

The effect of water vapor broadening on CH_4 flux measurements has been investigated (Luebken et al., 1991). A high-resolution TDLAS system was used to measure the broadening effect of water vapor on three CH_4 lines in the ν_4 fundamental. Over the range of H_2O variations expected in the atmosphere the effect on the CH_4 flux range from 3% for the least broadened line to 10% for the most broadened line. The authors recommend drying the sampled air as the most effective means of eliminating this effect.

An open-path tunable diode laser system for flux measurements has been reported (Anderson and Zahniser, 1991). The instrument combines a TDL with a 0.6-m baselength, open-path Herriott cell. The system operates at atmospheric pressure, which minimizes sampling errors and provides a fast time response on the order of 0.1 s. The trace gas sensing volume can be made coincident with a sonic anemometer, thus improving correlation with wind fluctuation measurements. The system was tested for CH_4 and O_3 (two gases whose pressure broadening is small enough to permit atmospheric pressure detection) fluxes in a marsh area. The detection limit of the system was reported to be 2 ppbv in 100 ms averaging time.

Stanton and co-workers (see Stanton and Hovde, 1992; Hovde and Stanton, 1993; Meyers et al., 1993) describe the first application of NIR DFB laser diodes for flux measurements. They used a GaAsP diode emitting near 1.65 μm, corresponding to the $2\nu_3$ band of CH_4, combined with 5 MHz wavelength modulation. An open-path (50 m total) Herriott cell was used.

Sensitivities in the ppmv range were sufficient to make measurement of both CH_4 and CO_2 in a landfill site.

5.8.3. Concentration and Flux Measurements from Aircraft

An airborne TDLAS system for the NASA Global Tropospheric Experiment (GTE) has been described (Hoell et al., 1984, 1985, 1987; Sachse and Hill, 1987; Sachse et al., 1987, 1991). The instrument evolved from one initially capable of high-precision measurements of CO with a several-second time response to a dual gas sensor for CO and CH_4 with two modes of operation: one capable of precisions of 0.1% for CH_4 and 1% for CO with response times of 5 s (corresponding to a horizontal resolution of 0.5 km), and a second mode suitable for eddy correlation flux measurements with response times of 60 ms made possible by pumping the air through the multipass cell at a flow rate of 350 L/min provided by a venturi pump operating off the aircraft engines. The White cell has a total path length of 20 m, and multiplexing was achieved with a two-color detector.

This system was first used for ground-based (Hoell et al., 1984, 1985) and airborne (Sachse et al., 1987; Hoell et al., 1987) intercomparisons of CO measurements with gas chromatographic (GC) techniques. Later the systems were used to determine the geographic distributions of CO and CH_4, and the high-precision mode was used to study plumes from forest and tundra fires, vertical distributions from convective storms, and long-range transport from urban plumes. Major sources of error included vibrations with resulting misalignment of the optics, temperature variations, and changing water vapor concentrations. Methods used to circumvent these problems are discussed by Sachse et al., (1991).

Eddy correlation flux measurements over tundra and northern wetlands have been made at several altitudes and extrapolated to the ground. The major source of error are noise from the buffeting of the aircraft that correlates with the eddies. Corrections had to be made for water vapor and heat fluxes.

As part of the GTE, NASA undertook an intercomparison of methods for measuring NO_2 and HNO_3 at the sub-ppbv level in very clean air of the boundary layer and free troposphere over the Pacific Ocean (Gregory et al., 1990a,b), and continental United States (Carroll et al., 1990; Ridley et al., 1990). The methods used for the NO_2 intercomparison were TDLAS (Schiff et al., 1990), laser-induced fluorescence (Bradshaw and Davis, 1982), and photolysis–chemiluminescence (Ridley et al., 1987, 1988), and for HNO_3 were TDLAS, a high-volume filter pack (Goldan et al., 1983), and denuder technique using NO chemiluminescence detection (Braman et al., 1982, 1986).

Figure 5.23 shows the comparison between the TDLAS NO_2 measurements and the average values for all three methods, which ranged from below

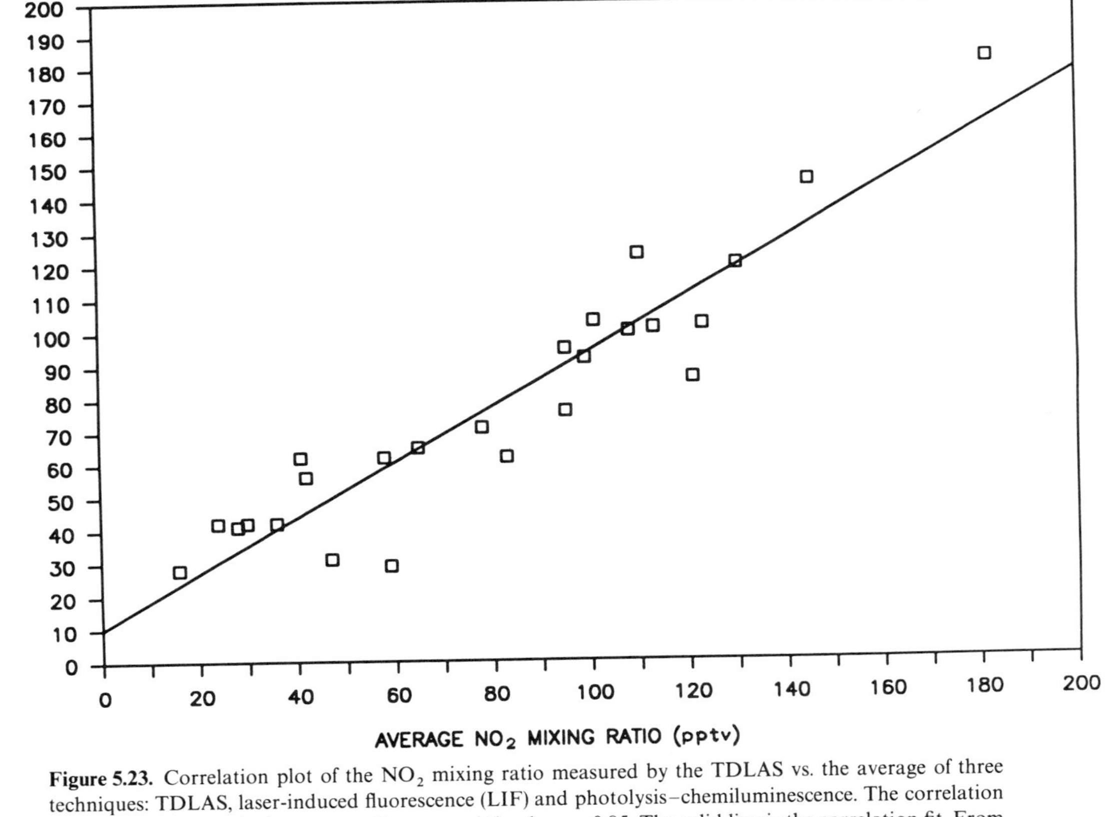

Figure 5.23. Correlation plot of the NO_2 mixing ratio measured by the TDLAS vs. the average of three techniques: TDLAS, laser-induced fluorescence (LIF) and photolysis–chemiluminescence. The correlation coefficient $R = 0.94$; the intercept = 10 pptv; and the slope = 0.85. The solid line is the correlation fit. From Gregory et al. (1990a).

the detection limits ($\sim$ 10 pptv) to $\sim$ 200 pptv. These results demonstrate that the TDLAS is capable of measuring NO_2 at very low concentration in the relatively harsh aircraft environment.

The TDLAS system described in Section 5.7 was used on board a DeHaviland Twin Otter aircraft to measure CH_4 flux over the Hudson Bay Lowland as part of the Northern Wetland Project (Mackay et al., 1991). The results of these first measurements with this instrument are ambiguous owing to the strong mechanical turbulence encountered associated with the low level (50 m) of the flight pattern, which added noise to the high-frequency component of the signal. The longer frequency components gave flux measurements of $\sim 40\,\mathrm{mg \cdot m^{-2} \cdot day^{-1}}$, which are consistent with chamber and TDLAS flux measurements (Edwards et al., 1993) made during this study from a tower in the area. Changes are now being made to the airborne instrument to increase its mechanical stability and in the data acquisition system to improve the sensitivity of the system.

5.8.4. Measurements from Ships

The first application of a TDLAS system on board a ship was the measurement of hydrogen chloride downwind of incinerator ships used to destroy highly chlorinated organic waste (Weitkamp, 1987). Concentration measurements were made at distances up to 10 km from the source with an inlet 20 m above sea level. The results of this investigation were used to assess the environmental impact of using waste incinerators at sea.

TDLAS systems have been used on ships making long scientific cruises (Harris et al., 1989a). Measurements were made of NO_2, HCHO, H_2O_2, and CO in the marine boundary layer over the tropical and equatorial eastern Atlantic Ocean during the period November 1987 through January 1988. Detection limits based on the reproducibility of the background spectra were 25 pptv for NO_2 and 70 pptv for HCHO for 70 s averaging times and 400 pptv for H_2O_2 in 120 s averaging time. The precision for CO ambient measurements was 1–2% obtained for 40 s averaging time. A complete set of measurements required 5 min.

Analysis of the data showed that all trace gases studied with the exception of NO_2 tended to increase between 25 °N and the intertropical convergence zone (ITCZ) because of biomass burning on the African continent during that period. The mean concentration of H_2O_2 in the boundary layer was found to be 2.2 ppbv. HCHO concentrations varied in the range 350–500 pptv, and NO_2 values were less than 50 pptv most of the time. Increases in the concentrations of all the gases except NO_2 were found as a result of biomass buring on the African continent. Strong correlation was found between O_3 and CO north of the ITCZ, with a linear regression slope of 0.07 (70 pptv O_3/1 ppbv CO).

5.8.5. Measurements from Balloons and Rockets

A balloon-borne TDLAS capable of in situ measurement of NO, NO_2, and HNO_3 in the stratosphere has been described by Hastie and Miller 1985). A Cassegrain optical arrangement was used to combine the beams of four laser diodes through a 48-m White cell onto a single detector. Frequency multiplexing was used to separate the signals. Detection limits were 0.5 ppbv for NO and NO_2 and 0.8 ppbv for HNO_3 under stratospheric conditions, with a response time of 3 s. The detection of NO_2 at 13.4 ± 5.2 ppbv at an altitude of 31.4 km in Manitoba, in July 1983, was the first stratospheric TDLAS measurement.

An alternative, open-path approach was developed by the group at the Jet Propulsion Laboratory (Menzies et al., 1983) and has been used to make a variety of excellent stratospheric measurements. The key to the system is a retroreflector suspended 500 m below the balloon gondola, providing an optical path of 1 km. A He-Ne laser co-aligned with the diode laser beam was used with a TV camera and tracking system to keep the beams aligned with the retroreflector during typical pendulum oscillations. Time multiplexing, as described in Section 5.5.3, was used to bring the beams of four diode lasers sequentially into the optical path. The vertical resolution of the measurements is 0.2 km. The sensitivity of the system is in the sub-ppbv level for most gases in the 20–40 km region. Species measured include the following: NO, NO_2, CH_4, H_2O, and CO_2 (Webster and May, 1987); HNO_3, O_3 and N_2O (Webster et al., 1990a); HCl (May and Webster, 1989); and $^{13}CH_4$ and $OC^{18}O$ (Webster and May, 1992).

A rocket-borne TDLAS system designed to measure altitude profiles of trace gases in the mesosphere has been described by Luebken (1991). Two TDLs in a single liquid nitrogen Dewar are used to measure CO_2 and H_2O simultaneously. The White cell is designed to permit the laser beams to traverse outside the shock fronts of the rocket during most of each of the flights planned for the summer of 1994.

5.8.6. Isotope Ratio Measurements

Measurement of stable isotope ratios in atmospheric gases provides a powerful tool for identifying the sources of these gases, whether of biogenic or anthropogenic origin (Wahlen and Yoshinari, 1985; Yoshinari and Wahlen, 1985; Cicerone and Omerland, 1988; Stevens and Engelkemeir, 1988). Very high-precision mass spectrometric (MS) techniques have been developed to determine these ratios with precisions of fractional parts per thousand. However, MS analysis is costly and time consuming, since a great deal of chemical preparation is generally required to concentrate and separate the gas and to convert it quantitatively into a gas that can be handled unequivocally. For example, a GC is used to separate isotopic forms of CH_4 ($^{12}CH_4$, $^{13}CH_4$, CH_3D) from CO, CO_2, and other hydrocarbons. The methane isotopes are then oxidized to CO_2, and MS is used to measure the $^{13}CO_2/^{12}CO_2$ ratio. The measurement of a single sample takes several hours. Moreover, since the equipment is quite bulky it cannot be used in the field and is therefore incapable of performing real-time measurements.

The TDLAS affords an excellent opportunity to measure isotope ratios without requiring gas conversion and, in principle, could be used in the field for continuous real-time measurements. The resolution is more than adequate to distinguish rotational lines from the isotopic species (J. K. Becker et al., 1992). In fact, the wavelength separations between lines from the same rotational levels of $^{12}CH_4$ and $^{13}CH_4$ are generally greater than the scanning range of a single diode. The major problem yet to be overcome is that of dynamic range. Since the $^{13}CH_4/^{12}CH_4$ ratio, for example, is 0.011, measurements of ratios to 0.1 parts per thousand accuracy with the same rotational lines would require a precision of 10^{-6}. To overcome these problems lines from different rotational states that lie close together can be used, but in this case the temperature must be carefully controlled since the ratio of the absorptions will then be a strong function of temperature.

A TDLAS was used in the laboratory (Wahlen and Yoshinari, 1985) to measure the $N_2{}^{18}O/N_2{}^{16}O$ ratio in ambient air samples and in air samples from nitrification of waste water, soils, and stack gas from coal-burning power plants. Neighboring lines in the ν_3 band at 1225 cm^{-1} were selected for the two isotopic species so that their intensities were comparable. The precision of the ratio measurements were ± 0.4 parts per thousand. To attain this sensitivity preconcentration of the samples was employed.

The open-path stratospheric instrument described in Section 5.8.5 was used by Webster and May (1992) to measure the $^{13}CH_4/^{12}CH_4$ ratio and $OC^{18}O$ at an altitude of 30 km. They report enrichment factors that are not in agreement with more precise MS measurements. It is not clear how the large dynamic range problem was overcome to achieve the accuracy that they claimed to have achieved.

A TDLAS system for methane isotope measurements has been described (Schupp et al., 1992). A triple optical path was used to simultaneously measure and compare the spectra of a methane sample, a $^{13}CH_4/^{12}CH_4$ standard, and a $^{13}CH_4$ pure sample in three single-pass cells (36 cm) at a total pressure of 30 mbar. The system uses an absorption line pair of $^{13}CH_4$ and $^{12}CH_4$ at 3007 cm^{-1}. All cells were temperature stabilized to ± 0.1 K. Interferences fringes were suppressed by linearly polarizing the laser beam and by using Brewster windows for the gas cells and detectors.

Under laboratory conditions Schupp and colleagues (1992) were able to demonstrate an accuracy in the $\delta^{13}C$ of 1 part per thousand for a sample size of 10 μmol with a measurement time of a few minutes. This is still 20 times less accurate than the best that can be achieved by MS. The use of multipass cells should result in considerable improvement in the accuracy of the TDLAS method.

5.8.7. Automobile Exhaust

The incomplete combustion of fuel in an engine results in the formation of a complex mixture of gaseous and particulate exhaust. Real-time monitoring of exhaust gas components provides important information on the dependence of the exhaust gas composition on engine operating conditions and on the efficiency and behavior of catalytic converters. The TDLAS technique has been applied extensively to real-time measurements in the exhaust gases of diesel and automobile engines (Harris et al., 1987; Lach et al., 1989; Wolf et al., 1989; Klingerberg and Winckler, 1987; Riedel et al., 1991, 1992; Nitzschke and Wolf, 1991), and recently for the determination of internal combustion engine oil consumption (Carduner et al., 1991, 1992; Mackay et al., 1993b).

Measurements of NO_2, HNO_3, and HCHO have been made in diesel exhaust (Harris et al., 1987) using the TAMS-150. Mixing ratios (corrected for flow dilution) in the raw exhaust were 40–800 ppbv for HNO_3, and 1–30 ppmv for both NO_2 and HCHO. Furthermore, the data indicate that the observed HNO_3 and HCHO are present in the exhaust stream immediately after dilution and are not produced by subsequent reactions in the dilution stream.

Instrument response time in these measurements was 5 min, as dictated by the data acquisition rate. At the high concentrations found in engine exhaust high sensitivity is not required, but rapid response >1 Hz is required if changes in species concentration are to be followed during test procedures designed to mimic on-road changes in acceleration and load conditions. The data acquisition rate was improved to provide 30 s response time for a study of SO_2 mixing ratios in automobile exhaust. A response time of 30 s is equivalent to the residence time of the gas in the absorption cell of the TAMS-150 for the flow rates available in these measurements as determined by the engine test conditions.

The TDLAS was applied for the determination of internal combustion engine oil consumption (Carduner et al., 1991, 1992). Oil consumption by internal combustion engines is a measure of the engine performance with impact on engine emissions. It also plays an important role in the customer's perception of engine quality. Monitoring oil consumption by TDLAS is achieved by measuring the concentration of SO_2 in the exhaust gas, produced by the oxidation of sulfur, a natural constituent of oil. Special sulfur-free fuel was used in these measurements. Sulfur dioxide was monitored via its IR absorption in the $1350\,cm^{-1}$ region. Interferences due to other exhaust gas species (CO, CO_2, CH_4, NO, NO_2) were less than 10% for mixtures typical of exhaust gas composition. Interference from water was more significant (equivalent to 100 ppbv SO_2 at 10% water) and was compensated for by mathematical correction of the data.

Figure 5.24 shows the variation in SO_2 measured in the raw automobile exhaust at two engine speeds with and without exhaust gas recirculation; no preconditioning of the exhaust sample was required. Transient SO_2 emissions are obvious.

Application of the EMS (environmental monitoring system) with its more rapid response has permitted even more detailed probing of the behavior of selected gases under rapidly varying engine conditions. Measurements of the transient behavior of CH_4 were made at the Ford Motor Company (Mackay et al., 1993b) along with measurements made simultaneously using FTIR. Figure 5.25 shows the good correlation between the two techniques and demonstrates the faster response (0.2 compared to 3 s) of the EMS system compared to the FTIR even when the EMS data was averaged into 3-s blocks to coincide with the FTIR sampling period. At 0.2-s response times (Fig. 5.26), additional transient emissions of CH_4 were identified that were not detectable by the FTIR.

Simultaneous measurements of CO, CO_2, NO, NO_2, HCHO, and methanol in car exhaust with a multicomponent TDLAS system [DIOLA (a diode laser instrument)] have been reported (Klingerberg and Winckler, 1987; Lach et al., 1989). The DIOLA system uses also Pb-salt diode lasers for multicomponent automobile exhaust measurements. The main features of the system are (a) the use of a polychromator arrangement to combine the beams from six different diode lasers each operated at a different temperature and current setting, and (b) the operation of the lasers in the pulsed mode. The sensitivity of the system is a few ppmv, with a time response of 0.5 s.

A fast exhaust gas analysis system has been developed for dynamic car engine testing (Riedel et al., 1991). The system uses pulse-operated diode lasers in combination with a gas sampling system designed to achieve a rise time of < 3 ms, with detection limits of 100 and 20 ppmv for NO and CO, respectively, with a 20-cm path length. The system allowed the resolution of single engine strokes and the monitoring of dynamic engine operation changes. The system was also operated in conjunction with a 100-m White cell for high-sensitivity real-time car exhaust analysis (Riedel et al., 1992; Nitzschke and Wolf, 1991).

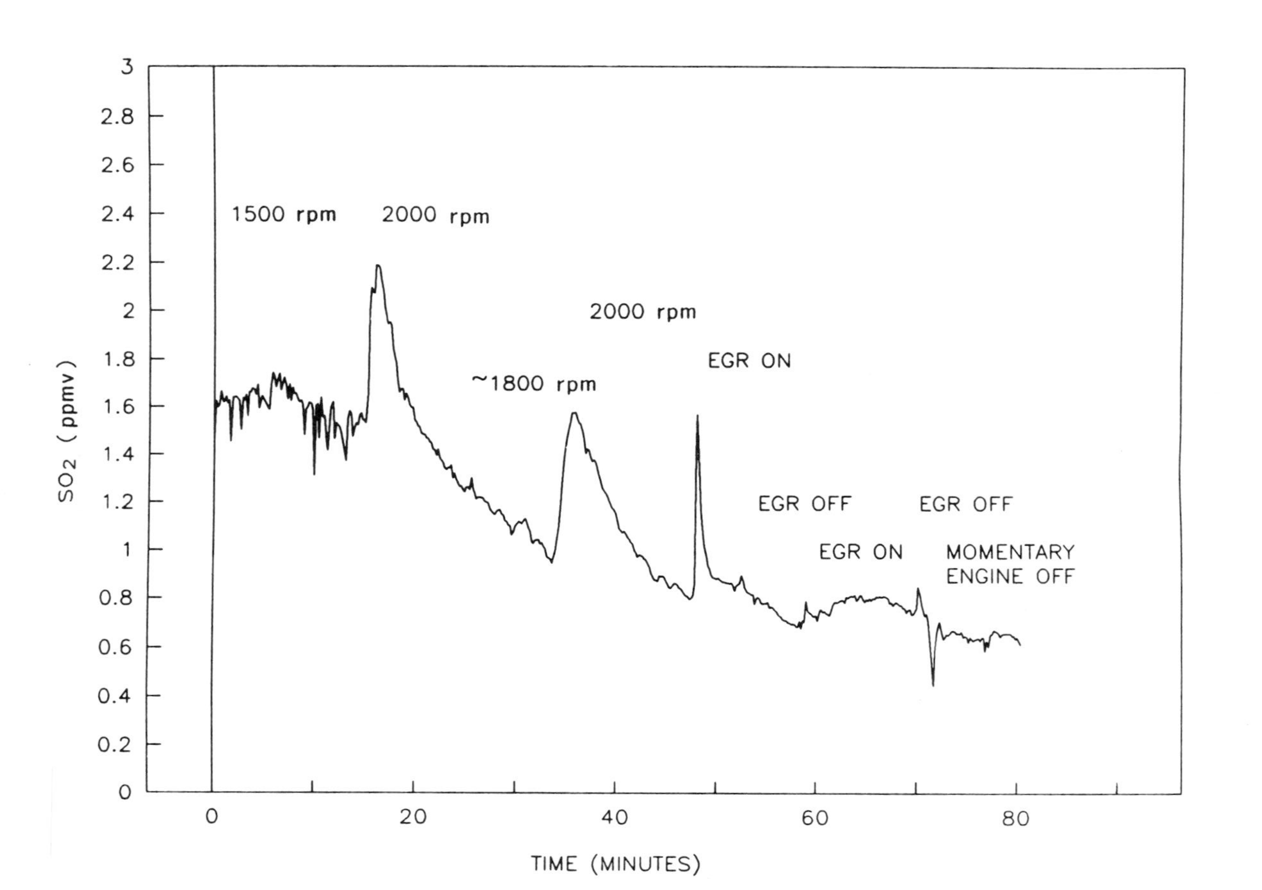

Figure 5.24. TDLAS response to SO_2 in automobile exhaust at 1500 and 2000 rpm with and without exhaust gas recirculation (EGR). The engine speed was not controlled in this test, and at 30 min into the test the engine speed had drifted down to ~ 1800 rpm, at which time it was increased to 2000 rpm. By the end of the

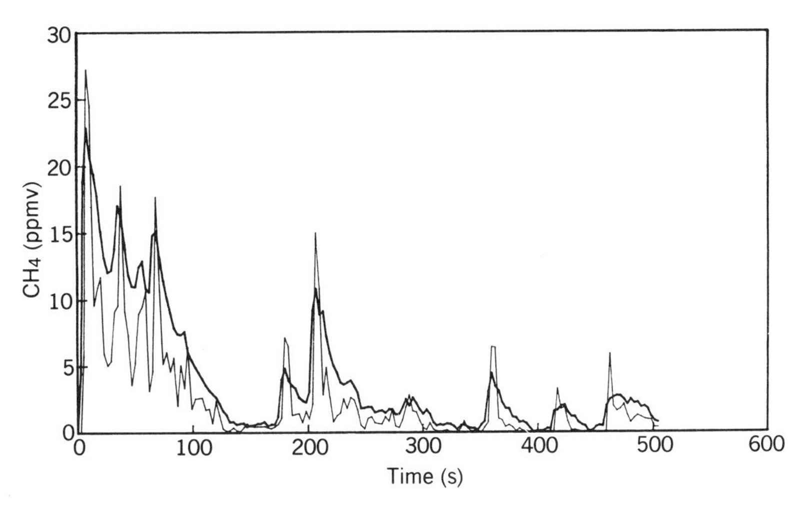

Figure 5.25. Time series plot for the variation of methane (CH_4) in engine exhaust as the automobile performed a standard test procedure from a cold start as measured by TDLAS (light line) and FTIR (heavy line). The response time of the FTIR was about 3 s and the TDLAS data, obtained at 5 Hz, was averaged into 3 s blocks to match the FTIR acquisition period. The FTIR data was kindly provided by Dr. D. Schuetzle, Ford Motor Company, Dearborn, Michigan. From Mackay et al. (1993b).

Trace components such as HCHO, ethane, and methanol were measured with a detection limit in the ppbv range and a response time of 1 s. First tests of the system in a dynamometer facility with a CVS (constant volume sampling) dilution tunnel have been reported (Riedel et al., 1992).

5.8.8. Pollution and Other Applications

TDLAS systems have been applied to measurements of gases in a large variety of environments where concentrations of polluting gases are higher than those encountered in the applications discussed above. A few examples will illustrate the versatility of the technique.

The excess ammonia (NH_3) in the exhaust of a coal power plant equipped with a Denox catalyst was analyzed by TDLAS (Wolf and Riedel, 1987). The exhaust gas was drawn into a sampling cell at 35 mbar, maintained at 350 °C. The detection limits and precision was reported to be <0.5 ppmv, with a response time of a few seconds.

Stanton and co-workers (see Silver et al., 1991) have reported the use of TDLAS for the measurement of trace concentrations of ammonia in an entrained flow reactor at 1225 K. The use of a Herriott cell combined with wavelength modulation, etalon fringe suppression, and rapid scanning result-

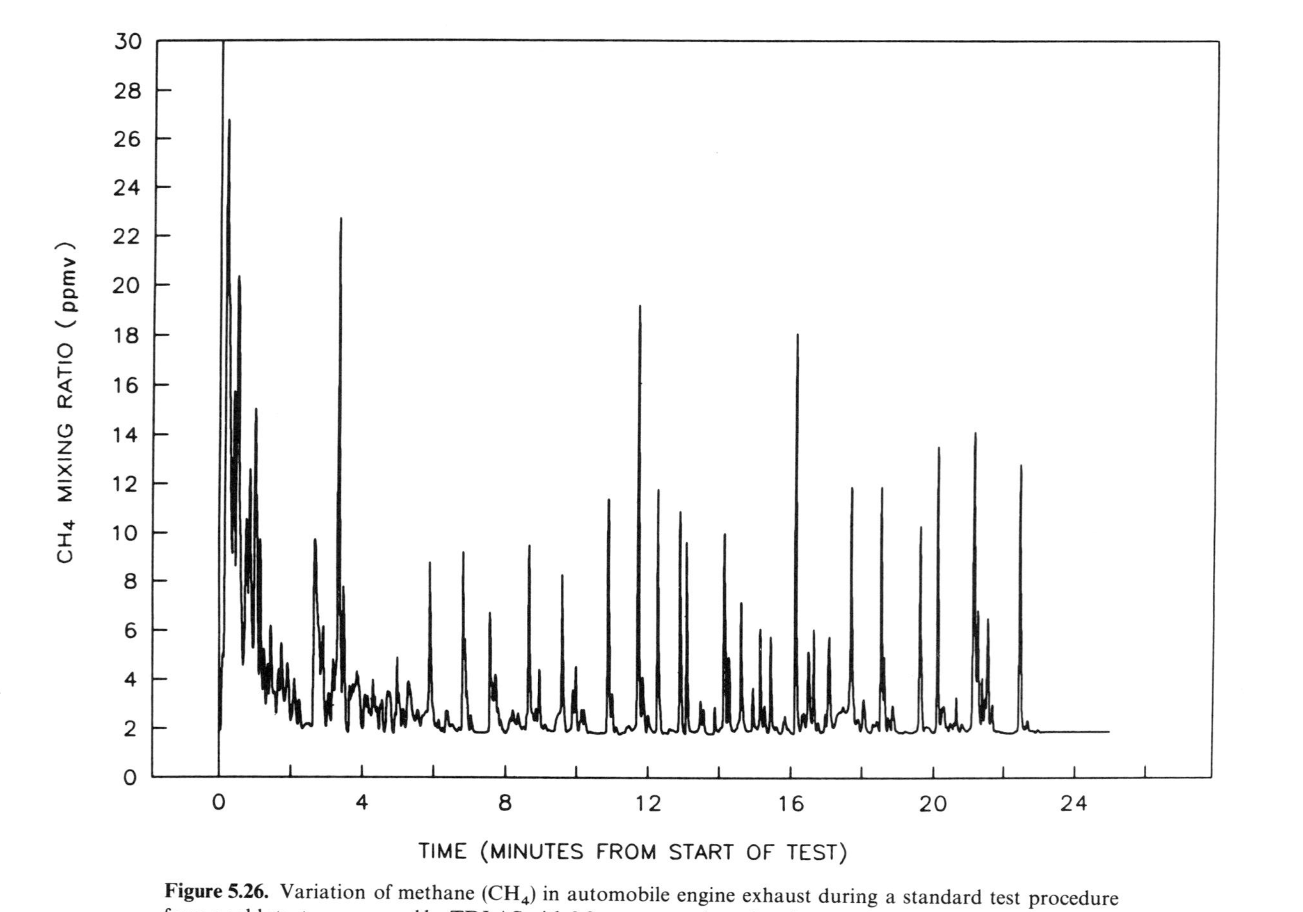

Figure 5.26. Variation of methane (CH_4) in automobile engine exhaust during a standard test procedure from a cold start as measured by TDLAS with 0.2-s response time, showing transient behavior in the methane

ed in a sensitivity of 40 ppbv for NH_3, several orders of magnitude better than could be achieved by traditional techniques.

Measurements of NO and CO in postflame gases of an atmospheric pressure flat flame burner in the temperature range 2000–2100 K have been reported (Hanson et al., 1977; Hanson and Falcone, 1978; Falcone et al., 1983). K. H. Becker and co-workers (1992) investigated the N_2O and CH_4 emissions from various small furnaces operated with charcoal, gas, oil, and wood as fuel and from a coal-fired power plant with a circulating atmospheric fluidized bed to reduce NO_x emissions. The N_2O emissions from small furnaces were comparable to those of conventional power plants (<12 ppmv); however, N_2O emissions from the full-scale plant with fluidized bed were much larger (75–180 ppmv). Methane emissions were found to depend strongly on the fuel used and the combustion conditions. Methane concentrations in excess of 1000 ppmv were measured for wood and pit coal furnaces with reduced air supply.

TDLAS systems were used to measure C_2H_6, CH_4, and H_2O vapor in cigarette smoke (Forrest, 1980) and CO, CO_2, (Stepanov et al., 1992), and NH_3 (Lachish et al., 1987) in human breath. The systems had sufficient sensitivity to provide real-time analyses.

Partridge and Curtis (1989) used a mobile open-path TDLAS system to measure air pollution and emissions inside and around industrial buildings and over landfill sites. Retroreflectors and plane mirrors were used to guide the laser beam along the desired path and thence to the detector, which permitted the entire location to be monitored for high-emission sources. The sensitivity was on the order of several ppbv for a path length of 250 m. The system was used to measure CO, CH_4, C_2H_4, and NH_3 at landfills, coke ovens, and coal-processing and chemical plants.

Brassington (1989) used a TDLAS for the measurement of HCl in the plume of a coal-fired power station and for ambient measurements of NH_3 at a mountain site in England. The system used a White cell (75 m; 50 mbar) to achieve detection limits of 0.4 ppbv for HCl and 0.1 ppbv for NH_3 with a 1 s instrumental time constant.

TDLAS systems have also been used in the laboratory to study the kinetics of atmospheric chemical processes (Bechara et al., 1989; K. H. Becker et al., 1990; J. K. Becker et al., 1992; Fox et al., 1988; Hjorth et al., 1989; Worsnop et al., 1992).

A laboratory study (Bechara et al., 1989; K. H. Becker et al., 1990) used a TDLAS system with a 200-m path length White cell along with other instruments to follow the production of hydrogen peroxide in the photooxidation of a number of terpenes and alkenes in the presence of other atmospheric constituents, e.g., H_2O, SO_2, CO, and NO. The increased H_2O_2 production in the presence of H_2O suggested that this reaction could contribute to the production of hydroperoxides and be a significant contributor to the decline of forests in western Europe and North America.

The same laboratory also investigated the chemical mechanism leading to the formation of N_2O in combustion processes (K.H. Becker et al., 1992). The kinetics of the reaction $NCO + NO \rightarrow$ products was studied, and the branching ratio for N_2O formation determined to be 0.25 at 300 K.

A kinetic study of the interaction of NO_2 and NO_3 (Hjorth et al., 1989) was made using TDLAS to measure NO. FTIR was to monitor, simultaneously, NO_2, N_2O_5, and HNO_3. The reactions occurred within a White cell at reduced pressure.

A TDLAS system was used (Worsnop et al., 1992) to study homogeneous and heterogeneous atmospheric reactions. The system employed rapid sweep integration at 300 kHz and an astigmatic Herriott cell with a 100-m path length. Peak heights and areas were determined by curve fitting using non-linear least square methods. Discharge flow tube kinetic studies were made with HO_2 radicals by use of isotopic labeling. Vapor pressure measurements of nitric acid and water over prototypical stratospheric aerosol surfaces and the heterogeneous uptake of HCl, H_2O_2, and N_2O_5 on aqueous and sulfuric acid droplets were also studied with this system.

5.9. SUMMARY: PRESENT STATUS AND FUTURE DEVELOPMENTS

In this chapter we have attempted to describe how tunable diode laser absorption spectroscopy can be applied to measurements of clean and polluted air. Pb-salt diodes emit in the 2–20 μm region, where most gases of atmospheric interest have their strong fundamental rotational–vibrational absorptions, providing the basis for a universal system. The high monochromaticity of these diodes have made TDLAS the most specific method available for unequivocal identification of atmospheric species, a particularly useful property for complex mixtures in polluted air or in automobile stack emissions and for serving as a comparison standard for other, less definitive methods. The combination of modulation and long path techniques have permitted sensitivities in the parts per trillion range, required for measurements in remote clean air or for minor but important constituents in complex mixtures. The rapid tuning rate of these diodes, permitting real-time measurements in the fractions of second range, has been exploited for measurements from fast-flying aircraft, for eddy correlation flux measurements, and for study of transients in automobile exhaust. Interesting advances are being made in the use of TDLAS systems to perform real-time isotope ratio measurements.

Theoretical improvements of several orders of magnitude in sensitivity by using high-frequency modulation techniques have not yet been realized. To accomplish that objective, better laser diodes need to be developed—a need that is being addressed by several suppliers. Considerable improvements have been made in the tuning range of commercially available diodes that have permitted the use of liquid N_2 cryostats or Dewars. A miniature He cryostat cooling system has also recently been developed that weighs 7 kg and requires only 75 W of electrical energy.

Early TDLAS systems were quite large, expensive, and required skilled operators. Commercial systems are now available that can fit in a standard instrument rack or small aircraft and can be operated automatically, without attendance for long periods of time. A technician can operate the system without supervision after about 2 weeks of training. The price of these systems has also declined significantly.

The main disadvantage of the TDLAS for atmospheric measurements is the relatively small tuning range, which limits the number of species that can be measured by the same diode. This limitation can be mitigated to some extent by the use of multiplexing techniques that permit simultaneous operation of a number of diodes. The tuning range of modern laser diodes is continuously being extended and is now sufficient in many cases to encompass strong absorption lines from four or five gases.

The high specificity of TDLAS depends on the molecule possessing resolved rotational–vibronic structure. Roughly speaking, this limits the method to molecules having less than 10 atoms or those having symmetrical structures. Extension to larger, asymmetric molecules has not yet been utilized, although it is possible that fast-scan techniques can be used to measure larger molecules that do not have completely resolved rotational–vibrational spectra.

Another limitation is that the best specificity and sensitivity is achieved when measurements are made at subatmospheric pressure. This necessitates the use of sampling cells that can introduce surface effects for reactive and unstable molecules. The pressure broadening of some molecules such as N_2O, CH_4, O_3, HCl, NO, HNO_3, NH_3 is, however, sufficiently small that measurement at atmospheric pressure is possible with open paths.

Increased interest is being shown in the application of laser diodes made of III–V compounds to gas measurements. The attractive features of these diodes are that they operate at or near room temperature, and are relatively inexpensive and of relatively high quality and power owing to their widespread use in communication and consumer electronic industries. They operate in the NIR in the range 0.78–1.6 μm, a spectral region that corresponds to overtone and combination bands of some molecules. Absorption coefficients of these bands are some 2–3 orders of magnitude lower than those in the fundamental, MIR region. But thanks to the quality and power of these diodes, very-high-frequency modulation techniques can be used to regain some of the sensitivity so that, at present, detection limits in the parts per billion can be achieved, approaching 1–10% of the sensitivity achievable from the more complex and more expensive MIR systems. In addition, both the optical and electronic components are more readily available, less expensive, and much smaller, which permits construction of relatively inexpensive compact instruments.

Another advantage is that pressure broadening is not as great a problem in this spectral region, components to operate which permits at atmospheric pressure either in closed or open-path optical systems. A disadvantage is that not as many molecules absorb in this spectral region, although this can sometimes be an advantage in analyzing complex mixtures. There are a number of manufacturers making these devices, resulting in an increasing variety and spectral coverage and a decrease in price.

It is unlikely that the NIR diodes will ever completely replace the Pb-salt MIR diodes. Rather, the two spectral regions will provide complementary systems. For a certain limited number of species where ultrahigh sensitivity is not required, the NIR systems will provide advantages of size, simplicity, and cost. For other species requiring a more universal and sensitive system, the Pb-salt MIR diodes will continue to provide a superior and highly specific measurement device. Improvements both in quality and performance of the Pb-salt diodes and in the hardware and software to operate the system will no doubt result in continuing enhancement in sensitivity, reliability, and utility of these systems as well as further cost reduction.

REFERENCES

Abramovitz, M., and Stegun, I. A. (1970). *Handbook of Mathematical Functions.* Dover, New York.

Altmann, J., Baumgart, R., and Weitkamp, C. (1981). Two-mirror multipass absorption cell. *Appl. Opt.* **20**, 995–999.

Anderson, S. M., and Zahniser, M. S. (1991). Open-path tunable diode laser absorption instrument for eddy correlation flux measurements of atmospheric trace gases. In *Measurement of Atmospheric Gases* (H. I. Schiff, Ed.), SPIE **1433**, 167–178.

Anlauf, K. G., Schiff, H. I., Wiebe, H. A., Mackay, G. I., and Mactavish, D. C. (1988). Measurement of atmospheric nitric acid by the filter method and comparisons with the tunable diode-laser and other methods. *Atmos. Environ.*, **22**, 1579–1586.

Armstrong, J. A., and Smith, A. W. (1965). Intensity fluctuations in GaAs laser emission. *Phys. Rev.*, **140**, A155–A164.

Bachor, H. A., and Manson, P. J. (1990). Practical implications of quantum noise. *J. Mod. Opt.*, **37**, 1727–1740.

Bauerecker, S., Cammenga, H. K., Taucher, F., Weitkamp, C., and Michaels, W. (1992). Enclosive flow cooling: Concept of a new method for simplifying complex molecular spectra. In *Monitoring of Gaseous Pollutants by Tunable Diode Lasers*, (R.Grisar, H.Boettner, M. Tacke, and G. Restelli, Eds.), pp. 291–302. Kluwer Academic, Dordrecht, The Netherlands.

Bechara, J., Becker, K. H., and Brockmann, K. J. (1989). Studies on the spectroscopy and chemistry of some atmospheric hydroperoxides with a tunable diode laser. In *Physico-Chemical Behaviour of Atmospheric Pollutants* (G. Restelli and G. Angeletti, Eds.), pp. 27–31. Kluwer Academic, Dordrecht, The Netherlands.

Becker, J. K., Sauke, B. T. B., and Loewenstein, M. (1992). Stable isotope analysis using tunable diode laser spectroscopy. *Appl. Opt.*, **31**, 1921–1927.

Becker, K. H., Brockmann, K. J., and Bechara, J. (1989). Tunable diode measurements of CH_3OOH absorption cross-sections near 1320 cm^{-1}. *Geophys. Res. Lett.*, **16**, 1367–1370.

Becker, K. H., Brockmann, K. J., and Bechara, J. (1990). Production of hydrogen peroxide in forest air by reaction of ozone with terpenes. *Nature (London)*, **346**, 256–258.

Becker, K. H., Bechara, J., and Brockmann, K. J. (1991). Investigation of the chemistry of atmospheric peroxides. In *Statuskolloqium fuer Luftverunreinigung and Waldschaeden*, unpublished final report, Res. Proj. FP 15, Ministry of Environment (MURL) of North Rhine–Westphalia, Duesseldorf, p. 41.

Becker, K. H., Huedepohl, R., Kurtenbach, R., and Wiesen, P. (1992). Formation and emission of atmospheric greenhouse gases in the combustion of fossil fuels: Monitoring by diode laser spectroscopy. In *Monitoring of Gaseous Pollutants by Tunable Diode Lasers* (R. Grisar, H. Boettner, M. Tacke, and G. Restelli, Eds.), pp. 13–20. Kluwer Academic, Dordrecht, The Netherlands.

Besson, J. M., Butler, J. F., Calawa, A. R., Paul, W., and Rediker, R. H. (1965). Pressure-tuned PbSe diode laser. *Appl. Phys. Lett.*, **7**, 206–208.

Bjorklund, G. C. (1980). Frequency-modulation spectroscopy—A new method for measuring weak absorptions and dispersions. *Opt. Lett.*, **5**, 15–17.

Bomse, D. S., Stanton, A. C., and Silver, J. A. (1992). Frequency-modulation and wavelength modulation spectroscopies—Comparison of experimental methods using a lead-salt diode laser. *Appl. Opt.*, **31**, 718–731.

Bradshaw, J., and Davis, D. D. (1982). Sequential two-photon laser induced fluorescence: A new method for detecting atmospheric trace levels of NO. *Opt. Lett.*, **7**, 224–226.

Braman, R. S., Shelley, T. J., and McClenny, W. A. (1982). Tungstic acid for preconcentration and determination of gaseous and particulate ammonia and nitric acid in ambient air. *Anal. Chem.*, **54**, 358–364.

Braman, R. S., de la Cantera, M. A., and Han, Q. X. (1986). Sequential selective hollow tube preconcentration and chemiluminescence analysis system for nitrogen oxide compounds in air. *Anal. Chem.*, **58**, 1537–1541.

Brassington, D. J. (1989). Measurements of atmospheric HCl and NH_3 with a mobile tunable diode laser system. In *Monitoring of Gaseous Pollutants by Tunable Diode Lasers* (R. Grisar, G. Schmidtke, M. Tacke, and G. Restelli, Eds.), pp. 16–24. Kluwer Academic, Dordrecht, The Netherlands.

Britov, A. D., Karavaev, S. M., Kalyuzhnaya, G. A., Kurbatov, A. L., Maksimovskii, S. N., and Sivachenko, S. D. (1977). Mode tuning in pulse lead–tin chalcogenide laser diodes. *Sov. J. Quantum Electron.*, (*Engl. Transl.*) **7**, 1138–1140.

Butler, J. F., Calawa, A. R., Phelan, R. J., Jr., Harman, T. C., Strauss, A. J., and Rediker, R. H. (1963). PbTe diode laser. *Appl. Phys. Lett.*, **5**, 75–77.

Butler, J. F., Linden, K. J., and Reeder, R. E. (1983). Mesa-stripe $Pb_{1-x}Sn_xSe$ tunable diode-lasers. *IEEE J. Quantum Electron.*, **QE–19**, 1520–1525.

Calawa, A. R., Dimmock, J. O., Harman, T. C., and Melngailis, J., (1969). Magnetic field dependence of laser emission in $Pb_{1-x}Sn_xSe$ diodes. *Phys. Rev. Lett.*, **23**, 7–10.

Cappellani, F., Melandrone, G., and Restelli, G. (1987). Post detection data handling techniques for application in derivative monitoring. In *Monitoring of Gaseous Pollutants by Tunable Diode Lasers* (R. Grisar, H. Preier, G. Schmidtke, and G. Restelli, Eds.), pp. 51–60. Reidel, Dordrecht, The Netherlands.

Carduner, K. R., Colvin, A. D., Scheutzle, D., and Mackay, G. I. (1991). Use of the tunable diode laser in determination of internal combustion engine oil consumption. In *Measurement of Atmospheric Gases* (H. I. Schiff, Ed.), SPIE **1433**, pp. 190–201.

Carduner, K. R., Colvin, A. D., Leong, R. Y., Schuetzle, D., Mackay, G. I., Karecki, D. R., and Schiff, H. I. (1992). Application of tunable diode laser spectroscopy to the real-time analysis of engine oil economy. *Environ. Sci. Technol.*, **26**, 930–934.

Carlisle, C. B., and Cooper, D. E. (1989a). An optical isolator for mid-infrared diode lasers. *Opt. Commun.*, **74**, 207–210.

Carlisle, C. B., and Cooper, D. E. (1989b). Tunable diode laser frequency modulation spectroscopy using balanced homodyne detection. *Opt. Lett.*, **14**, 1306–1308.

Carlisle, C. B., Cooper, D. E., and Preier, H. (1989). Quantum noise-limited FM spectroscopy with a lead-salt diode laser. *Appl. Opt.*, **28**, 2567–2576.

Carroll, M. A., Hastie, D. R., Ridley, B. A., Rodgers, M. O., Torres, A. L., Davis, D. D., Bradshaw, J. D., Sandholm, S. T., Schiff, H. I., Karecki, D. R., Harris, G. W., Mackay, G. I., Gregory, G. L., Condon, E. P., Trainer, M., Hübler, G., Monyzka, D. D., Madronich, S., Albritton, D. L., Singh, H. B., Beck, S. M., Shipham, M. C., and Bachmeier, A. S. (1990). Aircraft measurements of NO_x over the eastern Pacific and Continental United States and implications for ozone production. *J. Geophys. Res.* **95D**, 10, 205–10, 233.

Cassidy, D. T. (1988). Trace gas detection using 1.3 μ InGaAsP diode laser transmitter modules. *Appl. Opt.*, **27**, 610–614.

Cassidy, D. T., and Reid, J. (1982a). Atmospheric-pressure monitoring of trace gases using tunable diode-lasers. *Appl. Opt.*, **21**, 1185–1190.

Cassidy, D. T., and Reid, J. (1982b). Harmonic detection with tunable diode lasers. 2 tone modulation. *Appl. Phys. B* **29**, 279–285.

Cassidy, D. T., and Reid, J. (1982c). High-sensitivity detection of trace gases using sweep integration and tunable diode lasers. *Appl. Opt.* **21**, 2527–2530.

Chou, N. Y., Sachse, G. W., Wang, L. G., and Gallagher, T. F. (1989). Optical fringe reduction technique for FM laser spectroscopy. *Appl. Opt.*, **28**, 4973–4975.

Cicerone, R. J., and Omerland, R. S. (1988). Biogeochemical aspects of atmospheric methane. *Global Biogeochem. Cycles* **2**, 299–328.

Cooper, D. E., and Martinelli, R. U. (1992). Near-infrared diode lasers monitor molecular species. *Laser Focus World*, **11**, 133–146.

Cooper, D. E., and Warren, R. E. (1987). 2-tone optical heterodyne spectroscopy with diode lasers—Theory of line shapes and experimental results. *J.Opt. Soc. Am.*, **B4**, 470–480.

Cooper, D. E., and Watjen, J. P. (1986). 2-tone optical heterodyne spectroscopy with a tunable lead-salt diode laser. *Opt. Lett.*, **11**, 606–608.

Cooper, D. E., Riris, H., and van der Laan, J. E. (1991). Frequency modulation spectroscopy for chemical sensing of the environment. In *Measurement of Atmospheric Gases* (H. I. Schiff, Ed.), SPIE **1433**, 120–127.

Edwards, C. G., Neumann, H. H., den Hartog, G., Thurtell, G. W., and Kidd, G. (1993). Eddy correlation measurements of methane using a TDL at the Kinosheo Lake tower site during the Northern Wetland Study (NOWES). *J.Geophys. Res.* (to be published).

Eng, R. S., and Ku, R. T. (1982). High resolution linear laser absorption spectroscopy— Review. *Spectrosc. Lett.*, **15**, 803–929.

Eng, R. S., Mantz, A. W., and Todd, T. R. (1979). Low frequency characteristics of Pb-salt semiconductor lasers. *Appl. Opt.*, **18**, 1088–1091.

Eng, R. S., Butler, J. F., and Linden, K. J. (1980). Tunable diode-laser spectroscopy—An invited review. *Opt. Eng.*, **19**, 945–960.

Fach, A., Boettner, H., and Tacke, M. (1992). Embossed grating DFB-DH lead chalcogenide diode lasers. In *Monitoring of Gaseous Pollutants by Tunable Diode Lasers* (R. Grisar, H. Boettner, M. Tacke, and G. Restelli, Eds.), pp. 63–68. Kluwer Academic, Dordrecht, The Netherlands.

Falcone, P. K., Hanson, R. K., and Kruger, C. H. (1983). Tunable diode laser absorption measurements of nitric oxide in combustion gases. *Combust. Sci. Technol.*, **35**, 81–89.

Feit, Z., Kostyk, D., Woods, R. J., and Mak, P. (1990). Molecular-beam epitaxy-grown PbSnTe PbEuSeTe buried heterostructure diode lasers. *IEEE Photonics Tech. Lett.*, **2**, 860–862.

Feit, Z., Kostyk, D., Woods, R. J., and Mak, P. (1991). Single-mode molecular-beam epitaxy grown PbEuSeTe/PbTe buried-heterostructure diode lasers for CO_2 high-resolution spectroscopy. *Appl. Phys. Lett.*, **58**, 343–345.

Feit, Z., Kostyk,D., Woods, R. J., and Mak, P. (1992). Recent developments in MBE grown PbEuSeTe/PbSnTe diode lasers for high resolution spectroscopy. In *Monitoring of Gaseous Pollutants by Tunable Diode Lasers* (R. Grisar, H. Boettner, M. Tacke, and G. Restelli, Eds.), pp. 105–110. Kluwer Academic, Dordrecht, The Netherlands.

Fischer, H., and Tacke, M. (1991). High-frequency intensity noise of lead-salt diode lasers. *J.Opt. Soc. Am.*, **B8**, 1824–1830.

Fischer, H., Wolf, H., Halford, B., and Tacke, M. (1991). Low frequency amplitude noise characteristics of lead-salt diode lasers fabricated by molecular-beam-epitaxy. *Infrared Phys.*, **31**, 381–385.

Forrest, G. T. (1980). Tunable diode laser measurements of methane, ethane and water vapor in cigarette smoke. *Appl. Opt.*, **19**, 2095–2096.

Fox, D. L., Stockburger, L., Spicer, C. W., Weathers, W., Hansen, L. D., Kleindienst, T. E., Mackay, G. I., Edney, E. O., Eatough, D. J., Mortensen, F., Shepson, P. B., and Schiff, H. I. (1988). Intercomparison of nitric acid diffusion denuder methods with tunable diode laser absorption spectroscopy. *Atmos. Environ.*, **22**, 575–585.

Fried, A., Berg, W. W., and Sams, R. (1984). Application of tunable diode-laser absorption for trace stratospheric measurements of HCl—Laboratory results. *Appl. Opt.*, **23**, 1867–1880.

Fried, A., Drummond, J. R., Henry, B., and Fox, J. (1990). Reduction of interference-fringes in small multipass absorption cells by pressure modulation. *Appl. Opt.*, **29**, 900–902.

Fried, A., Drummond, J. R., Henry, B., and Fox, J. (1991a). A tunable diode laser spectrometer for high precision concentration and ratio measurements of long lived atmospheric gases. In *Measurement of Atmospheric Gases.* (H.I. Schiff, Ed.), SPIE **1433**, 145–156.

Fried, A., Drummond, J. R., Henry, B., and Fox, J. (1991b). Versatile integrated tunable diode-laser system for high-precision: Application for ambient measurements of OCS. *Appl. Opt.*, **30**, 1916–1932.

Fried, A., Henry, B., Parrish, D. D., Carpenter, J. R., and Buhr, M. P. (1991c). Intercomparison of tunable diode laser and gas filter correlation measurement of ambient CO. *Atmos. Environ.*, **254A**, 2277–2284.

Gallagher, T. F., Kachru, R., Gounand, F., Bjorklund, G. C., and Lenth, W. (1982). Frequency modulation spectroscopy with a pulsed dye laser. *Opt. Lett.*, **7**, 28–30.

Gehrtz, M., Bjorklund, G. C., and Whittaker, E. A. (1985). Quantum limited laser frequency modulation spectroscopy. *J.Opt. Soc. Am.*, **B2**, 1510–1525.

Gehrtz, M., Lenth, W., Young, A. T., and Johnson, H. S. (1986). High frequency modulation spectroscopy with a lead-salt diode laser. *Opt. Lett.*, **11**, 132–134.

Glenar, D., Kostiuk, T., Jennings, D. E., Buhl, D., and Mumma, M.J. (1982). Tunable diode laser hetrodyne spectrometer for remote observations near 8 μ. *Appl. Opt.*, **21**, 253–259.

Goldan, P. D., Kuster, W. C., Albritton, F. C., Fehsenfeld, P., Connell, S., Norton, R. B., and Huebert, B. J. (1983). Calibration and tests of the filter collection method for measuring clean air, ambient levels of nitric acid. *Atmos. Environ.*, **17**, 1355–1364.

Gregory, G. L., Hoell, J. M., Carroll, M. A., Ridley, B. A., Davis, D. D., Bradshaw, J., Rodgers, M. O., Sandholm, S. T., Schiff, H. I., Hastei, D. R., Karecki, D. R., Mackay, G. I., Harris, G. W., Torres, A. L., and Fried, A. (1990a). An intercomparison of airborne nitrogen-dioxide instruments. *J. Geophys. Res.*, **95D**, 103–127.

Gregory, G. L., Hoell, J. M., Huebert, B. J., Van Bramer, S. E., Level, P. J., Vay, S. A., Marinaro, R. M., Schiff, H. I., Hastie, D. R., Mackay, G. I., and Karecki, D. R.

(1990b). An intercomparison of airborne nitric acid measurements. *J. Geophys. Res.*, **95D**, 10,089–10,102.

Grisar, R., Preier, H., Schmidtke, G., and Restelli, G., Eds. (1987). *Monitoring of Gaseous Pollutants by Tunable Diode Lasers*, Reidel, Dordrecht, The Netherlands.

Grisar, R., Schmidtke, G., Tacke, M., and Restelli, G., Eds. (1989). *Monitoring of Gaseous Pollutants by Tunable Diode Lasers*, Kluwer Academic, Dordrecht, The Netherlands.

Grisar, R., Boettner, H., Tacke, M., and Restelli, G., Eds. (1992). *Monitoring of Gaseous Pollutants by Tunable Diode Lasers*, Kluwer Academic, Dordrecht, The Netherlands.

Hanson, R. K., and Falcone, P. K. (1978). Temperature measurement technique for high temperature gases using a tunable diode laser. *Appl. Opt.*, **17**, 2477–2480.

Hanson, R. K., Kuntz, P. A., and Kruger, C. H. (1977). High resolution spectroscopy of combustion gases using a tunable IR diode laser. *Appl. Opt.*, **16**, 2045–2048.

Harris, G. W., Iguchi, T., Mackay, G. I., Schiff, H. I., and Schuetzle, D. (1987). Measurement of NO_2 and HNO_3 in diesel exhaust-gas by tunable diode-laser absorption spectrometry. *Environ. Sci. Technol.*, **21**, 299–304.

Harris, G. W., Burrows, J. P., Klemp, D., and Zenker, T. (1989a). Measurement of trace gases in the remote maritime boundary layer. In *Monitoring of Gaseous Pollutants by Tunable Diode Lasers* (R. Grisar, G. Schmidtke, M. Tacke, and G. Restelli, Eds.), pp. 68–76. Kluwer Academic, Dordrecht, The Netherlands.

Harris, G. W., Iguchi, T., Schiff, H. I., Mackay, G. I., and Mayne, L. K. (1989b). Measurements of formaldehyde in the troposphere by tunable diode-laser absorption spectroscopy. *J. Atmos. Chem.*, **8**, 119–137.

Harward, C. N., and Hoell, J. M. (1979). Optical feedback effects on the performance of $Pb_{1-x}Sn_xSe$ semiconductors lasers. *Appl. Opt.*, **18**, 3978–3983.

Hastie, D. R., and Miller, M. D. (1985). Balloon-borne tunable diode laser absorption spectrometer for multispecies measurement in the stratosphere. *Appl. Opt.*, **24**, 3694–3701.

Hastie, D. R., Mackay, G. I., Iguchi, T., Ridley, B. A., and Schiff, H. I. (1983). Tunable diode laser systems for measuring trace gases in tropospheric air. *Environ. Sci. Technol.* **17**, 352A–364A.

Hayward, J. E., Cassidy, D. T., and Reid, J. (1989). High sensitivity transient spectroscopy using tunable diode lasers. *Appl. Phys. B* **48**, 25–29.

Heikes B. G., Kok, G. L., Walega, J. G., and Lazrus, A. L. (1987). H_2O_2 and SO_2 measurements in the lower troposphere over the eastern United States during fall. *J. Geophys. Res.*, **92D**, 915–913.

Henry, C. H. (1982). Theory of the linewidth of semiconductors. *IEEE J. Quantum Electron.*, **QE-18**, 259–264.

Hering, S. V., Hicks, B. B., Lawson, D. R., Horrocks, J., Maclean, A., Mackay, G. I., Knapp, K. T., Grosjean, D., John, W., Sutton, R., Wall, S., Solomon, P. A., Stedman, D. H., Spicer, C. W., Tuazon, E. C., Womack, J. D., Wiebe, A., Winer, A. M., Paur, R. J., Peake, E., Mitchell, W. J., Ondo, J., Possanzini, M., Perrino, C., Pleasant, M.,

Pierson, W. R., Sickles, J. E., Schiff, H. I., Ellestad, T. G., Ellis, E. C., Febo, A., Eatough, D. J., Eatough, N. L., Cass, G. R., Allegrini, I., Appel, B. R., Anlauf, K. G., Biermann, H. W., Braman, R. S., and Brachaczek, W. (1988). The nitric-acid shootout—Field comparison of measurement methods. *Atmos. Environ.*, **22**, 1519–1539.

Herriott, D., and Schulte, H. J. (1965). Folded optical delay lines. *Appl. Opt.*, **4**, 883–889.

Herriott, D., Kogelnik, H., and Kompfner, R. (1964). Off–axis paths in spherical mirror interferometers. *Appl. Opt.*, **3**, 523–526.

Hicks, B. B., and McMillen, R. T. (1988). On the measurement of dry deposition using imperfect sensors and in non-ideal terrain. *Boundary-Layer Meteorol.*, **42**, 79–94.

Hicks, B. B., Matt, D. R., and McMillen, R. T. (1989). A micrometeorological investigation of surface exchange of O_3, SO_2, and NO_2: A case study. *Boundary-Layer Meteorol.*, **47**, 321–326.

Hinkley, E. D. (1970). High-resolution infrared spectroscopy with a tunable diode laser. *Appl. Phys. Lett.*, **16**, 351–354.

Hinkley, E. D., and Freed, C. (1969). Direct observation of the Lorentzian line shape as limited by quantum phase noise in a laser above threshold. *Phys. Rev. Lett.*, **23**, 277–280.

Hinkley, E. D., and Kelley, P. L. (1971). Detection of air pollutants with tunable diode lasers. *Science* **171**, 635–639.

Hinkley, E. D., Ku, R. T., and Kelley, P. L. (1976). Techniques for detection of molecular pollutants by absorption of laser radiation. *Top. Appl. Phys.* **14**, 237–295.

Hirota, E. (1985). High resolution spectroscopy of small molecules. In *Vibrational Spectra and Structure* (J. R. Durig, Ed.), Vol 14, p. 1–67. Elsevier, Amsterdam.

Hjorth, J., Cappellani, C., Nielsen, C., and Restelli, G. (1989). Application of tunable diode lasers to laboratory studies of atmospheric chemistry: Kinetics of the reaction $NO_3 + NO_2 \rightarrow NO + NO_2 + O_2$. In *Monitoring of Gaseous Pollutants by Tunable Diode Lasers* (R. Grisar, G. Schmidtke, M. Tacke, and G. Restelli, Eds.), pp. 205–218. Kluwer Academic, Dordrecht, The Netherlands.

Hoell, J. M., Gregory, G. L., McDougal, D. S., Sachse, G. W., Hill, G. F., Condon, E. P., and Rasmussen, R. A. (1984). An intercomparison of carbon monoxide nitric oxide and hydroxyl measurement techniques: Overview of results. *J. Geophys. Res.*, **85D**, 819–825.

Hoell, J. M., Gregory, G. L., McDougal, D. S., Sachse, G. W., Hill, G. F. Condon, E. P., and Rasmussen, R. A. (1985). An intercomparison of carbon monoxide measurement techniques. *J. Geophys. Res.* **90D**, 881–889.

Hoell, J. M., Gregory, G. L., McDougal, D. S., Sachse, G. W. Hill, G. F. Condon, E. P., and Rasmussen, R. A. (1987). Airborne comparisons of carbon monoxide measurement techniques. *J. Geophys. Res.* **92D**, 2009–2019.

Horn, D., and Pimentel, G. C. (1971). 2.5 km low-temperature multi-reflection cell. *Appl. Opt.* **10**, 1892–1898.

Hovde, D. C., and Stanton, A. C. (1993). Fast response instrumentation for methane flux measurements: An open path near infrared diode laser sensor. *J. Atmos. Chem.* (to be published).

Hsieh, H. H., and Fonstad, C. G. (1980). Liquid phase epitaxy grown PbSnTe distributed feedback laser diodes with broad continuous single mode tuning range. *IEEE J. Quantum Electron.* **QE-16**, 1039–1044.

Iguchi, T. (1986). Modulation waveforms for 2nd-harmonic detection with tunable diode lasers. *J. Opt. Soc. Am.* **B3**, 419–423.

Janik, G. R., Carlisle, C. B., and Gallagher, T. F. (1986). Two-tone frequency modulation spectroscopy. *J. Opt. Soc. Am.* **B3**, 1070–1074.

Jennings, D. E. (1978). Diode laser spectra of CCl_2F_2 near 10.8 μ: Air-broadening effects. *Geophys. Res. Lett.* **5**, 241–244.

Jennings, D. E. (1980). Absolute line strengths in ν_4, $^{12}CH_4$: A dual-beam diode laser spectrometer with sweep integration. *Appl. Opt.* **19**, 2695–2700.

Johnson, J. T., Wienhold, F. G., Burrows, J. P., and Harris, G. W. (1991). Frequency modulation spectroscopy at 1.3 μ using InGaAsP lasers: A prototype field instrument for atmospheric chemistry research. *Appl. Opt.* **30**, 407–413.

Kikuchi, K. (1989). Effect of $1/f$-type FM noise on semiconductor laser linewidth residual in high power limit. *IEEE J. Quantum Electron* **QE-25**, 684–688.

Kingston, R. H. (1978). Detection of optical and infrared radiation. *Springer Ser. Opt. Sci.* **10.**

Kittel, C. (1971). *Introduction to Solid State Physics*, 4th ed. Wiley, New York.

Kleindienst, T. E., Tejada, S. B., Arnts, R. R., Dasgupta, P. K., Dong, S., Nero, C. M., Mayne, L. K., Shepson, P. B., Schiff, H. I., Lind, J. A., Mackay, G. I., Kok, G. L., and Lazrus, A. L. (1988a). An intercomparison of formaldehyde measurement techniques at ambient concentration. *Atmos. Environ.* **22**, 1931–1939.

Kleindienst, T. E., Shepson, P. B., Schiff, H. I., Mayne, L. K., Nero, C. M., Hodges, D. N., Hwang, H., Lazrus, A. L., Lind, J. A., Kok, G. L., Mackay, G. I., Dasgupta, P. K., and Arnts, R. R. (1988b). Comparison of techniques for measurement of ambient levels of hydrogen peroxide. *Environ. Sci. Technol.* **22**, 53–61.

Klingerberg, H., and Winckler, J. (1987). Multicomponent automobile exhaust measurements. In *Monitoring of Gaseous Pollutants by Tunable Diode Lasers* (R. Grisar, H. Preier, G. Schmidtke, and G. Restelli, Eds.), pp. 108–115. D. Reidel, Dordrecht, The Netherlands.

Koga, R., Kosaka, M., and Sano, H. (1981) Improvement of etalon-fringe immunity in diode-laser derivative spectroscopy. *Mem. Sch. Eng. Okayama Univ.* **16**, 21–30.

Kosichkin, Y. V., Nadezhdinskii, A. I., and Stepanov, E. V. (1990). Diode laser spectroscopy of collisional broadening in the spectra of polyatomic molecules. *J. Quant. Spectrosc. Radiat Transfer* **43**, 499–509.

Kruse, P. W. (1980). The photon detection process. *Top Appl. Phys.* **19**, 2–69.

Ku, R. T., Hinkley, E. D., and Sample, J. O. (1975). Long-path monitoring of atmospheric carbon monoxide with a tunable diode laser system. *Appl. Opt.*, **14**, 854–861.

Kuritsyn, Yu A. (1986). Infrared absorption spectroscopy with tunable diode lasers. In *Laser Analytical Spectrochemistry* (V. S. Letokhov, Ed.), pp. 152–228. Adam Hilger, Bristol.

Kuznetzov A. I., Loukianov, D. M., and Martin, H. (1992)., High speed data acquisition and control system for diode laser spectroscopy and diode laser based analytical applications. In *Tunable Diode Laser Applications* (A. I. Nadezhdinskii and A. M. Prokhorov, Eds.), SPIE **1724**, pp. 128–135.

Lach, G., Luf, H., and Winckler, J. (1989). Functional testing of a multicomponent diode laser spectrometer (DIOLA) in comparison with conventional technology. In *Monitoring of Gaseous Pollutants by Tunable Diode Lasers* (R. Grisar, G. Schmidtke, M. Tacke, and G. Restelli, Eds.), pp. 46–60. Kluwer Academic, Dordrecht, The Netherlands.

Lachish, U., Adler, E., Elhanany, U. and Rotter, S. (1987). Tunable diode-laser based spectroscopic system for ammonia detection in human respiration. *Rev. Sci. Instrum.* **58**, 923–927.

Lawson, D. R., Biermann, H. W., Tuazon, E. C., Winer, A. M., Mackay, G. I., Schiff, H. I., Kok, G. L., Dasgupta, P. K., and Fung, K. (1990). Formaldehyde measurement methods evaluation and ambient concentrations during the Carbonaceous Species Methods Comparison Study. *Aerosol Sci. Technol.*, **12**, 64–76.

Lenth, W. (1983). Optical heterodyne spectroscopy with frequency and amplitude modulated semiconductor lasers. *Opt. Lett.*, **8**, 575–577.

Lenth, W. (1984). High-frequency heterodyne spectroscopy with current-modulated diode-lasers. *IEEE J. Quantum Electron.*, **QE-20**, 1045–1050.

Lenth, W., Ortiz, C., and Bjorklund, G. C. (1981). Pulsed frequency modulation spectroscopy as a means for fast absorption measurements. *Opt. Lett.* **6**, 351–353.

Lenth, W., Gehrtz, M., and Young, A. T. (1985). High-frequency modulation spectroscopy with tunable GaAs and lead salt diode-lasers. *J. Opt. Soc. Am.*, **A2**, 99–99.

Li, H., and Abraham, N. B. (1989). Analysis of the noise spectra of a laser diode with optical feedback from a high finesse etalon. *IEEE J. Quantum Electron.*, **QE-25**, 1782–1793.

Linden, K. J. (1985). Single-modes, short cavity, Pb-salt diode-lasers operating in the 5, 10, and 30 μ spectral regions. *IEEE J. Quantum Electron.*, **QE-21**, 391–394.

Linden, K. J., and Reeder, R. E. (1984). Operation of cleaved-coupled cavity Pb-salt diode lasers in the 4–5μ spectral region. *Appl. Phys. Lett.*, **44**, 377–379.

Liu, S. C., Trainer, M., Carroll, M. A., Huebler, G., Montzka, D. D., Norton, R. B., Ridley, B. A., Walega, J. G., Atlas, E. L., Heikes, B. G., Huebert, B. J., and Warren, W. (1992). A study of the photochemistry and ozone budget during MLOPEX. *J. Geophys. Res.*, **97D**, 463–471.

Lo, W. (1984). Cleaved coupled cavity lead-salt diode lasers. *Appl. Phys. Lett.* **44**, 1118–1119.

Lo, W., and Partin, D. L. (1984). Overview of tunable diode laser technology. In *New Lasers for Analytical and Industrial Chemistry* (A. Bernhardt, Ed.), SPIE **461**, 5–10.

Luebken, F. J. (1991). Maserati—A new rocketborne tunable diode laser experiment to measure trace gases in the middle atmosphere. *Proc. ESA Symp. Eur. Rocket Balloon Programs*, 10th, Mandelieu-Cannes, France, 1991, ESA SP-317, pp. 99–104.

Luebken, F.J ., Eng, R., Karecki, D. R., Mackay, G. I., Nadler, S., and Schiff, H. I. (1991). The effect of water vapour broadening on methane eddy correlation flux measurements. *J. Atmos. Chem.*, **13**, 97–108.

Mackay, G. I., Anlauf, K., Schiff, H. I., and Wiebe, A. (1988). Measurements of NO_2, H_2CO and HNO_3 by tunable diode-laser absorption-spectroscopy during the 1985 Claremont Intercomparison Study. *Atmos. Environ.*, **22**, 1555–1564.

Mackay, G. I., Mayne, L. K., and Schiff, H. I. (1990). Measurements of H_2O_2 and HCHO by tunable diode-laser absorption-spectroscopy during the 1986 Carbonaceous Species Methods Comparison Study in Glendora, California. *Aerosol Sci. Technol.*, **12**, 56–63.

Mackay, G. I., Karecki, D. R., and Schiff, H. I. (1991). Tunable diode laser systems for trace gas measurements. In *Measurement of Atmospheric Gases* (H. I. Schiff, Ed.), SPIE **1433**, 104–119.

Mackay, G. I., Karecki, D. R., and Schiff, H. I. (1993a). Tunable diode laser absorption measurements of H_2O_2 and HCHO during the Mauna Lao Observatory Photochemistry Experiment II Field Study, Oct. 5–23, 1991. In *Optical Methods in Atmospheric Chemistry* (H. I. Schiff and U. Platt, Eds.), SPIE **1715**, 126–137.

Mackay, G. I., Karecki, D. R., Carduner, K. R., and Schiff, H. I. (1993b). In preparation.

May, R. D., and Webster, C. R. (1989). In situ stratospheric measurements of HNO_3 and HCl near 30-km using the balloon-borne laser in situ sensor tunable diode laser spectrometer. *J. Geophys. Res.* **94D**, 6343–6350.

McElroy, C. T., Goldman, A., Fogal, P. F., and Murcray, D. G. (1990). Heterodyne spectrophotometry of ozone in the 9.6 μ band using a tunable diode laser. *J. Geophys. Res.* **95D**, 5567–5575.

McManus, J. B., and Kebabian, P. L. (1990). Narrow optical interference fringes for certain setup conditions in multipass absorption cells of the Herriott type. *Appl. Opt.*, **29**, 898–900.

Menzies, R. T. (1976). Laser heterodyne detection techniques. In *Laser Monitoring of the Atmosphere*. (E. D. Hinkley, Ed.), pp. 297–353. Springer, New York.

Menzies, R. T., Hinkley, E. D., and Webster, C. R. (1983). Balloon-borne diode laser absorption spectrometer for measurements of stratospheric trace species. *Appl. Opt.*, **22**, 2655–2664.

Meyers, T. P., Hovde, D. C., Stanton, A. C., and Matt, D. R. (1993). Micrometeorological measurements of methane emission rates from a sanitary landfill. *J. Atmos. Chem.* (to be published).

Mohebati, A., and King, T. A. (1988). Remote detection of gases by diode-laser spectroscopy. *J. Mod. Opt.*, **35**, 319–324.

Muecke, R., Werle, P., Slemr, F., and Pretl, W. (1991). Comparison of time and frequency multiplexing techniques in multicomponent FM spectroscopy. In *Measurement of Atmospheric Gases* (H. I. Schiff, Ed.), SPIE **1433**, 136–144.

Muertz, M., Schaefer, M., Schneider, M., Wells, J. S., Urban, W., Schiessl, U., and

Tacke, M. (1992). Line narrowing and frequency control of lead-salt diode lasers by optical feed back. In *Monitoring of Gaseous Pollutants by Tunable Diode Lasers*, (R. Grisar, H. Boettner, M. Tacke, and G. Restelli, Eds.), pp. 191–202. Kluwer Academic, Dordrecht, The Netherlands.

Nadezhdinskii, A. I., and Prokhorov, A. M. (1992). *Tunable Diode Laser Applications* SPIE **1724**.

Nadezhdinskii, A. I., and Stepanov, E. V. (1991). Diode laser spectroscopy of atmospheric pollutants. In *Measurement of Atmospheric Gases* (H. I. Schiff, Ed.), SPIE **1433**, 202–210.

Nadler, S., Mackay, G. I., Karecki, G. I., and Schiff, H. I. (1993). A compact tunable diode laser spectrometer for environmental monitoring. In *Optical Methods in Atmospheric Chemistry* (H. I. Schiff and U. Platt, Eds.), SPIE **1715**, 194–199.

Nillsson, O., and Buus, J. (1990). Linewidth and feedback sensitivity of semi-conductor lasers. *IEEE J. Quantum Electron.*, **QE-26**, 2039–2042.

Nitzschke, E., and Wolf, H. (1991). Dynamic exhaust gas measurement with diode lasers. *Motortech. Z.* **52**, 362–368.

Olivero, J. J., and Longbothum, R. L. (1977) Empirical fits to the Voigt linewidth: A brief review. *J. Quant. Spectrosc. Radiat. Transfer* **17**, 233–236.

Partin, D. L., (1988). Lead salt quantum effect structures. *IEEE J. Quantum Electron.*, **QE-24**, 1716–1726.

Partin, D. L., and Lo, W. (1981). Low threshold current lead-telluride diode lasers grown by molecular beam epitaxy. *J. Appl. Phys.* **52**, 1579–1582.

Partin, D. L., Majkowski, R. F., and Thrush, C.M. (1984). Stripe geometry lead-telluride diode lasers grown by molecular beam epitaxy. *J. Appl. Phys.*, **55**, 678–682.

Partridge, R. H., and Curtis, I. H. (1989). Long-path diode laser measurement of industrial air pollution. In *Monitoring of Gaseous Pollutants by Tunable Diode Lasers*, (R. Grisar, G. Schmidtke, M. Tacke, and G. Restelli, Eds.), pp. 3–15. Kluwer Academic, Dordrecht, The Netherlands.

Preier, H. (1979). Recent advances in lead chalcogenide diode lasers. *Appl. Phys.* **20**, 189–206.

Preier, H., and Riedel, W. (1974). NO spectroscopy by pulsed $PbS_{1-x}Se_x$ diode lasers. *J. Appl. Phys.* **45**, 3955–3958.

Preier, H., Bleicher, M., Riedel, W., Pfeiffer, H., and Maier, H. (1977). Peltier cooled PbSe double-heterostructure lasers for IR gas spectroscopy. *Appl. Phys.*, **12**, 277–281.

Preier, H., Feit, Z., Fuchs, J., Kostyk, D., Jalenak, W., and Sproul, J. (1989). Status of lead salt diode lasers development at spectra-physics. In *Monitoring of Gaseous Pollutants by Tunable Diode Lasers*, (R. Grisar, G. Scmidtke, M. Tacke, and G. Restelli, Eds.), pp. 85–102. Kluwer Academic, Dordrecht, The Netherlands.

Reich, M., Schieder, H., Clar, H. J., and Winnewisser, G. (1986). Internally coupled Fabry-Perrot interferometer for high precision wavelength control of tunable diode lasers. *Appl. Opt.*, **25**, 130–135.

Reid, J., and Labrie, D. (1981). 2nd-harmonic detection with tunable diode lasers—Comparison of experiment and theory. *Appl. Phys. B* **26**, 203–210.

Reid, J., Shewchun, J., Garside, B. K., and Ballik, E. A. (1978). High sensitivity detection employing tunable diode lasers. *Appl. Opt.* **17**, 300–306.

Reid, J., El-Sherbiny, M., Garside, B. K., and Ballik, E. A. (1980). Sensitivity limits of a tunable diode laser spectrometer with application to the detection of NO_2 at the 100 pptv level. *Appl. Opt.* **19**, 3349–3353.

Reid, J., Sinclair, R. L., Grant, W. B., and Menzies, R. T. (1985). High-sensitivity detection of trace gases at atmospheric pressure using tunable diode lasers. *Opt. Quantum Electron.*, **17**, 31–39.

Restelli, G., and Cappellani, F. (1985). Calibration technique for IR-laser 2nd-derivative monitoring of some trace gases in tropospheric air. *Appl. Opt.* **24**, 2480–2481.

Ridley, B. A., Carroll, M. A., and Gregory, G. L. (1987). Measurements of nitric oxide in the boundary layer and free troposphere over the Pacific Ocean. *J. Geophys. Res.* **92D**, 2025–2048.

Ridley, B. A., Carroll, M. A., Gregory, G. L., and Sachse, G. W. (1988). NO and NO_2 in the troposphere: Technique and measurements in regions of a folded tropopause. *J. Geophys. Res.* **93D**, 15,813–15,830.

Ridley, B. A., Shetter, J. D., Gandrud, B. W., Salas, L. J., Singh, H. B., Carroll, M. A., Hübber, G., Albritton, D. L., Hastie, D. R., Schiff, H. I., Mackay, G. I., Karecki, D. R., Davis, D. D., Bradshaw, J. D., Rodgers, M. O., Sandholm, S. T., Torres, A. L., Condon, E. P., Gregory, G. L., and Beck, S. M. (1990). Ratios of peroxyacetyl nitrate to active nitrogen observed during aircraft flights over the eastern Pacific Ocean and continental United States. *J. Geophys. Res.* **95D**, 10,179–10,192.

Riedel, W. J., Grisar, R., Klocke, U., Knothe, M., Wolf, H., Schottka, P., Bessey, E., and Pelz, N. (1992). Analysis of trace gas components in automotive exhaust gas. In *Monitoring of Gaseous Pollutants by Tunable Diode Lasers*, (R. Grisar, H. Boettner, M. Tacke, and G. Restelli, Eds.), pp. 319–324. Kluwer Academic, Dordrecht, The Netherlands.

Riedel, W. J., Klocke, U., and Wolfe, H. (1991). Time-resolved exhaust gas analysis by infrared laser spectroscopy. *Proc. ISATA Int. Symp. Automot. Technol. Autom., 24th* Florence, Italy, pp. 289–295.

Sachse, G. W., and Hill, G. F. (1987). Aircraft based sensor for fast response measurements of atmospheric trace gases. In *Monitoring of Gaseous Pollutants by Tunable Diode Lasers*, (R. Grisar, H. Preier, G. Schmidtke, and G. Restelli, Eds.), pp. 68–69. Reidel, Dordrecht, The Netherlands.

Sachse, G. W., Hill, G. F., Wade, L. O., and Perry, M. G. (1987). Fast response, high-precision carbon monoxide sensor, using a tunable diode laser absorption technique. *J. Geophys. Res.* **92D**, 2071–2081.

Sachse, G. W., Collins, J. E., Jr., Hill, G. F., Wade, L. O., Burney, L. G., and Ritter, J. A. (1991). Airborne tunable diode laser sensor for high-precision concentration and flux measurements of carbon monoxide and methane. In *Measurement of Atmospheric Gases*, (H. I. Schiff, Ed.), SPIE **1433**, 157–166.

Schiff, H. I., Ed. (1991). *Measurement of Atmospheric Gases* SPIE **1433**.

Schiff, H. I., and Mackay, G. I. (1989). Tunable diode laser absorption spectrometry as a reference method for tropospheric measurements. In *Monitoring of Gaseous Pollutants by Tunable Diode*, (R. Grisar, G. Schmidtke, M. Tacke, and G. Restelli, Eds.), pp. 36–46. Kluwer Academic, Dordrecht, The Netherlands.

Schiff, H. I., Harris, G. W., and Mackay, G. I. (1987). Measurement of atmospheric gases by laser absorption spectrometry. *ACS Symp. Ser.*, **349**, 274–288.

Schiff, H. I., Karecki, D. R., Harris, G. W., Hastie, D. R., and Mackay, G. I. (1990). A tunable diode laser system for aircraft measurements of trace gases. *J. Geophys. Res.* **95D**, 147–153.

Schiff, H. I., Mackay, G. I., Karecki, D. R., and Nadler, S. D. (1991). *A Small Tunable Diode Laser Absorption Spectrometer for CH_4 Flux Measurements from a small Aircraft*. Am. Geophys. Union, Baltimore, MD.

Schiff, H. I., Mackay, G. I., Karecki, D. R., Pisano, J. T., and Nadler, S. D. (1992). Measurements of H_2O_2 in the Grand Canyon by tunable diode laser spectrometry during the Navajo generating station visibility study. In *Monitoring of Gaseous Pollutants by Tunable Diode Lasers*, (R. Grisar, H. Boettner, M. Tacke, and G. Restelli, Eds.), pp. 21–30. Kluwer Academic, Dordrecht, The Netherlands.

Schupp, M., Bergamaschi, P., and Harris, G. W. (1992). Measurements of the $^{13}C/^{12}C$ ratio in methane using a tunable diode laser spectrometer. In *Monitoring of Gaseous Pollutants by Tunable Diode Lasers*, (R. Grisar, H. Boettner, M. Tacke, and G. Restelli, Eds.), pp. 343–352. Kluwer Academic, Dordrecht, The Netherlands.

Shani, Y., Katzir, A., Tacke, M., and Preier, H. M. (1989). $Pb_{1-x}Sn_xSe/Pb_{1-x-y}Eu_ySn_xSe$ corrugated diode lasers. *IEEE J. Quantum Electron.* **25**, 1828–1844.

Shinohara, K. (1989). Developments of lead-chalcogenide lasers at Fujitsu. In *Monitoring of Gaseous Pollutants by Tunable Diode Lasers*, (R. Grisar, G. Schmidtke, M. Tacke, and G. Restelli, Eds.), pp. 77–84. Kluwer Academic, Dordrecht, The Netherlands.

Silver, J. A., (1992). Frequency modulation spectroscopy for trace species detection: Theory and comparison among experimental methods. *Appl. Opt.* **31**, 707–717.

Silver, J. A., and Stanton, A. C. (1987). Airborne measurements of humidity using a single mode Pb-Salt laser. *Appl. Opt.* **26**, 2558–2566.

Silver, J. A., and Stanton, A. C. (1988). Optical interference fringe reduction in laser absorption experiments. *Appl. Opt.* **27**, 1914.

Silver, J. A., Bomse, D. S., and Stanton, A. C. (1991). Diode-laser measurements of trace concentrations of ammonia in an entrained-flow coal reactor. *Appl. Opt.*, **30**, 1505–1511.

Slemr, F., Harris, G. W., Hastie, D. R., Mackay, G. I., and Schiff, H. I. (1986). Measurement of gas phase hydrogen peroxide in air by tunable diode laser spectroscopy. *J. Geophys. Res.* **91D**, 5371–5376.

Spaink, H. A., Lub, T. T., and Smit, H. C. (1990). Improvement of the performance of a diode laser spectrometer by using sweep averaging and numerical data-processing. *Anal. Chim. Acta.*, **241**, 83–94.

Spanger, B., Schiessl, U., Lambrecht, A., Boettner, H., and Tacke, M. (1988). Near room temperature operation of DH–PbSrSe infrared diode lasers using MBE growth techniques. *Appl. Phys. Lett.*, **53**, 2582–2583.

Spears, D. L. (1983). IR detectors: Heterodyne and direct. *Springer Ser. Opt. Sci.*, **39**, 278–286.

Spilker, G., Daddato, R., Schiessl, U., Lambrecht, A., and Tacke, M. (1992). Linewidth and noise of lead chalcogenide diode lasers. In *Monitoring of Gaseous Pollutants by Tunable Diode Lasers*, (R. Grisar, H. Boettner, M. Tacke, and G. Restelli, Eds.), pp. 85–92. Kluwer Academic, Dordrecht, The Netherlands.

Stanton, A., and Hovde, C. (1992). Near infrared diode lasers measure greenhouse gases. *Laser Focus World* **8**, 117–120.

Stepanov, E. V., Zasavitskii, I. I., Moskalenko, K. L., and Nadezhdinskii, A.I. (1992). Application of tunable diode lasers for human expiration diagnostics. In *Monitoring of Gaseous Pollutants by Tunable Diode Lasers*, (R. Grisar, H. Boettner, M. Tacke, and G. Restelli, Eds.), pp. 353–370. Kluwer Academic, Dordrecht, The Netherlands.

Stevens, C. M., and Engelkemeir, A. (1988). Stable carbon isotopic composition of methane from some natural and anthropogenic sources. *J. Geophys. Res.* **93D**, 725–733.

Svelto, O. (1989). *Principles of Lasers*, 2nd ed., Plenum, New York.

Tacke, M. (1989). Recent results in lead salt laser developments at the MPI. In *Monitoring of Gaseous Pollutants by Tunable Diode Lasers*, (R. Grisar, G. Schmidtke, M. Tacke, and G. Restelli, Eds.), pp. 103–118. Kluwer Academic, Dordrecht, The Netherlands.

Thompson, G. H. B. (1980). *Physics of Semi-conductor Laser Devices*. Wiley, New York.

Uehara, K., and Tai, H. (1992). Remote detection of methane with a 1.66 μ diode laser. *Appl. Opt.* **31**, 809–814.

Verma, S. B., Ullman, F. G., Billesbach, D., Clement, R. J., and Kim, J. (1992). Eddy correlation measurements of methane flux in a northern peatland ecosystem. *Boundary-Layer Meteorol.* **58**, 289–304.

Wahlen, M., and Yoshinari, T. (1985). Oxygen isotope ratios in N_2O from different environments. *Nature (London)* **313**, 780–782.

Walega, J. G., Iguchi, T., Mackay, G. I., Schiff, H. I., Shetter, R. E., and Stedman, D. H. (1984). Comparison of a chemiluminescent and a tunable diode laser absorption technique for the measurement of nitrogen oxide, nitrogen oxide, and nitric acid. *Environ. Sci. Technol.* **18**, 823–826.

Wall, D. L. (1991). Advances in tunable diode laser technology for atmospheric monitoring applications. In *Measurement of Atmospheric Gases*, (H.I. Schiff, Ed.), SPIE **1433**, pp. 94–103.

Wang, L. G., Riris, H., Gallagher, T. F., and Carlisle, C. B. (1988). Comparison of approaches to modulation spectroscopy with GaAlAs semiconductor lasers—Application to water vapor. *Appl. Opt.* **27**, 2071–2077.

Webster, C. R. (1985). Brewster plate spoiler—A novel method for reducing the amplitude of interference fringes that limit tunable laser absorption sensitivities. *J. Opt. Soc. Am.* **B2**, 1464–1470.

Webster, C. R., and May, R. D. (1987). Simultaneous in situ measurements and diurnal variations of NO, NO_2, HNO_2, CH_4, H_2O and CO_2 in the 40 to 26 km region using an open-path tunable diode laser spectrometer. *J. Geophys. Res.* **92D**, 11,931–11,950.

Webster, C. R., and May, R. D. (1992). In situ stratospheric measurements of CH_4, $^{13}CH_4$, N_2O, and ^{18}OCO using the Bliss tunable diode laser spectrometer. *Geophys. Res. Lett.* **19**, 45–48.

Webster, C. R., Menzies, R. T., and Hinkley, E. D. (1988). Infrared laser absorption: Theory and applications. In *Laser Remote Chemical Analysis* (R. M. Measures, Ed.), pp. 163–272. Wiley, New York.

Webster, C. R., May, R. D., Toumi, R., and Pyle, J. A. (1990a). Active nitrogen partitioning and the nighttime formation of N_2O_5 in the stratosphere: Simultaneous in situ measurements of NO, NO_2, HNO_3, O_3, N_2O, using the Bliss diode laser spectrometer. *J. Geophys. Res.* **95**, 13,851–13,866.

Webster, C. R., Sander S. P., Beer, R., May, R. D., Knollenberg, R. G., Hunten, D. M., and Ballard, J. (1990b). Tunable diode laser IR spectrometer for in situ measurements of the gas phase composition and particle size distribution of Titan's atmosphere. *Appl. Opt.* **29**, 907–917.

Weitkamp, C. (1984). Calibration of diode laser 2nd-derivative modulation spectrometry with a reference cell. *Appl. Opt.* **23**, 83–86.

Weitkamp, C. (1987). Measurements of hydrogen chloride in the marine atmosphere with a tunable diode laser. In *Monitoring of Gaseous Pollutants by Tunable Diode Lasers*, (R. Grisar, H. Preier, G. Schmidtke, and G. Restelli, Eds.), pp. 17–28. Reidel, Dordrecht, The Netherlands.

Welford, D., and Mooradian, A. (1982). Observation of linewidth broadening in GaAlAs diode lasers due to electron number of fluctuations. *Appl. Phys. Lett.* **40**, 560–562.

Werle, P., and Slemr, F. (1991). Signal to noise ratio analysis in laser absorption spectrometers using optical multipass calls. *Appl. Opt.* **30**, 430–434.

Werle, P., Slemr, F., Gehrtz, M., and Brauchle, C. (1989a). Quantum limited FM-spectroscopy with a lead-salt diode laser. *Appl. Phys. B* **49**, 99–108.

Werle, P., Brauchle, C., Slemr, F., and Gehrtz, M. (1989b). Wideband noise characteristics of a lead-salt diode laser—possibility of quantum noise limited TDLAS performance. *Appl. Opt.* **28**, 1638–1642.

Werle, P., Josek, K., and Slemr, F. (1991). Application of FM spectroscopy in atmospheric trace gas monitoring: Study of some factors influencing the instrument design. In *Measurement of Atmospheric Gases*, (H.I. Schiff, Ed.), SPIE **1433**, pp. 128–135.

Werle, P., Muecke, R., and Slemr, F. (1992). Development of a prototype IR-FM absorption spectrometer: Design criteria and system performance. In *Monitoring of Gaseous Pollutants by Tunable Diode Lasers*, (R. Grisar, H. Boettner, M., Tacke, and G. Restelli, Eds.), pp. 169–182. Kluwer Academic, Dordrecht, The Netherlands.

White, J. [U] (1942). Long optical paths of large aperture. *J. Opt. Soc. Am.*, **32**, 285–288.

White, J. U. (1976). Very long optical paths in air. *J. Opt. Soc. Am.*, **66**, 411–416.

Whittaker, E. A., Shum, C. M., Grebel, H., and Lotem, H. (1988). Reduction of residual amplitude modulation in frequency modulation spectroscopy by using harmonic frequency modulation. *J. Chem. Phys.*, **5**, 1253–1256.

Wolf, H., and Riedel, W. J. (1987). NH_3 measurements in power plants with denox installations. In *Monitoring of Gaseous Pollutants by Tunable Diode Lasers* (R. Grisar, H. Preier, G. Schmidtke, and G. Restelli, Eds.), pp. 120–126. Reidel, Dordrecht, The Netherlands.

Wolf, H., Grisar, R., Klocke, U., Riedel, W. J., and Wissler, R. (1989). Dynamic car exhaust gas analysis using IR tunable diode lasers. In *Monitoring of Gaseous Pollutants by Tunable Diode Lasers* (R. Grisar, G. Schmidtke, M. Tacke, and G. Restelli, Eds.), pp. 61–67. Kluwer Academic, Dordrecht, The Netherlands.

Worsnop, D. R., Nelson, D. D., and Zahniser, M. S. (1993). Chemical kinetic studies of atmospheric reactions using tunable diode laser spectroscopy. In *Optical Methods in Atmospheric Chemistry* (H. I. Schiff and U. Platt, Eds.), SPIE **1715**, 18–33.

Yariv, A. (1981). *Quantum Electronics*. Wiley, New York.

Yoshinari, T., and Wahlen, M. (1985). Oxygen isotope ratios in N_2O from nitrification at a wastewater treatment facility. *Nature* (*London*) **317**, 349–350:

Zasavitskii, I. I., Kosichkin, Y. V., Nadezhdinskii, A. I., Stepanov, E. V., Tishchenko, A. Y., and Shotov, A. P. (1984). Investigation of the broadening of absorption lines of molecular gases by methods of pulsed diode laser spectroscopy. *Sov. J. Quantum Electron Engl. Transl.* **14** (12), 1615–1620.

CHAPTER

6

GAS MEASUREMENT IN THE FUNDAMENTAL INFRARED REGION

PHILIP L. HANST and STEVEN T. HANST

Infrared Analysis, Inc., Anaheim, California 92801

6.1. INTRODUCTION

6.1.1. Gases and Vapors

We consider here the measurement of substances in the gaseous state—gases and vapors. By *gases* we mean those compounds that at room temperature have a vapor pressure higher than 1 atm. Examples are CO_2, CO, NO_2, NO, methane, dimethyl ether, and arsine. By *vapors* we mean the gaseous phases of compounds that normally are liquids or solids (room temperature vapor pressure less than 1 atm). Examples here are water, methanol, pentane, diethyl ether, and naphthalene. At room temperature the group of compunds that may be measured as vapors is very large, and when the temperature is raised it becomes even larger. When the detection limit is extended down to 10^{-9} atm, a large percentage of monomeric materials are detectable as gases. However, most polymers, metals, salts, sugars, and other crystalline materials do not have sufficient vapor pressure for detection as gases. Most organic molecules that are part of living systems will never be detectable as gases because they will decompose before they are vaporized.

6.1.2. Infrared Signatures

All polyatomic molecules and heteronuclear diatomic molecules absorb infrared radiation. The absorption changes the molecular rotation and vibration. The pattern of absorption therefore depends on the physical properties of the molecule such as the number and type of atoms, the bond angles, and the

---

Air Monitoring by Spectroscopic Techniques, Edited by Markus W. Sigrist. Chemical Analysis Series, Vol. 127.
ISBN 0-471-55875-3

bond strengths. This means that each spectrum differs from all others and may be considered the molecular "signature." Monatomic gases such as radon and homonuclear diatomic molecules such as oxygen and nitrogen do not have infrared bands and therefore must be measured by noninfrared means. Diatomic molecules such as NO, CO, HCl, and HF have a single major band that is an array of individual lines, each with a width of about 0.2 cm^{-1}. Linear polyatomic molecules like CO_2, N_2O, and C_2H_2 also show arrays of individual lines. Nonlinear polyatomic molecules like O_3, SO_2, NH_3, H_2CO, CH_4, and H_2O have many apparent "lines" that are in fact small bundles of lines, with the widths of the bundles varying from 0.2 cm^{-1} to many cm^{-1}. For larger polyatomic molecules at atmospheric pressure there are so many lines overlapping each other that the spectral features are broad and smooth, except for occasional "spikes." Figure 6.1 shows the principal absorption bands of five molecules that are important as air pollutants. The shapes of all these bands are different from each other, and each band is centered at a different frequency in the infrared spectrum. These spectra illustrate the effects of molecular size and shape.

6.1.3. The Fundamental Infrared Region

The spectral region 3700–500 cm^{-1} includes almost all the important absorption bands of gaseous molecules. This is the fundamental infrared region where the rotation and vibration of the molecules give rise to infrared absorption. It has been fashionable lately to call this region the middle infrared, to distinguish it from the near infrared and the far infrared. In the context of measuring gases, in our opinion however, middle infrared is not an appropriate term because there is nothing of any value on either side of the fundamental region. HF is the only important molecule that has a strong band at frequencies higher than 3700 cm^{-1}. At higher frequencies there are only overtones, which have extremely small absorption coefficients. At lower frequencies there may be good strong rotation lines, but because of extremely strong absorption by water vapor they are quite inaccessible. For these reasons we choose here to retain the traditional term, the *fundamental infrared region.*

6.1.4. Instrumentation Improvements and Detection Limit Extensions

Infrared spectrometers have been widely used for more than 40 years, and during that time there have been many improvements in instrumentation and technique. In recent years, the rate of instrument improvements has increased and prices have come down markedly while instrument performance has gone up. Improvements in detectors and computers have especially influenced the increase in gas measurement capability. It is now practical to measure gases

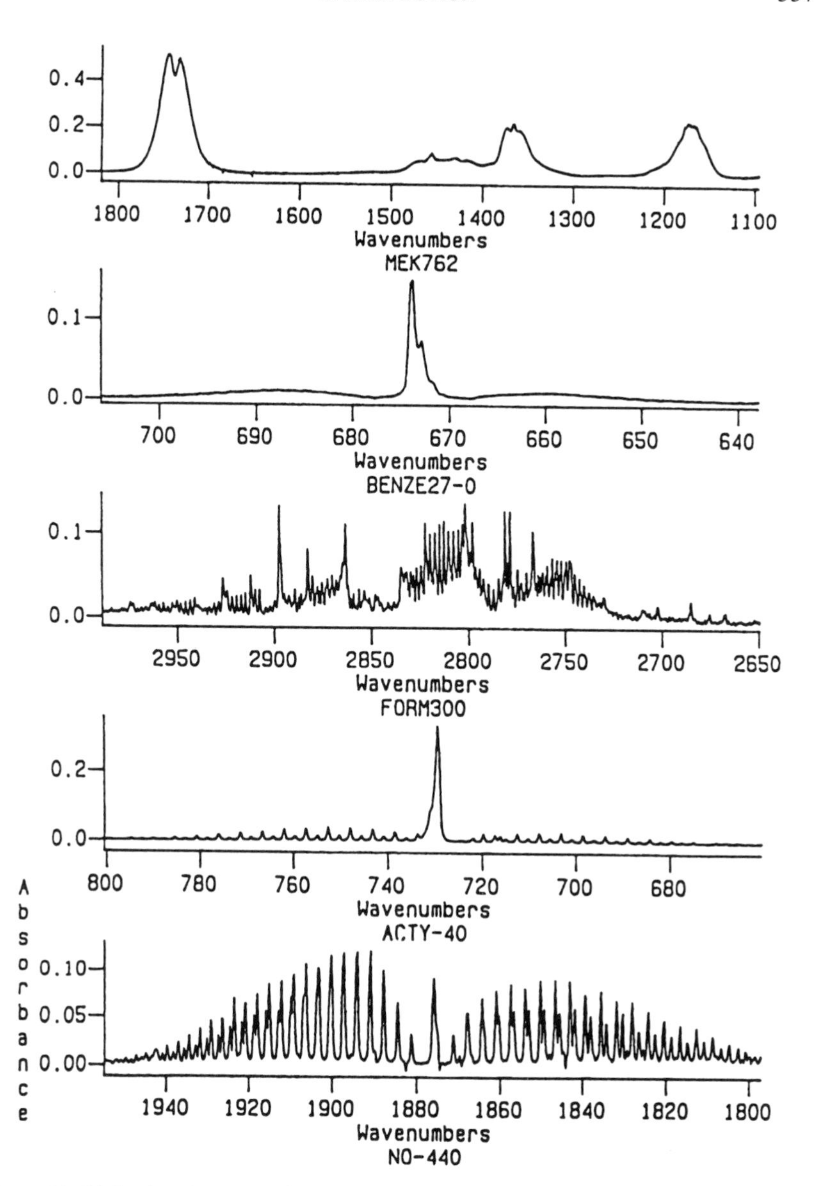

Figure 6.1. Molecular signatures (from bottom to top: nitric oxide, a heteronuclear diatomic molecule; acetylene, a 4-atom linear molecule; formaldehyde, a 4-atom nonlinear molecule; benzene, a 12-atom symmetrical molecule; and methyl ethyl ketone, a 13-atom nonsymmetrical molecule. Resolution = $0.5\ \text{cm}^{-1}$.

over the range of partial pressures from above 1 atm to less than 10^{-9} atm (Hanst, 1978). The limit of detectability is determined by path length and signal-to-noise ratio in the spectrum. With a Fourier transform spectrometer and a good nitrogen-cooled photodetector, a few minutes of scan time can give a signal-to-noise ratio of 10^4 or higher. Currently available software allows detection and quantitative measurement of trace gases whose strongest spectral features are at the noise limit. The software also allows removal of spectral interference from water and other compounds without introducing noise into the spectrum. The low detection limits for gases that are obtained with current instrumentation are presented below.

6.1.5. Detection Limits

It can be assumed that one can detect any spectral line or band that is approximately as high in the spectrum as the noise level. Detectability is increased in direct response to reduction of the noise. The noise level is a function of the source intensity, the number of scans, the spectral resolution, the types of optical filters used, the stability of the optical components, and especially the quality of the detector itself. Twenty years ago, in a Fourier transform infrared (FTIR) experiment at a folded path of several hundred meters, a signal-to-noise ratio of 2000 was obtained (Hanst et al., 1973). With present day equipment and a 100-m White cell, a signal-to-noise ratio of 10,000 is readily obtained.

There are two ways to lower the detection limit for a given compound: (1) lower the noise level in the spectrum, and (2) increase the size of the absorption features.

The noise level is made lower by choosing a low-noise detector and by bringing to that detector a maximum number of photons while avoiding detector saturation. The best available detectors today are the photoconductors that operate at liquid nitrogen temperature, like mercury–cadmium–telluride (HgCdTe). These detectors usually produce a noise level in the spectrum that is about 100 times lower than the noise level produced by detectors that operate at room temperature. The rate at which photons are delivered to the detector is kept high by having a bright source, an efficient Fourier transform spectrometer, and an optical system that conserves the energy. In designing systems for air analysis, this energy conservation factor is sometimes not given enough emphasis.

An additional way to lower the noise level in the spectrum is to use longer measurement times and co-add more interferometer scans. It is difficult to make up for lost photons in this way, however, because the lowering of the noise level is not proportional to the extension of the time. The lowering is only proportional to the square root of the time extension. If for some reason the rate of delivery of photons to the detector is reduced by a factor of 10, the noise level in the spectrum will go up by a factor of 10. If one sought to restore the lower noise level by extending the measurement time, a 100-fold extension would be required, changing a 10-min measuring time to 17 h.

For maximum spectral detail and maximum line heights the spectrometer should fully resolve the spectral lines. When one is working at 1 atm total pressure, a resolving power of $0.125\,cm^{-1}$ is appropriate because pressure-broadened line widths are about $0.2\,cm^{-1}$ [*full width at half-maximum intensity* (fwhm)]. For most instruments, however, going to higher resolution increases the noise in the spectrum at a faster rate than it increases the height of the lines. A compromise in the matter of resolution is required. A value of $0.5\,cm^{-1}$ seems a good choice.

The main way to increase the size of the spectral features is to lengthen the optical path. The principal limitation on path lengthening is that it should not waste the available photons. An additional problem is that when the path gets longer than about 100 m, absorption by water vapor and carbon dioxide precludes the use of three important bands of atmospheric pollutants: the NO_2 band near $1600\,cm^{-1}$, the SO_2 band near $1360\,cm^{-1}$, and the benzene band at $674\,cm^{-1}$. The importance of these three bands makes it worthwhile to keep the total path at 100 m or less.

When the optical path is at 100 m, the resolution is at $0.5\,cm^{-1}$ and the spectrum noise level is at 10^{-4} AU (absorbance units) nearly every important gas can be measured at ppb (parts per billion) levels. Minimum detection limits under these conditions were calculated from the atlas of spectra (Hanst and Hanst, 1992). These are listed in Table 6.1.

An important condition for validity of Table 6.1. is that the water and carbon dioxide lines must be subtracted from the spectrum without introducing noise. In the past this condition was almost impossible to meet, but now it can be met. The modern software has made it possible to match the line positions and line widths of two spectra with a high degree of precision. The software also has made it possible to create reference spectra showing deviations from the absorbance law that are the same as the deviations in the sample spectrum. (For a detailed discussion of subtraction problems, see Section 6.2.8, below.) Digitized spectra for subtraction of interferences of water, carbon dioxide, and other compounds are available from Infrared Analysis, Inc.

The column of minimum detection limits (MDLs) in Table 6.1 shows that under the conditions discussed most compounds are detectable down to the ppb level. Some compounds have MDLs much below 1 ppb. Note the extremely low numbers for CF_4 and SF_6.

Some of the strongest bands of the compounds have not been used in preparing the table because of interferences from water or carbon dioxide. All

Table 6.1. Minimum Detection Limits When Impurities Are Measured in Air, for Compounds in the Infrared Analysis, Inc. Atlas of Digitized Quantitative Reference Spectra

Compound	Frequency (cm^{-1})	MDL[a] (ppb)	Remarks
Acetaldehyde	2900–2600	6.0	Broad band with some fine structure; open region
Acetic acid	642	1.0	Spike
Acetone	1218	4.0	Broad band with some structure
Acetonitrile	1080–1010	30.0	Array of lines
Acetyl chloride	1130–1090	1.0	Band with spikes
Acetylene	739	0.15	Spike
Acrolein	959	5.0	Spike
Acrylic acid	640	0.5	Spike
Acrylonitrile	954	3.0	Spike
Ammonia	970–920	1.0	Two bundles of lines
Aniline	782	10.0	Spike
Arsine	2126	2.0	Spike
Benzene	674	0.6	Sharp line; CO_2 must be subtracted
Boron trichloride	970–940	0.2	Strong band
Bromoform	1150	2.0	Smooth band
Bromomethane	1030–900	15.0	Array of lines
1,3-Butadiene	908	1.0	Spike
Butane	2968	1.5	C—H band; has a spike
2-Butanone	1171	3.0	Band about 40 cm^{-1} wide
Carbon dioxide	2363	0.4	MDL is 0.4 ppb if no other CO_2 is present; in air, the minimum detectable change in CO_2 would be about 50 ppb
Carbon monoxide	2200–2100	2.0	Array of lines
Carbon tetrachloride	800–790	0.2	Very strong band
Carbonyl sulfide	2080–2030	0.3	Broad band
Chlorobenzene	741	0.8	Spike
Chlorodifluoromethane	809	0.5	Spike
Chloroethane	995–950	9.0	Band with *P*–*Q*–*R* structure
Chloroform	780–760	0.5	Wide band, but strong
Chloromethane	732	5.0	Spike

Table 6.1. *(Continued)*

Compound	Frequency (cm^{-1})	MDL[a] (ppb)	Remarks
Chlorotrifluoromethane	1107	0.7	Spike
Crotonaldehyde	2900–2600	10.0	Broad bands in unique position
Cyclohexane	2934	0.5	Spike in C—H band
Cyclohexene	2936	1.0	Spikes in C—H band; also spikes in region 1160–650 cm^{-1}
Cyclopentene	698	1.2	Spike
Cyclopropane	3101	1.0	Sharp spike in C—H band
1,2-Dibromoethane	1195–1180	2.5	Band with characteristic shape
m-Dichlorobenzene	800–770	1.5	Band with spike
o-Dichlorobenzene	749	1.5	Spike
p-Dichlorobenzene	820	1.5	Spike
Dichlorodifluoromethane	1161	0.2	Strong spike
1,1-Dichloroethane	1060	3.0	Spike
1,2-Dichloroethane	740–710	2.0	Broad band
cis-1,2-Dichloroethylene	696	0.6	Spike
Dichloromethane	770–740	1.3	Band with some structure
1,2-Dichloro-tetrafluoroethane	1250–800	0.7	Group of strong bands
Diethylamine	1130–1160	3.0	Broad band
Diethyl ether	1150–1120	1.0	Smooth band
Diisopropyl ether	1400–1000	2.0	Group of bands
Dimethyl ether	1179	2.5	Band with *P–O–R* structure
1,1-Dimethylhydrazine	2814, 2775	4.0	Spikes
Dimethyl sulfide	2980–2900	8.0	Structure in C—H band
Dinitrogen pentoxide	1240, 740	1.0	Two strong bands
Ethane	3050–2880	1.5	Structure in C—H band
Ethanol	1090–1020	4.0	Broad band with some structure
Ethyl acetate	1270–1225	0.5	Broad band
Ethyl acrylate	1210–1180	0.5	Broad band
Ethylbenzene	698	4.0	Spike
Ethyl formate	1210–1150	0.5	Broad band with some structure

Table 6.1. *(Continued)*

Compound	Frequency (cm^{-1})	MDL[a] (ppb)	Remarks
Ethylene	950	1.0	Spike
Ethylene oxide	3066	0.5	Spike
Ethyl vinyl ether	1230–1200	2.0	Wide band with some structure
Fluorobenzene	754	0.5	Spike
Formaldehyde	2711, 2779	3.0	Two sharp spikes
Formic acid	1105	0.8	Spike
Furan	744	0.2	Spike
n-Hexane	2970	1.0	C—H band not specific; total hydrocarbon is often quoted in hexane equivalents
Hydrazine	1000–890	4.0	Band with complex fine structure
Hydrogen bromide	2700–2400	8.0	Array of lines
Hydrogen chloride	3050–2700	1.5	Array of lines
Hydrogen cyanide	712	0.4	*Q*-branch
Hydrogen fluoride	4200–3700	1.0	Array of lines
Hydrogen sulfide	1300–1200	400	Many lines, but extremely weak
Isobutane	2967	1.0	Spike in C—H band
Isobutanol	1060–1030	3.0	Broad band
Isobutylene	890	1.5	Spike
Isoprene	900	1.5	Two distinctive spikes near 900 cm^{-1}
Isopropanol	1420–900	7.0	Group of bands
Mesitylene	836	3.0	Spike
Methane	3018	2.0	Spike and fine structure in C—H band
Methanol	1033	1.5	Spike
Methyl acetate	1265–1230	1.0	Broad band with some structure
Methyl acrylate	1220–1180	1.0	Broad band with some structure
Methylamine	820–720	2.5	Band with structure
2-Methyl-2-butene	2969	5.0	C—H band not specific; use spike at 890 cm^{-1} for identification

Table 6.1. *(Continued)*

Compound	Frequency (cm^{-1})	MDL[a] (ppb)	Remarks
3-Methyl-1-butene	2970	2.0	C—H band not specific; use feature at 913 cm^{-1} for identification
Methyl formate	1210	1.5	Spike
Methyl methacrylate	1215–1150	1.2	Two-hump broad band
Methyl nitrite	860–790	2.0	Band with structure
2-Methylpentane	2967	1.0	C—H band not specific
3-Methylpentane	2967	1.0	C—H band not specific
2-Methyl-1-pentene	2970	2.5	C—H band not specific; use feature at 892 for identification
2-Methyl-2-pentene	2973	2.0	C—H band not specific
4-Methyl-2-pentene	2967	1.5	C—H band not specific; use feature at 721 cm^{-1} for identification
Methyl vinyl ether	1400–800	1.0	Five different spikes
Methyl vinyl ketone	744	8.0	Spike
Nitric acid	896, 879	2.0	Spikes
Nitric oxide	1920–1870	4.0	Array of lines
Nitrobenzene	701	1.0	Spike
Nitroethane	900–850	10.0	Band with P–Q–R shape
Nitromethane	656	3.0	Spike
Nitrogen dioxide	1600	2.0	Several bundles of lines
Nitrous acid	852, 790	0.5	Q-branches
Nitrous oxide	2210	1.0	Array of lines
Ozone	1054	2.0	Band with structure
n-Pentane	2966	1.5	Difficult to distinguish from other large alkanes
1-Pentene	2968	2.0	C—H band strongest but not specific; use feature at 915 cm^{-1} for identification
2-Pentene	2970	2.0	C—H band strongest but not specific; use feature at 961 cm^{-1} for identification
Phosgene	864–830	0.3	Strong band with structure
Phosphine	992	10.0	Spike
Phosphorus trichloride	515–495	3.5	Smooth band

Table 6.1. *(Continued)*

Compound	Frequency (cm^{-1})	MDL[a] (ppb)	Remarks
Propane	2968	1.0	Spike in C—H band
Propionaldehyde	2850–2650	8.0	Broad bands in the open
Propionic acid	1114	2.0	Band 30 cm^{-1} wide
Propylene	913	2.0	Spike
Propylene oxide	3050	3.0	Spike
Styrene	695	2.0	Spike
Sulfur dioxide	1361	2.0	Spike; water must be carefully subtracted
Sulfur hexafluoride	950–940	0.04	Extremely strong band
1,1,1,2-Tetrachloroethane	980–700	1.3	Group of bands
Tetrachloroethylene	930–900	0.7	Smooth band
Tetrafluoromethane	1283	0.03	Extremely strong spike
Toluene	729, 694	2.0	Spikes
1,1,1-Trichloroethane	735–715	0.5	Smooth band
1,1,2-Trichloroethane	760–720	0.8	Broad band
Trichloroethylene	783	1.4	Spike
Trichlorofluoromethane	855–840	0.2	Very strong band
Trichlorotrifluoroethane	1240–800	0.6	Group of strong bands
Vinyl acetate	1240–1210	0.7	Broad band
Vinyl chloride	942, 896	2.0	Two spikes
Vinylidene chloride	869	1.0	Spike
Water	1700–1400	5.0	MDL is 5 ppb if no other water is present; in humid air, the minimum detectable change in water content would be 1000–2000 ppb
m-Xylene	768	2.0	Spike
o-Xylene	741	1.0	Spike
p-Xylene	795	2.0	Spike

[a] MDL is the minimum detection limit in parts per billion (10^{-9} atm) when resolution is 0.5 cm^{-1}, optical path is 100 m and spectrum noise level is at 10^{-4} AU (logarithm, base 10).

the carbonyl bands in the region 1800–1700 cm^{-1} have been left out, for example, because of water vapor interference.

In Table 6.1 a narrow and intense spectral feature has been designated "spike." These features occur in the majority of the spectra, and they are the main reason for the specificity and sensitivity of the infrared method of measurement. It is the families of spikes and families of lines that give us confidence that a spectrum can be identified when the maximum absorbance is equal to the peak-to-peak noise level.

It is noteworthy that the MDLs in Table 6.1 are similar to those published 20 years ago by Hanst et al. (1973). The reason for this is that the calculations then and now were based on similar experimental results obtained with an FTIR spectrometer system, HgCdTe detector, and multiple-pass cell. The main difference between then and now is that formerly we could not properly subtract the interferences but now we can.

The MDLs in our table are some 10 to 100 times lower than the MDLs recently published by Grant et al. (1992) for an FTIR system working over an open-air doubled long path totaling 400 m. The main source of the differences is the two different spectrum noise levels: 1.3×10^{-3} AU for Grant et al., and 1×10^{-4} AU for us. The reason for the two different noise estimates is that they were derived from experience with two radically different optical systems. The simple fact is that our measurement system with the multiple-pass cell is a high-throughput (low-noise) system whereas their measurement system with the long-path retroreflector technique is a low-throughput (high-noise) system.

An additional advantage for us is that we calculate for only a 100-m path in air. This lets us consider some important absorption bands that are not accessed in a 400-m path. This includes strong bands of NO_2, SO_2, and benzene. Finally, we have said that we can recognize a line or band when its height is equal to the peak-to-peak noise level in the spectrum, whereas they assumed their detectable band height to be 3.3 times greater than the noise.

6.2. MANIPULATING THE SPECTRA

6.2.1. Line Widths and Resolution Requirements

The fine structure in the spectra of gaseous molecules is of great value in identification, in discrimination against interferences and in quantitative measurements. High enough resolution should be used to take advantage of this fine structure. At atmospheric pressure and room temperature the widths of individual spectral lines are about 0.2 cm^{-1} (fwhm). To see all spectral detail therefore requires resolving power on the order of 0.1 cm^{-1}. Most commercial

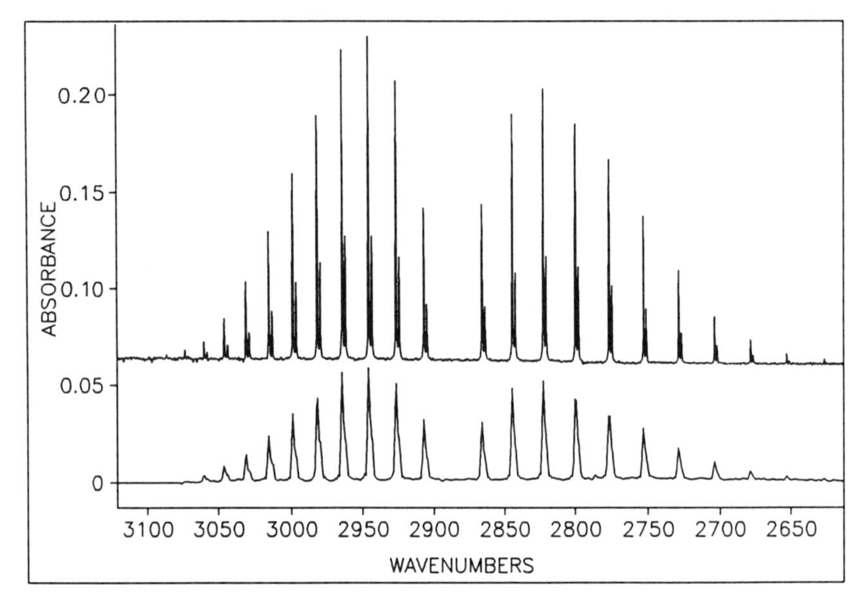

Figure 6.2. Hydrogen chloride at two different spectral resolutions: (lower) resolution = 2 cm^{-1} (upper) resolution = 0.5 cm^{-1}—same absorbance scale as the lower, with spectrum displaced 1-in. upward.

spectrometers, however, have a resolving power limit of 0.5, 1, or 2 cm^{-1}. A good example of the difference between instruments having limits of 0.5 cm^{-1} and those having limits of 2 cm^{-1} is given by the HCl spectra of Fig. 6.2. Here, the lines are four times taller when the resolution is four times better. Furthermore the isotopic doublets are clearly resolved at 0.5 cm^{-1} and not resolved at all at 2 cm^{-1}. At a resolving power of 0.1 cm^{-1}, the HCl lines would be another two or three times taller. However, there are some disadvantages to going all the way to 0.1 cm^{-1}. The data arrays are proportionately larger at the higher resolution, requiring more computer memory and longer computation times. Also, at the higher resolution the spectral noise is higher.

Thus it appears that trade-offs are in order. High resolving power is exchanged for a more favorable signal–to–noise ratio. In choosing resolving power, one should also consider what kinds of molecules are to be measured. Only the smaller and more symmetrical molecules show arrays of individual lines that would benefit from higher resolution. The heavier molecules have so many spectral lines that their overlap produces smooth bands with occasional "spikes." High resolution brings no benefits to the study of these heavier molecules. For measurement of a wide variety of molecular gases, a good compromise choice of resolution is 0.5 cm^{-1}.

6.2.2. Some Benefits from Pressure Reduction

At constant temperature and at pressures above about 0.5 atm, the absorption coefficient α at wavenumber v will depend on pressure P approximately as

$$\alpha_v = \frac{C_1 P}{(v - v_0)^2 + C_2 P^2} \tag{6.1}$$

where C_1 and C_2 are constants characteristic of the absorbing gas. At the center of the line ($v = v_0$), α will be inversely proportional to pressure. In the wings of the line, α will be directly proportional to pressure. Near the steeply

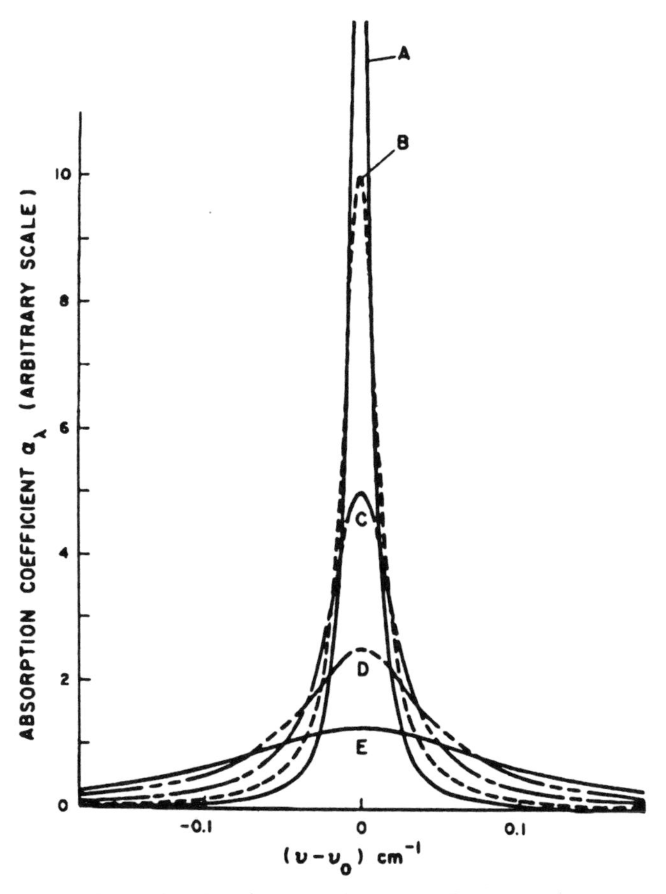

Figure 6.3. Line profile as a function of pressure. Curves A to E correspond to pressures of 48, 95, 190, 380, and 760 torr.

rising sides of the lines, α will vary with total pressure in a more complicated way, and as pressure is changed, α may go through a maximum.

The change of line profile with pressure is illustrated in Fig. 6.3, which was plotted from the formula above. A full width of 0.2 cm^{-1} at the half-maximum intensity points was chosen as being typical of absorption lines of gases at normal temperatures and pressure. The area under the plotted spectral lines depends on the transition probability of the vibration–rotation change and therefore remains constant as long as the number of molecules in the light beam remains constant.

The extent to which an absorption band changes with pressure reduction depends on the spacing of the individual absorption lines. When there are hundreds of lines per wavenumber, as may be the case with some of the larger pollutant molecules, the absorption band will not develop a fine-line structure until pressure is reduced to a few hundredths of an atmosphere. When there are about 10 lines per wavenumber, as is the case with some of the lighter

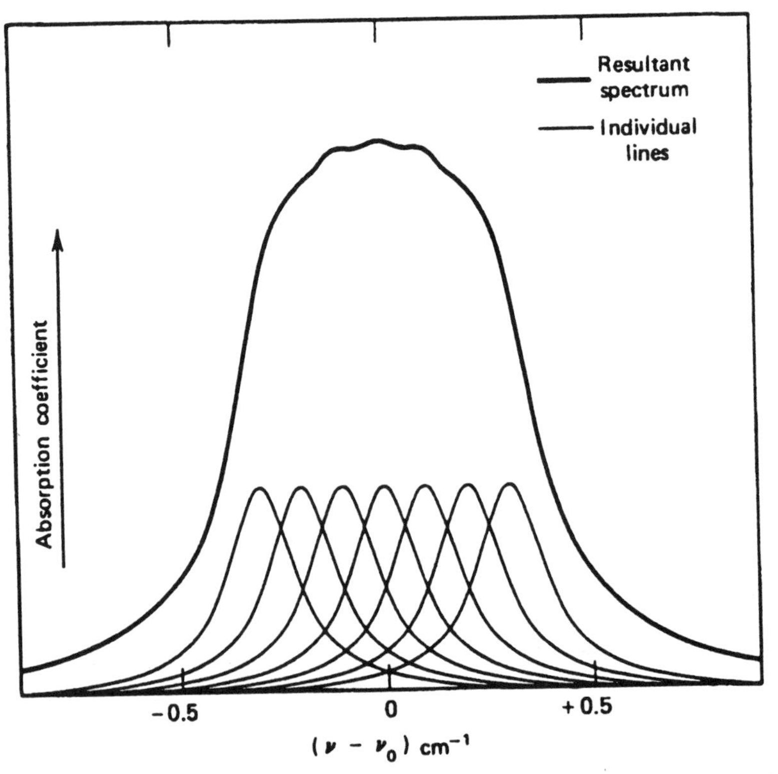

Figure 6.4. Spectral lines at atmospheric pressure, with their resultant absorption band.

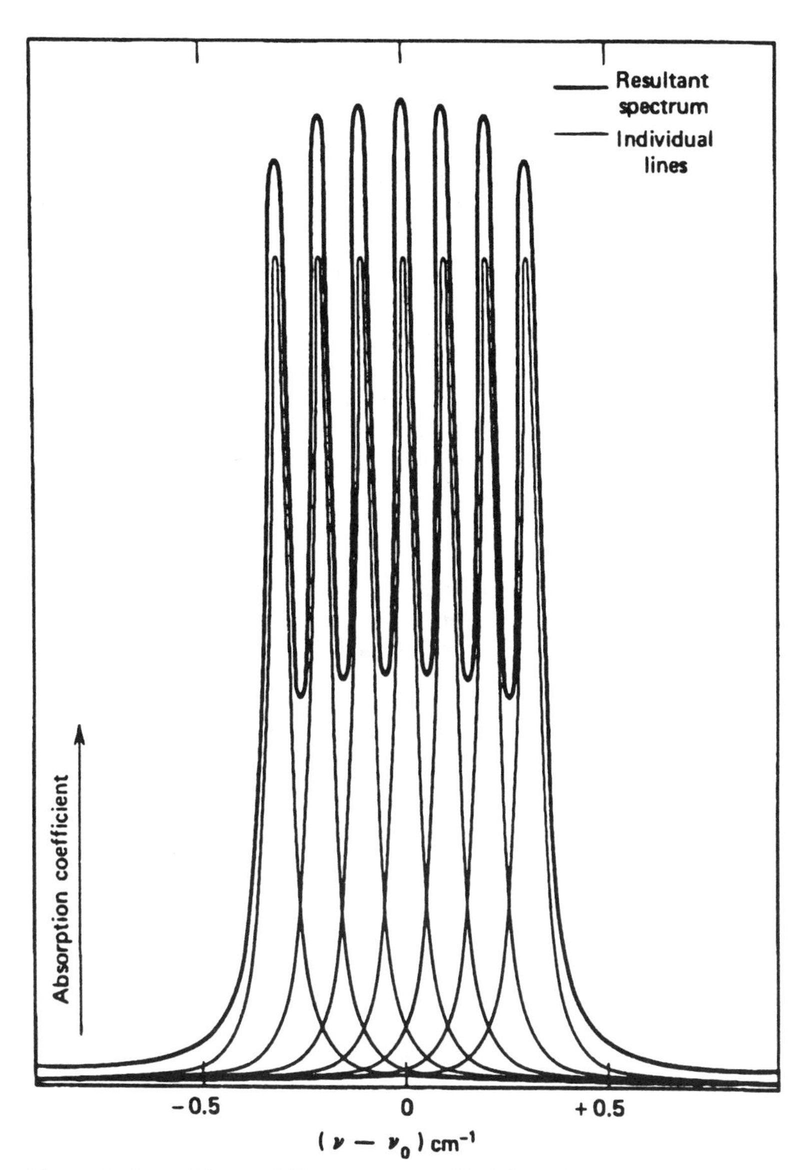

Figure 6.5. Spectral lines at 0.25-atm pressure, with their resultant absorption band.

pollutant molecules, the spectrum will be continuous at 1 atm pressure but will break into individual lines with only a fourfold reduction in pressure. This is illustrated in Figs. 6.4 and 6.5, which show seven lines spaced 0.1 wavenumber apart. In the atmospheric pressure case there is a single absorption feature almost a wavenumber in width. The second case (Fig 6.5) shows the same seven lines with pressure reduced to 0.25 atm and the light path made four times longer. The integrated absorption seen by an instrument of low resolution would still be the same after pressure reduction, but the true spectrum would have acquired the indicated fine structure, which can be used to advantage by some instrument designs.

The spectrum of nearly every gaseous air pollutant can thus be broken into a fine-line structure by pressure reduction. The method will fail only when the lines are closer together than their Doppler width, which in the infrared vibration–rotation spectrum is on the order of one-thousandth of a wavenumber.

6.2.3. Contending with the Overlapping of Absorption Bands

When there is a small number of gases in a sample, severe interferences between the spectra are not likely. The reason for the infrequency of interferences is that the strong spectral features that are used for quantitative analysis are relatively narrow compared to the width of the whole spectrum and they fall at many different spectral locations. The worst cases of interference arise in the study of mixtures of compounds with similar molecular structures, like mixtures of paraffinic hydrocarbons or of aliphatic acids.

The main case of interference encountered in air studies is the interference between the spectrum of water and the spectra of the trace gases. The water in the air is so much higher in concentration than any other infrared-absorbing molecule that a spectrum of an atmospheric sample seems to have water lines everywhere. The problem of the water lines is much more severe for the lower atmosphere than it is for the upper atmosphere.

Figure 6.6 shows an example of removing interfering water lines by forming a ratio between the polluted air spectrum and a reference spectrum. These spectra were recorded with a path of 81 m, air pressure 150 torr, and spectral resolution 1 cm^{-1}. The contaminated air of the upper spectrum contained about 1 ppm NO_2 and 10 ppm NO. The reference spectrum was obtained with a sample of pure compressed air to which water vapor had been added. The scale–expanded ratio plot in the lower part of the figure shows a rather complete cancellation of the water lines.

An interesting facet of Fig 6.6 is the presence of extraneous pen lines in the ratio plot. It can be seen that these erratic pen excursions take place at the centers of the water lines. The reason is that there is practically no transmitted energy at the line centers in either the sample spectrum or the reference

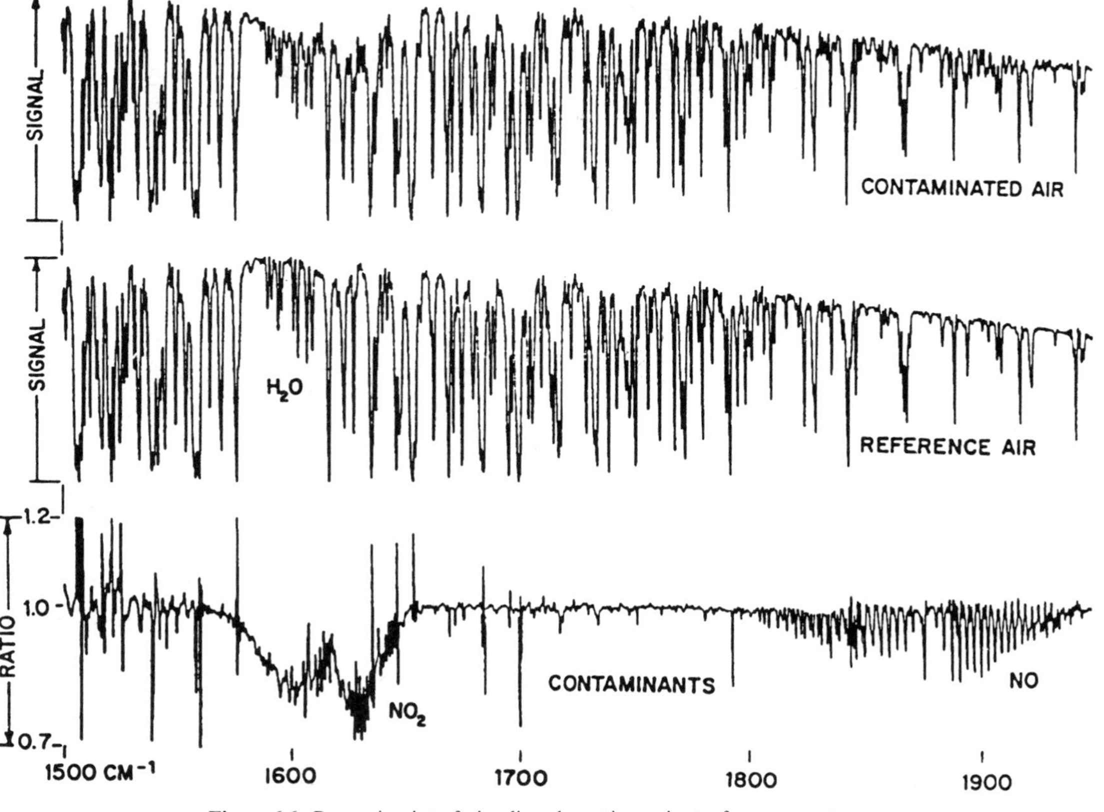

Figure 6.6. Removing interfering lines by ratio against reference spectrum.

spectrum. The instrument takes the ratio of noise over noise, which may fall at random somewhere between zero and a very large number. This noise-over-noise problem becomes more serious as resolution is increased. The spectra in Fig 6.6 were recorded in 1971, using an early version of the digital computer plotting programs that did not allow for the noise problem. Subsequently, the plotting programs have been altered to avoid noise. In ratio plotting the computer now passes over those points in the spectrum where the energy level in the reference beam falls to a specified low value. This experience illustrates the opportunities of spectrum enhancement that are inherent in the computer manipulation of the digitized specta.

Perhaps the worst case of severe interference arises when one is measuring impurities in an infrared-absorbing medium, such as in propane or in nitric acid. In these cases one should try to remove the bands of the principal constituents. However, one must keep in mind that no measurements can be made in regions where no infrared energy is transmitted. If the analytical band of the trace component falls in a region where the principal component is absorbing all of the energy, then the sample must be diluted or the optical path must be shortened. The most sensitive detection of the trace component occurs when the principal component absorbs 63% $(1 - 1/e)$ of the energy.

The examples given in subsequent sections of this chapter include many cases of removal of interfering bands by subtraction of one spectrum from another.

6.2.4. Measuring Gases with Less Than Full Resolution

In quantitative analysis of gases, errors will arise when the resolution bandwidth is greater than the actual spectral line widths. We consider here the magnitude of error as a function of the width differences. We also discuss rules for minimizing or avoiding the error. With a low noise level in the spectrum, it is possible to do quantitative analysis over a wide range of concentrations, using only one reference spectrum for each compound measured. When noise levels are higher, a family of reference spectra may be needed.

We start by deriving the logarithmic law. When radiation passes through a thin layer of an absorbing medium (dx), the reduction of intensity $(-dI)$ is proportional to the intensity (I), the concentration of absorber (C), and the absorption coefficient (α):

$$-dI = \alpha CI\,dx \tag{6.2}$$

For sample thickness (L) and incident intensity (I_0), the above equation integrates to

$$-\log I/I_0 = \alpha CL \tag{6.3}$$

Where αCL is called the absorbance, and Eq. (6.2) is the absorbance law (Beer–Lambert law); I/I_0 is the transmittance.

Across a single absorption line the variation of α with frequency (ν) gives the line shape. This shape changes with total pressure, but when α is integrated over a whole line, the integral is constant, independent of total pressure.

If an instrument can reveal the true shape of spectral features, the true absorbance is revealed and it always obeys Eq. (6.2). Frequently, however, the instrumental resolving power is not high enough to reveal the true spectrum and one sees only an apparent absorbance. Apparent absorbance frequently does not obey Eq. (6.3). Basically, this happens because the transmittance saturates, or "bottoms out." This is illustrated in Fig. 6.7.

The figure applies to one of the line pairs of HCl with the concentration–pathlength product at 10000 ppm·m and a total sample pressure of 1 atm. Figure 6.7a shows the line pair with its nearly correct line widths of 0.2 cm^{-1} and the nearly correct absorbance values. (If the absorbance values were exactly right, the two maxima would have the ratio 3:1, because that is the isotopic ratio of chlorine-35 to chlorine-37.) Figure 6.7b shows the actual transmittance values for the sample. Figure 6.7c shows the transmittance values that would be measured by an instrument with resolution of 1 cm^{-1}. In Fig. 6.7d the lines are shown with the absorbance values calculated from Fig 6.7c. The striking difference between parts a and of Fig. 6.7 is the result of the bottoming of the transmittance lines and the instrumental smoothing to 1 cm^{-1}.

In Fig. 6.7 we have shown how incomplete resolution creates deviations from the absorption law. In Fig. 6.8 we show a full spectrum illustration of these deviations. This is an absorbance spectrum of a sample of air containing 1% HCl gas in a 1-m absorption cell. The resolution bandwidth was 0.5 cm^{-1}, which is approximately three times greater than the actual line width (fwhm). The spectrum illustrates our case because of isotopic doubling of the lines. The ratio of chlorine-35 to chlorine-37 is approximately 3:1. If the absorption law is being followed, the intensities of the two components of each line pair should also have the ratio of 3:1. This ratio holds true for the weak lines at the edges of the band, but as one goes up the band to the stronger lines the ratio becomes much smaller.

It thus appears that correct measurements are more easily carried out with low absorbance than with high absorbance. The magnitude of possible errors is illustrated in Fig. 6.9 by three calculated absorbance curves for an observed spectral feature (a single line, a bundle of lines, a Q-branch, or a whole band). We designate the wavenumber width of the resolution element as W and the wavenumber width of the spectral feature as V. If W is equal to or smaller then V, the spectral feature will be seen with its true maximum absorbance, which will always be proportional to the concentration of the absorbing

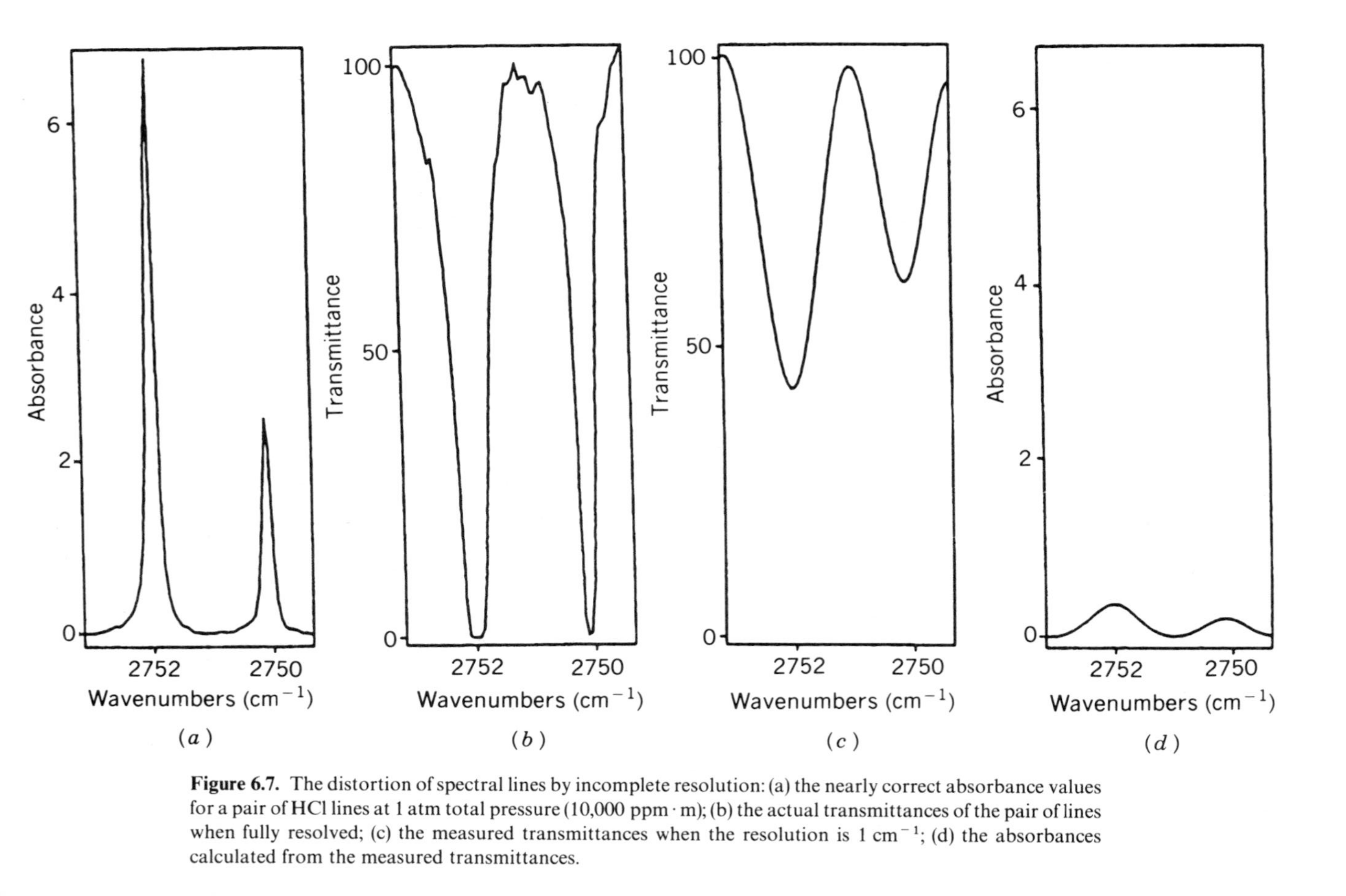

Figure 6.7. The distortion of spectral lines by incomplete resolution: (a) the nearly correct absorbance values for a pair of HCl lines at 1 atm total pressure (10,000 ppm · m); (b) the actual transmittances of the pair of lines when fully resolved; (c) the measured transmittances when the resolution is 1 cm^{-1}; (d) the absorbances calculated from the measured transmittances.

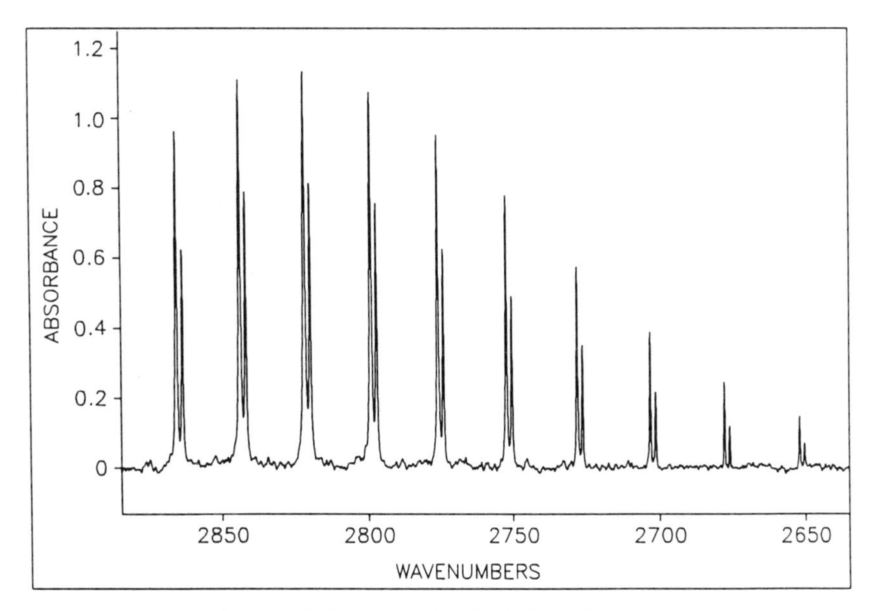

Figure 6.8. Hydrogen chloride (1%) in a 1-m cell.

molecule. The absorbance curve is then a straight line. If W is equal to $2V$, the maximum absorbance is only half as great as in the first case and furthermore the absorbance plot curves over and flattens out at high concentrations of the molecule. If W equals $4V$, the maximum absorbance is only one-fourth as great as when W equals V and the absorbance plot curves over and flattens even sooner.

In the case of the straight-line absorbance plot, only one calibration spectrum is needed to allow quantitative analysis over the full range of absorbance. In the cases of the curved absorbance plots, precise quantitative analysis requires multiple reference spectra. However, if measurements are confined to the low-absorbance region, a single reference spectrum may suffice for any of the three cases. In Fig. 6.9, two straight lines have been added to show the absorbance plots that are assumed by the use of only one calibration spectrum of maximum absorbance 0.1. The shaded areas show the errors incurred because of the actual curvature of the absorbance plots. For absorbance values below 0.1 the shaded areas indicate a small positive error—the concentrations calculated will be slightly higher than they should be. At the absorbance value of 0.1 there is no error. At absorbance values above 0.1 the error assumes negative values that increase with increasing absorbance until at high absorbance the calculated concentrations are far lower than they should be. If the absorbance of the calibration line is only 0.05, the positive errors in the

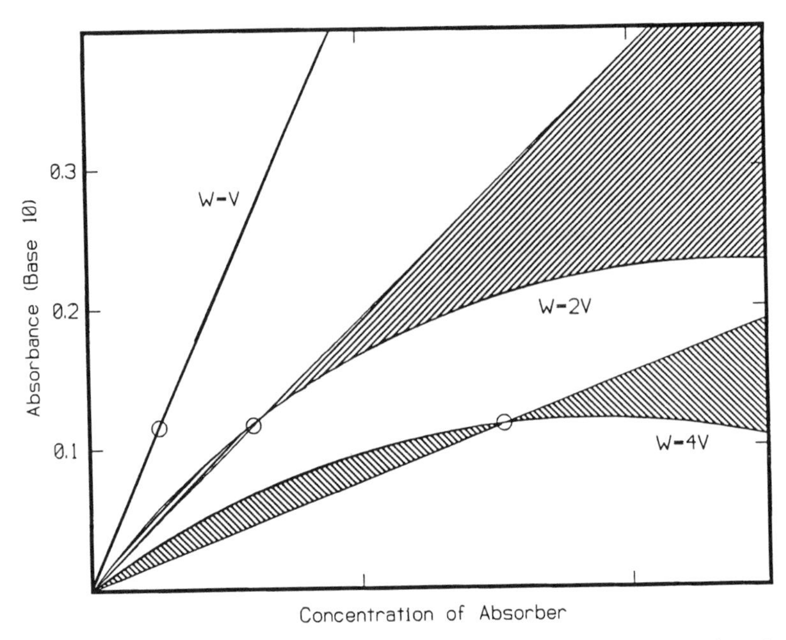

Figure 6.9. Absorbance curves and measurement error. W, Wavenumber width of resolution element; V, wavenumber width of spectral feature.

low absorbance region are smaller, and nearly correct measurements can be made at absorbance values all the way down to the spectrum noise level.

The above considerations suggest that the following points be observed when one is conducting quantitative analysis of gases:

1. The absorption law is most likely to fail for small molecules that have individual spectral lines, because it is probable that the resolution bandwidth will be larger than the spectral line width.
2. The absorption law is not likely to fail for heavy gaseous molecules that do not have fine structure in their infrared bands.
3. As resolution is increased, the absorbance errors are decreased.
4. The total sample pressure should be kept high by the addition of inert gas. The total pressure should be the same for both the sample spectrum and the reference spectrum.
5. The sample spectrum and the reference spectrum must both be recorded with the same spectral resolution.
6. When lines are not fully resolved, it is advantageous to record both the sample spectrum and the reference spectrum with small maximum

absorbance. Only one reference spectrum is then required for quantification over a wide range of values.

7. In measuring gases, the signal-to-noise ratio in the spectrum should be kept as high as possible, thus maximizing the range over which valid measurements can be made. In many cases, a signal-to-noise ratio of 10^4 will allow valid concentration measurements over a range of 10^3, using only one reference spectrum.

6.2.5. Reference Spectra and Subtraction Rules

The reference spectra have three important functions: (1) to assist in identifying the compound being measured; (2) to indicate the amount of compound present in the sample; and (3) to permit removal of the spectrum of the compound in order to allow identification and quantification of other compounds. Each reference spectrum should have a lower noise level than the sample spectrum. Each reference spectrum should also have the same pattern of deviations from the Beer–Lambert absorbance law that is exhibited by the sample spectrum (see Section 6.8).

The subtraction process for air requires one low-noise spectrum of the sample plus a low-noise reference spectrum of each molecule to be measured. The subtractions are carried out either interactively, with the operator viewing the computer screen, or automatically, with the computer operating within specified wavenumber intervals. Discussed below are some rules for obtaining correct subtraction results:

1. *Use high resolution.*

Since most of the molecules being measured have detailed infrared spectra with narrow absorption features, the spectrometer resolution should be high. The higher the resolution, the higher the apparent absorption coefficients and the wider the range of absorbances that may be used for quantification.

For air at total pressure of 1 atm the individual spectral lines have a width of about 0.2 cm^{-1}. To resolve this is beyond the capacity of most commercial spectrometers. Partial resolution must therefore be tolerated.

There are a number of moderately priced spectrometers that have maximum resolution of 0.5 cm^{-1}. Such a spectrometer is a good choice for air analysis. Although it will not show all spectral detail, it will allow successful spectral subtraction and quantification over a wide range of concentrations.

2. *Work with spectral features of low maximum absorbance.*

Subtraction and quantification work best at low apparent absorbances. How low the absorbance should be depends on the spectral resolution and on total sample pressure. With resolution of 0.5 cm^{-1} and 1-atm pressure, single spectral lines will obey the Beer–Lambert law at absorbances below 0.1 AU. Above 0.1 AU the apparent absorption coefficients begin to decrease and quantification becomes more difficult. With a family of reference spectra that accounts for the deviations from the Beer–Lambert law, one can work up to 1 AU, but the work will be less convenient and less accurate than work below 0.1 AU.

3. *Use nitrogen-cooled photodetectors.*

The noise level in the spectrum determines the range of concentrations over which the spectral subtraction will be valid. The lower the noise, the lower the detection limits and the wider the quantification range. A spectrometer with a room temperature detector might yield a noise level of 0.01 AU. This will restrict spectral subtraction and quantification to the 100-fold range from 0.1 to 1 AU. A nitrogen-cooled photodetector produces a noise level about 100 times lower than is obtained from a room-temperature detector. This will drop the noise level from 0.01 to 0.0001 AU, yielding a concentration range for spectral subtraction and quantification that is 100 times wider than before.

One can always lower the noise level in a spectrum by co-adding more scans. However, the reduction of noise is only proportional to the square root of the number of scans. To bring the spectrum noise level obtained with a room-temperature DTGS (deuterated triglycine sulfate) detector down to that obtained with a nitrogen-coolded HgCdTe detector would require a 10,000-fold increase in the number of scans. One scan with HgCdTe is equivalent to 10,000 scans with DTGS.

4. *Match the absorbance scales between sample and reference spectra.*

In a spectrum that covers a wide range of absorbances, with spectral details not fully resolved, only the weak spectral features will obey Beer–Lambert's law. If the concentration–path length products differ substantially between the sample and the reference spectra, the subtraction will be successful only for the weak spectral features. At the stronger lines and bands, the spectra are distorted and the logarithmic absorption law is not followed. In an atmospheric spectrum these distortions will occur especially for water and CO_2.

In the case of gases that will appear in the sample at ppb concentrations, a problem arises in getting low enough absorbances to match the sample absorbance. If the path is 200 m and the concentration is 1 ppb, the concentration–path length product is 0.2 ppm·m. This is several hundred times smaller than the concentration–path length products in the reference spectra. In doing a quantitative subtraction, the large mismatch of absorbances yields a very low subtraction factor. Then the steps in subtraction are too large for good quantification.

The software provides a way out of the problem. The spectrum can be divided by any chosen number. In the division, the relation between the absorption bands and the noise remains unchanged. Only the absorbance scale is reduced. Dividing a 300 ppm · m spectrum by 1500 yields a 0.2 ppm · m reference spectrum. For the case we are considering, the subtraction factor then is near unity and the ability to quantify is restored.

5. *Use noise-free reference spectra.*

In a subtraction, the noise in the reference spectrum is transferred to the sample spectrum. When multiple subtractions are carried out it is especially important that the reference spectra be low in noise. There is normally no problem with subtracting the spectra of components at ppb levels because spectra that have been reduced by division are very low in noise. The main danger of noise introduction comes when one is subtracting the spectra of the major components H_2O, CO_2, CH_4, N_2O, and CO. Removing these involves five separate subtractions, so the noise level of each reference spectrum should be at least five times lower than the noise level of the sample spectrum.

6. *Correct for frequency shifts between sample and reference spectra.*

Frequency shifts seem to be an inescapable part of Fourier transform spectroscopy. The spectra often differ in frequency by one sampling point. With resolution of 0.5 cm^{-1}, the frequency shift would be 0.25 cm^{-1}. In subtracting a spectrum with much fine structure, the frequency shift causes a spurious pattern that can obscure the underlying bands.

The Spectra-Calc software carries the remedy for this frequency shift problem through the program called SHIFT. The data array may be shifted any number of sample points in either direction. With this program the sample and reference spectra may be compared and brought into registration.

7. *Correct for resolution differences between sample and reference spectra.*

Resolution differences also cause spurious patterns to arise during spectral subtraction. The smoothing programs of Spectra-Calc provide the remedy for resolution mismatch. Whichever of the two spectra has the higher resolution should be de-resolved to match the resolution of the other spectrum.

6.2.6. An Example of the Effects of Mismatches in Resolution and Frequency

Figure 6.10 shows subtraction examples for the main CO_2 band seen in the air within an unpurged Digilab spectrometer. At the bottom are two spectra that were obtained at resolution of 0.5 cm^{-1}. These spectra were imported into a Spectra-Calc computer program and manipulated within Spectra-Calc. Each was put in absorbance form by using an artificial background spectrum

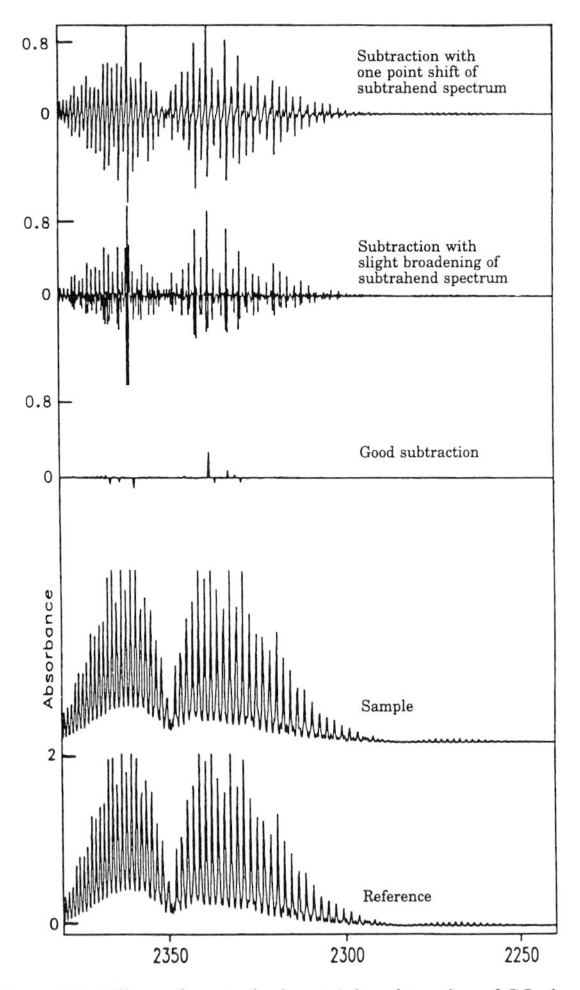

Figure 6.10. Effects of spectral mismatch in subtraction of CO_2 band.

created from the ZAP[1] function of the program. Subtraction of one spectrum from the other then gave the difference spectrum marked *Good Subtraction.* Just above this is the difference spectrum obtained after the reference spectrum had its lines slightly broadened from 0.5 cm^{-1} to about 0.6 cm^{-1} (Spectra-Calc smoothing program). At the top of Fig. 6.10 is the difference spectrum obtained after the reference spectrum had been shifted by one data point (0.25 cm^{-1}). The absorbance scales on the figure show the magnitude of the apparent "noise" resulting from the two types of spectral mismatch.

Figure 6.11 is a second illustration involving the water band of the two spectra. Again it is seen that slight line broadening and frequency shift create highly damaging subtraction patterns.

6.2.7. Preparing the Reference Spectra

Digitized quantitative reference spectra may be purchased from Infrared Analysis, Inc. The person doing the spectroscopic analysis, however, may choose to record his/her own reference spectra. For most compounds a 10-cm single-pass absorption cell may be used. The intensity of the infrared absorption depends on the total number of molecules in the path of the infrared radiation. At a fixed total pressure, trace gas concentration and optical path length are interchangable. Thus a 10-cm cell containing air with a trace gas at a partial pressure of 10^{-4} atm will give the same spectrum as a 10-m cell with air containing the trace gas at a partial pressure of 10^{-6} atm. The reference gas should be mixed with an inert gas to the same total pressure as in the sample being analyzed.

In the case of water vapor, a full interchange between concentration and path length may be prevented by the low vapor pressure of water at room temperature. The water vapor reference spectrum may therefore need to be recorded through the same long path cell that is used in recording the sample spectrum. The water vapor should be mixed in pure air or nitrogen. If the long-path cell is evacuable, a simple way to prepare the reference sample is to pump out the cell and then allow pure air to flow from a tank through a wet tube and into the cell. The single-beam spectrum of the wet air is then divided by the single beam spectrum of the evacuated cell, and the absorbance spectrum is calculated.

A different method is used when the long-path cell can only be flushed with air. Air or nitrogen from a tank is flushed through until the spectrum shows that the room air has been displaced. A single-beam spectrum is then recorded. Next, a water vaporizer is used to blow humidity into the cell. The spectrum is

[1]ZAP means remove or eliminate. The ZAP function of Spectra-Calc replaces a designated portion of a spectrum with a straight line.

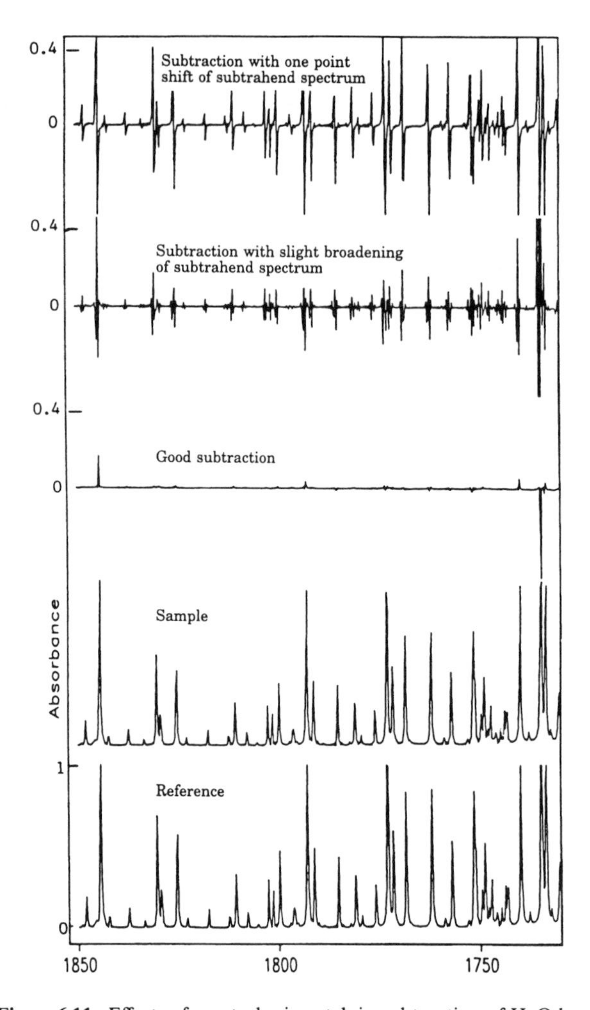

Figure 6.11. Effects of spectral mismatch in subtraction of H_2O band.

recorded again and the absorbance spectrum is calculated. The spectrum of the dry clean air (or nitrogen) is used in calculating the absorbance spectrum of the sample

A third way to prepare the humidified reference air is to draw on two large plastic bags of clean air, one containing water and the other dry. The proportion of air drawn from each bag will determine the relative humidity of the reference air.

6.2.8. Automatic Subtraction or Interactive Subtraction?

Automatic subtraction can be relied upon in cases where the spectra of the gases to be measured are noninterfering. If one knows in advance what the spectrum looks like and is therefore assured of what is being measured, it is safe to proceed. When one is probing an unknown situation, however, it is necessary to project the unknown spectrum and the reference spectrum together on a monitor screen so that the subtraction may be observed. When subtraction is done with lever control or with the plus and minus keys of a computer, it is called interactive subtraction. In gas analysis, interactive subtraction gives a feeling of assurance that the right compound is being measured and the right answer is being obtained. It is a great benefit to observe the subtraction of the pollutant bands on the screen while the interfering bands remain unchanged. Furthermore, the operator can determine any possible error by visually observing oversubtraction and undersubtraction. During the use of automated data analysis programs it is often beneficial to use interactive subtraction routines as a check to verify the accuracy of the automated quantitative analysis. In many cases the simplicity, accuracy, and speed of the interactive subtraction routines will make the automated analysis unnecessary.

6.3. SAMPLE HANDLING AND COLLECTION TECHNIQUES

6.3.1. Flow Sampling

A sample that may be acquired continuously, such as air or a process stream, may be flowed through the absorption cell. This flow method is preferred if there is a likelihood that the gases to be measured will react with the walls or other parts of the absorption cell. In this case it is likely that an equilibrium will be reached where absorption and desorption are balanced out.

The flow method requires an outlet at one end of the absorption cell and an inlet at the other end. For cells that have their front end (or field mirror end with transfer optics) close to or inside the sample compartment, it may be necessary to have the inlet and outlet valves both on the rear end of the cell. Infrared Analysis, Inc. places a flow tube inside the cell, running from a rear end valve to the front end of the cell. The sample is then made to flow in one pass from the rear end to the front end and then out the tube.

6.3.2. "Grab" Sampling

The main alternative technique is *grab sampling*, where a small sample is brought to the cell in a container and is either pushed into the cell or allowed to flow in under its own pressure. When using the grab sample method, if it is observed that some compounds are decreasing in concentration while the sample is in the cell, it may be possible to get the correct starting concentration by extrapolating the concentration curve back to zero time.

For grab sampling, Infrared Analysis, Inc. provides glass containers that may be evacuated in the laboratory and taken to the sampling site where the sample is allowed to flow in under its own pressure.

6.3.3 Cryogenic Sample Concentration

A third sampling technique is the cryogenic concentration method in which trace gases are separated from the air before being placed in the absorption cell. This is feasible with stable gases such as hydrocarbons and halocarbons. This separation allows great gains in detection sensitivity. Separation is not feasible for gases that are reactive, such as ozone and peroxides, or for gases that are highly soluble in water, such as HCl and HNO_3.

An application of the cryogenic method in the measurement of fluorocarbons, carbon tetrachloride, carbonyl sulfide, and other atmospheric trace gases has been described by Hanst et al. (1975a). A simple collection technique was used. Air was driven by a pump into a closed vessel immersed in liquid nitrogen. Condensation took place until the vessel was full, after which the oxygen and nitrogen were pumped off, leaving a residue of ice, solid CO_2, and condensed trace gases. When the container with condensate was attached to the infrared absorption cell and warmed to room temperature, the trace gases flowed into the cell in a highly concentrated state.

The method depends on the vapor pressure–temperature relations as shown in Fig. 6.12. The most volatile pollutant in the air is CO, as shown by the vapor pressure line at the top of the figure. The CO will condense and distill with the oxygen and nitrogen without separation.

Methane is next down the scale in volatility. At liquid nitrogen temperature its vapor pressure is approximately 0.01 atm. This is low enough for separation. Toward the end of the distillation, when the nitrogen and oxygen are nearly gone, a methane-rich fraction distills. Test spectra have shown that during removal of the methane-rich fraction there is no appreciable removal of carbon dioxide, nitrous oxide, or other minor constituents of the residue. Even ethylene remains in the cryocondenser during methane removal.

Ethylene is the most volatile of the nonmethane hydrocarbons, with a vapor pressure of about 2×10^{-6} atm at liquid nitrogen temperature. Nevertheless, spectra recorded during distillation have shown that ethylene is fully retained in the residue as long as the distillation is not carried very far beyond the methane removal stage.

Acetylene, carbon dioxide, nitrous oxide, halogenated compounds, and all

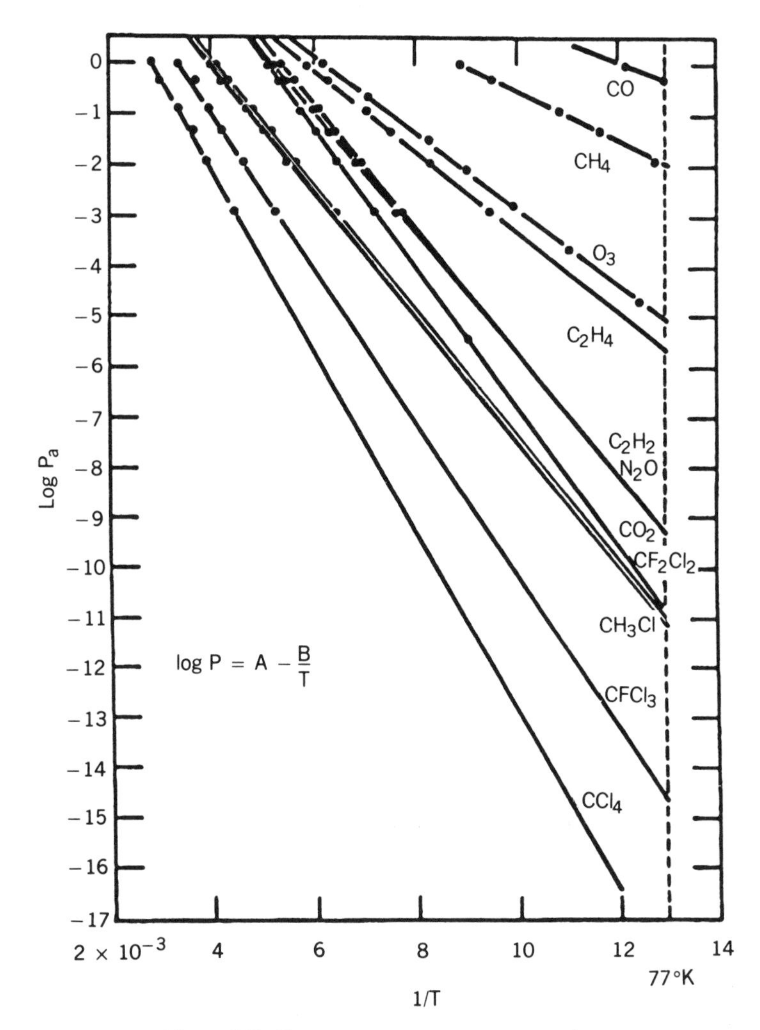

Figure 6.12. Vapor pressure vs. temperature plots.

other minor molecular constituents of the air do not vaporize from the cryocondenser during the distillation of nitrogen and oxygen.

Nitrous oxide serves as an inherent calibration of the degree to which the concentrations of trace gases have been increased by the condensation and distillation process. N_2O occurs everywhere in the lower atmosphere at a mixing ratio of 3×10^{-7}. Since the vapor pressure of N_2O at liquid nitrogen temperature is less than 10^{-9} atm, all the N_2O remains with the other trace gases until final removal from the cryocondenser. Because of its inertness, the

N_2O behaves in handling like the hydrocarbons, halocarbons, and other inert pollutants. When the infrared spectrum is recorded, the strength of the N_2O bands in the spectrum tells how much gain in pollutant concentration was obtained through the condensation. In other words, the N_2O bands give the spectrum "amplification factor."

In the early trials of the technique, the collected residue was vaporized from the cryocondenser into the infrared absorption cell with no attempt to remove CO_2. This produced an infrared spectrum dominated by CO_2 lines that tended to obscure the bands of other trace gases. One way to minimize the interference is to plot a ratio of the spectrum of the residue over a spectrum of pure CO_2. Unfortunately it has been found that in such a plot the strong absorption causes some parts of the spectrum to be lost and others to be rendered unduly noisy. It is better to remove the CO_2, provided that other trace gases are not removed. Sodium hydroxide has been found effective in removing CO_2 without disturbing most of the other gases. Halocarbons, hydrocarbons, nitrous oxide, and carbonyl sulfide, for example, are not taken up by the NaOH during CO_2 removal.

A second limitation of CO_2 is that it holds down the spectrum amplification factor. If the collected residue with CO_2 present is placed in an infrared absorption cell at a total pressure of 1 atm, the gas will consist mainly of CO_2 whose concentration has been raised from its normal 0.00033 atm to 1 atm—an amplification factor of 3000. If the CO_2 is removed, however, the N_2O becomes the most prevalent gas in the mixture, allowing a maximum amplification factor of 3.3×10^6. The maximum amplification factor obtained in experiments so far as been approximately 1×10^6.

When the collected residue is placed in an infrared cell large enough to hold it all, it is the path-to-volume ratio of the cell that determines the strength of the infrared bands, rather than the total path length. Small compact cells are more efficient than large cells.

Figure 6.13 shows an example of the detection power of the Fourier transform spectrometer when it is used in conjunction with the cryogenic concentration technique and a miniaturized multiple-pass cell. The cell in this case was not our smallest. It was an intermediate size, with a path of 1100 cm folded inside a volume of 200 cm^3, giving a path-to-volume ratio of 5.5. The spectrum between 700 and 1200 cm^{-1} shows identifiable bands of 11 species of pollutant, sampled on a clear day, October 22, 1976, outside of Research Triangle Park, North Carolina. In addition, the spectra showed the naturally occurring nitrous oxide. Carbon dioxide was removed from the gas mixture by NaOH. The amplification factor indicated by the strength of the N_2O band is approximately 16,000. It is therefore calculated that approximately 3 m^3 of air were condensed and distilled to yield the residue that produced the spectrum.

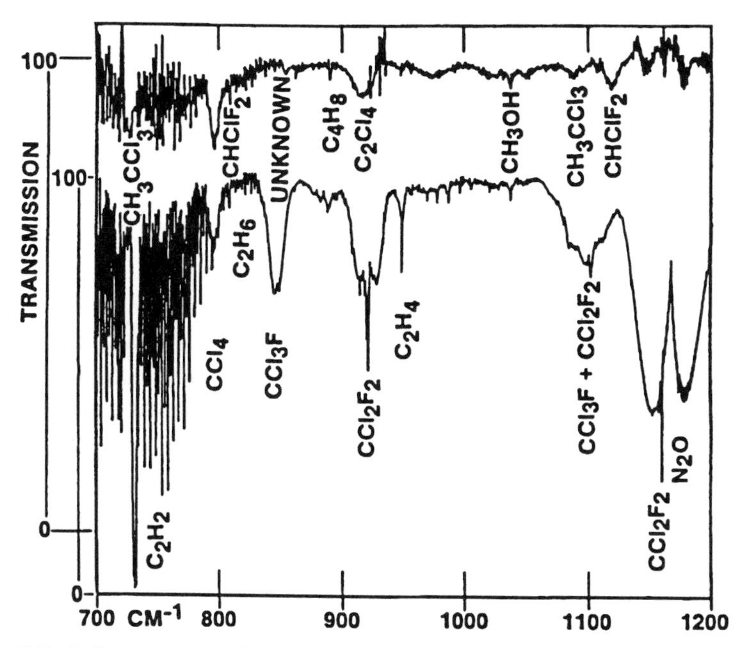

Figure 6.13. Pollutant spectra obtained with the aid of the cryogenic concentration technique. From Hanst (1978).

Figure 6.13 also shows the great value of the spectrum subtraction routines that are available with the Fourier transform spectrometer system. In the lower spectrum the dominant bands obscure the weaker bands. In the upper spectrum, however, the dominant bands of acetylene, trichlorofluoromethane, dichlorodifluoromethane, ethylene, and nitrous oxide have been subtracted, revealing the weaker bands of other pollutants and allowing them to be measured. Table 6.2 gives the calculated mixing ratios for the gases in the

Table 6.2. Mixing Ratio of Gases in Air Sample[a]

N_2O	3×10^{-7}[b]	CCl_4	1.2×10^{-10}
C_2H_6	4×10^{-9}	CH_3CCl_3	1×10^{-10}
C_2H_2	2×10^{-9}	C_2Cl_4	1×10^{-10}
C_2H_4	4×10^{-11}	CH_3OH	4×10^{-11}
CCl_2F_2	2.6×10^{-10}	$CHClF_2$	4×10^{-11}
CCl_3F	1.7×10^{-10}	C_3H_8	2×10^{-11}

[a] Data from Hanst (1978); see Fig. 6.13.
[b] Assumed.

original air sample. It is obvious from the table that an extremely high degree of analytical sensitivity is obtained when the FTIR method is combined with appropriate methods of sample preparation. The example also shows that in many aspects of air analysis, gas chromatographic separations are not required. The subtraction technique will generally suffice for the unraveling of complex spectra.

6.3.4. Using a "Marker" Compound to Calibrate the Sample Collection

It has been shown above that N_2O is an inherent calibration gas for the cryogenic sample concentration. In engine exhaust samples, carbon dioxide may also serve as a calibration gas. If a mixture of air and exhaust gas is captured, the carbon dioxide concentration will tell what the relative proportions are. A gasoline-burning engine, for example, will convert almost all the oxygen of its air intake into CO_2 and H_2O. Since the air enters with an approximate composition of 20% O_2, 77% N_2, 2% H_2O, and 0.04% CO_2, the exhaust composition will be about 72% N_2, 12% CO_2, and 15% H_2O. Then, if the sample measures 6% CO_2, for example, one knows that there was dilution by a factor of 2 during sample gathering. If the sample measures 1% CO_2, one calculates a dilution factor of 12. Sample gathering can therefore be done casually and qualitatively. A bus exhaust can be sampled by a person standing at the bus stop as the bus pulls out. Car exhaust can be sampled by the driver in a moving car. If the fuel economy of the vehicle is known (miles per gallon, or liters per kilometer), one can convert the pollutant concentrations to mass rate of emissions (grams per kilometer).

6.3.5. Multiple-Reflection Cells

The measurement of gaseous air pollutants by infrared absorption has greatly benefited from the use of multiple-reflection long-path cells. The size and configuration of the cells have varied with the type of pollution under study. Gases that cannot easily be separated from the air such as the peroxides and the acid anhydrides require a long-path cell. Preferably the cell should be placed in the polluted atmosphere. In this case the cell is allowed to have a very large volume because the whole ambient atmosphere is available as a sample.

Gases that can be separated from the air (for example, by the cryogenic method discussed in Section 6.3.3, above) can be placed in a cell in a highly concentrated state. In this case it is the path-to-volume ratio rather than the total path length that determines the strength of the infrared bands. Small, compact cells are more efficient than large cells and will minimize the amount of air that must be processed to prepare the pollutant concentrate.

The three-mirror multiple-pass cell known as the White cell was described in a paper entitled "Long Optical Paths of Large Aperture" (White, 1942). Although a number of improvements have been developed since then, including some by the original author (White, 1976), the basic three-mirror system is still widely used. The cell is diagrammed in Fig. 6.14, showing the minimum unit of four passes.

The operation of the cell is described by following the light from the source into the cell and through at least four passes (see Figs. 6.14 and 6.15). The light from the source is initially focused into a real image in the entrance aperture of the cell. In Fig. 6.15 that is designated the zeroth image. After passing through the zeroth image, the beam diverges and is collected by one of the two objective mirrors. This is a spherical mirror situated two focal lengths from the image so that it refocuses the image, inverted, on the lower part of the opposite single mirror, called the field mirror, which is another spherical mirror of the same focal length. The first image is marked 2—for two passes. The field mirror is aimed so that the reflected diverging beam falls entirely on the second objective mirror. This is then aimed to form another image (marked 4) above the center line of the field mirror alongside the zeroth image. If this image falls symmetrically opposite the first image (numbered 2), the beam will be returned to the first objective at the required small angle with the intput beam, so that

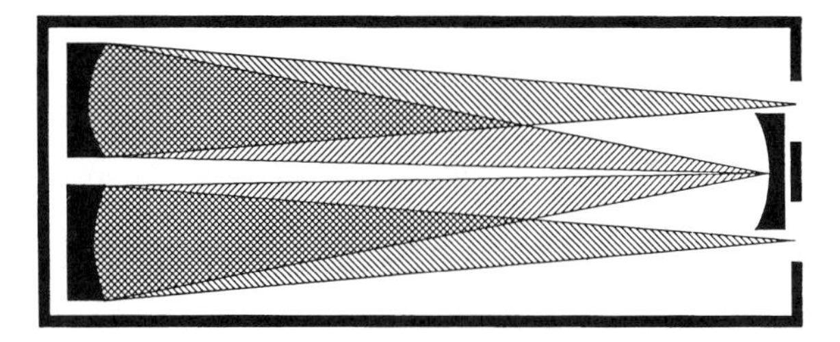

Figure 6.14. White optical system; basic set of four passes.

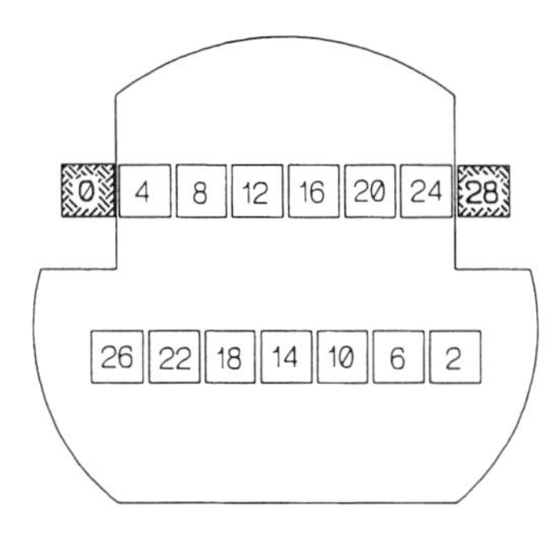

Figure 6.15. Placement of images on field mirror.

all the energy is again collected and returned and there will be at least four more passes through the system.

The number of images allowed in the row depends on the placement of the first image in the lower part of the mirror. If it falls exactly on the vertical center line, no more than four passes are possible. The farther to the right it falls, the greater the number of passes allowed. The number of passes is equal to the number of the exit image. In practice, the number of passes is determined by counting the number of images in the bottom row and multiplying by 4.

In order to have as much energy as possible going through the system, an enlarged image of the source is usually projected into the entrance aperture and the exit image is demagnified when the transmitted light is fed into the detection system.

Frequently, the only major source of energy loss in the system is the absorption at the mirror surfaces. If R is the reflectivity of the mirrors, R^n is the fractional amount of energy transmitted by the cell after n reflections. Some sample calculations show the magnitude of this loss. The traditional choice for mirror coating has been gold, for which $R = 0.98$, approximately. In a 52-pass gold-coated mirror system, the fraction of energy transmitted is 0.98^{52}, or 35%. With 104 passes, the fraction of energy transmitted is 0.98^{104}, or 12%. A better choice of mirror coating is protected silver, which can have R as high as 0.995. Then the 52-pass cell will transmit 0.995^{52}, or 77%, and the 104 pass cell will transmit 0.995^{104}, or 59%. The recent availability of the silver coatings considerably enhances the applicability of the multiple-pass cell to trace gas analysis.

A modification of the White system that increases its energy throughput was described by Horn and Pimentel (1971). They created two additional rows of images and doubled the number of passes by means of a retroreflecting pair of mirrors at the normal exit port. This increased the amount of available mirror surface at the in-focus end of cell. That allowed the use of a more collimated beam, thus increasing the energy throughput without enlarging the main collecting mirrors.

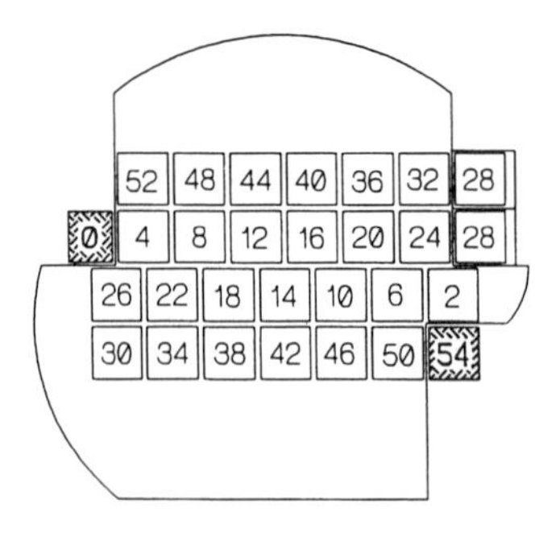

Figure 6.16. Placement of images of field mirror and retroreflecting mirrors in the Horn and Pimentel (1971) system.

Figure 6.16 shows the four rows of images and the retroreflecting mirror pair. The numbering is carried on from that used in Fig. 6.15, but rather than the last image in a typical White cell exiting the system (image 28) it is redirected by way of the retroreflecting mirror to the upper portion of the oversized field mirror, and two additional rows of images fall onto the field mirror. If the mirror has an exit cutout across from the input cutout as shown in the bottom right-hand corner of Fig. 6.16, the last pass (number 54) will exit the optical system well separated from the input beam. Using this system with special high-quality mirror coatings, Horn and Pimentel (1971) were able to work at 254 reflections along a 10-m base path, for a total path of 2.54 km.

We have carried the Horn and Pimentel concept one step further by putting a second retroreflecting pair of mirrors on the input side of the field mirror (Infrared Analysis, Inc., 1992). This creates six rows of images. In this case the field mirror shape can be the "inverted mushroom," as is used in the simple White cell. Figure 6.17 shows the distribution of images in the six-row case when the total number of passes is 84. For atmospheric studies the image array in Fig. 6.17 will allow an efficient and simple energy transfer between a Fourier transform spectrometer and a large long-path cell. If the mirror is cut from a 30-cm-diameter blank (which is a common size), the multiple-pass cell will directly accept nearly all the output energy from the FTIR instrument. There will be no need to have any transfer optics between the interferometer and the cell.

Additional improvements in the design of the multiple-reflection cell have recently been described by Ritz et al. (1993).

6.3.6. The Advantage of a Multiple-Pass Cell over a Single or Double Long Path

The title of White's 1942 paper, "Long Optical Paths of Large Aperture," emphasized the main point that he was trying to make, namely, that the cell can capture and conserve nearly all the energy that can be projected from an

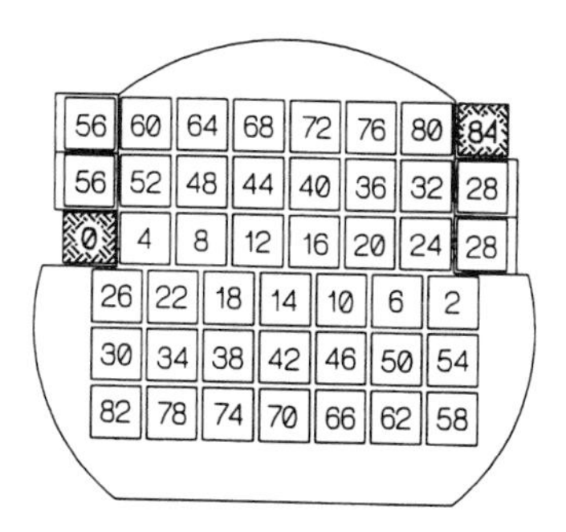

Figure 6.17. Image array in "tripled–up" White cell of Infrared Analysis, Inc.

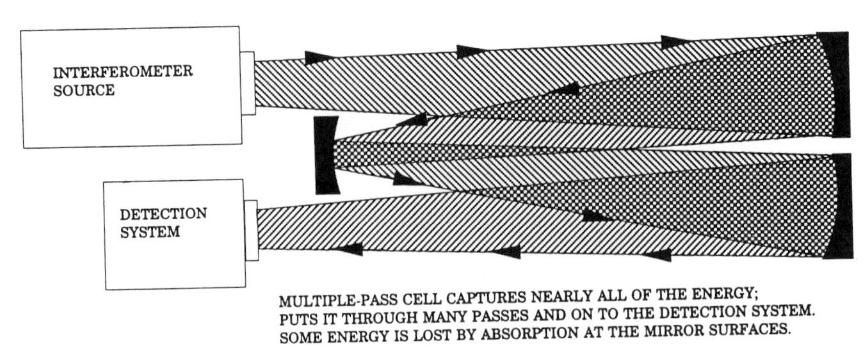

Figure 6.18. Air analysis system with multiple-pass cell.

optical source. The energy conservation is the result of the refocusing that occurs on every other pass, as was depicted in Fig. 6.15. The White cell has been used with success in atmospheric studies at total optical paths of more than a kilometer (Hanst et al., 1982; Tuazon et al., 1978). The use of the White cell is diagrammed in Fig. 6.18.

When one is using a single or doubled optical path at such long path lengths, most of the transmitted energy is usually lost because available mirrors are not large enough to capture it all. With a double path there is a double loss. Transmitted energy misses the retroreflector, and returned energy misses the collecting mirror. The losses are high because at each end the energy spreads in two dimensions. The use of doubled long path is illustrated in Fig. 6.19.

Compared below are a multiple-pass cell and a two-pass retroreflector system, each with practical-sized optical components. The total paths considered are 1 km and 200 m. We find that for the 1-km case the throughput of a multiple-pass cell exceeds the throughput of a double-long-path system by a factor of 312. For the 200-m case the cell throughput is greater by a factor of 27.

First, we consider establishing a 1-km path. To work at $0.5\,cm^{-1}$, a typical FTIR system would have its source aperture (Jacquinot stop) about 1.5 mm in diameter. A mirror would collect the source radiation at about $f/1$ and project through the interferometer a collimated beam about 5 cm in diameter. Of course, the beam would not be truly collimated; that is only possible with a point source. The beam would spread at the angle subtended by the source, as seen from the mirror surface. The spread angle, in radians, is equal to the source diameter divided by the focal length of the collimating mirror. In the case we are considering, this angle would be about 30 mrad.

If this beam were allowed to pass directly into a multiple-pass cell with a 10-m base path, the beam would spread to about 350-mm diameter at the first

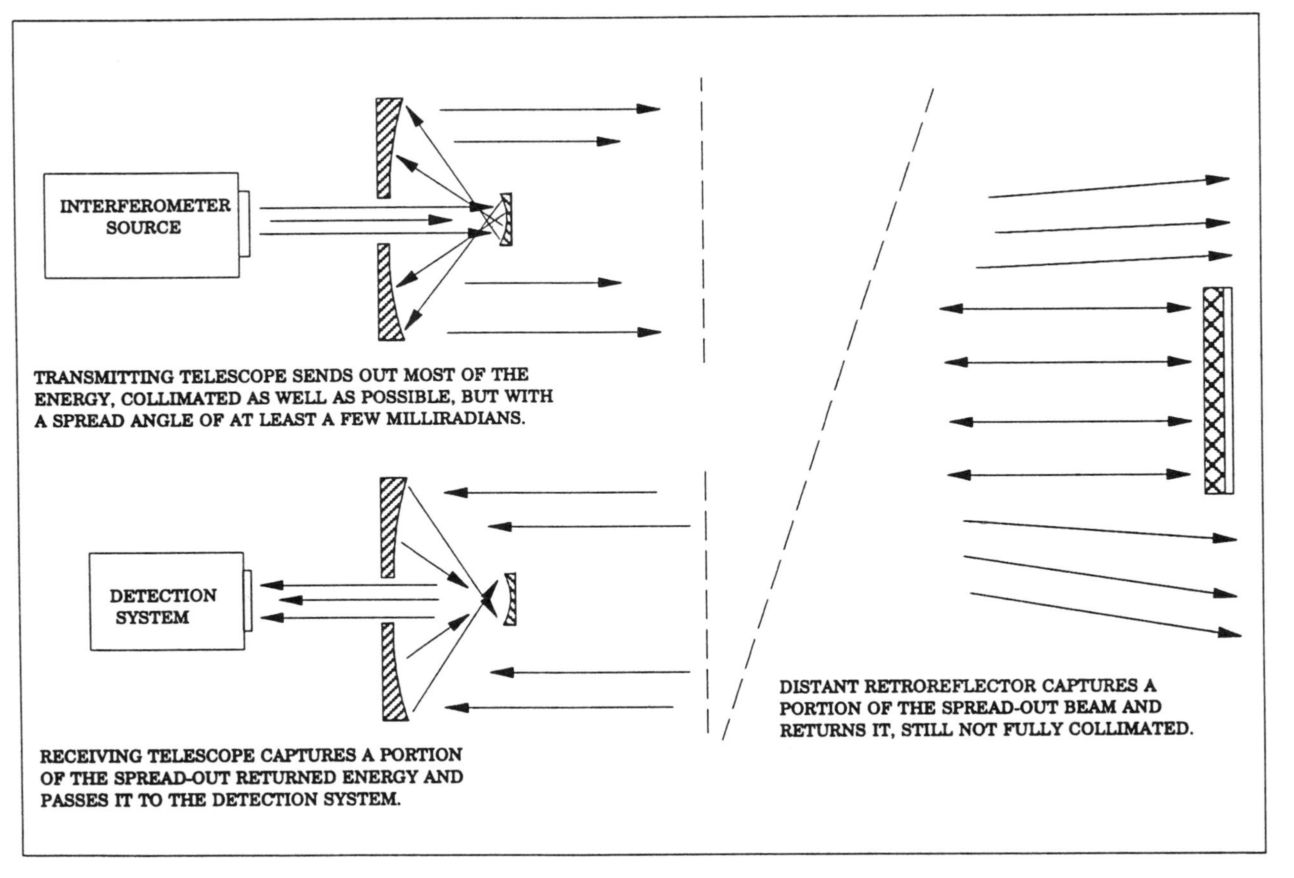

Figure 6.19. Air analysis system using doubled long path.

objective mirror. If this were a 35-cm-diameter mirror, nearly all the energy would be captured and focused onto the field mirror. If the image areas on the field mirror are in the form of squares, 40 mm on a side, then 82% of the energy would be folded into the multiple-pass mirror system. Let the cell be a tripled-up version set for 96 passes. The field mirror could be fashioned in two pieces, each cut from a 35-cm blank. The retromirrors used for path tripling could be inexpensive flats. The only energy losses in the cell would be due to absorption by the mirror surfaces. With high-quality protected silver, the loss would be about 0.5% per reflection. For 96 passes the cell throughput would be 62%. Since 82% of the original energy was accepted by the system, the total system throughput would be slightly over 50%.

When one is establishing a two-pass 1-km system, it is first necessary to reduce the 30-mrad spread angle of the beam emanating from the interferometer. This can be done with a two-mirror concentric system having a 5-cm-diameter focusing mirror and a 30-cm-diameter projection mirror. The beam spread would then be reduced to about 5 mrad. Were the beam then projected to a retroreflector 0.5 km away, it would spread until it covered a circle 2.8 m in diameter. To capture all the energy would require a huge retroreflector of about 6.2 m^2 area. Let us assume a more practical size of a square 0.5 m on a side. This reflector would have an area of 0.25 m^2 and would return about 4% of the energy; that is, 96% would be lost, as depicted in Fig. 6.19. The returned energy would continue to spread. How much it would spread and in what manner would depend on the nature of the retroreflector. If the retroreflector were of high optical quality, perhaps the beam spread would be only half as great as it was on the way out. Let us assume this to be the case. Then the beam would be about 1.5 m in diameter at the receiver. Using a 30-cm-diameter receiving telescope would allow the capture of about 4% of the returned energy (another 96% loss). The total system throughput has thus been calculated to be approximately 0.16%. The 50% throughput calculated for the 1-km multiple-pass cell was 312 times greater than this.

Let us now consider an example more favorable to the two-pass system, namely, a system with a 200-m total path. In this case the multiple-pass cell could be scaled down in size, using a 3-m base path and 15-cm-diameter objective mirrors. The throughput of the multiple pass cell is calculated in a manner similar to that used for the 1-km cell, with the result being 60%.

In the two-pass 200-m system the retroreflector would only be 100 m away from the source and the beam of radiation at the retroreflector would be about 1 m in diameter. The square retroreflector 0.5 m on a side would return about 32% of the energy. At the receiver, the beam cross section would be about 1 m^2 and the 30-cm-diameter receiving telescope would capture about 7% of the returned energy. The total throughput would then be about 2.2%. The calculated multiple-pass cell throughput was about 27 times greater than this.

The main consequence of the energy loss in the infrared system would be a corresponding loss of the ability to detect and measure. This happens in direct proportion: if the number of photons received per second is cut in half, the noise in the spectrum is doubled; if the noise doubles, the amount of compound required for detection also doubles. Infrared instrument designers therefore go to great lengths to maintain a high energy throughput. They know that the instrument with the highest throughput gives the lowest MDLs.

Fence-line pollution monitoring is a concept that is in vogue at present, with plans being developed for setting up long single or double optical paths along the edge of an industrial property. The above considerations indicate, however, that such a monitoring application is not especially practical. The energy losses in a long fence-line optical path would be excessive, and as a result the MDLs would be much higher than desired. It would be better to monitor the fence-line with multiple sensing systems that are compact and therefore have high-energy throughput and low MDLs. Better yet would be to use a compact, highly sensitive measuring system that is mobile. With a mobile system, the person making the measurement is not at the mercy of the direction of the wind. If the pollution plume were to move, the measurement system could also be moved.

6.4. FIELD APPLICATIONS

6.4.1. Optical Paths Through the Atmosphere

The infrared method has been used for many years in the measurement of atmospheric trace gases. There are three main methods of measuring the infrared absorption by the atmospheric gases, as illustrated in Fig. 6.20.

Figure 6.20a shows the case of looking at the sun from an observing point on the earth's surface. For measurements not concerned with the lower troposphere, an observation point at high altitude is chosen. The higher the observatory, the less is the interference from water vapor. High flying aircraft have also served as the observing platform, for example at the National Center for Atmospheric Research in Boulder, Colorado (Mankin et al., 1979).

The second method of recording atmospheric spectra, shown in Fig. 6.20b, is to observe the setting or rising sun from a balloon or satellite that is above most of the atmosphere. This is called atmospheric limb spectroscopy. In these cases the equivalent optical paths can be hundreds of kilometer-atmospheres (1 km · atm is a 1-km path at sea level). The lowest point in the optical path, called the tangent altitude, depends on how far down the system is pointed. For a viewing direction 94° off the vertical, for example, from an instrument at 30-km altitude, the tangent altitude is about 15 km and the equivalent optical

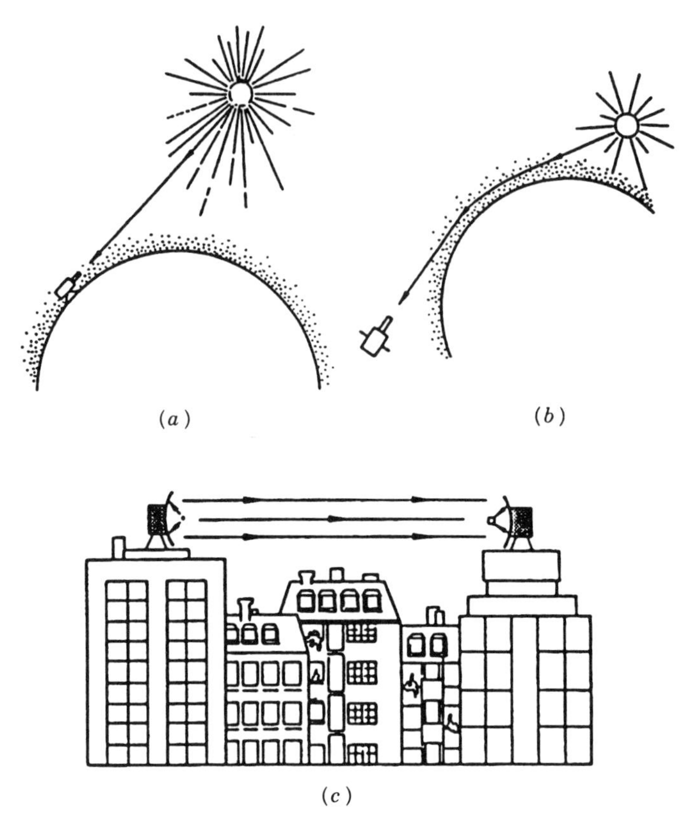

Figure 6.20. Three methods of obtaining a long optical path through the atmosphere. (See the text.)

path is about 100 km · atm. The major advantage of sighting along a high altitude path to the sun is that very little water vapor exists in the stratosphere and the water lines do not dominate the spectrum as they do at lower altitudes. A disadvantage of this method is that observations can be made only at sunrise and sunset.

The observation method depicted in Fig. 6.20c is that which has been used in the urban smog. In this case the path is horizontal, between an infrared radiation source and the detection system. The source may be at one end of the optical path and the detector at the other end, as shown in Fig. 6.20c, or the source and detector may be together at one end, with a reflector at the other end. A single reflection or many reflections may be used. The three-mirror multiple-pass system developed by White (1942) has been used more than any other, and White's system has been of fundamental importance in developing an understanding of the atmospheric chemistry. This sytem was diagrammed in Fig. 6.14, where the basic set of four passes is indicated. Many additional sets of four passes can be added. When one is using a long horizontal path in the urban atmosphere, water vapor absorption is extremely strong, eliminating some spectral regions completely and showing at least some lines in all other regions. One must learn to recognize the presence of the ultra-trace gases by their small perturbations on the ever-present water vapor spectrum.

Most of the molecules that were detected and measured in the atmosphere by infrared absorption up to 1984 are listed in Table 6.3. The approximate mixing ratios (ratio of the number of molecules of the trace gas to the number of molecules of all gases) and comments on the distribution of the gases and references to literature reports are also given. These are molecules that were seen in situ in ambient air. Many additional molecules, mainly industrial pollutants, have been identified by infrared after being captured and taken into the laboratory.

Atmospheric measurements have, of course, continued and there are many additional reports in the literature. An especially valuable group of high-resolution spectra was obtained in measurements from the U.S. space shuttle. These results are described, for example, in papers by Zander et al. (1986), Rinsland et al. (1986a, b), and Park et al. (1986).

Infrared methods are also used to confirm and calibrate chemical methods of atmospheric analysis. Chemical methods can be extremely sensitive, but they also are notoriously subject to interference. In many instances, reports of chemical measurements of atmospheric pollution have been disproved or corrected by the application of infrared methods.

To account for the strong differences in the spectra seen along the various optical paths, Fig. 6.21 shows the amounts of trace gases in terms of darkened areas. Starting with the lower atmosphere case, at the bottom of the figure, one sees that a 2-km path at sea level may show an amount of water vapor equivalent to a 20-m path through 1 atm of pure water vapor (steam). The amount of CO_2 is much smaller, being equivalent to 0.66 m· atm. The ultra-trace gases in which one is interested are so much lower in concentration that they cannot fairly be represented by areas on the graph. In most cases, the area of a dot is too large. The middle of the figure shows that on the path from a balloon to the sun, CO_2 is dominant, with the amount of water vapor being even smaller than the amount of ozone. The upper part of the figure shows that when the sun is observed from a high-altitude location (mountaintop), the amounts of water vapor and carbon dioxide in the path are comparable.

6.4.2. Some Examples of Atmospheric Measurements

For the case of observing the sun from a point on the earth's surface, an early and noteworthy example was the work carried out at Jungfraujoch,

Table 6.3. Molecules That Have Been Detected and Measured in the Air by Infrared Absorption

Molecule	Location of Bands or Lines (cm^{-1})	Approximate Mixing Ratio	Comments	Some References
H_2O	Everywhere in spectrum	10^{-6} in stratosphere; 10^{-2} in troposphere	Interferes with detection of nearly everything else.	All spectra
CO_2	2380, 670	3×10^{-4}	Being a linear, symmetrical molecule, CO_2 has a relatively simple spectrum that does not interfere seriously in the detection of other compounds.	All spectra
CH_4	3020, 1305	1.5×10^{-6}	Quite uniformly distributed, but in urban areas slightly higher than in others.	Most spectra
N_2O	2220, 1280	3×10^{-7}	Uniformly distributed; not significantly higher in urban areas than in nonurban areas. May be used as an internal standard, allowing ratios of other compounds do be calculated in reference to N_2O.	Most spectra
CO	2146	2×10^{-6}	Higher in Northern Hemisphere than in Southern; in urban areas CO can be 10–100 times higher than in nonurban areas.	Migeotte et al. (1956); Zimmerman et al. (1978)
O_3	1050	10^{-6} in stratosphere; 10^{-7} in lower troposphere	Shows in all solar spectra; O_3 is beneficial in stratosphere but poisonous in urban areas.	Most spectra

NO	1900	10^{-6} in stratosphere; 10^{-10} in clean troposphere; 10^{-7} in urban smog	Difficult to see because of H_2O interference, but seen in urban smog and stratosphere.	Farmer (1985); Hanst et al. (1982); Fontanella et al. (1975); Blatherwick et al. (1980); Coffey et al. (1981)
NO_2	2920, 1620	10^{-6} in stratosphere; 10^{-9} in clean troposphere; 10^{-7} in urban air	Lines of strong band near 1620 cm^{-1} can be seen in some solar spectra. In urban smog these lines are obscured by water vapor, but one can see the lines of the weaker band near 2920 cm^{-1}.	Goldman et al. (1978); Coffey et al. (1981); Fontanella et al. (1975); Blatherwick et al. (1980); Hanst et al. (1982); Camy-Peyret et al. (1983)
N_2O_5	1240	10^{-9} in stratosphere(?)	N_2O_5 is undoubtedly formed in the urban smog, but it hydrolyzes and photolyzes. It has not yet been detected in smog but has been seen in sunrise measurements from the space shuttle.	Farmer (1985)
HNO_2	791	10^{-9} in urban smog	Nitrous acid has a well-known infrared spectrum, but it can be more easily measured by its UV absorption; 5×10^{-9} atm of HNO_2 have been seen in urban smog early in the morning. The compound is photolyzed by sunlight.	Perner and Platt (1979); Hanst et al. (1982)
HNO_3	879, 967	10^{-9} in clean troposphere; 5×10^{-9} in stratosphere; 10^{-8} in urban areas	End product of nitrogen oxide oxidation. Washed out of troposphere, Prominent in stratospheric spectra.	Coffey et al. (1981); Hanst et al. (1982); Murcray et al. (1969)

Table 6.3. *(Continued)*

Molecule	Location of Bands or Lines (cm^{-1})	Approximate Mixing Ratio	Comments	Some References
NH_3	932, 967	10^{-8}–10^{-9}	NH_3 is seen near farms and feed lots. Spectra show it to be absent from urban smog. It has not been seen in the stratosphere	Tuazon et al. (1978); Hanst et al. (1982)
$ClNO_3$	780, 809, 1293	10^{-9}	Partial reservoir for atmospheric chlorine.	Murcray et al. (1977)
HCl	2924, 2926	10^{-9} in stratosphere; not seen in troposphere	Sink for Cl in stratosphere. Washes out of troposphere.	Farmer et al. (1976, 1980); Williams et al. (1976b) Ackerman et al. (1976)
HF	4029, 4174	10^{-10} in stratosphere; 10^{-8} near fertilizer plant.	HF is also of concern in aluminum production facilities.	Herget (1979); Farmer et al. (1980)
HCOOH	1105	10^{-8} in urban smog; 10^{-9} in upper troposphere	Product of oxidation of organic compounds.	Hanst (1971); Hanst et al. (1982); Goldman et al. (1984)
H_2CO	2780, 2870	10^{-8} in urban smog; 10^{-10} in upper atmosphere	Product in oxidation of methane and other organic compounds.	Jouve (1980); Hanst et al. (1982)
HCN	3270, 3290	10^{-10}	Detected in stratosphere.	Coffey et al. (1981); Smith and Rinsland (1985)
Chloro fluorocar-bons (CFCs)	Various	10^{-9}	See text.	
Non-methane hydro carbons	2850–3000	10^{-7} in smog; 10^{-9} in stratosphere	See text.	

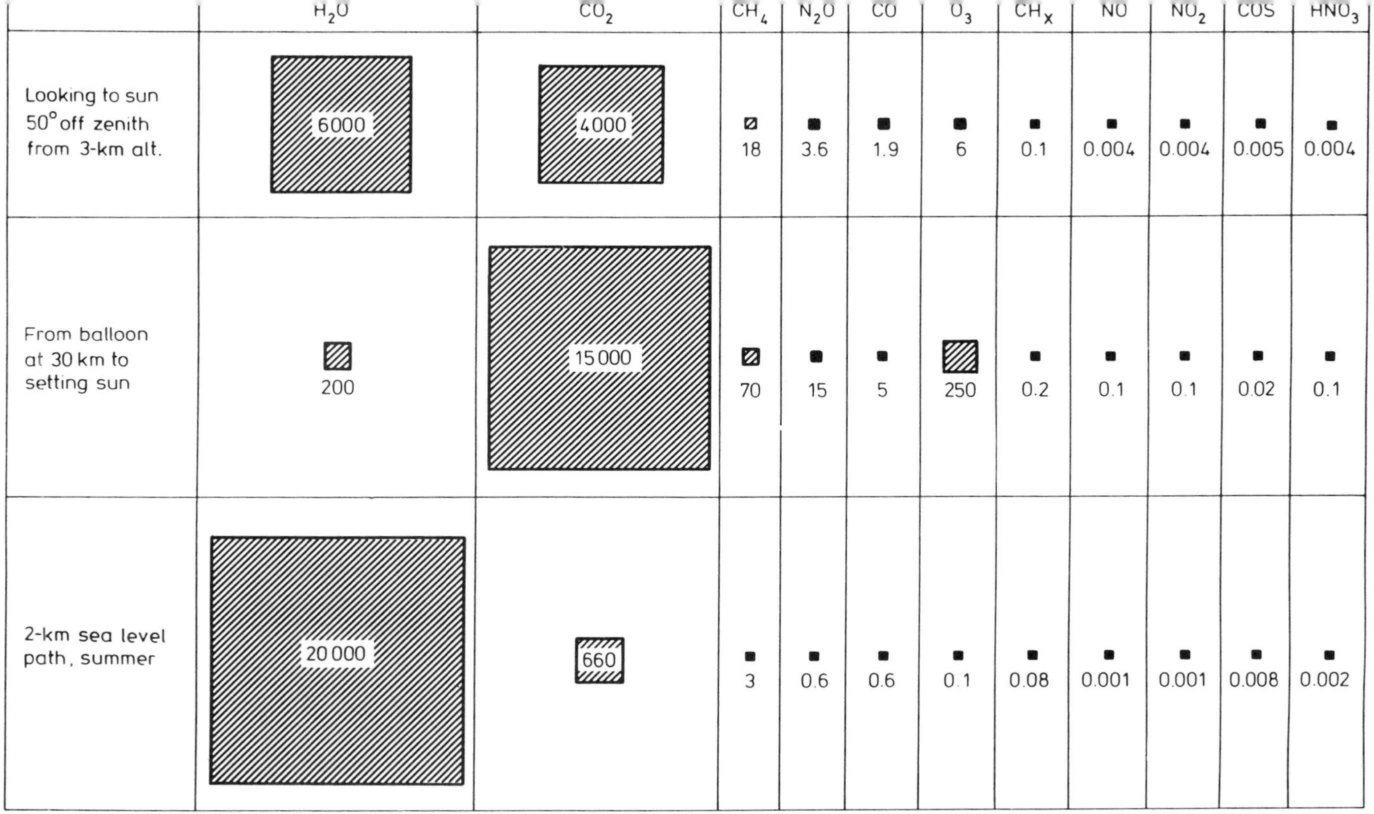

	H_2O	CO_2	CH_4	N_2O	CO	O_3	CH_x	NO	NO_2	COS	HNO_3
Looking to sun 50° off zenith from 3-km alt.	6000	4000	18	3.6	1.9	6	0.1	0.004	0.004	0.005	0.004
From balloon at 30 km to setting sun	200	15 000	70	15	5	250	0.2	0.1	0.1	0.02	0.1
2-km sea level path, summer	20 000	660	3	0.6	0.6	0.1	0.08	0.001	0.001	0.008	0.002

Figure 6.21. Proportions of atmospheric trace gases seen in various viewing geometries. (Numbers are the path length–concentration products in millimeter–atmospheres.)

Switzerland, in the early 1950s and reported as an atlas of the solar spectrum (Migeotte et al., 1956). In that work a large grating spectrometer was used from a laboratory at an altitude of about 3580 m. When the viewing direction was 45° off the vertical, the equivalent path was 7.3 km·atm.

The aforementioned atlas demonstrates an important aspect of spectral observations: they are a permanent record of the condition of the atmosphere, and this record may continue to be useful in latter years. An example of this is the work of Rinsland and collaborators in which they calculated carbon monoxide and methane concentrations from a 1951 spectrum and compared them to the higher concentrations that existed in 1985 (Rinsland and Levine, 1985; Rinsland et al., 1985).

Spectra recorded in the open atmosphere also serve to confirm or deny the presence of trace gases that may be indicated by nonspectroscopic methods. In the case of ethane, for example, gas chromatographic methods have indicated a background atmospheric mixing ratio on the order of $1-2 \times 10^{-9}$. If this is true, some evidence of ethane lines should appear in the 1951 spectra. Figure 6.22 shows a portion of the 1951 spectrum in which the positions of ethane lines are indicated by the arrows pointing downward from above the spectrum; solid arrows mark ethane lines that can be seen, and dotted arrows mark lines that are hidden by the methane and water lines. An absorption coefficient of $20\,\text{cm}^{-1}\cdot\text{atm}^{-1}$ was obtained for these ethane lines from a laboratory spectrum recorded with pressure and resolution conditions similar to those that prevailed during the recording of the solar spectrum. The measurement conditions were as follows: 11,746-ft (~3580-m) altitude; grating spectrometer; coelostat; thermocouple detector; 13-cycle chopper; pressure at laboratory, 480 mm; temperature, −20 to 6 °C; date, April 15, 1951, 6:43 a.m.

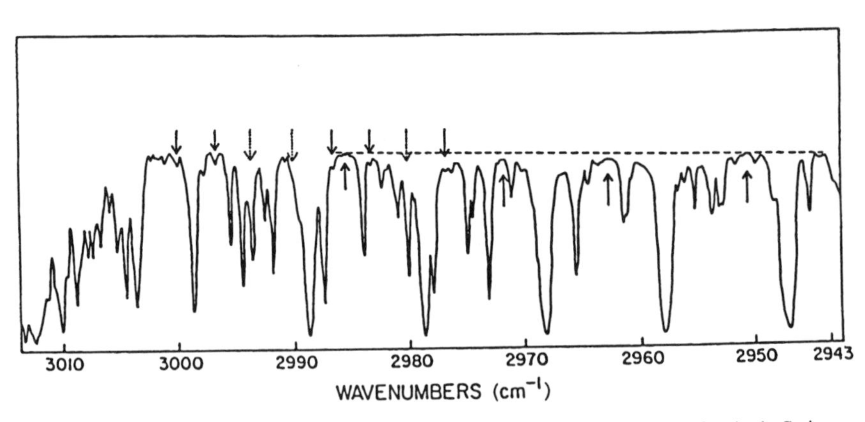

Figure 6.22. A portion of the solar spectrum from Migeotte et al. (1956); Jungfraujoch, Switzerland, 11,746-ft (~3580-m) altitude.

to 7:23 a.m.; sun height, 19°36′ to 26°18′; (center) secant of zenith distance, 2.56; band pass 0.26 cm^{-1}; temperature, 15 °C; relative humidity, 38%. The geometric data supplied with the solar spectrum indicate a distance–air mass product of 13.3 km · atm. Use of this number and a value of 4% for the depths of the stronger ethane lines then indicates the average ethane mixing ratio along the line of sight to be approximately 1.5×10^{-9}.

One can also detect in the spectrum of Fig. 6.17 a broad continuum absorption due to organic C—H groups. This is indicated in the figure by the four arrows that point upward. The intensities at the two inner arrows are reduced by some 2–3% from the intensities at the two outer arrows.

As an example of what has been done from aircraft, Fig. 6.23 shows the carbonyl sulfide lines seen among the carbon dioxide lines in the spectral region 2050–2060 cm^{-1}. This spectrum was obtained by Mankin et al. (1979) using an airborne Fourier transform spectrometer and solar tracking apparatus. In Fig. 6.23, asterisks mark the COS lines used for analysis. The upper curve was from solar elevation 9.2°; the lower curve, from solar elevation 6.5°. These lines indicate a COS mixing ratio of about 5×10^{-10}.

Many recordings of the solar spectrum from balloons have been presented by D. Murcray and his colleagues at the University of Denver. These observations have been carried out over a span of 20 years and have had an especially important impact on the understanding of atmospheric composition and

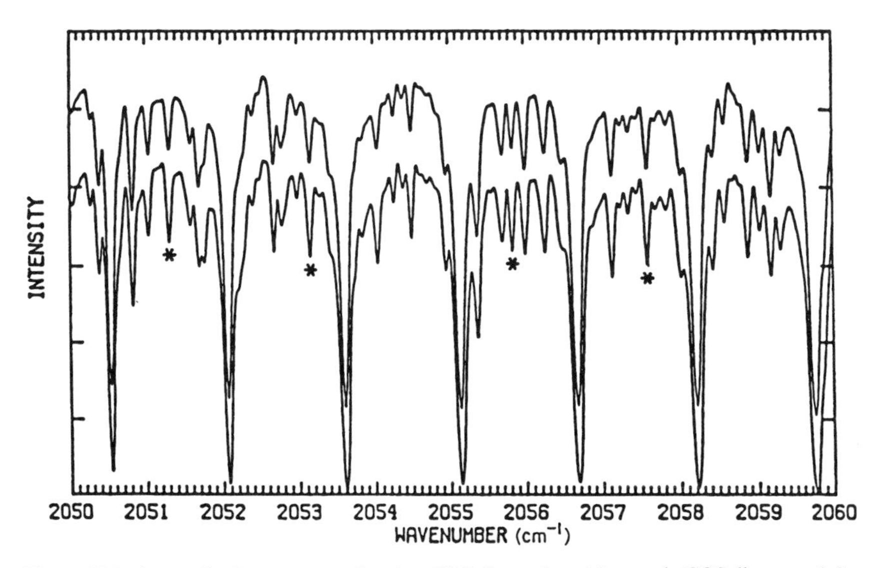

Figure 6.23. Atmospheric spectrum showing COS lines. Asterisks mark COS lines used for analysis. From Mankin et al. (1979).

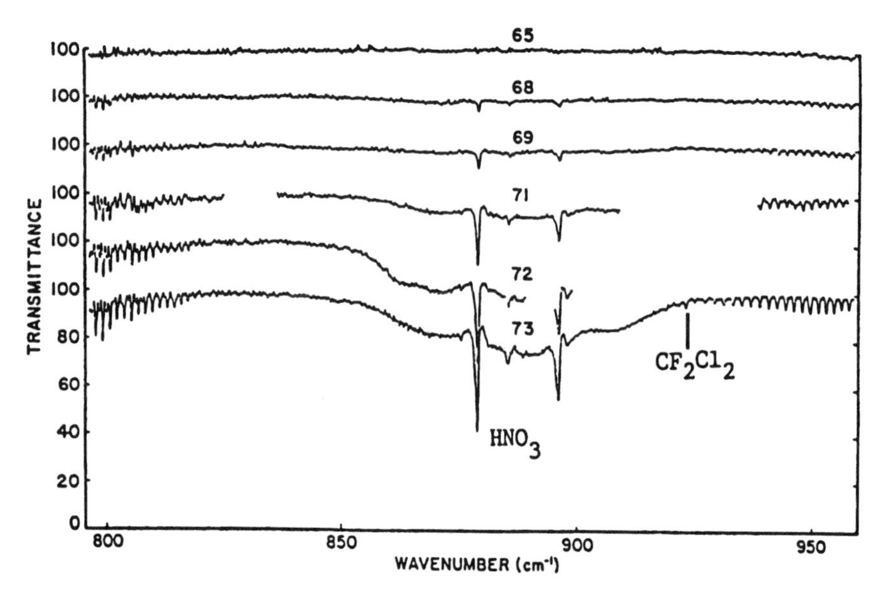

Figure 6.24. Solar spectrum published by Murcray et al. (1969), recorded from a balloon at 30.5 km, with observations directed 94.1° from the vertical.

chemistry. The balloon-borne instruments usually observed the sun from an altitude of about 30 km. Figure 6.24 shows a spectrum recorded in 1968 (Murcray et al., 1969). The main feature of this spectrum is the nitric acid band, with peaks at 879 and 896 cm^{-1}. However, the spectrum also offers an example of how one can derive valuable information from a spectrum years after it was recorded. In 1968 people were not aware of the accumulation of chlorofluorocarbons (CFCs) in the atmosphere; but in 1975, when concern arose over the danger of CFCs leading to stratospheric ozone depletion, the 1968 spectrum was reexamined and the absorption feature of CF_2Cl_2 at 921 cm^{-1} was detected. This is indicated by the arrow in Fig. 6.24.

Further spectra of the halogenated pollutants were obtained in 1975 by the University of Denver group (Williams et al., 1976a,b). Although the CFC threat to stratospheric ozone had been outlined by Molina and Rowland (1974) on the basis of knowledge of the atmospheric chemistry of the compounds involved, it was still necessary to confirm their predictions by physical measurements. The spectra of Fig. 6.25 made that confirmation in a most convincing way. Comparison of the heights of the HNO_3 peaks to the CF_2Cl_2 peak in Figs. 6.19 and 6.20 indicated that between 1968 and 1975 the ratio of CF_2Cl_2 to HNO_3 in the stratosphere increased by fivefold or more.

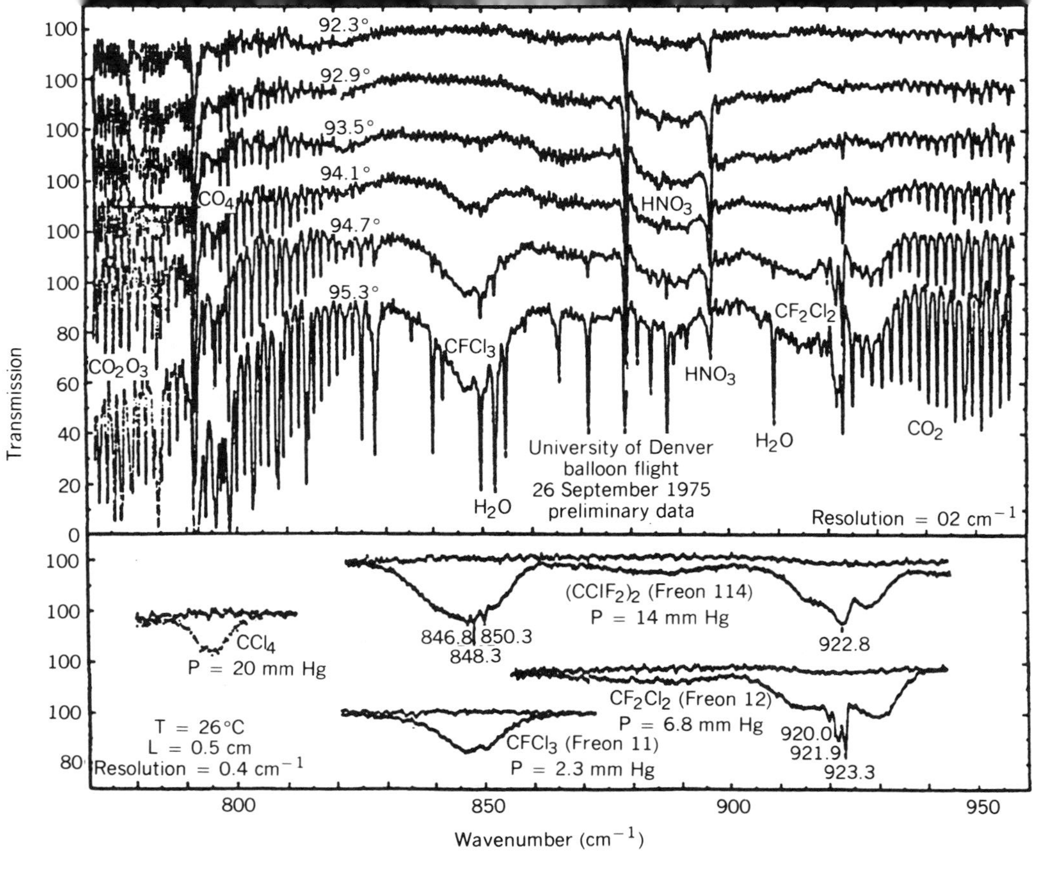

Figure 6.25. Solar spectrum published by Williams et al. (1976a), showing the presence of CFCs in the stratosphere.

An example of the value of "upper limit" calculations from atmospheric spectra is also drawn from the work of the University of Denver group. In discussions of the late 1970s on chlorine–ozone interactions the suggestion arose that much of the chlorine in the stratosphere might be tied up in the compound chlorine nitrate. No one had measured $ClNO_3$ in the air, but laboratory chemistry showed that the compound might well exist. There were not enough kinetic data available to allow accurate predictions, but some kineticists were saying that the $ClNO_3$ could be the major sink for Cl in the stratosphere and that predictions of ozone depletion by CFCs should be scaled down. However, when one examined spectra obtained by balloon-borne instruments (as in Fig. 6.20) and compared them to a laboratory spectrum of $ClNO_3$, no evidence of $ClNO_3$ was seen.

It was concluded that the mixing ratio of $ClNO_3$ in the stratosphere had to be below about 1×10^{-9} (Murcray et al., 1977). Subsequently, solar spectra recorded from balloons showed a weak absorption feature near 1292 cm^{-1} that is probably due to $ClNO_3$ in the 20 to 30-km altitude region. In this case the mixing ratio was estimated to be in the range of 6×10^{-10}–1×10^{-9} (Murcray et al., 1979).

An example of using ratio techniques to bring out information is given by the formic acid measurement of Goldman et al. (1984). There is a problem in seeing formic acid clearly in stratospheric spectra because of interference from ozone lines. Goldman and colleagues solved this problem by making a ratio of solar spectra taken at low and high tangent altitudes. The lower altitude spectrum has much more air mass and hence more formic acid. There is less ozone at lower altitudes, however, so the authors were able to find a low-altitude spectrum and a high-altitude spectrum in which the ozone lines were similar in strength. In the ratio, the ozone dropped out, revealing the formic acid. These results are shown in Fig. 6.26. The upper and lower spectra, **a** and **b**, were taken from an altitude of 33.5 km. For spectrum **a**, the zenith angle was 94.71° and the tangent altitude was 14.7 km. For spectrum **b**, the zenith angle was 95.58° and the tangent altitude was 7.7 km. The heavy dashed line shows a smoothed ratio of **a** over **b**. Three water lines marked with a W are evident, as well as the formic acid *Q*-branch marked F.

The horizontal optical path at ground level has been used mainly in the study of urban and industrial pollution. An example of a long single path is seen in the work of Herget (1979). He used a quartz-iodine lamp as light source and projected the radiation along a path on the order of 1–2 km to a receiving telescope and Fourier transform spectrometer. When his observations were in the vicinity of an oil refinery, species detected included ethylene, propylene, and sulfur dioxide (Fig. 6.27).

In the vicinity of a fertilizer plant he detected hydrogen fluoride (Fig. 6.28). Each of these spectra shows the column amount of trace gas distributed along

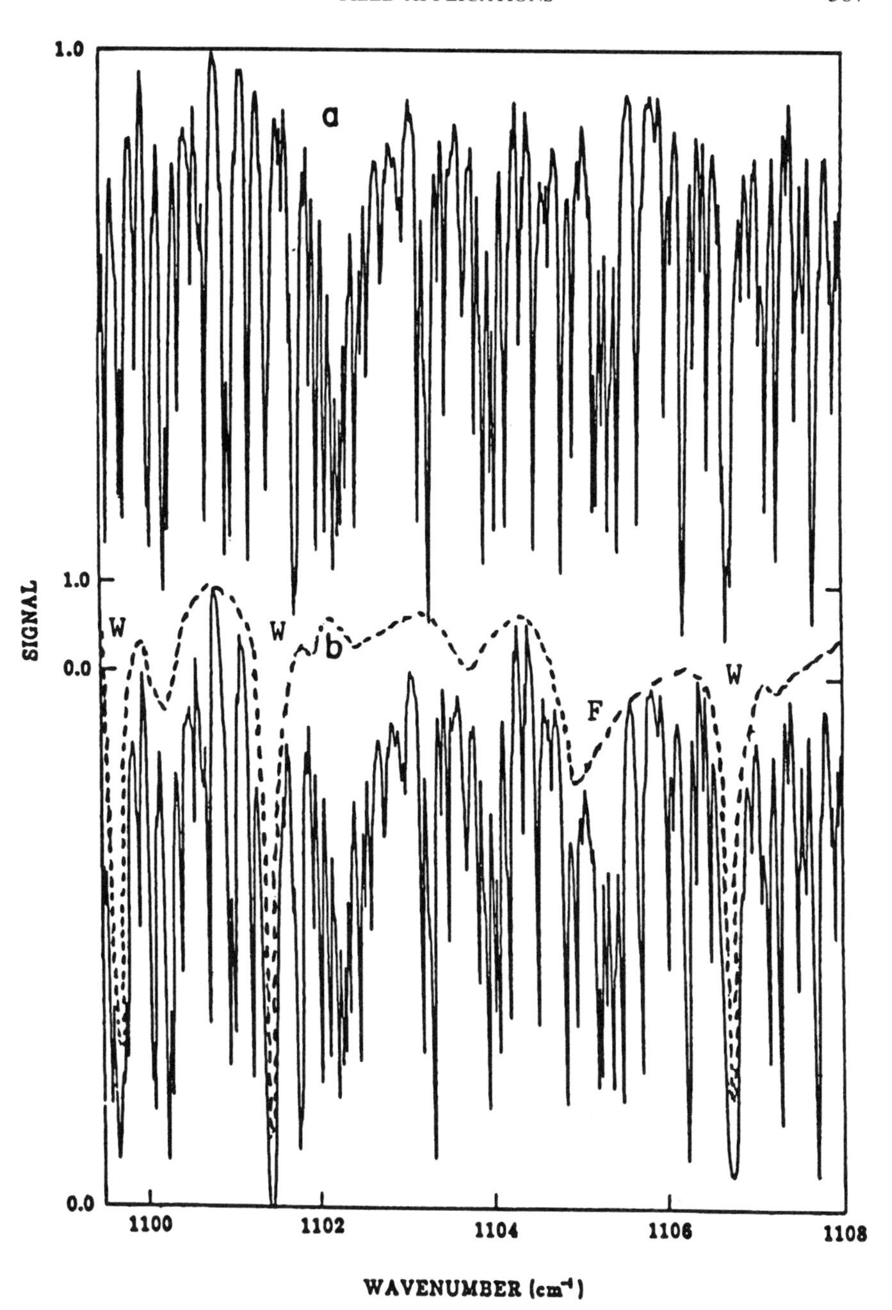

Figure 6.26. Solar absorption spectra published by Goldman et al. (1984).

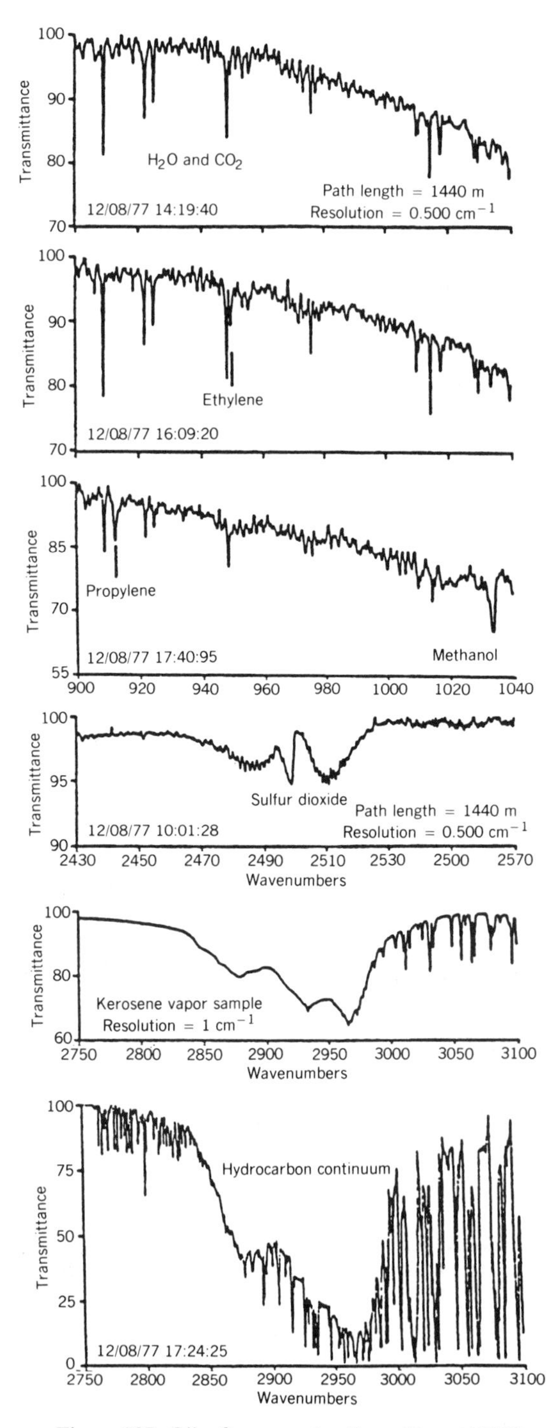

Figure 6.27. Oil refinery spectra. From Herget (1979).

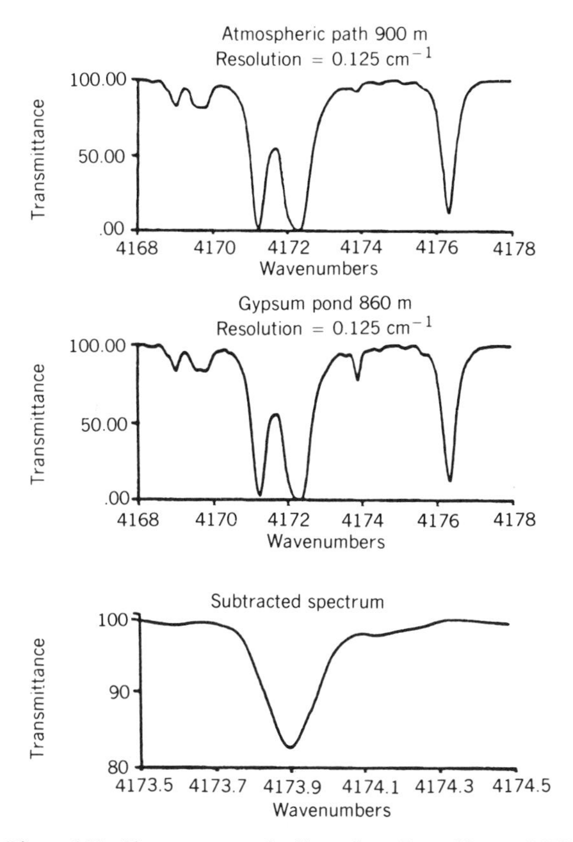

Figure 6.28. Air spectra near fertilizer plant. From Herget (1979).

the whole path. Such a single long path can be used to monitor the total pollution crossing the boundary of an industrial site.

One of us (P.L.H.) has participated in several programs of measurement of ambient air pollution by long-path absorption spectroscopy. The first of these took place in 1956 and used a spectrometer with sodium chloride prism; the last was in 1980 and used a high-resolution Fourier transform spectrometer. In each succeeding program, improvements in technique increased the number of pollutants measured and lowered the detection limits. In all cases, the long optical path was folded between the mirrors of a White cell. Table 6.4 lists conditions and some results of the various measurement programs. Figure 6.29 shows a portion of the spectra seen with the 1956 instrumentation. The resolution and signal-to-noise ratio in the spectra of 1956 were not good, but we did show for the first time that in situ spectroscopy could mea-

Table 6.4. Long-Path Infrared Studies of California Smog

Year and Location	Type of Instrument	Details of Long-Path Cell	Species Detected and Monitored	References
1956, S. Pasadena	NaCl prism spectrometer	240-m White cell, 3-m base	Ozone, PAN, CO, hydrocarbons	Scott et al. (1957)
1973, Pasadena	FTIR, 1-cm^{-1} resolution	400-m White cell, 9-m base	Acetylene, CO, ethylene, formic acid, ozone, methane, methanol, CF_2Cl_2,CCl_4, PAN	Hanst et al. (1975b)
1976, Riverside	FTIR, 0.25-cm^{-1} resolution	1-km cell, 8-mirror type, 23-m base	Hydrocarbons, PAN, formic acid, formaldehyde, nitric acid	Tuazon et al. (1978)
1980, Los Angeles	FTIR, 0.25-cm^{-1} resolution	1-km White cell, 23-m base	NH_3, CO, H_2CO, HCOOH, CH_3OH, HNO_3, NO, NO_2, ozone, PAN, hydrocarbons	Hanst et al. (1982)

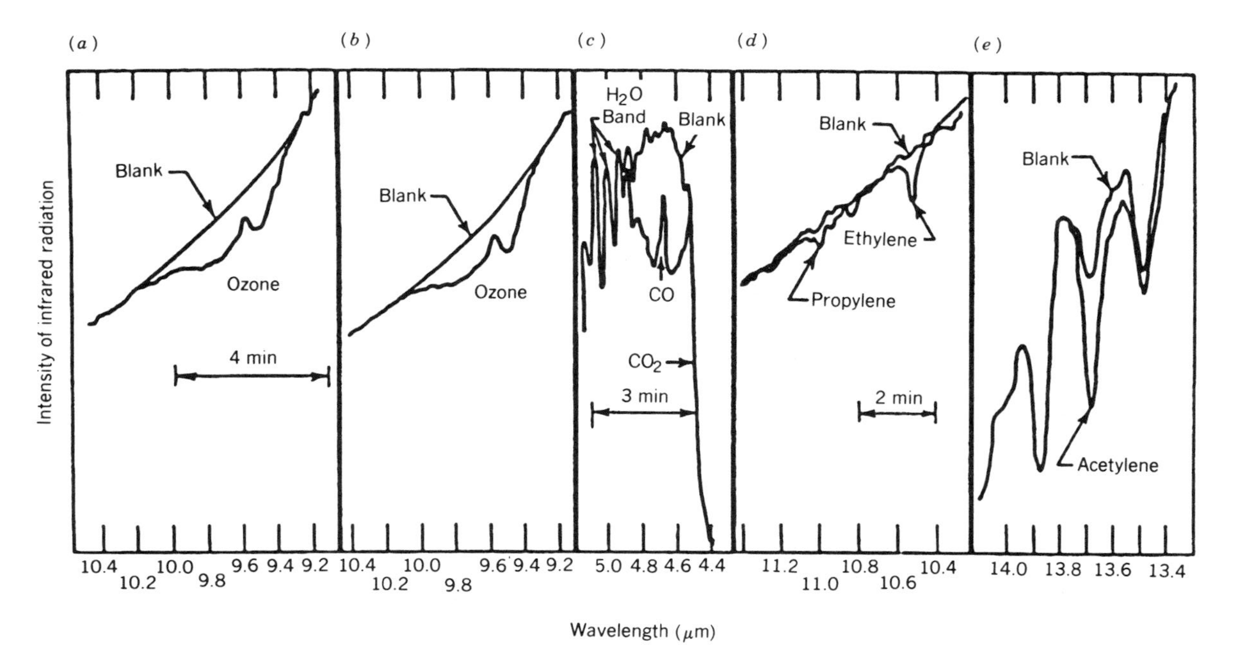

Figure 6.29. Infrared bands of pollutants in South Pasadena, California, 1956: (a) ozone, 0.4 ppm, 3 p.m. Sept. 28, 1956 (Slit, 500 μm; path, 336 m); (b) synthetic ozone, 1.34 ppm (slit 500 μm; path, 120 m); (c) carbon monoxide; 40 ppm, 8:30 a.m. Dec. 17, 1956 (slit, 100 μm; path, 300 m); (d) ethylene and propylene, 8:35 a.m. Dec. 17, 1956 (slit, 400 μm; path, 300 m); (e) acetylene, 8:35 a.m. Dec. 17, 1956 (slit, 800 μm; path, 300 m).

sure the ozone and other pollutants. In looking at more recent spectra, however, we can see that some of the absorption we attributed to hydrocarbons was really due to unresolved water lines.

6.4.3. Measurement of Polluted Urban Air, with Grab Sampling

Air samples have been analyzed by means of a Digilab FTS-40 spectrometer system, using a compact 7.2-m multiple-pass absorption cell and a "briefcase" sample collection system.

The Infrared Analysis, Inc. long-path minicell was used, mounted in the sample compartment of the spectrometer. When the sample compartment cover was lowered, a rectangular plexiglass plate with valves fit into the opening in the top of the cover. This allowed a purge to be maintained while the air samples were being passed in and out of the cell through the two vacuum valves.

Glass sample containers were used to transport polluted air samples to the laboratory for analysis. These were glass vessels with a Teflon vacuum value on each end. Since the absorption cell volume was only 0.5 L, the sample container volumes were chosen to be only about 0.7 L. Three sample containers were carried in a briefcase. The sample containers were evacuated in the laboratory and carried to the site in their case. At the sampling site a valve was opened and the sample flowed in under its own atmospheric pressure. When a sample was returned to the laboratory, it was allowed to flow into the evacuated absorption cell by way of a long evacuated transfer tube. The volume of the transfer tube was about 2 L, so that when the air sample had come to pressure equilibrium some 16% of the gas was in the cell, about 64% was in the transfer tube, and 20% was still in the sample bottle. Laboratory air was then allowed to flow into the end of the sample container away from the transfer tube, and the sample gas in the transfer tube was compressed into the absorption cell. Gas mixing down the long transfer tube was not fast enough for the room air to reach the cell. Vacuum was the only tool used to move the sample, and glass was the only surface the sample encountered. A schematic diagram of this configuration is shown in Figure 6.30.

An example of the use of the air measurement system in the analysis of captured samples of New York City air is shown in Fig 6.31–6.34. Figure 6.31 (lower part) shows the full infrared spectrum of an air sample captured near the Manhattan end of the Brooklyn Bridge on Friday, Nov. 27, 1987. The air sample was captured by opening the evacuated glass container while driving down the street. The main absorption bands in this transmittance plot are due to atmospheric water and carbon dioxide. The other trace gases contributed only very weak bands that are just barely visible in this transmittance plot. The signal-to-noise ratio in this spectrum is higher than 1000:1, however, and when

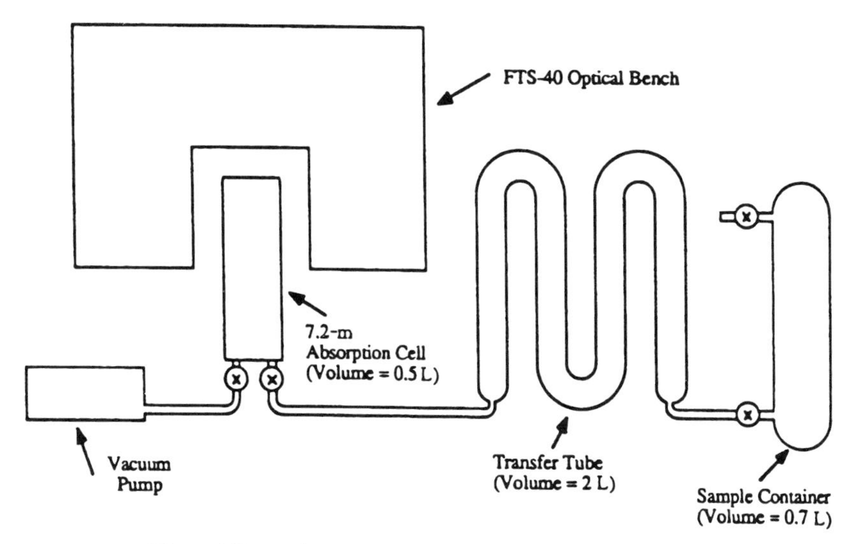

Figure 6.30. Equipment configuration for grab sample analysis.

the plot is amplified, as in Figs. 6.32–6.34, the absorption bands of the other trace gases are easy to see.

Water and CO_2 spectra stored in the computer may be subtracted from the sample spectrum before the bands of the other trace gases are measured. The upper spectrum of Fig. 6.31 shows the result of subtracting most of the water interference from the lower spectrum. A complete water removal is usually not feasible, but the example shows that nearly complete removal is sufficient to reveal hidden bands of the trace gases.

Figure 6.32 shows an absorbance plot of the C—H stretching region of the spectrum, after water subtraction. Above the air spectrum are reference spectra of methane and non-methane hydrocarbons, along with the amounts calculated from the air spectrum.

Figure 6.33 shows the CO–N_2O region of the spectrum with some of the CO_2 lines. The reference spectra and the calculated amounts are also shown.

Figure 6.34 shows the important spectral region that in air analysis is usually hidden by the main water band. Here one can see under the remains of the water band the absorptions due to NO, NOCl, NO_2, and HNO_3.

The amounts calculated from the spectrum were surprisingly high for such a heavily populated urban area. Normally these nitrogen-containing pollutants would be some 10–100 times lower in concentration than this. It is presumed that what happened near the Brooklyn Bridge on Nov. 27, 1987, was that a plume of pollution came down to ground level. In a sample captured

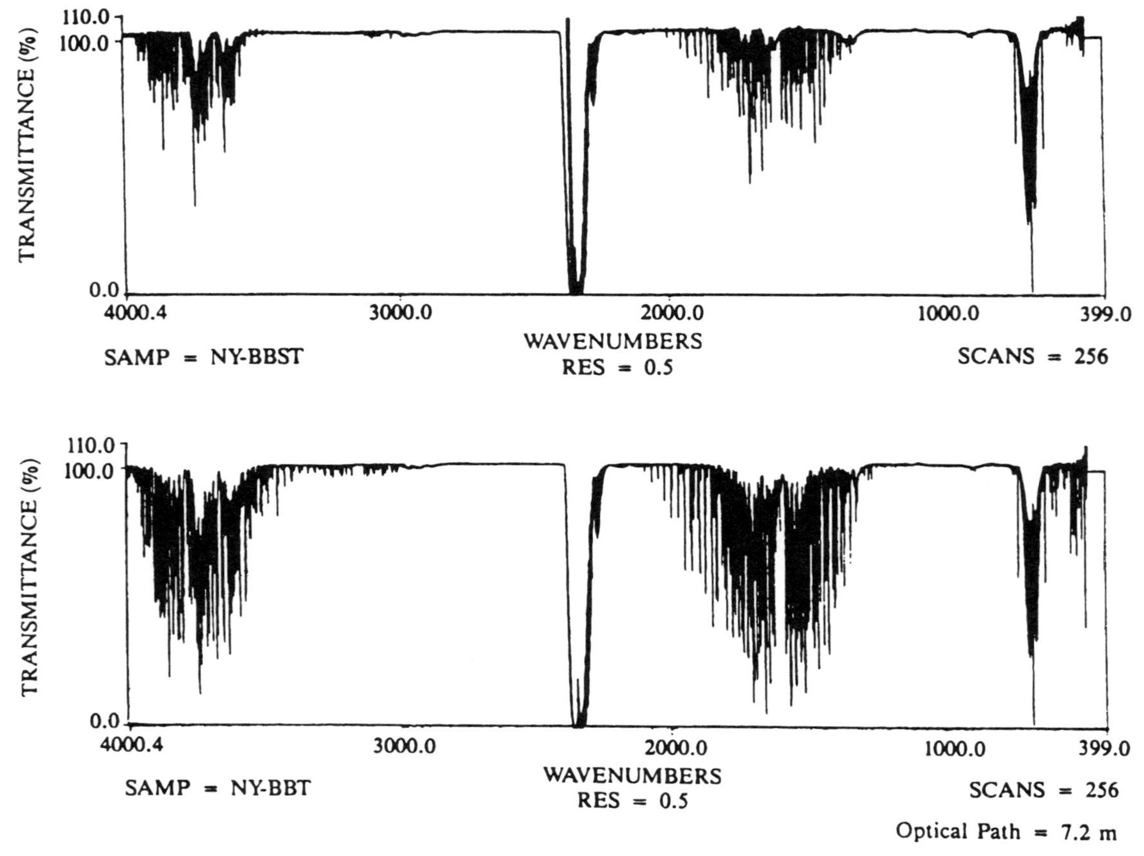

Figure 6.31. Transmittance plots of air sampled near Brooklyn Bridge, Nov. 27, 1987: (lower) full spectrum; (upper) spectrum with water subtracted.

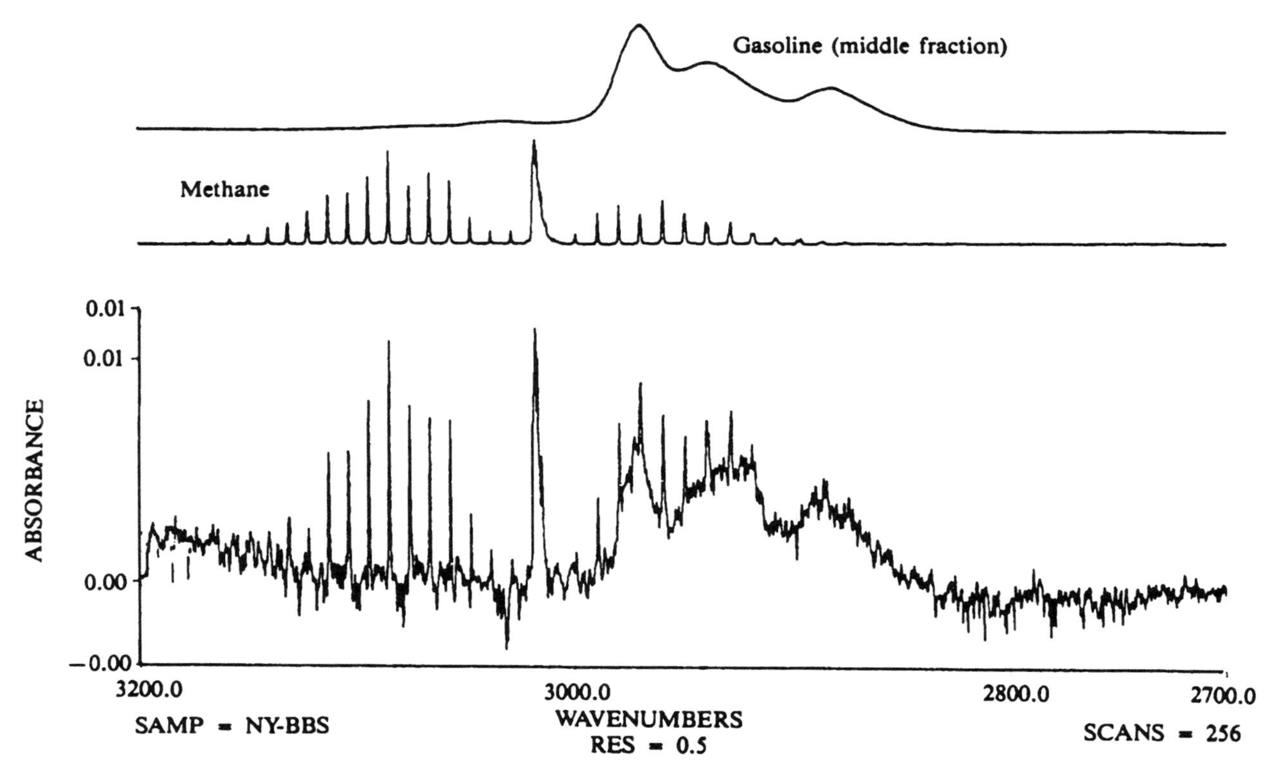

Figure 6.32. Air near Brooklyn Bridge, 1 p.m. Nov. 27, 1987: water subtracted; 7.2 m. Air spectrum shows 2.0 ppm methane and 0.6 ppm non-methane hydrocarbons, calculated as hexane.

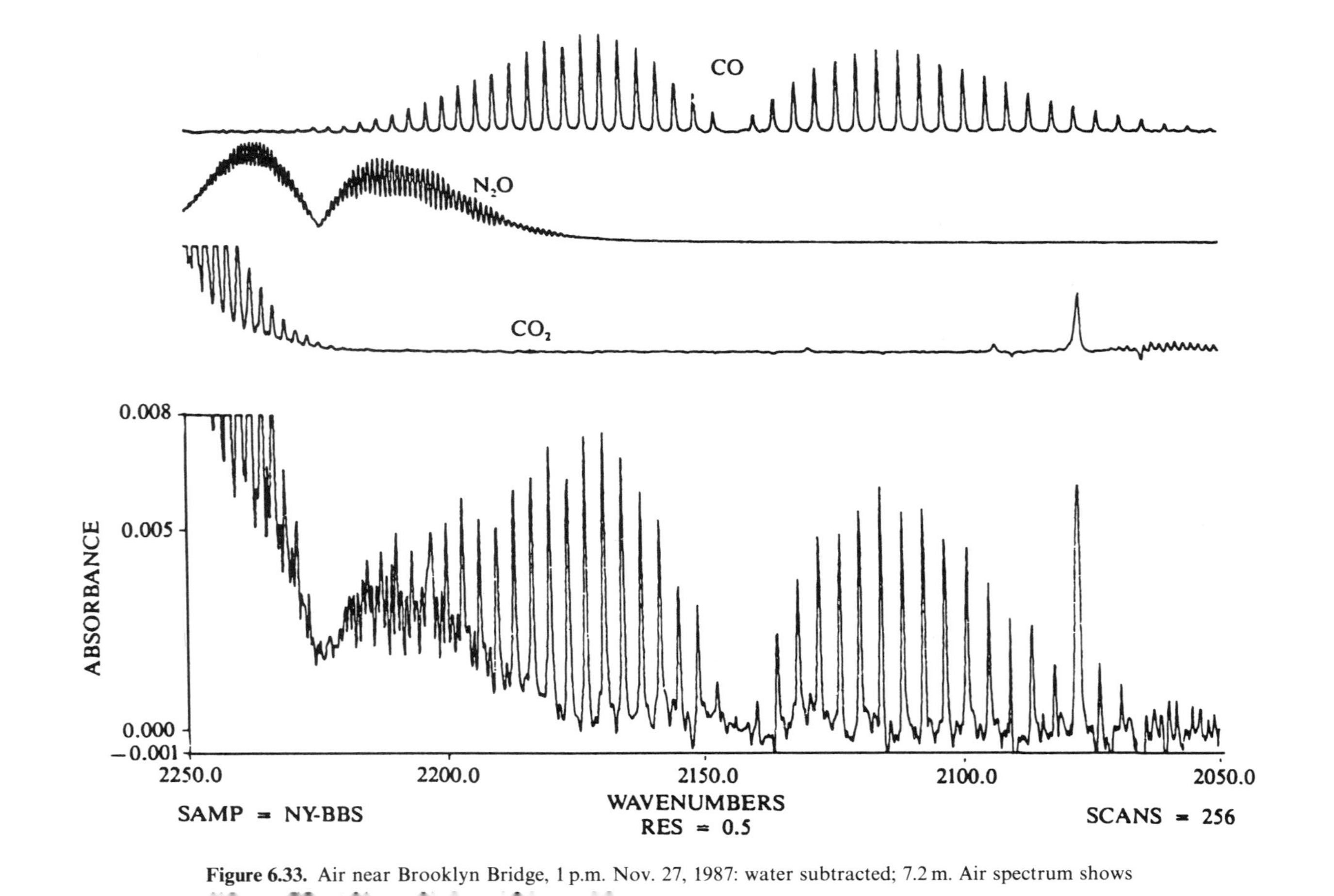

Figure 6.33. Air near Brooklyn Bridge, 1 p.m. Nov. 27, 1987: water subtracted; 7.2 m. Air spectrum shows

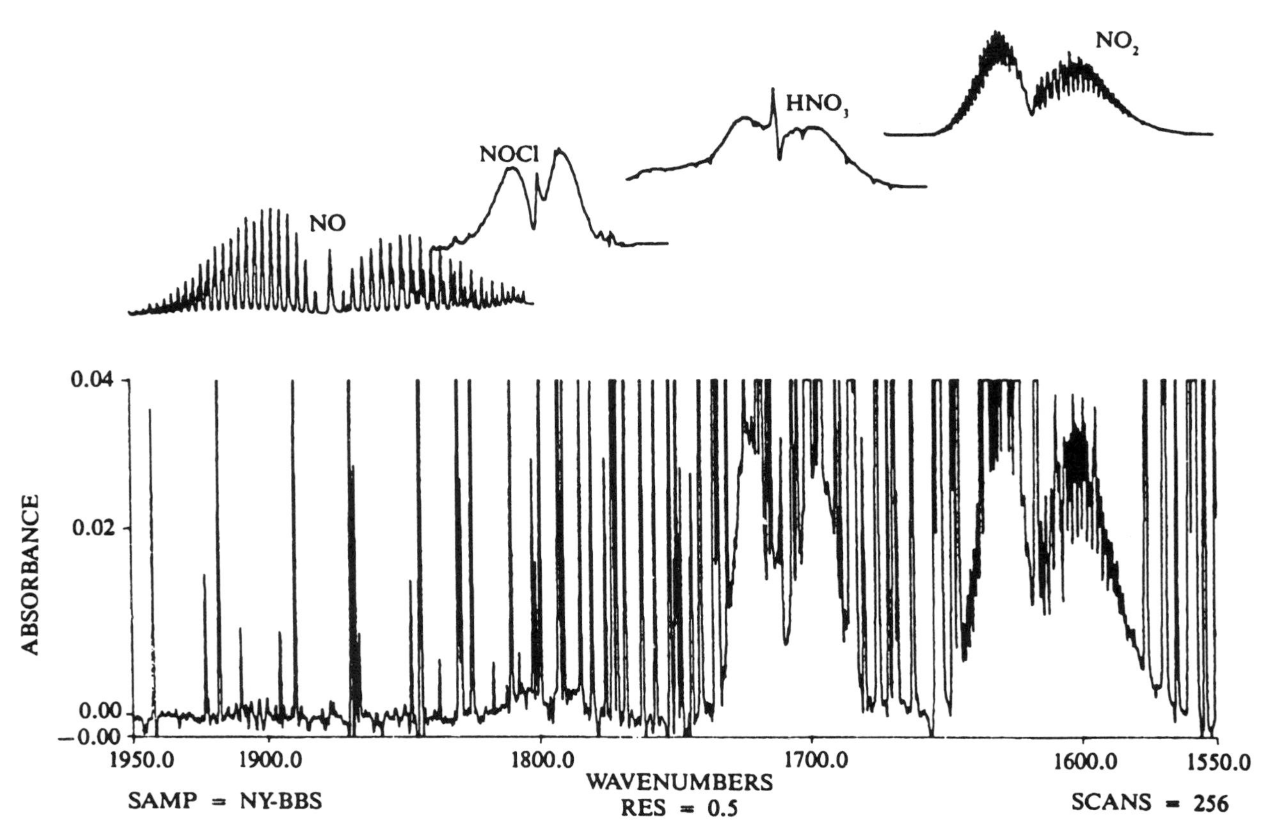

Figure 6.34. Air near Brooklyn Bridge, 1 p.m. Nov. 27, 1987: water subtracted; 7.2 m. Air spectrum shows 0.6 ppm NO, 0.07 ppm NOCl, 3.7 ppm HNO_3, and 2.6 ppm NO_2.

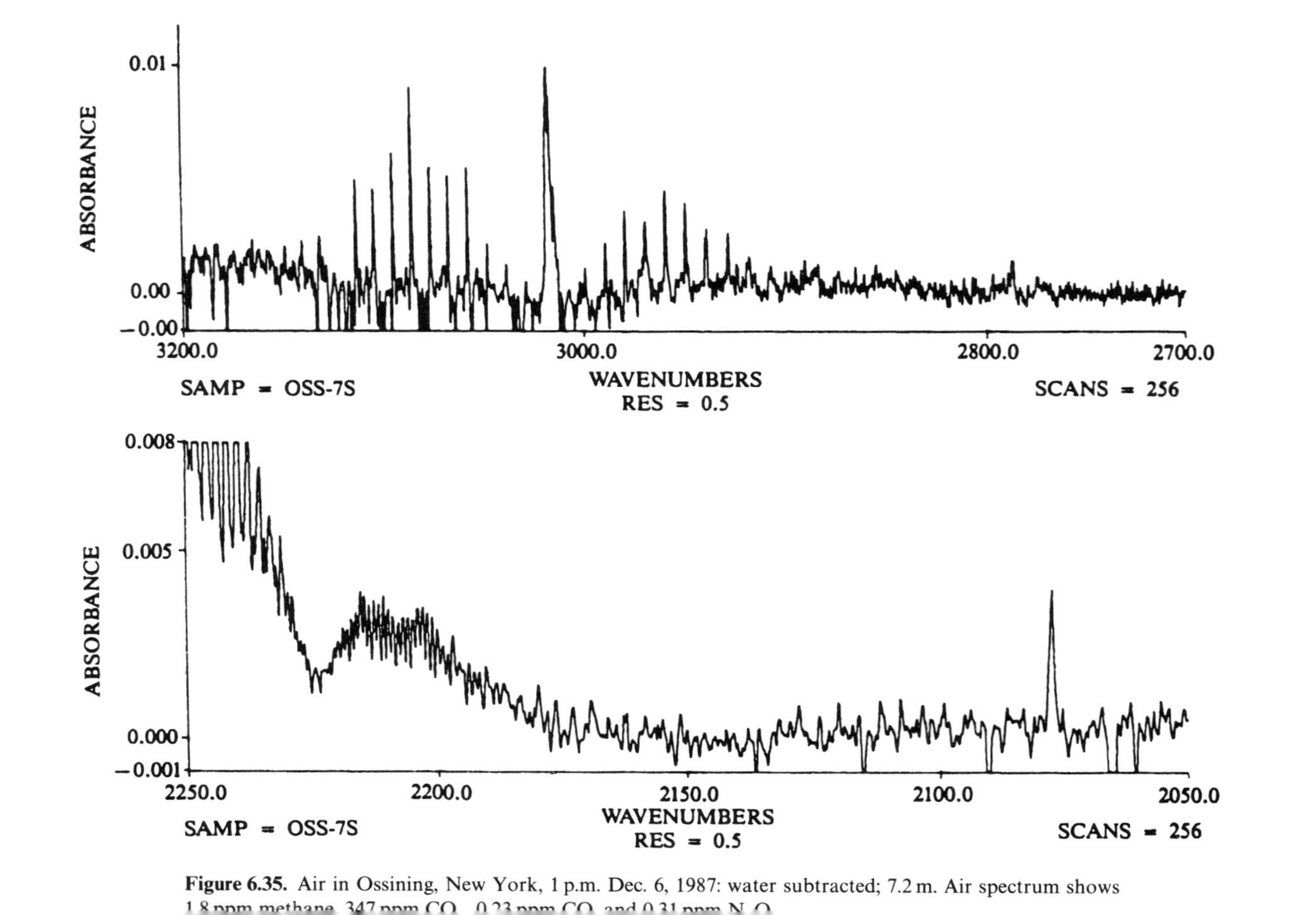

Figure 6.35. Air in Ossining, New York, 1 p.m. Dec. 6, 1987: water subtracted; 7.2 m. Air spectrum shows 1.8 ppm methane, 347 ppm CO_2, 0.23 ppm CO, and 0.31 ppm N_2O.

in Manhattan on 42nd Street about 15 min later, high concentrations of NO_2 and HNO_3 were also observed. Measurements on samples captured 10 days later at the same locations did not show abnormal amounts of the nitrogen-containing pollutants.

The quantitative analysis reported here was done with the aid of the Digilab subtraction routine. To measure a compound, the sample spectrum and the reference spectrum were displayed and the interactive subtraction was carried out. When the band being observed was removed from the sample spectrum the amount was readily obtained by multiplying the ppm-meters of the reference spectrum by the subtraction factor and then dividing by the path-length.

To verify the validity of the method of sample manipulation and the accuracy of the reference spectra, measurements were also made on clean air in the town of Ossining, about 30 miles north of New York City on the Hudson River. The Ossining air sample was captured on an exceptionally clear day, just after passage of a weather front. Clean "background" air is known to contain approximately 350 ppm CO_2, 1.7 ppm CH_4, 0.3 ppm N_2O, and 0.2 ppm CO. Portions of the Ossining spectrum are shown in Fig. 6.35, along with the calculated amounts of trace gases. These values are close to the expected background values.

6.4.4. An Example of Atmospheric Analysis by Open-Path Infrared Spectroscopy

6.4.4.1. Introductory Remarks

The work described here is a continuation of long-path infrared studies of the atmosphere that have been carried on intermittently for many years. Previous work using long folded optical paths is described in papers by Tuazon et al. (1978) and Hanst et al. (1982). These references contain additional citations of previous work.

In earlier work, an enclosed optical path was nearly always used. An exception was the outdoor long-path studies by Herget (1979), who used a single long path between large 30-in. transmitting and receiving telescopes. We report below on open-path studies using a three-mirror multiple-pass optical system (White cell), with total optical paths up to 1 km. At a chosen path length, a multiple-pass cell can transmit and refocus the same amount of energy as can a large pair of telescopes, even though the cell mirrors are much smaller than the telescope mirrors. This is the principal advantage of the three-mirror cell, as was stated in the title of White's original paper in 1942: "Long Optical Paths of Large Aperture."

6.4.4.2. Measurement Technique

A Digilab FTS-40 spectrometer was used, working with a spectral resolution of 0.5 cm^{-1}. The radiation from the scanning interferometer was projected out the side port of the spectrometer into a three-mirror multiple-pass cell with a 22.5-m base path. The two collecting mirrors were semicircular in shape (D-mirrors). They were cut from a single round mirror of 12-in. diameter. These mirrors were mounted on a pedestal placed in the open air. A wooden cover over the mirrors shaded them from the sun. The field mirror was of 12-in. width and was situated close up against the spectrometer. The cell mirrors were coated with silver, protected with a ceramic overcoating. This type of coating has reflectivity higher than 99% throughout the infrared region. The coating is resistant to tarnish and other types of corrosion. It is superior to gold both in reflectivity and durability.

The arrangement of optical components is diagrammed in Fig. 6.36. The helium–neon laser radiation centered in the infrared beam was found to be bright enough to be seen on the silver mirrors even in daylight. This red light was used for alignment and path length verification. After the infrared radiation had been passed through the long-path cell, it was captured and directed to the detector by a three-mirror transfer optics system mounted on a base plate in the sample compartment. The spectrometer could be returned to normal use merely by removing the transfer optics plate and moving aside the plane mirror that coupled the infrared beam out the side port. Following are some of the matters considered in the choice of the particular operating conditions:

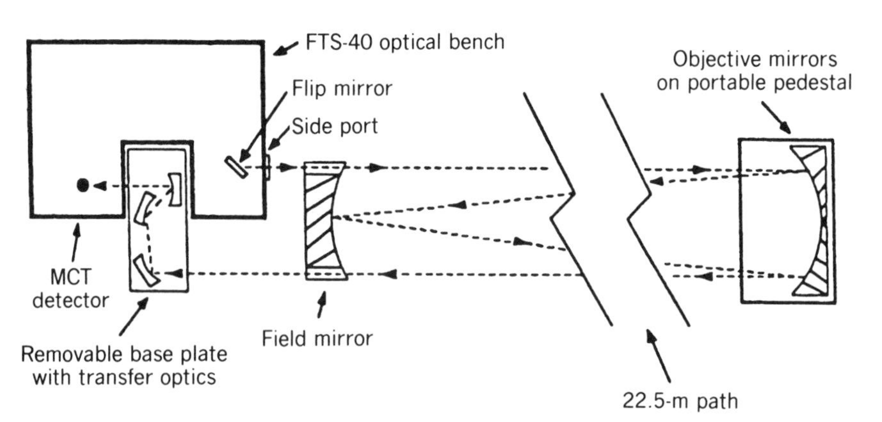

Figure 6.36. Optical system for open-air long-path infrared spectroscopy. *Key*: MCT = mercury cadmium telluride (HgCdTe).

a. Open Path. If the path is open, there are no wall effects. Reactive compounds are properly measured. Photochemical equilibria are not disturbed. If the air is moving, the trace gas measurements are averaged over all the air mass that moves through the optical path during the time of scanning. Previously, one of us (P.L.H.) has made an effort to surround the light path with an enclosure— a pipe or some kind of tunnel. One reason for this was to allow recording of a background spectrum when the light path was evacuated or flushed with nitrogen. Another reason was to stabilize the air sample so that refraction effects due to air turbulence would not cause excessive noise in the spectrum. A third reason was to permit filling the light path with water vapor and pure tank air so a water reference spectrum could be obtained. It now appears that, if necessary, these reasons for enclosing the light path can be ignored.

b. Lack of Noise from Air Turbulence. When there are many cell traversals in an open path, air turbulence causes the red laser beam to move about randomly. The laser beam image at the cell exit shows jitter and pulsations. Probably the infrared image has similar variations. The spectrum, however, does not seem to show noise from this image movement. Probably, this discrimination against the "seeing noise" is due to the high frequencies at which the scan modulates the infrared signal. These modulation frequencies are in the kilohertz range, whereas the "seeing noise" frequencies are probably some 10–100 times lower. Low-frequency modulations would appear as noise only if one were working in the far infrared.

c. Water Vapor. The concentration of water vapor in the air will be 10^5–10^6 times higher than the concentrations of the trace gases being measured. Absorption by water vapor therefore dominates the infrared spectrum. It is customary for spectroscopists either to remove the water vapor from the optical path or to prepare a background spectrum with the same amount of water absorption as in the sample spectrum. Unfortunately, working at kilometer path lengths complicates preparation of background spectra of water vapor. With the other infrared-absorbing gases, including CO_2, one can fill a relatively small absorption cell with a high pressure of the compound and match the absorption in the kilometer path of air. This is not possible with water because of its limited vapor pressure. To make a water reference spectrum for a kilometer path, one needs to fill the whole path with water vapor mixed with pure air or nitrogen. This was attempted in previous studies, but not here.

Subtracting the water lines does not add any information to the spectrum; it just makes it easier to read the information that is there. The main thrust of our present work is to show that one can read the spectrum directly for trace gases without removing any water lines.

d. Resolution. As mentioned in Section 6.2.1 the width of spectral lines is about $0.2\,cm^{-1}$ at atmospheric pressure. To see all the detail in the air spectrum therefore requires resolving power on the order of $0.1\,cm^{-1}$. Since the average laboratory FTIR system does not do that well, a lower resolution must be accepted. Other reasons for accepting lower resolution are that higher resolution requires more computer memory, leads to longer computation times, and produces more noise in the spectrum. For the present work, we have chosen resolution of $0.5\,cm^{-1}$. This allows one to utilize most of the detail in gas phase spectra while working with modest price instrumentation.

e. Choice of Pathlength. When working in a spectral region where water and CO_2 do not absorb strongly, like the region 1200–800 cm^{-1}, lengthening the optical path increases the measurement sensitivity. When working in a region of strong water and CO_2 absorption, however, the path may need to be shortened to allow transmission of enough energy for a measurement. For some molecules there is a choice between using a strong absorption band that falls in a region of heavy interference or a weak band that is in the clear. In the case of SO_2, our choice is to use the weak spectral feature at $1130\,cm^{-1}$ with a maximum path length. In the case of NO_2, our choice is to use the strong band at $1600\,cm^{-1}$ but to shorten the path to give about 30% transmittance at the measurement frequency.

f. Throughput Advantage of the Multiple-Pass Optical System. Radiation projected by an optical instrument spreads as it goes out. The intensity incident on a distant receiver decreases with the square of the distance from the source. The three-mirror multiple-pass cell brings the collecting mirror close to the source, even though the path is long. Thus one has a long path of large aperture, which gives high-energy throughput.

g. Convenience of the Multiple-Pass Optical System. The use of a multiple-pass cell allows the transmitter and the receiver to be together, as in the work described here, where they were part of a commercial spectrometer. The field mirror is mounted with the spectrometer. The objective mirrors are set up as a separate portable unit. A kilometer path can be set up in a single room—or in a parking lot or on a roof. If the spectrometer unit is in a van and the objective mirrors are on a tripod, the whole system can easily be transported from place to place.

h. Choice of Detectors. Any gain in detector sensitivity is equivalent to an increase in path length. In spectral regions of strong interference from water or carbon dioxide, one should adjust the path length to give about 30% transmittance; then the only way to increase the detection sensitivity is to increase the detector signal-to-noise ratio. A spectrometer system for trace gas analysis should be equipped with a nitrogen-cooled photodetector of the highest available sensitivity. In our work reported here, a "wide-band" HgCdTe detector was used. For compounds whose analytical bands fall in the high-frequency region like HF, HCl, and H_2CO, an indium antimonide (InSb) photodetector should be used.

i. The Background Spectrum. Computer manipulations of the data require that the spectra be in absorbance form. To make an absorbance plot, one needs a background spectrum. In gas studies the spectrum of the empty cell usually serves as background. In our aforementioned work, the multiple-pass cell was "shorted out" of the optical system in order to provide a background spectrum. A pair of plane mirrors did this. One mirror at the entrance to the cell sent the infrared beam across to the other mirror at the cell exit. This second plane mirror sent the beam on to the detector. When the multiple-pass cell spectra were used with these background spectra, the resultant absorbance plots were quite flat across the whole spectrum. This proves that the silver mirrors with their ceramic overcoating do not have any dips in reflectivity in the spectral regions used.

j. The Use of Digitized Quantitative Reference Spectra. In recent years the major improvements in infrared technique have come through advances in computer software. Thirty-five years ago when one of us (P.L.H.) was engaged in infrared studies of the Los Angeles smog, a transmittance spectrum was obtained from point-to-point hand measurements of the empty-cell spectrum and the sample spectrum. The measuring and replotting for one transmittance spectrum would take all afternoon. The computer now does this in a second or less.

The software capabilities now include automated quantitative analysis. For this one needs digitized quantitative reference spectra, which have not generally been available. Currently available collections of infrared spectra are mainly designed for identification, not quantitation. For quantitative analysis of gases, one needs to take into account the pressure-broadening effects and other line width effects that lead to deviation from the Beer–Lambert law. In order to avoid those deviations when working at modest resolution, the reference and sample spectra must be used only in the low-absorbance region (absorbance of 0.1 or less, for compounds whose spectra have single lines). For the present work a collection of quantitative reference spectra was prepared in digital form using the same instrument that was used for the long path studies. When these spectra are used at low absorbance, the logarithmic absorption law is always obeyed.

6.4.4.3. The Atmospheric Transmission

Shown in Figs. 6.37–6.40 are the transmittance spectra obtained at a 90-m path (four traversals) and a 720-m path (32 traversals). These spectra were

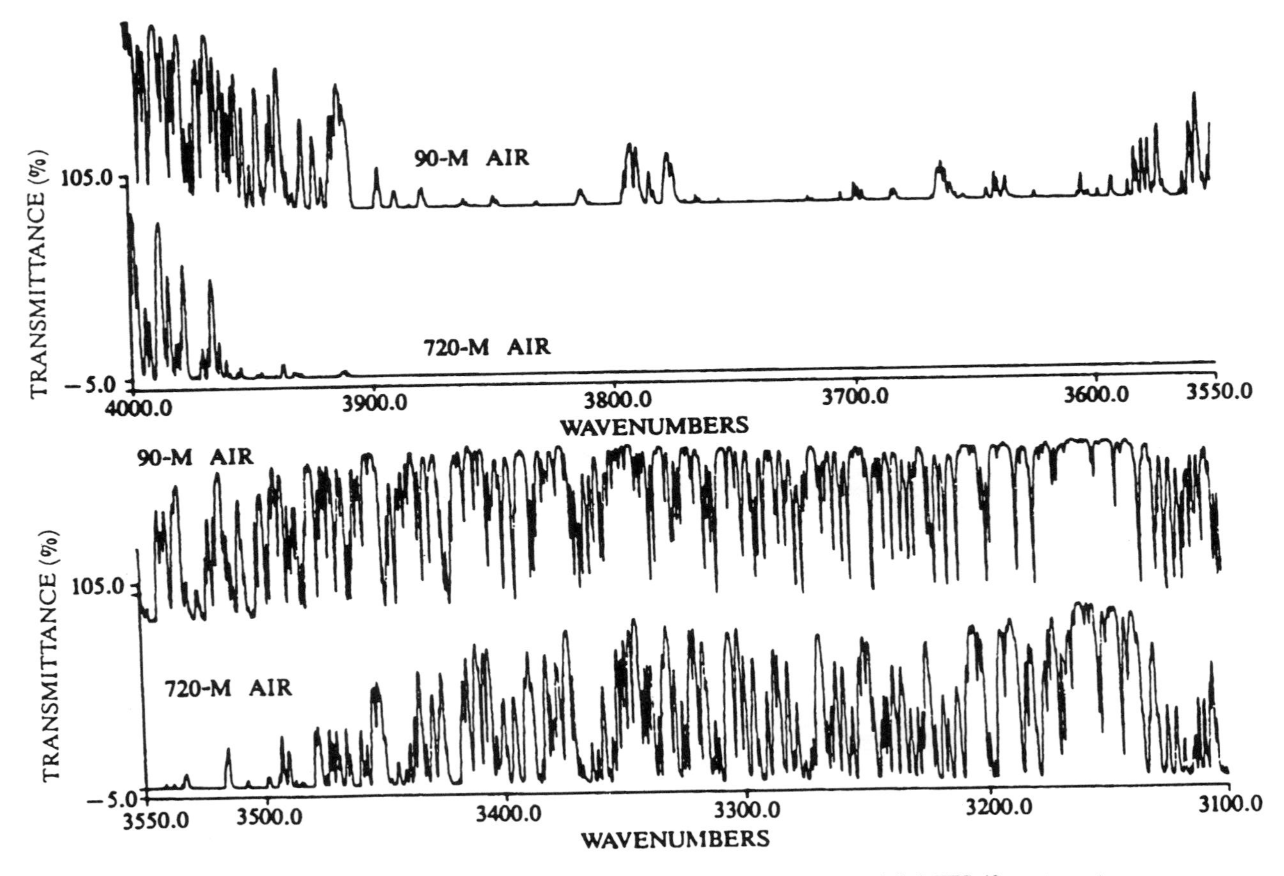

Figure 6.37. Infrared transmittance of air at Ossining, New York, July 3, 1988: Digilab FTS-40 spectrometer; HgCdTe detector; 0.5-cm^{-1} resolution; 30-min scan.

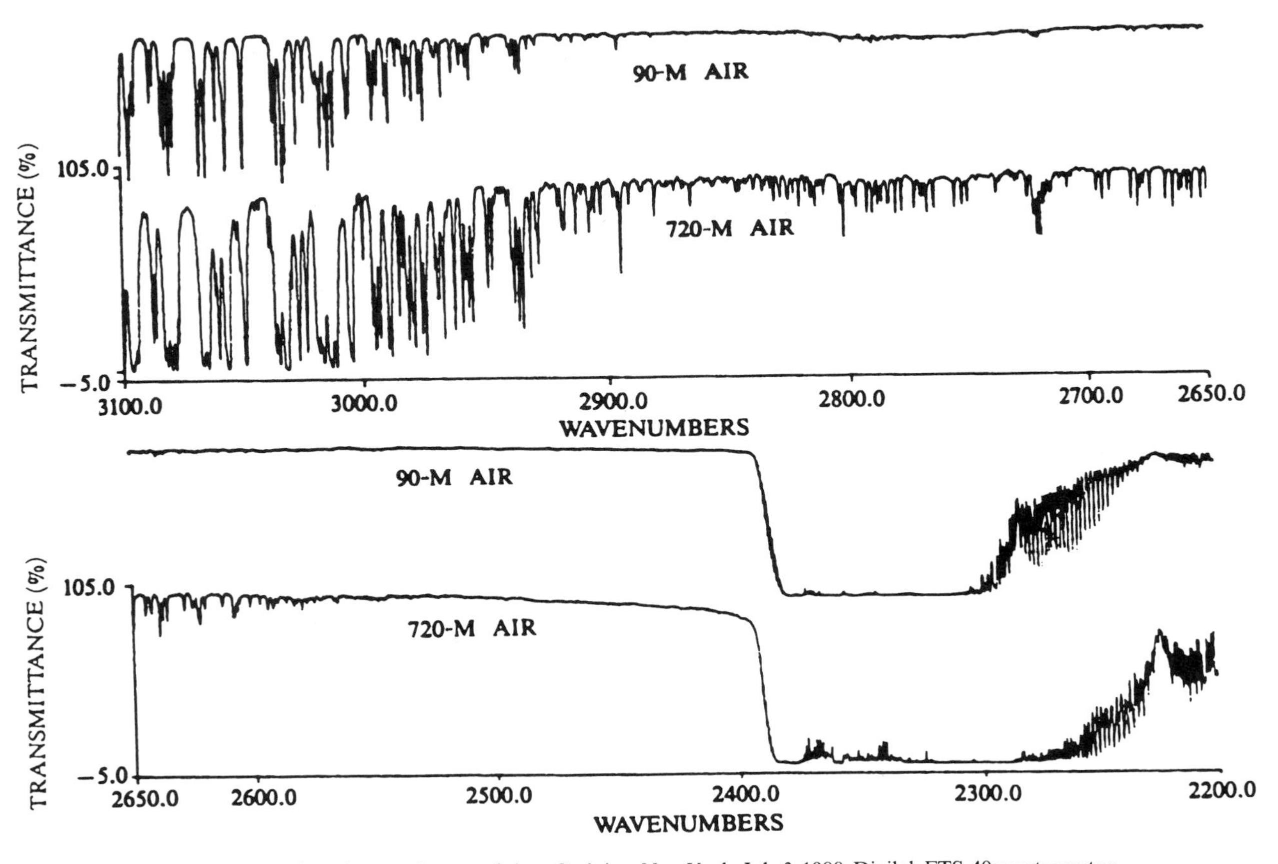

Figure 6.38. Infrared transmittance of air at Ossining, New York, July 3, 1988: Digilab FTS-40 spectrometer; HgCdTe detector; 0.5-cm^{-1} resolution; 30-min scan.

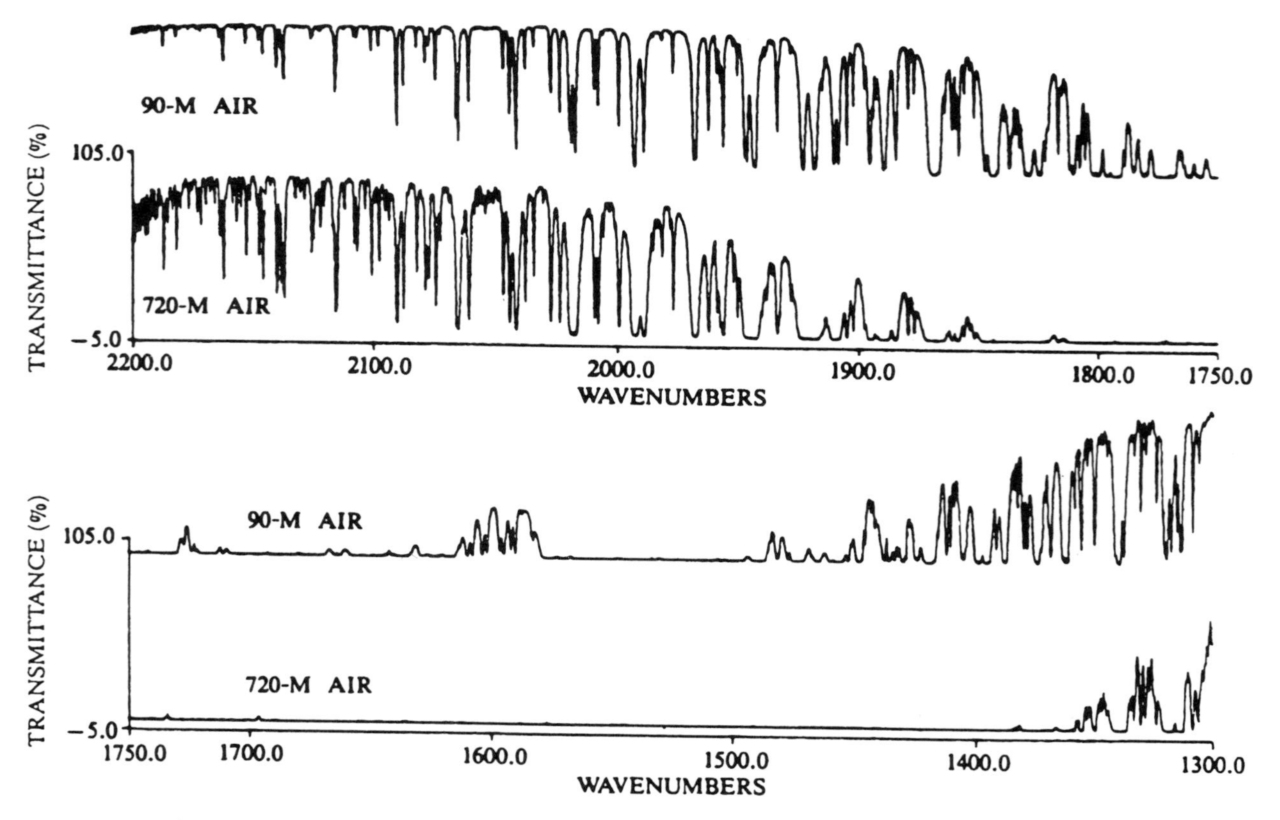

Figure 6.39. Infrared transmittance of air at Ossining, New York, July 3, 1988: Digilab FTS-40 spectrometer; HgCdTe detector; 0.5-cm^{-1} resolution; 30-min scan.

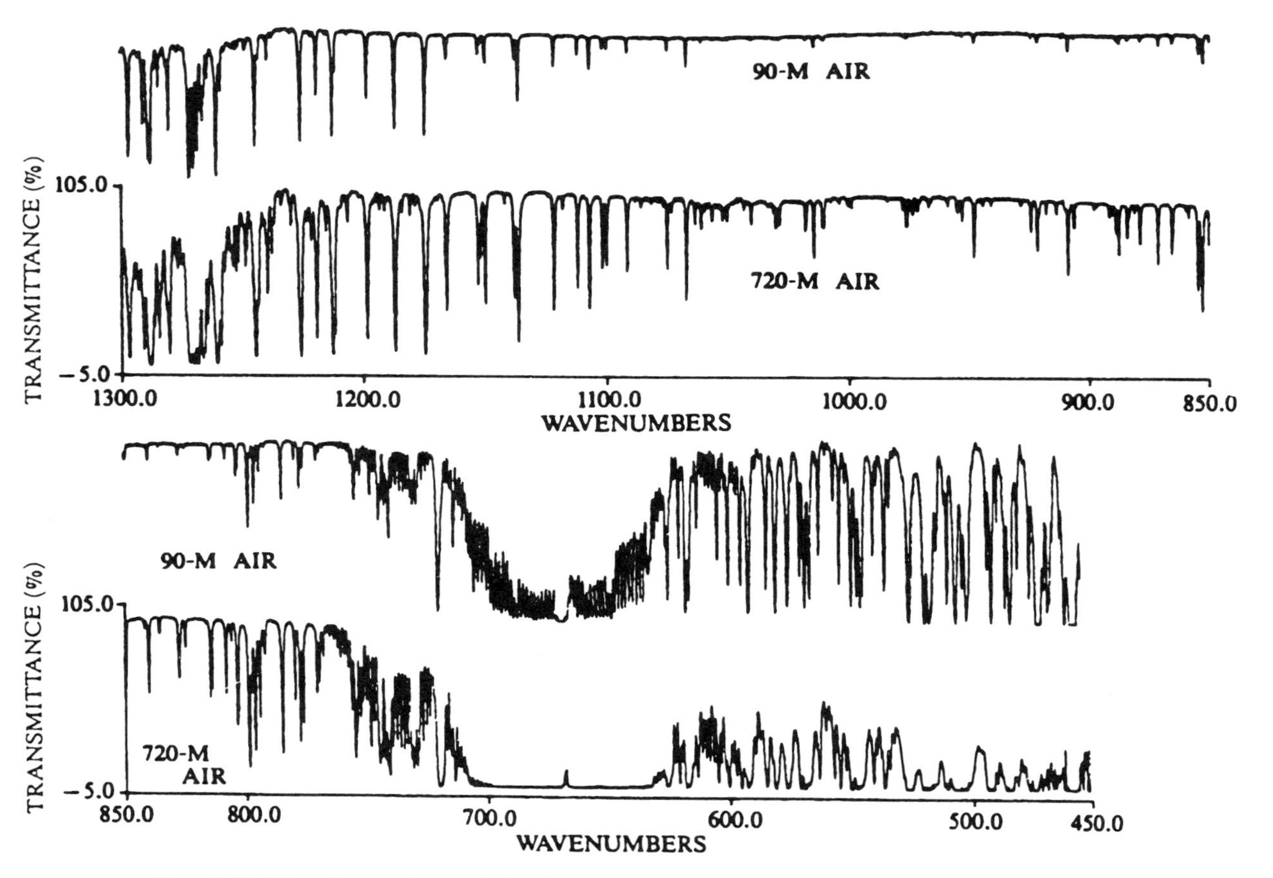

Figure 6.40. Infrared transmittance of air at Ossining, New York, July 3, 1988: Digilab FTS-40 spectrometer; HgCdTe detector; 0.5-cm^{-1} resolution; 30-min scan.

recorded on July 3, 1988, which was a warm but only moderately humid day. These spectra show which spectral regions are available for measurements and which are not.

In Fig. 6.37 we see that even for the shorter path the region between 3900 and 3550 cm^{-1} does not transmit enough energy to be of any use. This is the region of the OH bands. Thus alcohols and acids must be detected by bands other than those involving the OH stretch. Likewise, Fig. 6.39 shows that the carbonyl region between 1800 and 1610 cm^{-1} offers practically no energy, even at the 90-m path. The detection of carbonyl bands in open air studies is therefore out of the question. This is especially unfortunate because many strong and characteristic molecular bands fall in the carbonyl region. The region between 1580 and 1400 cm^{-1} is also practically useless, but that is not really serious because not very many significant bands fall there.

The plots show that the regions 3200–1800 cm^{-1} and 1400–700 cm^{-1} are the prime spectral regions for open-path gas measurements. There is also some transmission from 630 to 450 cm^{-1}, but there are not very many important bands that fall in that region. It will be shown that these open regions of the spectrum reveal bands and lines for almost every polyatomic and heteronuclear diatomic molecule in the air. When the revealed band is a strong one, the detection capability extends down to the level of a few ppb—or even lower. When the revealed bands are weaker, the detection limits are correspondingly higher.

6.4.4.4. *Examples of Trace Gas Measurement*

Carbon dioxide is a minor constituent of the atmosphere, but its normal mixing ratio of 350 parts CO_2 per million parts air (ppm) puts its concentration some 200 times higher than for the next most concentrated trace gas. The next gas is methane, at about 1.7 ppm. After methane comes nitrous oxide at 0.3 ppm and carbon monoxide at about 0.16 ppm. Figures 6.41 and 6.42 show some of the CO_2, CH_4, N_2O, and CO lines that may be used for direct measurements in an open atmospheric path. The lines are of course interspersed with water lines; but some are in the clear and can be used for quantitative measurements. These recommended lines are marked by arrows.

In order to verify our 720-m path length and also to test the validity of our quantitative reference spectra, we have calculated from the spectrum the concentrations of the four aforementioned naturally occurring molecules. To do this we used the Digilab software subtraction routine. To measure a compound, the sample spectrum and the reference spectrum were displayed and the interactive subtraction was carried out. When the band being measured was removed from the sample spectrum, the amount was readily calculated by multiplying the path length–concentration factor for the reference

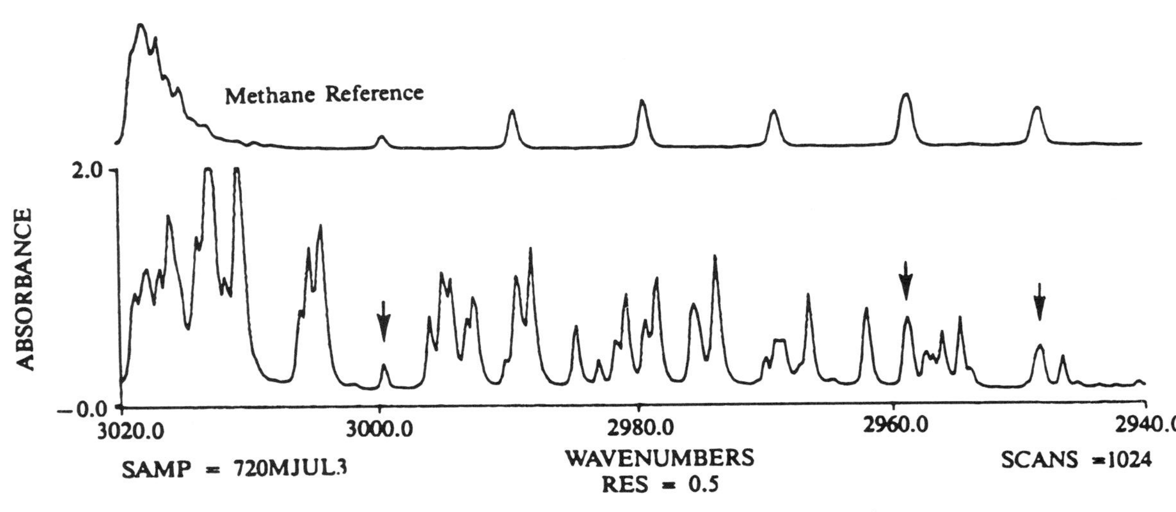

Figure 6.41. Identifying methane lines: 720-m air path; methane reference.

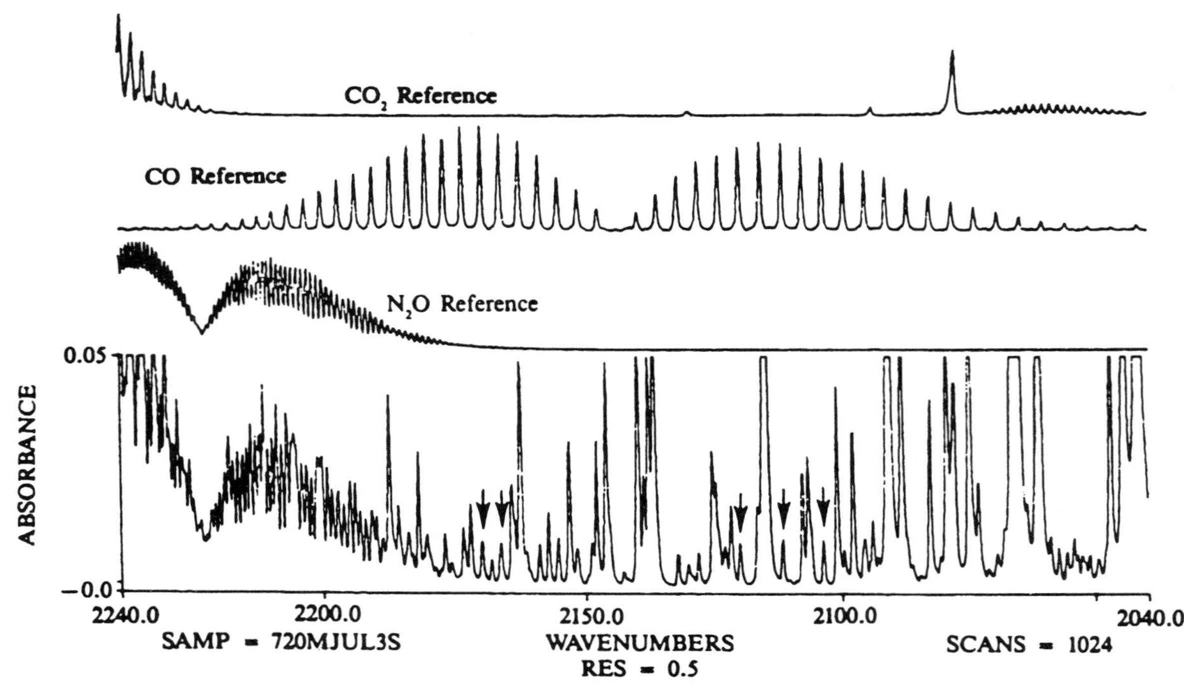

Figure 6.42. Identifying N_2O, CO, and CO_2 lines: 720-m open-air path length.

Compound	Background Amount in Unpolluted Air (ppm)	Measured Amount
CO_2	350	350
CH_4	1.7	1.9
N_2O	0.30	0.29
CO	0.16	0.27

spectrum by the subtraction factor and then dividing by the path length. The measured values are given in the accompanying tabulation. These values appear to be an adequate verification of our measurement method. It is especially important to obtain a nearly correct value for N_2O, which is not a pollutant. In previous studies, the N_2O concentration was never found to deviate measurably from its background value. The low measured value of carbon monoxide indicates that the air mass under study was not polluted by auto exhaust in any significant degree. In an urban area, the CO concentration will usually be about 10 times higher than the amount measured here. It appears that on July 3, 1988, a Sunday, Ossining was not in a plume of pollution moving north from New York City. Instead, it is presumed that the town was immersed in a clean air mass that had moved in from the west with the passage of a weather front during the previous day.

To further illustrate the detection of pollutant gases, we present Figs 6.43–6.52. In each figure the bottom plot is a portion of the atmospheric spectrum for July 3, 1988. Above is a reference spectrum of the constituent under consideration. The third spectrum presented in each figure is a synthetic spectrum obtained by adding together the atmospheric spectrum and the reference spectrum weighted to correspond to the presence of the indicated number of parts per billion (ppb) of the compound. These synthetic spectra therefore show the distortions of the clean air spectrum that are indicative of the presence of pollutants. Particular lines or bands that may be used for quantitative analysis are marked by arrows. These examples cover a number of important molecules, but the set is far from complete. In the future we expect to expand the set of examples to include halogen acids, nitrates, peroxides, nitriles, nitrosamines, alkenes, alkynes, hydrides, organometallics, and other groups of compounds. Discussion of individual cases follows:

a. Ammonia. Figure 6.43 (left) shows that ammonia has several lines and bands favorably located for detection. The two features marked in the figure are probably the best choices for quantitative analysis. Measurement sensitiv-

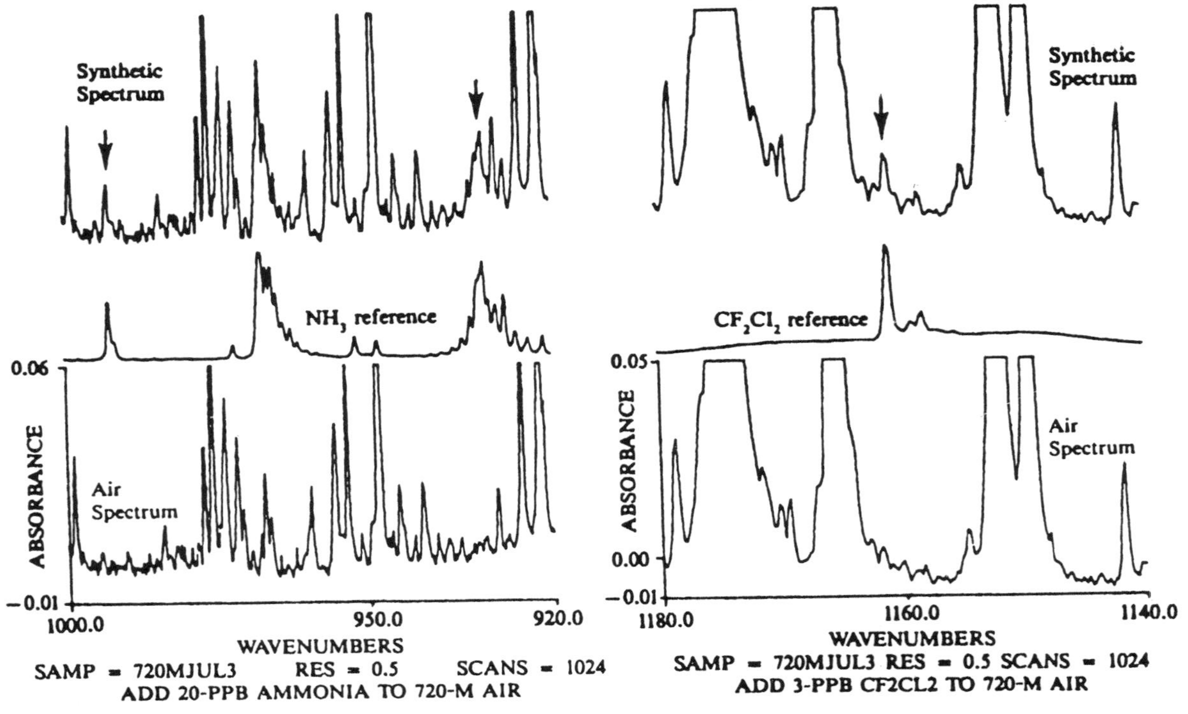

Figure 6.43. Ammonia and dichlorodifluoromethane.

ity is high. The spectrum of July 3, 1988, does not show any absorption attributable to ammonia. Comparing the synthetic spectrum with the real spectrum puts the detection limit at 1 or 2 ppb.

b. Dichlorodifluoromethane (Freon-12). Figure 6.43 (right) shows that there is a very strong spectral feature for CCl_2F_2 at approximately 1161 cm^{-1}. There are some weak N_2O lines in this region, which are barely resolved. We cannot see the CCl_2F_2 that is present in clean air at about 0.4 ppb. The synthetic spectrum shows that 3 ppb of the compound could easily be detected. Probably, higher spectral resolution would be helpful in detecting the compound at less than 1 ppb.

c. Benzene. The strongest feature in the benzene spectrum falls at 674 cm^{-1}. This is a region of strong CO_2 absorption, and therefore one might consider using a weaker benzene band that falls in a region with less interference. There is such a band centered at 1037 cm^{-1}. This is a case where it is best to choose the stronger band and minimize interference by backing off on path length. Figure 6.44 shows the case for 90 m of air. There is enough transmission between the CO_2 lines to allow detection of the benzene. The lower spectrum shows that the three absorbance minima centered around 674 cm^{-1} line up nicely in the absence of benzene. In the synthetic spectrum we see that 200 ppb of benzene clearly distorts the pattern in a way that can be used for quantitative measurement.

d. Butane. Methyl and methylene groups in organic molecules absorb in the region 3000–1850 cm^{-1}, the C—H stretch region. At a 720-m path the clean air spectrum shows a weak C—H band due to the organic matter. Figure 6.45 shows how the absorption in the atmospheric spectrum of July 3, 1988, was equivalent to the absorption by about 50 ppb of butane.

e. Ethylene. The strongest feature in the ethylene spectrum falls at 950 cm^{-1}, about 1 cm^{-1} to the side of a water line. At a resolution of 0.5 cm^{-1} a few ppb of ethylene will reveal their presence as a "shoulder" on the water line, as seen in Fig. 6.46 (left). In the clean air spectrum of July 3, 1988, there is a small shoulder at the bottom of the water line, but this is probably not due entirely to ethylene. There are also weak CO_2 lines in this region, one of which probably contributes to this shoulder. Higher resolution would separate the lines better.

f. Formaldehyde. The C—H stretch band of formaldehyde is rich in lines and falls on the low-frequency side of the C—H bands of most other molecules. At a 720-m path many weak water lines overlap the formaldehyde spectrum, as shown in Fig. 6.47. At least five formaldehyde lines fall between water lines and may be used for measurement. These are marked by arrows. The use of higher resolution and the use of an InSb detector would increase the measurement sensitivity for formaldehyde.

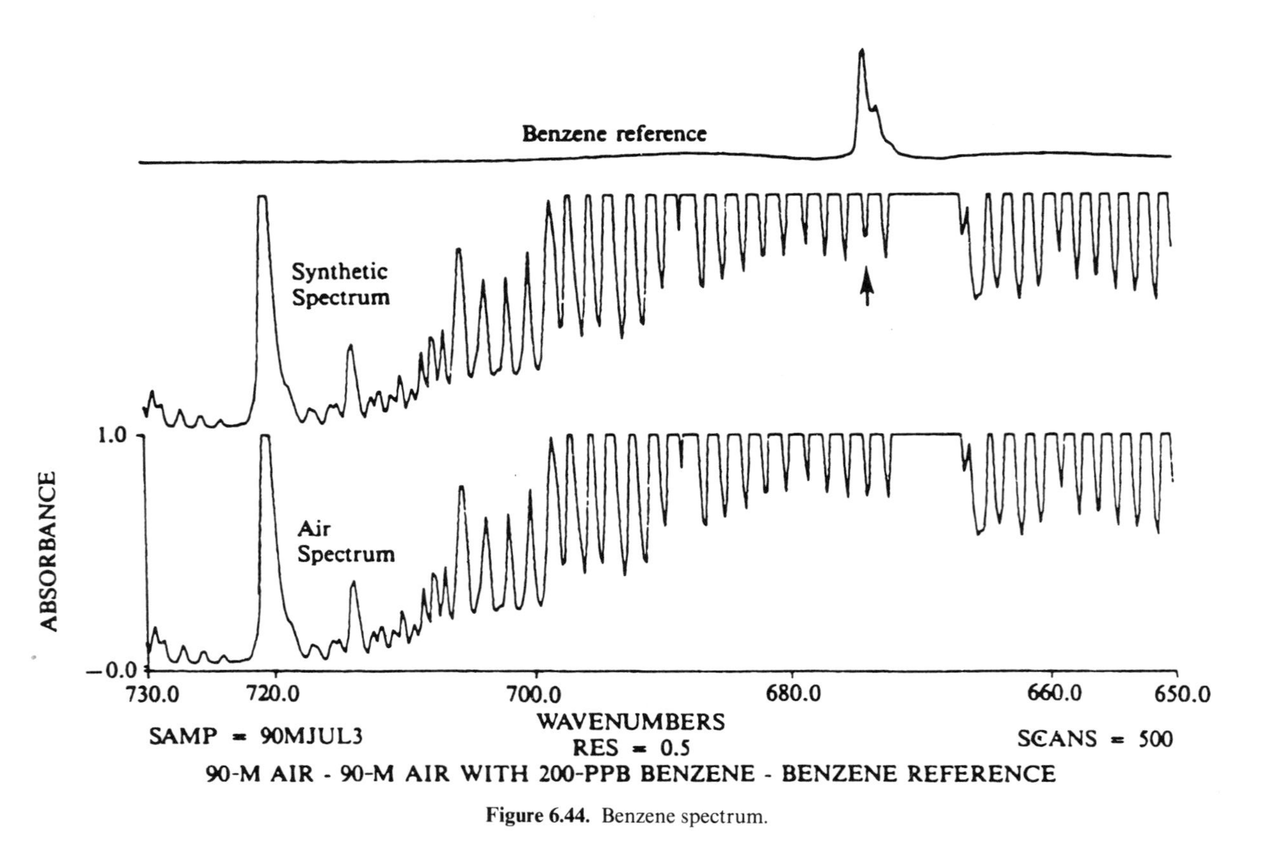

Figure 6.44. Benzene spectrum.

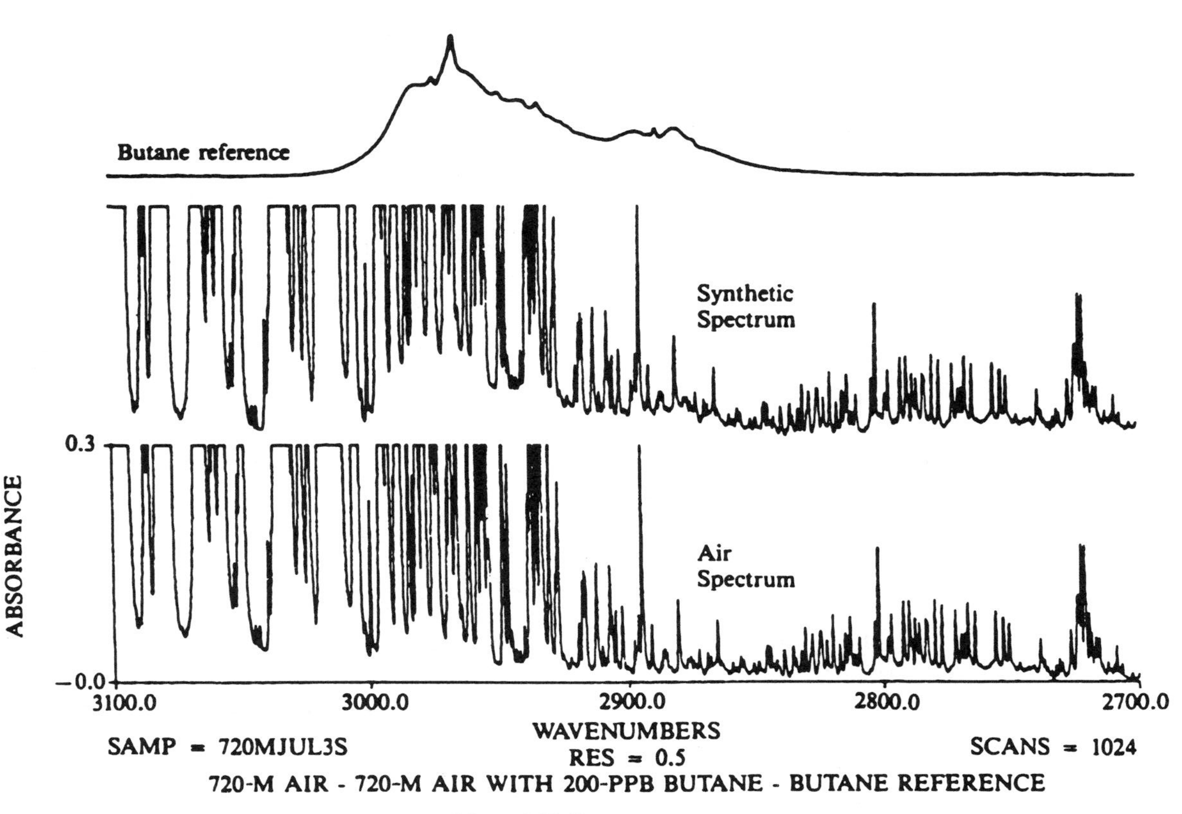

Figure 6.45. Butane spectrum.

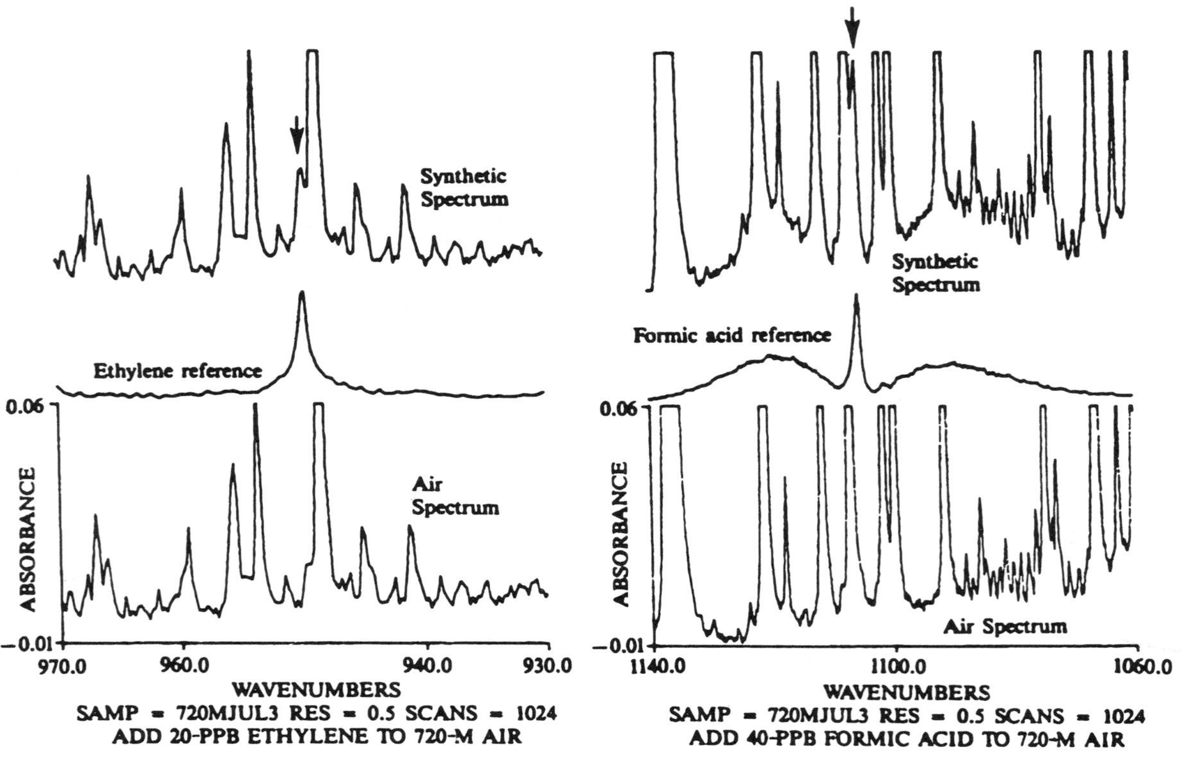

Figure 6.46. Ethylene and formic acid spectra.

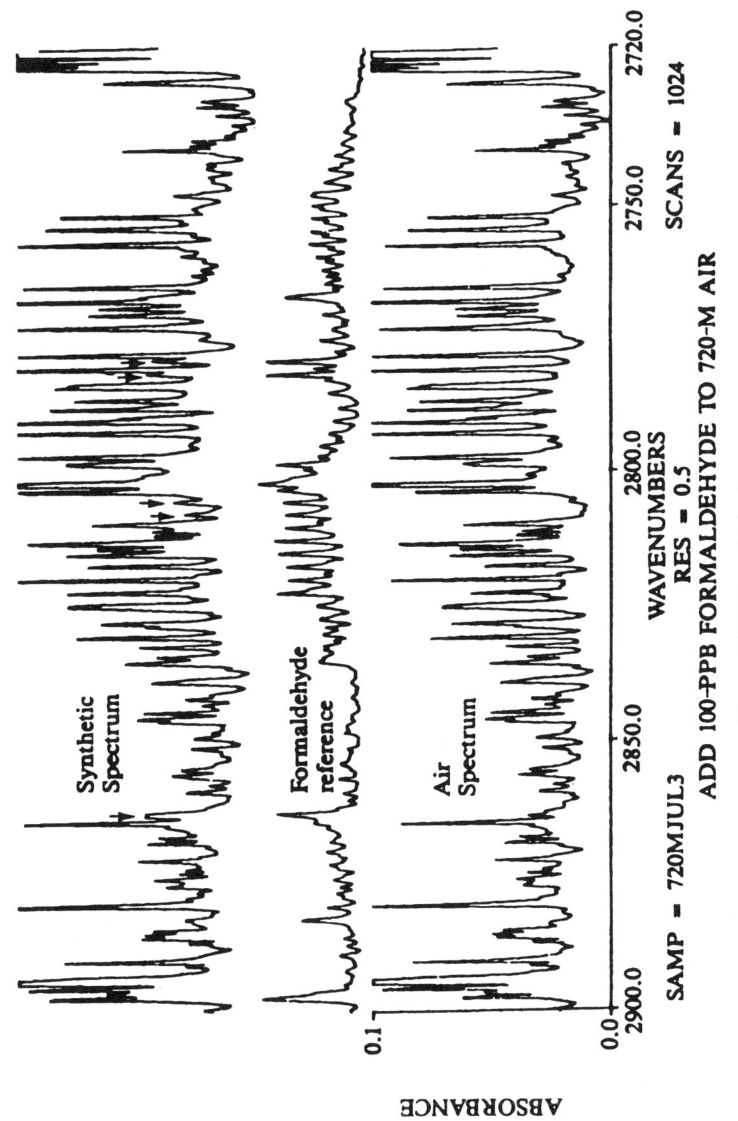

Figure 6.47. Formaldehyde spectrum.

g. Formic Acid. Formic acid reveals itself as a shoulder on a water line near 1105 cm^{-1} [Fig. 6.46(right)]. This molecule is a product of the atmospheric photochemistry. The 1105 cm^{-1} band is seen clearly in the spectra of the Los Angeles smog. It has also been observed in spectra recorded through the stratosphere.

h. Isoprene. Isoprene (C_5H_8) shows two strong spectral features near 900 cm^{-1}, one of which is in the clear [Fig. 6.48(left)]. It is important to be able to measure isoprene in the open air because this compound is released into the atmosphere in large quantities by living and decomposing vegetation.

i. Methanol. Methanol [Fig. 6.48(right)] is measured with good sensitivity by its sharp spectral feature at 1032 cm^{-1}. Because of the use of methanol in gasoline, the compound is now regularly detected in urban air.

j. Nitrogen Dioxide. NO_2 measurement is a case where it is necessary to back off on path length so that its major band may be seen through the water interference. This is shown in Fig. 6.49(left). At 90-m path there is a small "window" at 1600 cm^{-1} that shows some of the structure due to the NO_2 band. There is structure in the lower spectrum of the atmosphere on July 3, 1988, that corresponds to perhaps 30 ppb of NO_2. The use of higher spectral resolution would probably make the detection of NO_2 easier.

k. Nitric Oxide. NO, like NO_2, must be seen through water interference. With NO there is less interference, so the 720-m path has been used. The absorption coefficient for NO is much smaller than the absorption coefficient for NO_2, so that the detectability levels of the two molecules turn out to be about the same. Figure 6.49(right) shows the appearance of an NO "line" at 1900 cm^{-1}. This line is, in fact, an unresolved doublet, so here again is a case where higher resolution would benefit detection.

l. Nitric Acid. For nitric acid measurement we choose spectral features of moderate strength that fall in the region 900–880 cm^{-1} [Fig. 6.50(left)]. There are stronger features in the nitric acid spectrum, but they are not in the clear. The arrows indicate two features suitable for quantitative analysis. After seeing the distortion due to 20 ppb of HNO_3, one can look back at the "clean air" spectrum of July 3, 1988, and conclude that it contains absorption due to approximately 5 ppb of HNO_3.

m. Sulfur Dioxide. Unfortunately, the strongest SO_2 band, centered at 1360 cm^{-1}, is well into the main water band. In measuring SO_2 we must therefore rely on the relatively weak spectral features between 1200 and 1100 cm^{-1}. Bands useful for quantitative analysis are marked in Fig. 6.50(right). The amount of SO_2 used for the example was high:500 ppb. Increases in sensitivity are available by using a longer scanning time, a longer path, and a more sensitive "narrow-band" HgCdTe detector.

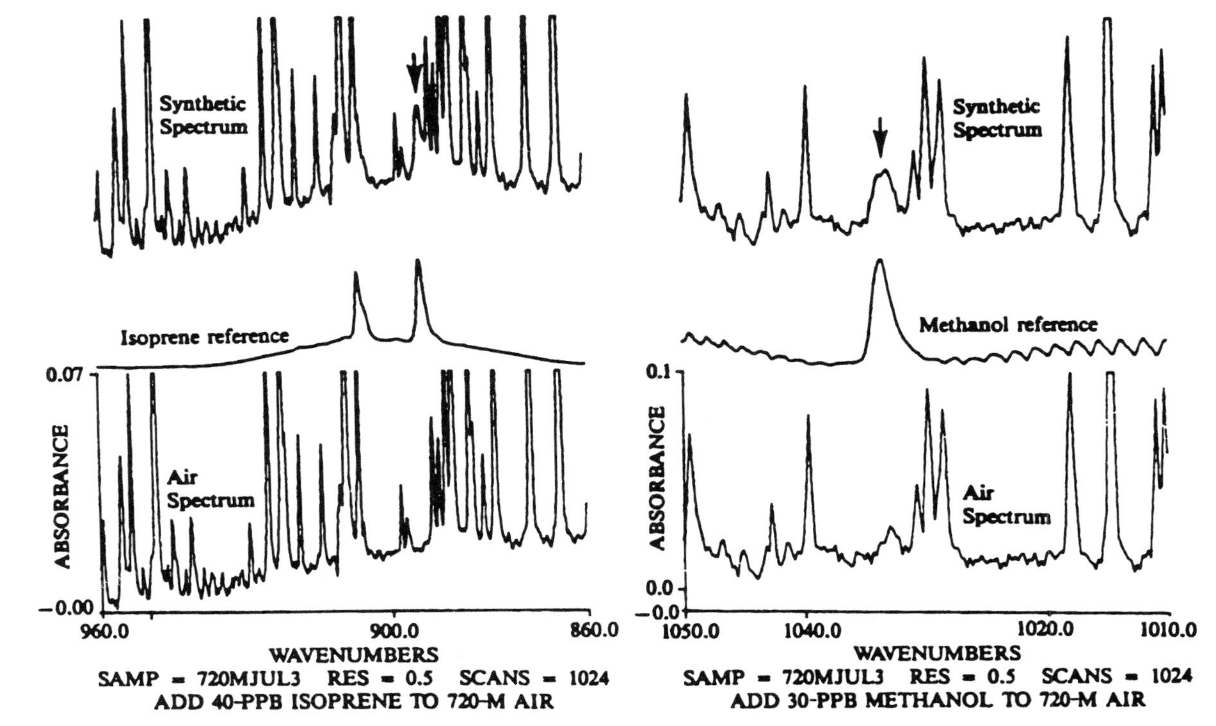

Figure 6.48. Isoprene and methanol spectra.

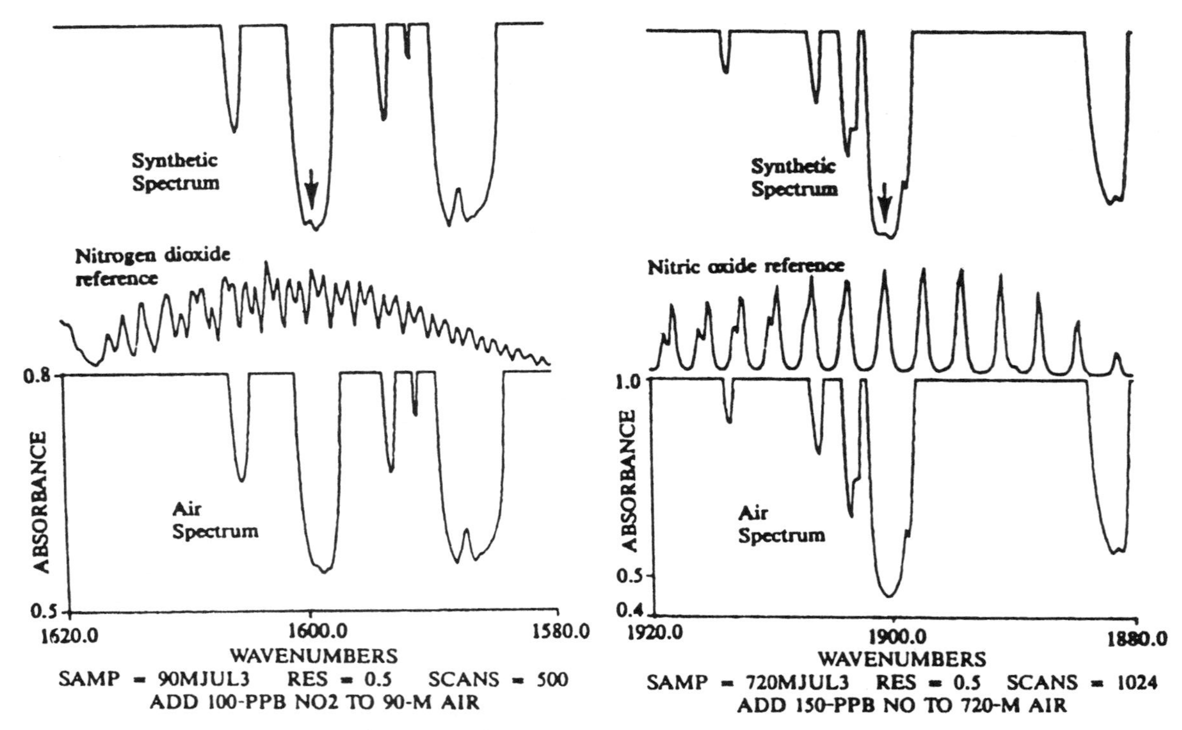

Figure 6.49. Nitrogen dioxide and nitric oxide spectra.

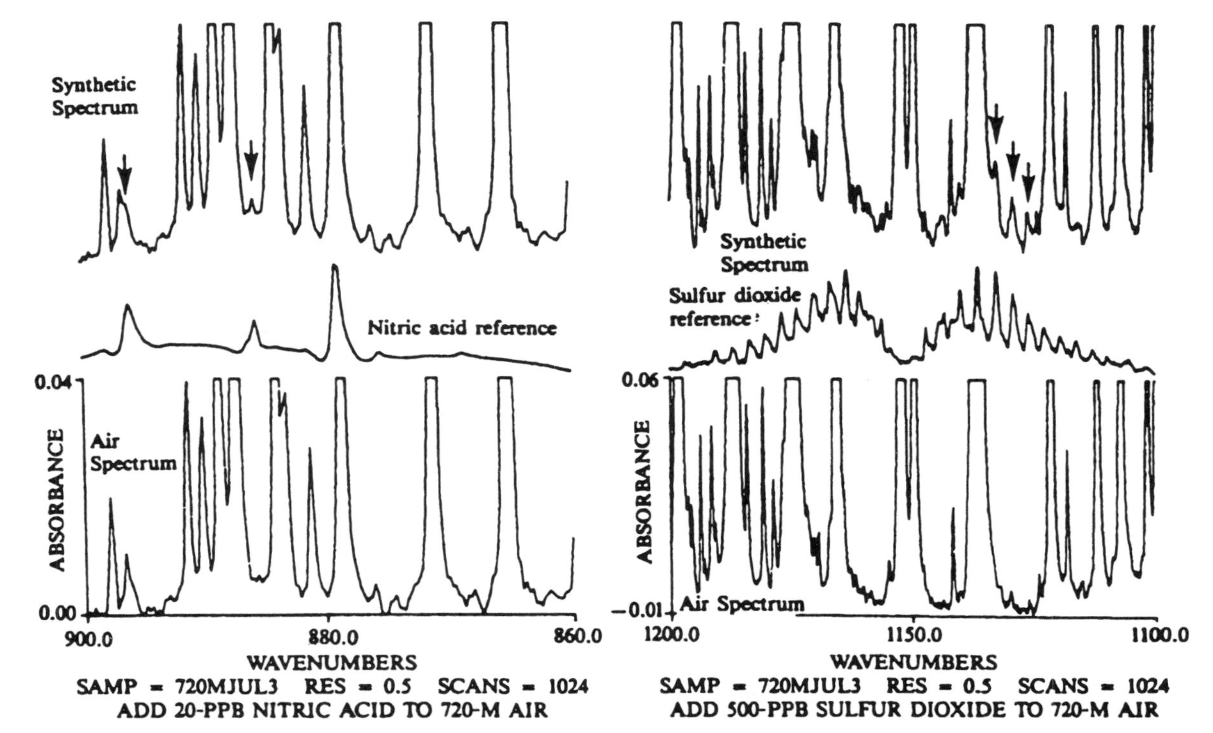

Figure 6.50. Nitric acid and sulfur dioxide spectra.

n. Ozone. Ozone detection is relatively straightforward, as shown in Fig. 6.51. The clean air spectrum of July 3, 1988, showed absorption due to about 30 ppb of ozone.

o. Acetone. The last example, for acetone (Fig. 6.52), shows that compounds with broad bands are detectable, but generally with less sensitivity than are compounds with sharp spectral features. Acetone is a poor case for infrared open-path work, because its very strong carbonyl band is hidden by water.

6.5. GENERAL ANALYTICAL APPLICATIONS

6.5.1. Laboratory Studies

Since the 1950s, long-path infrared absorption spectroscopy has played an important role in the identification and quantitative determination of trace pollutants in synthetic smog systems. Air pollution studies prior to 1970 have been reviewed previously by Hanst (1971). Among the significant contributions made with the technique in 1956 were the first spectroscopic proof of ozone formation in photochemical smog and the discovery of the highly toxic secondary pollutant peroxyacetyl nitrate (PAN), a compound unknown to chemists prior to its characterization in those studies (Stephens et al., 1956; Scott et al., 1957). These results were obtained with prism infrared spectrophotometers and multiple-pass absorption cells with path lengths of 40–400 m. To induce photochemical reactions the cells are surrounded by ultraviolet (UV) lamps. A typical laboratory setup is diagrammed in Fig. 6.53.

Other early applications of this method included studies of irradiated auto exhaust (Stephens et al., 1959), plant damage (Darley et al., 1959), hydrocarbon reactivity (Huess and Glasson, 1968), and ozone–olefin reactions (Leighton, 1961). More recent applications of the long-path infrared method included further investigations of ozone–olefin reactions (Niki et al., 1977b), as well as studies of peroxynitric acid (Niki et al., 1977a), oxidation of chlorinated ethylenes (Gay et al., 1976a), synthesis of peroxyacyl nitrates (Gay et al., 1976b), nitrous acid equilibria (Chan et al., 1976), and calibration techniques for ozone (Pitts et al., 1976).

Infrared detection, by virtue of its specificity and its in situ nondestructive and quantitative nature, is particularly useful in the measurement of organic oxygenates and nitrogenous compounds, including labile species. The development of FTIR spectrometers has extended the capabilities of systems like that shown in Fig. 6.53 far beyond the capabilities of the systems used in the early smog studies.

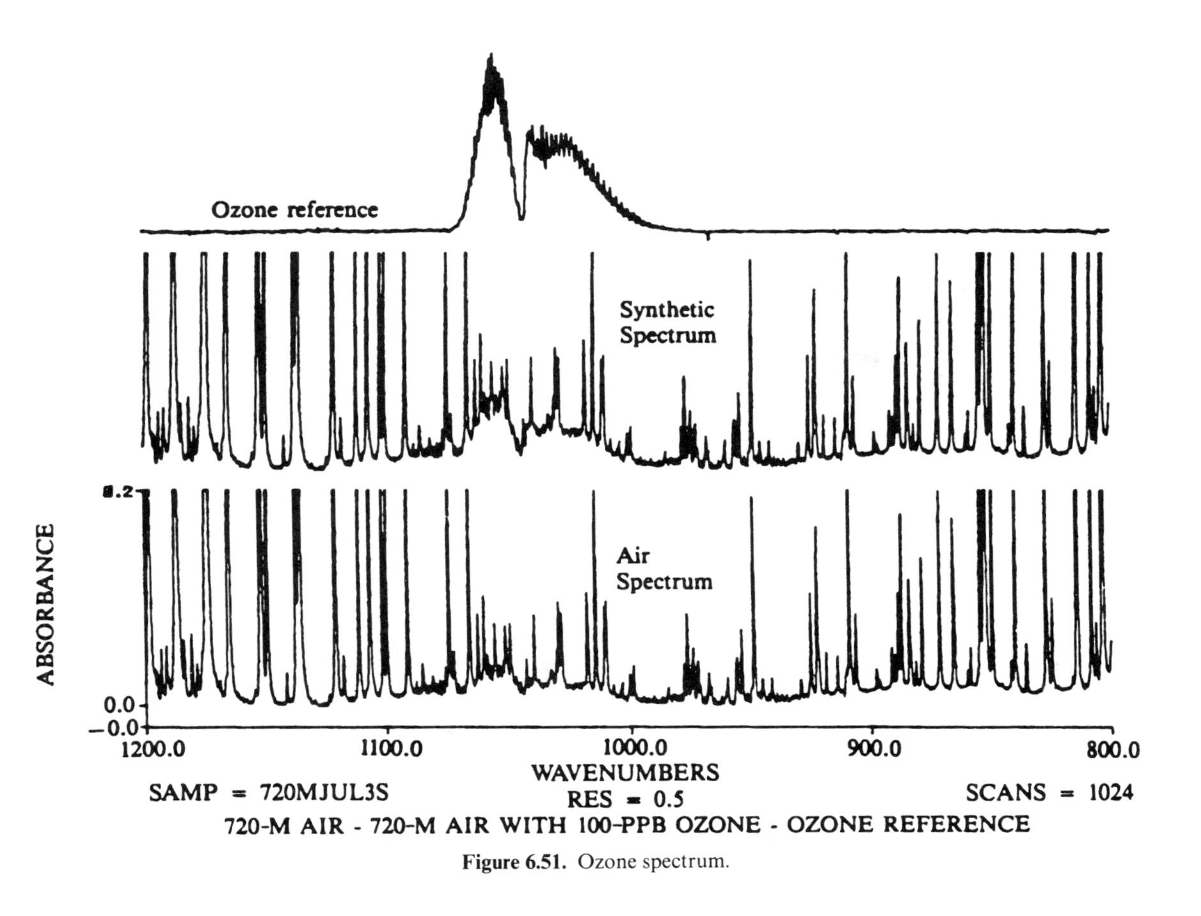

Figure 6.51. Ozone spectrum.

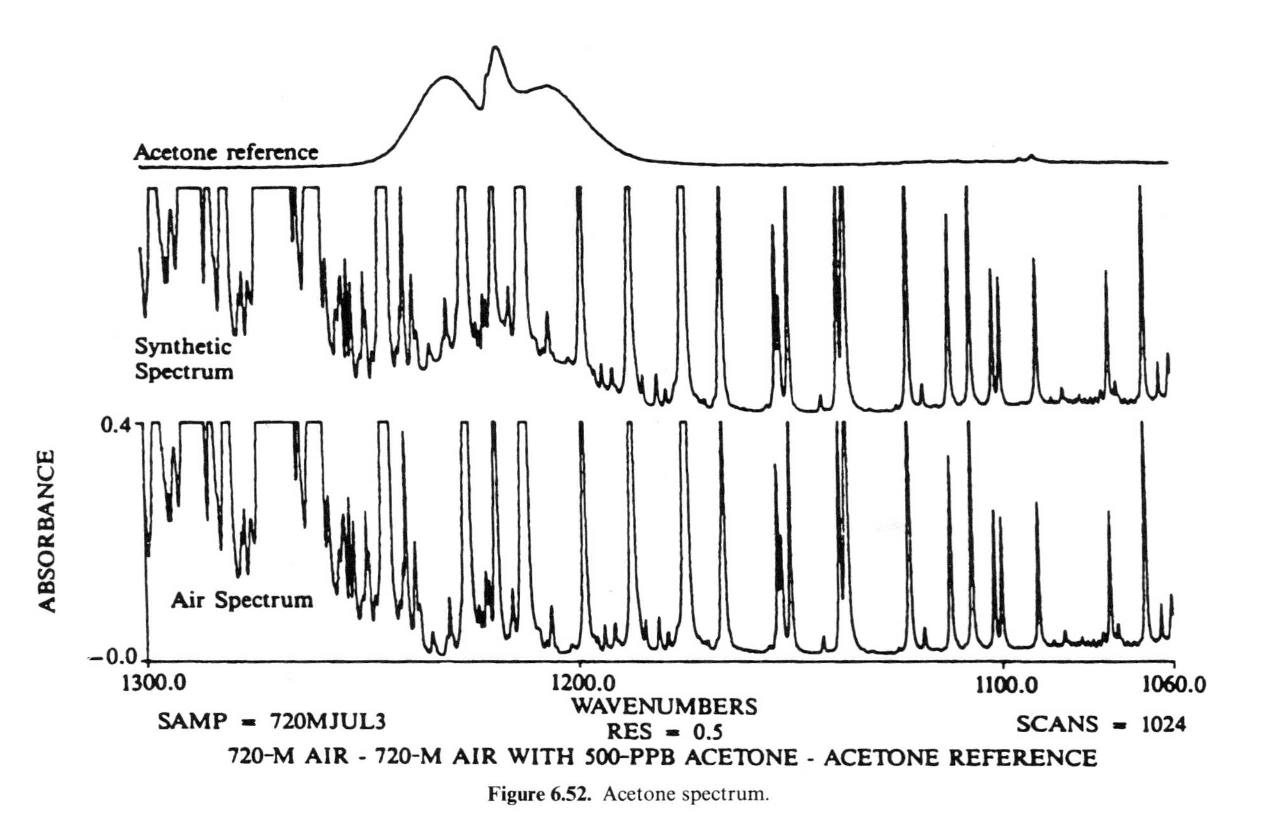

Figure 6.52. Acetone spectrum.

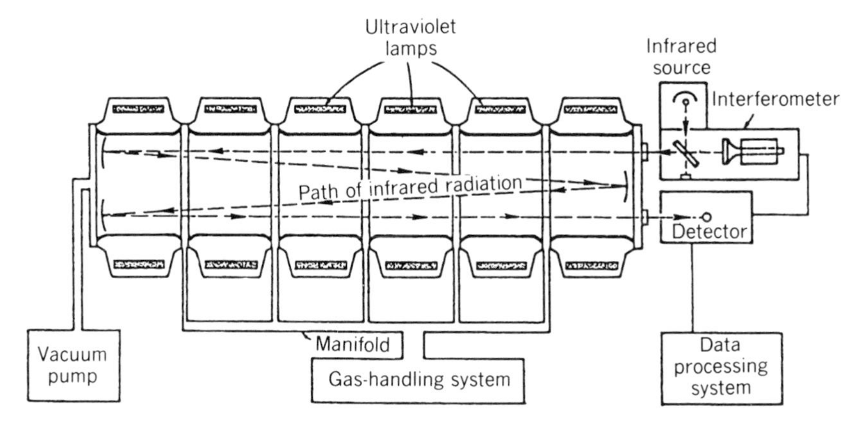

Figure 6.53. Photochemical reactor and infrared detection system. From Hanst and Gay (1977).

6.5.2. Measuring Impurities in Infrared Active Gases

Infrared absorption has long been used in research on the impurities in air. Since dry air is transparent in the infrared, its impurities show up clearly. Water vapor is the main constituent of air that absorbs infrared radiation. The water vapor in the air causes some interference in measuring the trace gases. Nevertheless, present technology can reveal most impurities in air at ppb mixing ratios. This was shown in Table 6.1 (see Section 6.1.5) in the MDL column.

Another important application of the infrared technique is the measurement of impurities in gases other than air. This category of gas measurement can be more difficult than measurements in air because, although air does not itself absorb infrared radiation, almost all other gases do. What can be seen as an impurity in an infrared-absorbing gas depends on the relative strengths of the infrared absorption by the principal component and the impurities. The contrast between the observed bands is the important thing. It is easy to see one compound in the presence of another if the main bands of the two compounds do not overlap and if there is much detailed structure in the two spectra. It is difficult to measure one compound in the presence of another if their bands are structureless and overlapping.

Simple molecules (diatomic, triatomic) will absorb only in restricted portions of the infrared spectrum, and measurement of their impurities will be easy. In many cases the detection limits will be the same as those listed in Table 6.1 for impurities in air. Absence of water vapor may allow some impurities to be measured with even lower detection limits than those in the table because bands in the region 1800–1400 cm^{-1} may be used.

Complex molecules like high-molecular-weight hydrocarbons, will absorb widely and strongly, making the impurity measurement more difficult. Available reference spectra do not usually show the absorption by large quantities of gas, so each case of impurity measurement will require the preparation of reference samples of the main component.

The ability to measure impurities will be greatly enhanced if an absolutely pure sample of the principal component is available for reference. When a low-noise spectrum of the pure compound is subtracted from the spectrum of the impure compound, the bands of the impurities are left standing alone. For success, it is required that enough infrared energy penetrate the sample to allow the recording of the spectrum: 10 m · atm of air is quite transparent in the fundamental infrared region, whereas 10 m · atm of a compound like butane would be almost totally opaque. To measure impurities in butane one would have to ensure that there was sufficient infrared transmission either by shortening the path or by diluting the butane with an infrared transparent gas. As mentioned in Section 6.7.3, the measurement sensitivity is generally greatest when the main component absorbs 63% of the infrared energy in the region being considered.

In all cases of measurement of trace gases, the detection sensitivity will depend on how well the measurement system performs. The higher the signal-to-noise ratio in the spectrum, the greater is the detection sensitivity. This brings us to a consideration of the brightness of the source, the throughput of the absorption cell, and the inherent signal-to-noise level of the detector. When these three factors are highest, the ability to measure impurities is maximized. The detector is especially important. For gas measurements, detectors that operate at room temperature should be avoided completely. A nitrogen-cooled detector like HgCdTe will give an impurity measurement sensitivity about 100 times greater than a room-temperature detector like DTGS. If with a specified scanning time and a DTGS detector the measurement limit for an impurity is deemed to be 0.01%, a HgCdTe detector used with the same scanning time will reduce that limit to 0.0001%. It is almost impossible to make up the difference by increasing the number of scans made with the DTGS detector. Signal-to-noise goes up only with the square root of the number of scans. One scan with a HgCdTe detector is equivalent to 10,000 scans with DTGS.

6.5.3. Measurement of Benzene and Other Aromatic Hydrocarbons

Aromatic hydrocarbon molecules have strong absorption features in their infrared spectra that can readily be used for identification and measurement. In this the aromatics are different from the paraffinic molecules of comparable molecular weight. With an infrared spectrum, one can identify and quantitatively measure methane, ethane, propane, butane, and isobutane, but for paraffins of five carbons or more the spectrum can only tell the total amount of paraffinic carbons without identifying individual molecules.

In the case of a mixture of aromatic hydrocarbons, however, the quantitative analysis is feasible, as indicated in Fig. 6.54. This figure shows the main absorption band of seven aromatic compounds that might be found as pollutants in the air. It is seen that the strong spikes in the spectra occur at different frequencies for each compound. These spikes are the spectral features that allow identification and quantitation of individual species in mixtures.

Figure 6.55 illustrates the measurement of benzene and toluene pollutants in room air. For this measurement a multiple-pass cell of 7.2-m total path was

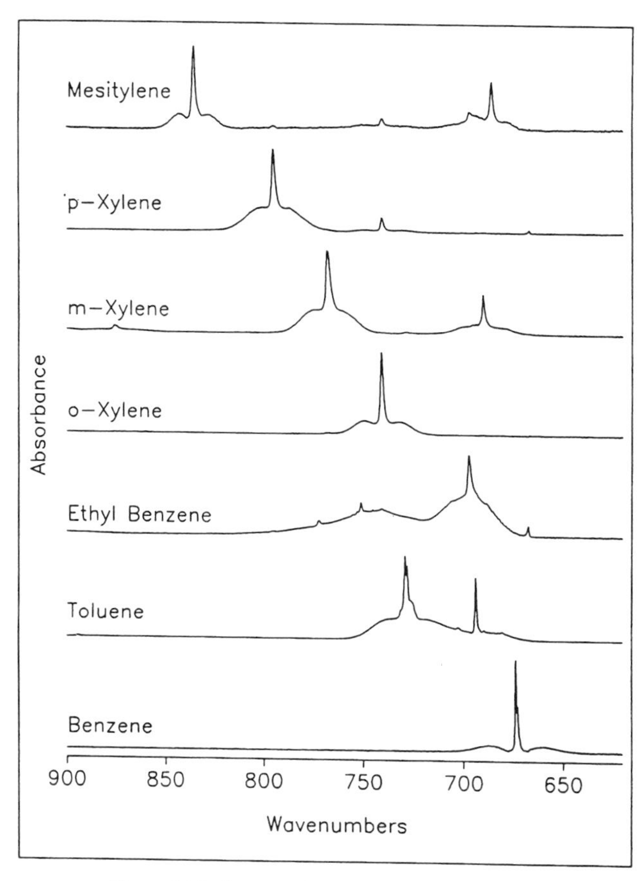

Figure 6.54. Spectra of aromatic hydrocarbons.

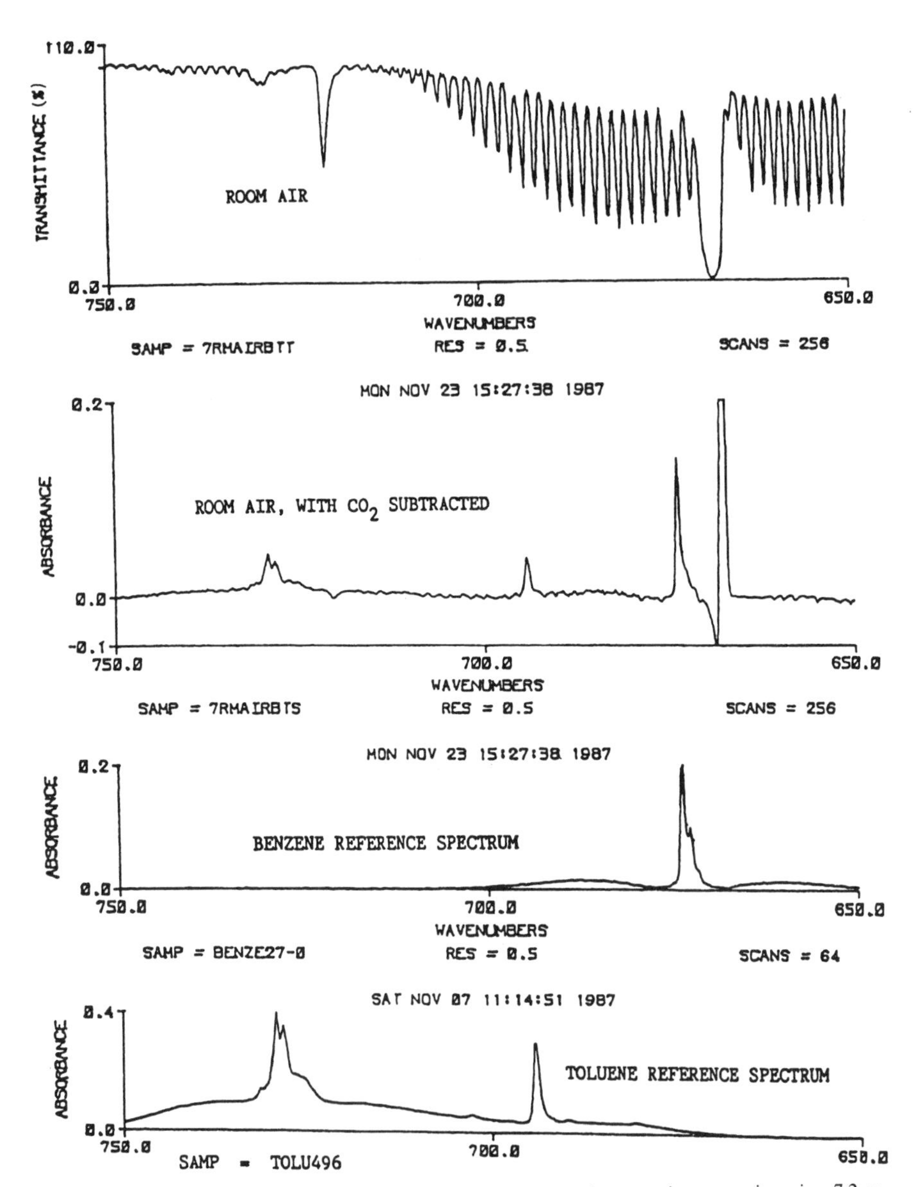

Figure 6.55. Measurement of 10.6-ppm toluene and 3.3-ppm benzene in room air, using 7.2-m absorption cell in a Digilab FTS-40 spectrometer.

mounted in the sample compartment of a Digilab FTS-40 FTIR spectrometer. The cell volume was only 0.5 L.

The upper plot of Fig. 6.55 shows the relevant region of the room air spectrum. Here the main absorption bands are due to the atmospheric CO_2 (transmittance plot). The second spectrum is an absorbance plot of the remnants of the spectrum after the CO_2 lines have been subtracted. The reference spectra of benzene and toluene are plotted below to confirm the identity of the spectral features. Using the reference spectra in the subtraction routine of the software allowed calculation of the amounts of pollutants, as shown.

It is estimated that for this absorption cell the detection limits are about 1 ppm (10^{-6} atm) for toluene and 0.1 ppm (10^{-7} atm) for benzene. For greater detection sensitivites a larger long-path cell should be used.

6.5.4. Water Analysis in the Gas Phase by FTIR

Infrared absorption spectroscopy has not been especially successful at measuring organic compounds dissolved in liquid water. The infrared radiation is attenuated so strongly by liquid water that the solution being studied is not allowed to be more than a few micrometers thick. At that small thickness, the concentration of the solutes must be relatively high for them to be detected. Another drawback is that compounds in aqueous solution do not have much detail in their spectra. Instead of showing peaks and lines, the spectra of the solvent and solutes are broad and nondescript. Measurements of solutes in liquid water by infrared absorption have really only been feasible when the solute mole fraction has been 10^{-2} or higher. Furthermore, the broad nature of the absorption bands has restricted the analysis of complex mixtures of solutes.

In the gas phase, things are very much different and sensitive measurements of complex mixtures are feasible. The infrared spectrum of water vapor is not really so intense, compared to the spectra of other molecules. This may seem contrary to the experience of infrared workers who put so much effort into purging their instruments; but in an instrument there may be 1 m of optical path with 10–15 torr of water vapor. In the direct analysis of vaporized water samples, the water vapor spectrum will appear even weaker than it appears in air, because the width of the spectral lines goes down nearly in proportion to the total sample pressure. The absorption is then concentrated at the line centers, and regions of transmission open up between lines that normally are overlapping. There is enough energy transmitted to reveal molecular bands beneath the water lines. The residual water absorption is then removed easily enough by subtraction methods.

Figure 6.56 shows the spectra of 5 torr of water vapor in a 6-m gas

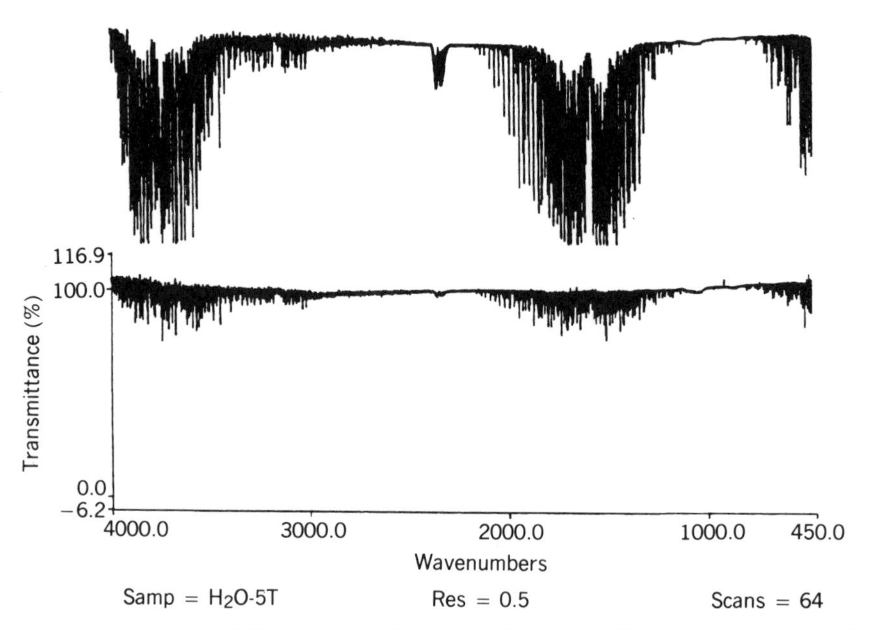

Figure 6.56. Five torr of H_2O in 6-m cell: (lower) without added N_2; (upper) with 1 atm of added N_2.

absorption cell, with and without pressure broadening by nitrogen. It is apparent that without the nitrogen all spectral regions transmit enough energy to allow the measurement of other gases that might be mixed with the water vapor. Even in a 100-m cell with 5 torr of water pressure, all spectral regions would be useful.

Handling gases is somewhat more complicated than handling liquids, which is one reason for the relative neglect of vapor phase infrared studies. The extra work is, however, not really great. Figure 6.57 shows the gas-handling apparatus used in our present work. Only a small droplet of aqueous solution, a few microliters in volume, was required for a measurement. This droplet was placed in an open-ended U-tube attached to the gas-handling manifold. The U-tube was then closed, and the droplet was frozen with liquid nitrogen. Next, the U-tube, the manifold, and the 6-m gas absorption cell were all evacuated. Then the valve connecting the U-tube to the manifold was closed, the liquid nitrogen was removed, and the frozen droplet was vaporized into the U-tube. The U-tube was large enough that the whole sample—water and solutes—was able to exist in the vapor phase at room temperature. Finally, the mixture of water vapor and solute vapors was allowed to flow into the evacuated absorption cell, with the total pressure being measured on the

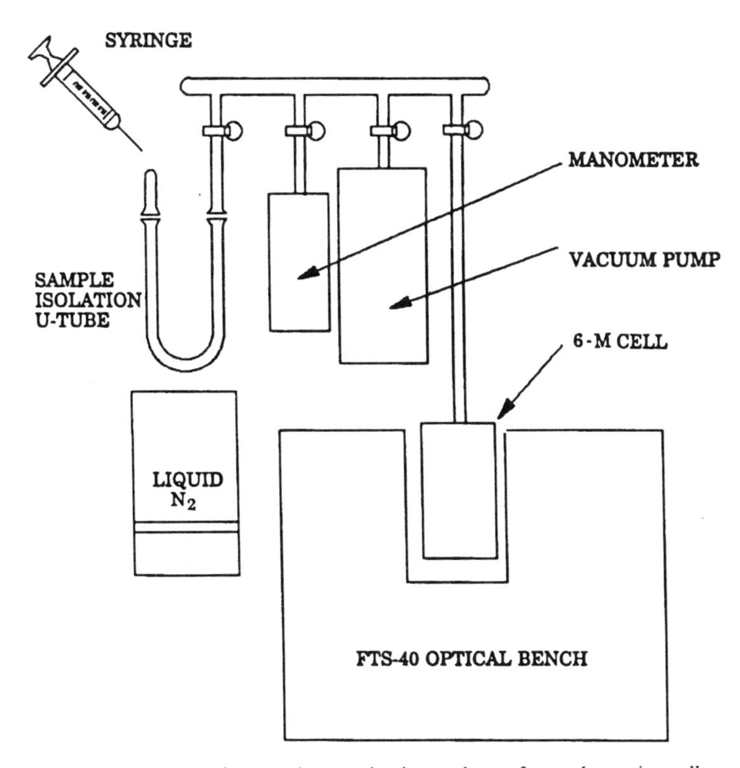

Figure 6.57. System for sample vaporization and transfer to absorption cell.

manometer. The spectrum was then recorded. For the background spectrum, a similar sample of pure water was used.

To demonstrate the sensitivity of the detection method, a drop of methanol and two drops of acetone were added to about 0.5 L of tap water. This was well stirred, and then, by syringe, 0.01 ml of the tainted water was placed in the sample isolation U-tube. When the sample was vaporized and allowed to flow into the manifold and absorption cell, about 10 torr of pressure was measured. The sample spectrum was plotted against a background spectrum of 10 torr of pure water, with the results shown in Fig. 6.58 (bottom). Above this sample spectrum are reference spectra of acetone and methanol. From the reference spectra the partial pressures calculated were as follows: methanol, 2.1×10^{-6} atm; acetone, 5.8×10^{-6} atm. Since 10 torr of water is about $13{,}000 \times 10^{-6}$ atm, the calculated mole fractions were 0.00016 for methanol and 0.00044 for acetone. From the size of the bands in the spectrum compared to the noise level, it is estimated that mole fractions 10 times smaller could have been measured. If a longer absorption cell had been used, the detection

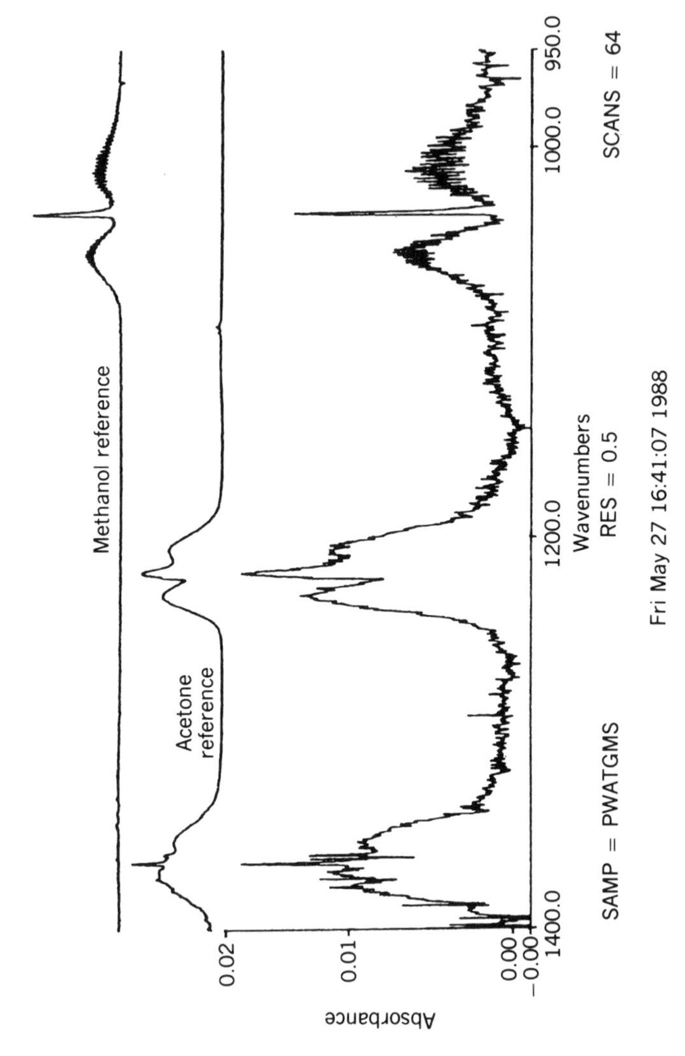

Figure 6.58. Five milligrams of polluted water in 6-m cell—water subtracted.

sensitivity would have been still greater. It is estimated that a 100-m cell would allow measurement of solutes in water at mole fractions of 10^{-6} or smaller.

As an additional check on the validity of the sample-handling method, the ethanol content was measured on three samples of alcoholic beverages. The resultant spectra showing the ethanol are presented in Fig. 6.59. The top spectrum was from gin, 94.6 proof. The total pressure of vaporized gin in the cell was 1.1 torr, and the partial pressure of ethanol, calculated by reference to ethanol spectrum published by the Coblentz Society, was about 0.5 torr. The middle spectrum in the figure was from vodka—only 80 proof—and the spectrum verifies that the alcohol content was slightly less than in the gin. The lower spectrum was from vermouth, 36 proof, and the alcohol content was correctly measured at the lower value. Thus, these measurements indicate that the sample-handling technique is appropriate for quantitative analysis.

It has been demonstrated that vaporizable organic compounds mixed with water may be measured by vaporizing the whole aqueous sample into an infrared absorption cell. For quantitative analysis, the sample handling should avoid recondensation of the water vapor. Interference from the infrared absorption by the water vapor is only a minor problem, easily handled in the background spectrum. By lengthening the infrared path, the sensitivity of the method may be increased to allow detection of solutes with mole fractions of 10^{-6} or smaller. Finally, it is expected that nonaqueous liquid and solid samples may be vaporized and investigated with the same sample-handling system used for aqueous solutions.

6.5.5. An Analysis of Indoor Air

The infrared absorption spectrum of the atmosphere is a combination of the spectra of all the heteronuclear diatomic molecules and the polyatomic molecules that are in the air sample. In the lower atmosphere the spectrum is dominated by the bands and lines of water vapor. The H_2O mixing ratio normally falls near 15,000 ppm.

In addition to water vapor, CO_2, CH_4, N_2O, and CO (see Section 6.4.3) clean air contains other trace gases that absorb infrared, but their concentrations fall in the ppb range. Pollutants in the air can have concentrations in the ppm to ppb range. Almost all pollutant gases are infrared active.

The spectrum of the air has all of the individual spectra piled on top of each other. Prior to development of modern spectroscopic software, it was quite impractical to try to separate the spectra from each other. The molecules that were low in concentration could only be detected if they had spectral features in the open—not overlaid with the spectral features of the water, CO_2, and other molecules. This limitation no longer exists. The spectral subtraction

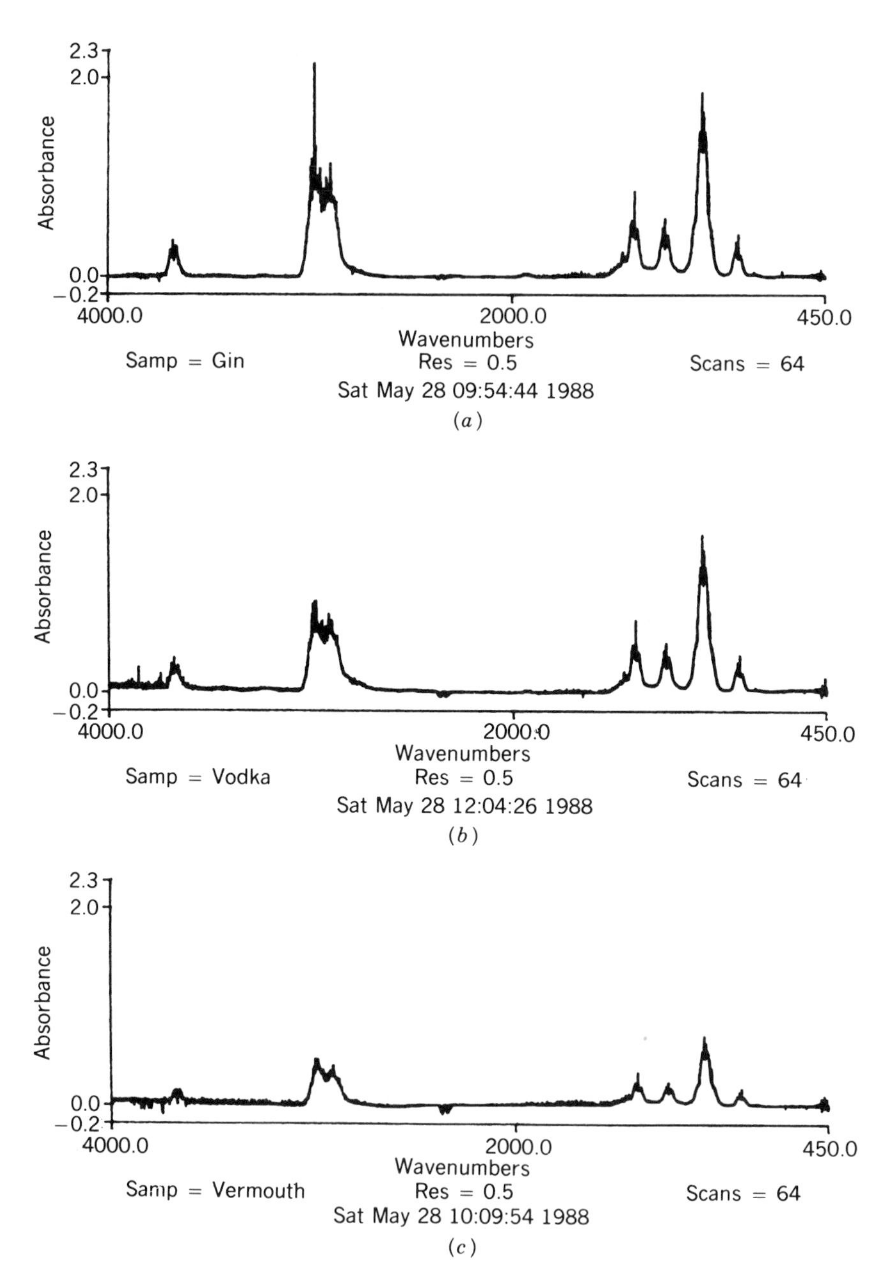

Figure 6.59. Testing for ethanol content of beverages in 6-m cell: (a) 1.1 torr of gin; (b) 1.1 torr of vodka; (c) 1.1 torr of vermouth.

routines of present computer programs make it possible to identify, separate, and measure the spectra of the individual molecules, one by one.

For example, an air sample was analyzed using a measurement system with the following components:

1. A Bomem MD-110 FTIR spectrometer.
2. A HgCdTe detector operating at liquid nitrogen temperature
3. An adjustable multiple-pass absorption cell of 3-m base path
4. A data station with personal computer and laser printer
5. The Spectra-Calc spectroscopic software from Galactic Industries Corp.
6. The Infrared Analysis, Inc. library of 130 digitized quantitative reference spectra

A sample spectrum of room air and a reference spectrum of vacuum were recorded in multiple-pass cell with 210 m of optical path. Resolution was 1 cm^{-1}; scanning time was 5 min. A portion of the resulting absorbance spectrum is shown in the upper part of Fig. 6.60. Below that is the absorbance spectrum of water in nitrogen, plotted with the vacuum spectrum as reference. The third spectrum is the remainder after subtraction of the water spectrum from the air spectrum. This remainder is mainly due to C—H stretch bands of hydrocarbons. Methane is first subtracted by means of the methane reference spectrum shown. The concentration of methane is obtained by multiplying the subtraction factor by the ppm-meters number from the reference spectrum and dividing it by the optical path length. Propane is recognized in the remaining absorption, and it is measured and subtracted. What then remains appears to be due to a mixture of hydrocarbons such as gasoline vapor. A hexane spectrum is used to remove this absorption, giving a hexane-equivalent value for the concentration of gasoline hydrocarbons.

Subtraction sequences for two other regions of the air spectrum are shown in Figs. 6.61 and 6.62. In general, the sequence of spectral subtractions should follow a decreasing order of constituent concentration. The five main polyatomic gases in the air should have their spectra removed first, in the order: (1) H_2O; (2) CO_2; (3) CH_4; (4) CO; and (5) N_2O. When these five are gone the lines and bands of the lesser trace gases are revealed. These can be identified by inspection or by a search program. When the constituents are identified, their reference spectra may be called up for the quantitative analysis.

In the present example, the measured amounts of trace gases were as shown in the tabulation on p. 439. There was also some broad unidentified absorption in the region 1200–800 cm^{-1}. Possibly this was due to the vapors of solvents and tapping fluid used in a machine shop adjoining the room where the sample was taken.

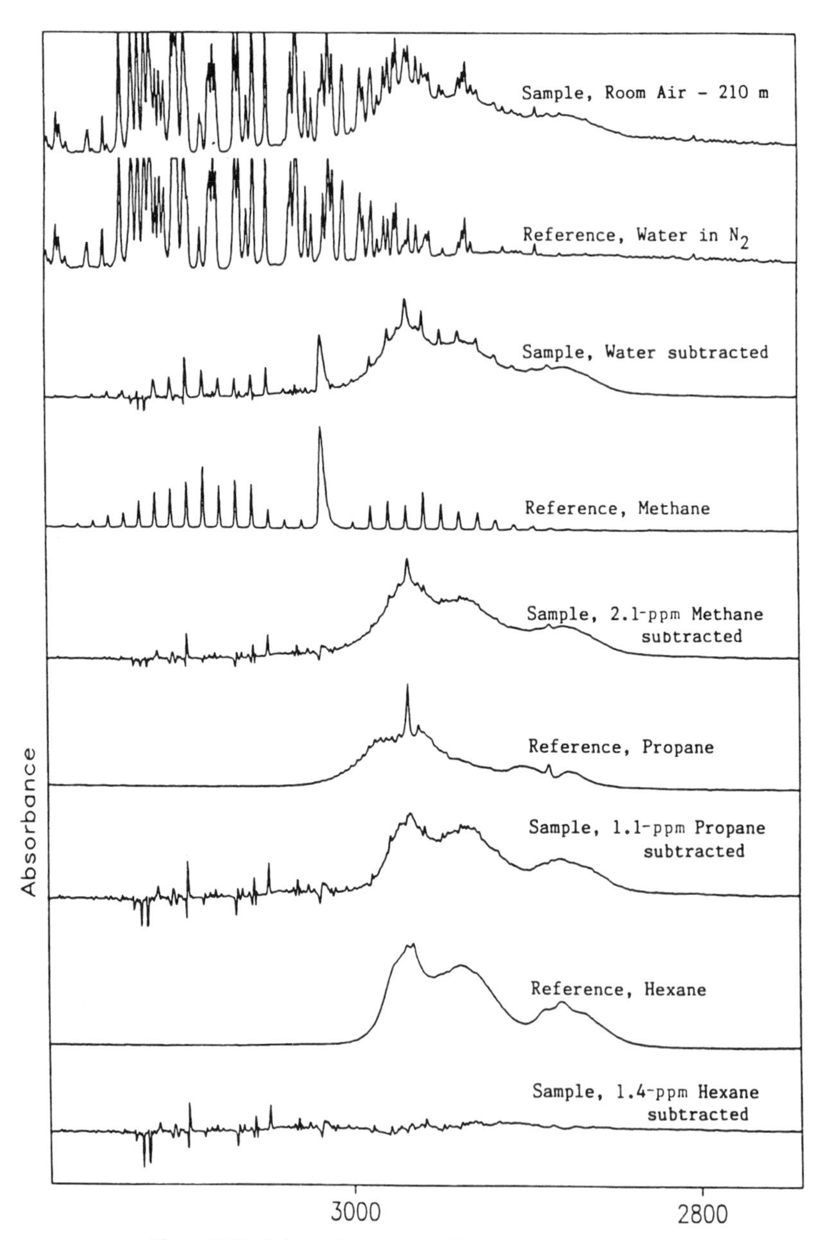

Figure 6.60. Subtraction sequence in analyzing air sample.

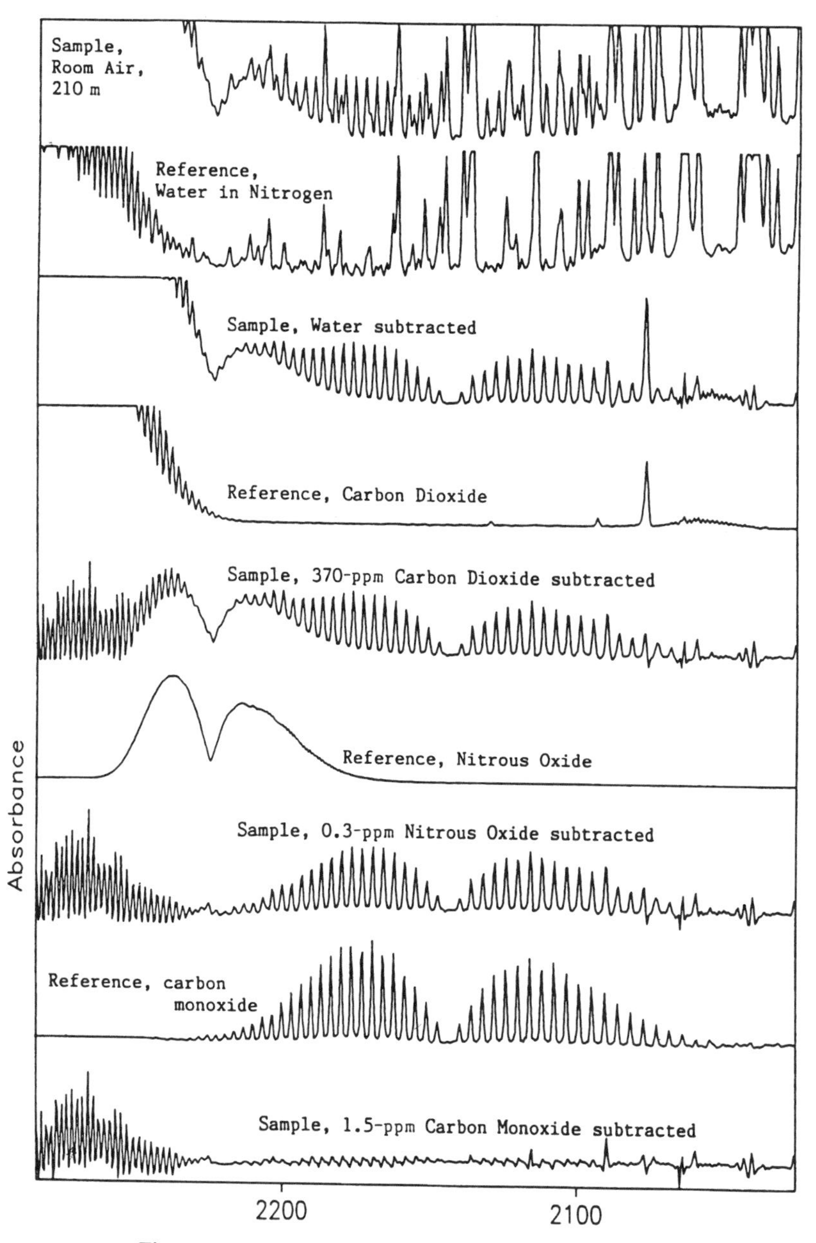

Figure 6.61. Subtraction sequence in analyzing air sample.

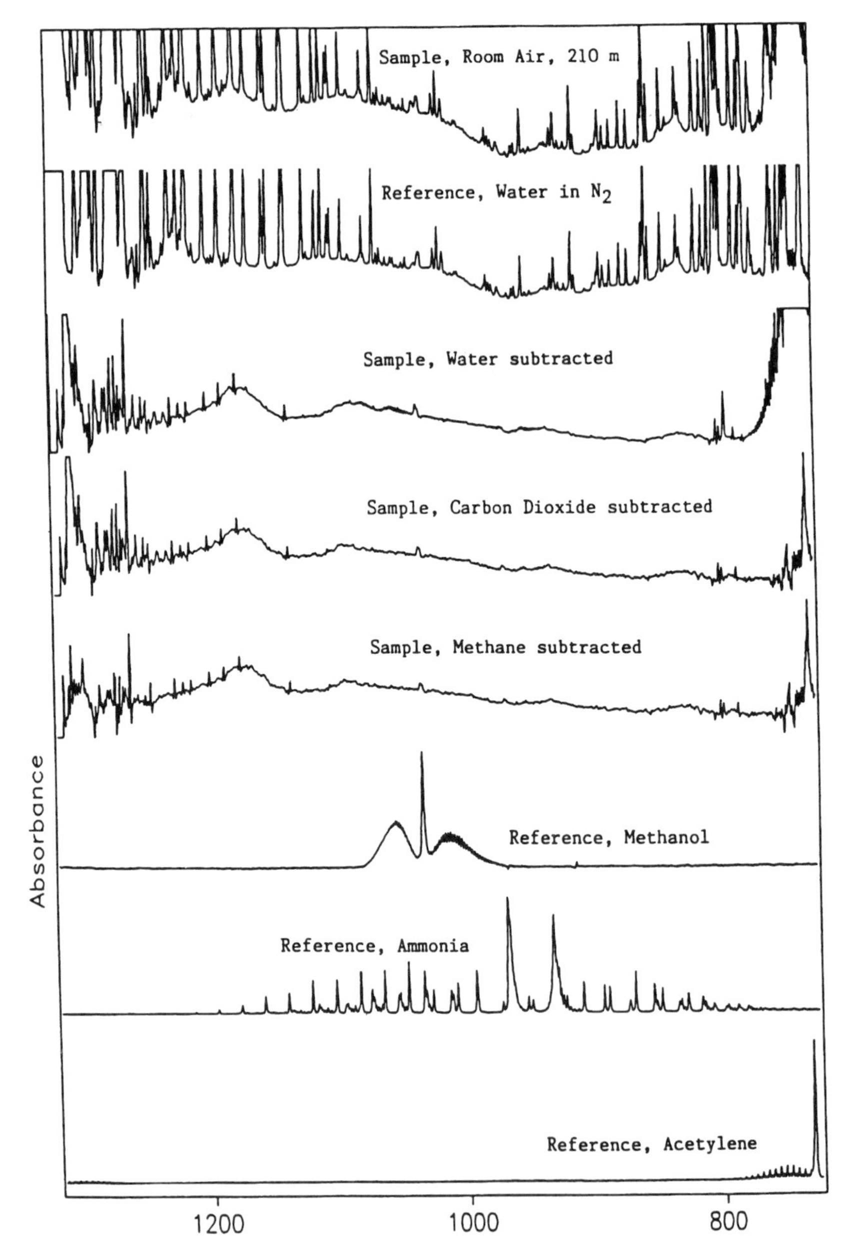

Figure 6.62. Subtraction sequence in analyzing air sample.

Trace Gases	Concentration (ppm)
Carbon dioxide	370
Methane	2.1
Propane	1.1
Nonmethane hydrocarbons C_5 and up, calculated as hexane	1.4
Nitrous oxide	0.3
Carbon monoxide	1.5
Methanol	0.035
Ammonia	0.009
Ethylene	0.007
Acetylene	0.038

In our example, the subtractions and analyses were done manually while the reference spectra and difference spectra were observed on the computer screen. However, programs for automating the subtraction and quantification have been written. The sequence of operations is approximately as follows: The computer directs the interferometer to make a chosen number of scans; then it co-adds these and transforms them to a spectrum. Next, the spectrum is divided by a prerecorded background and absorbance is calculated. The computer then subtracts the prerecorded water reference spectrum from the sample spectrum. The water concentration is calculated and stored. Next, a CO_2 reference spectrum is subtracted to remove CO_2 lines, and the CO_2 concentration is calculated and stored. Each additional trace gas spectrum is then subtracted, with the amounts being calculated and stored. The sequence of subtractions will start with compounds with strong bands, such as CH_4, non-methane hydrocarbon, carbon monoxide, nitrous oxide, and ozone, and work down to the compounds with the weakest bands. Finally, all concentrations are printed out and a new measurement sequence is started.

For air samples likely to have unidentified bands, visual inspection of the spectra and some manual manipulation of the reference spectra will still be required. Even when one is using an automatic monitoring program, it will occasionally be necessary to examine the spectrum visually to verify system performance.

6.5.6. Infrared Analysis of Engine Exhausts

Engine exhausts have been chemically analyzed by infrared absorption, using a Digilab FTS-40 Fourier transform spectrometer, a small multiple-pass absorption cell, and a set of digitized quantitative reference spectra. Exhaust samples were obtained from an automobile and a methanol-powered passenger bus by allowing the gases to flow into evacuated glass sample containers. Detected minor exhaust components included carbon monoxide, methane, formaldehyde, nitric oxide, nitrogen dioxide, methyl nitrite, methane, ethylene, acetylene, propylene, isobutylene, and total non-methane hydrocarbons. The instrumentation and technique used in this work eliminate any need for the analyst to prepare reference spectra or calibration curves. Using only a few liters of sample, this measurement method yields a detailed chemical composition, covering components that are present from percent levels down to ppm. If necessary, the sample may be captured in a casual and qualitative manner, because the spectrum gives an accurate carbon dioxide measurement against which all other component measurements can be normalized.

6.5.6.1. Introductory Remarks

The exhausts from internal combustion engines consist mainly of nitrogen, water, and carbon dioxide. Nitrogen oxides are also present, along with products of incomplete combustion such as carbon monoxide, formaldehyde, ethylene, acetylene and methane. There may also be unburned fuel.

The infrared method of analysis can reveal and measure quantitatively all the gases in the exhaust except for the oxygen and nitrogen. Infrared absorption has been used for exhaust analysis for many years. For example, an infrared study was reported by our group more than 30 years ago (see Stephens et al., 1959). Up to the present, however, the method has not been as fully successful as it could be. A principal difficulty with the infrared method has been that interferences by the spectra of water and carbon dioxide have obscured the lines and bands of many of the trace components. Another difficulty has been that past methods of recording the spectra for quantitative measurement have required establishment of nonlinear calibration curves.

New developments in spectroscopic instrumentation and software have very much reduced the problems of spectral interference and nonlinear calibrations. It is now possible to measure all of the important exhaust components simultaneously on a single small gaseous sample. The making of quantitative measurements has been simplified, and the need for calibration samples of the exhaust constituents has been eliminated.

Sample gathering is also simplified in this method, as a result of the completeness of the analysis. Not only are the minor exhaust components measured, but the concentrations of carbon dioxide and water are also obtained. Knowing the carbon dioxide concentration is especially useful because it tells what fraction of the sample was actually exhaust gas (see Section 6.3.4).

6.5.6.2. Measurement Method

The exhausts were analyzed by means of the FTS-40 Fourier transform spectrometer of Bio-Rad, Digilab Division, using the long-path minicell of Infrared Analysis, Inc. This adjustable absorption cell was set for 4.2-m total path. Spectral resolution was 0.5 wavenumbers. The detector was HgCdTe, wideband form.

The exhausts were captured in 5-L glass containers with a glass inlet tube that was inserted into the exhaust pipe of the vehicle. The sample containers were evacuated before sampling so that the exhaust gas flowed in under its own pressure. Similarly, the exhaust gas was allowed to flow under its own pressure from the sample container into the evacuated infrared absorption cell. Since the pressure equilibrated at less than 1 atm, purified tank air was then added to the cell to raise its total pressure to 1 atm. Figure 6.63 is a photograph of the long-path minicell and one of the sample containers. Sampling was well enough controlled that it was not necessary to use the CO_2 concentration as an indicator, as discussed above.

For all the samples, quantitative analyses were carried out with digitized quantitative reference spectra that were recorded under the same instrumental conditions as were used in running the sample spectra. The high-quality detector made it possible to work with small absorbances where calibration is linear and only one reference spectrum is required for each component. These reference spectra are stored in the computer and used for quantitation of all samples. To avoid errors due to line-broadening effects, all reference spectra were recorded while the trace gas was mixed in 1 atm of nitrogen or tank air.

The quantification came about through the use of Digilab's interactive subtraction routine, in which the sample spectrum and the reference spectrum are displayed together on the computer screen, and then by lever control (joystick) the spectrum of the compound is removed from the spectrum of the

Figure 6.63. Absorption cell (lower) and sample vessel (upper).

whole sample. The abilities of the human eye and brain to recognize small changes in a complex pattern make this interactive subtraction a very sensitive and reliable technique.

Each digitized reference spectrum has been made quantitative by including its concentration–path length product in its name. In obtaining this product, the gas partial pressure is expressed in microatmospheres and the optical path is expressed in meters. (At sea level, 1 μatm is equal to a volume mixing ratio of 1 ppm.) Concentration and path length are interchangable. If a methane partial pressure of 100 μatm were used to make a reference spectrum in a 1-m cell, the concentration–path length product would be 100 μatm·m. The name of the reference spectrum would then be METHA100. If 2000 μatm of methane gas were used in a 5-cm cell, the product would still be 100 μatm·m and the spectrum name would still be METHA100.

When the interactive subtraction has removed a band from the sample spectrum, the concentration of compound (C) is obtained by multiplying the subtraction factor shown on the computer screen (S) by the *concentration equivalent* of the reference spectrum, taking into account the path length used in the analysis. This concentration equivalent is simply the number (N) from the name of the quantitative spectrum divided by the optical path (P). In summary: $C = S(N/P)$. When P is expressed in meters, C is obtained in microatmospheres (ppm).

6.5.6.3. *Results of the Analyses*

Analyses were performed on a sample of exhaust obtained from a gasoline-powered 1983 Honda automobile, idling, and on samples from a passenger bus with methanol as fuel, idling and under heavy engine load. The automobile had a catalytic muffler, but the bus did not. Figures 6.64–6.66 are the complete infrared spectra for the samples that flowed from the sample containers into the 4.2-m absorption cell. In each case the equilibrium pressure was slightly less than 0.75 atm, so that tank air had to be added to raise the total pressure to 1 atm. These spectra are presented in transmittance form so that weak bands and strong bands can both be seen together.

The molecules responsible for the various infrared bands are marked on the spectra. In these spectra the carbon monoxide and methanol bands are too strong for quantitative measurement, so each sample was diluted 10-fold and a second spectrum was recorded. The resulting reduced carbon monoxide bands are presented in Fig. 6.67. Only carbon monoxide and methanol were measured from the diluted samples shown in Figs. 6.64–6.66. Figures 6.68–6.73 are derived from Figs. 6.63, 6.64, and 6.66, as discussed below.

Figure 6.68 is an absorbance plot of the high-frequency end of the spectrum of the Honda exhaust (replotted from Fig. 6.64). This is the spectral region of

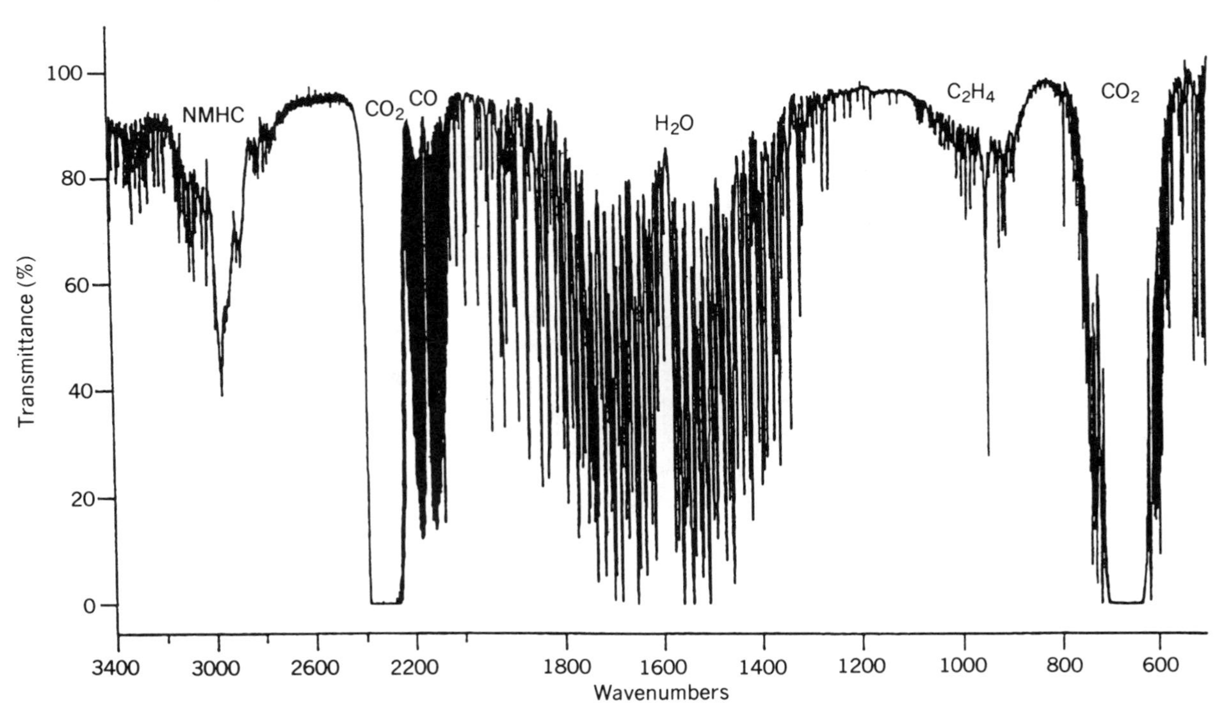

Figure 6.64. Exhaust from idling Honda car with catalytic muffler: 565 torr of exhaust gas and 200 torr of tank air in a 4.2-m cell.

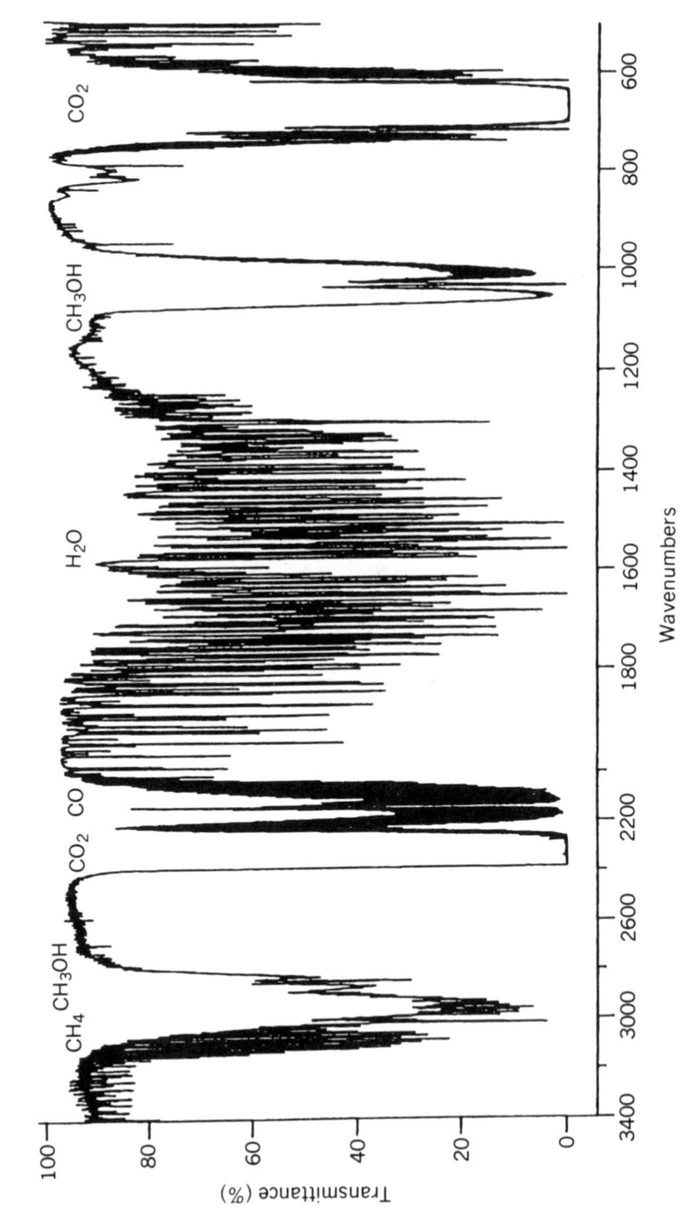

Figure 6.65. Exhaust from an idling passenger bus operating on pure methanol fuel without a catalytic muffler: 565 torr of exhaust gas and 200 torr of tank air in a 4.2-m cell.

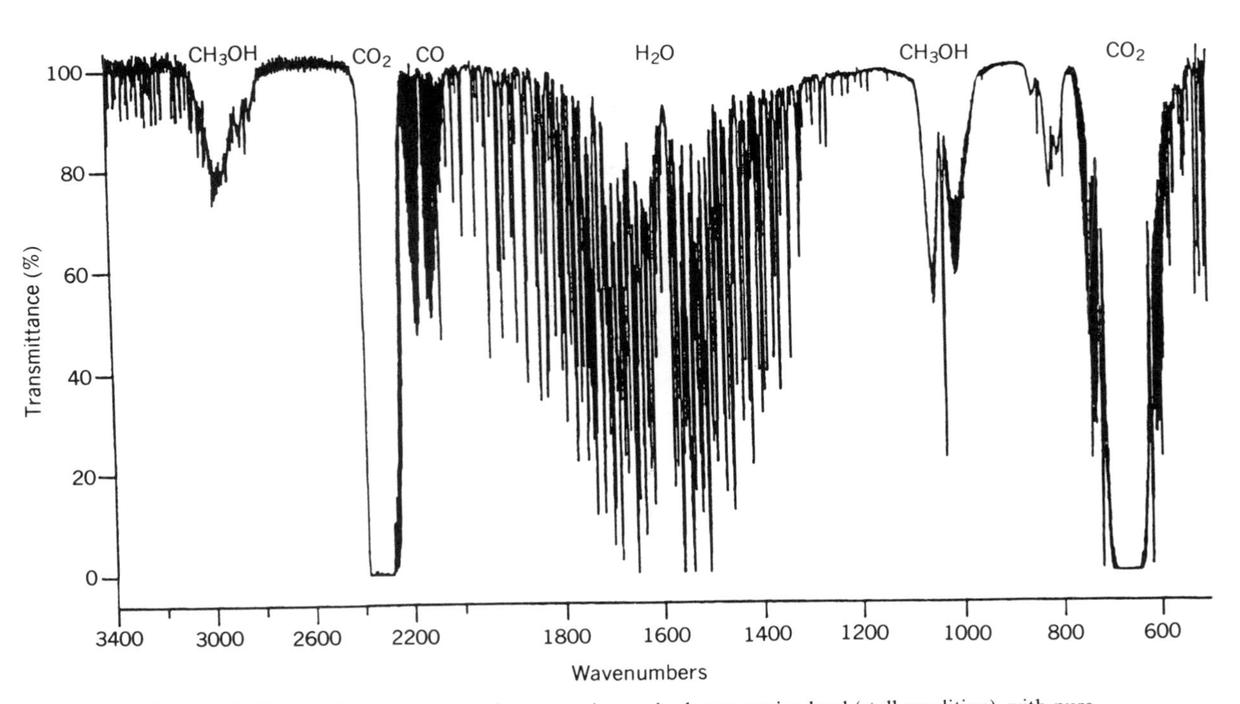

Figure 6.66. Exhaust from a passenger bus operating under heavy engine load (stall condition), with pure methanol fuel but without a catalytic muffler: 540 torr of exhaust gas and 215 torr of tank air in a 4.2-m cell.

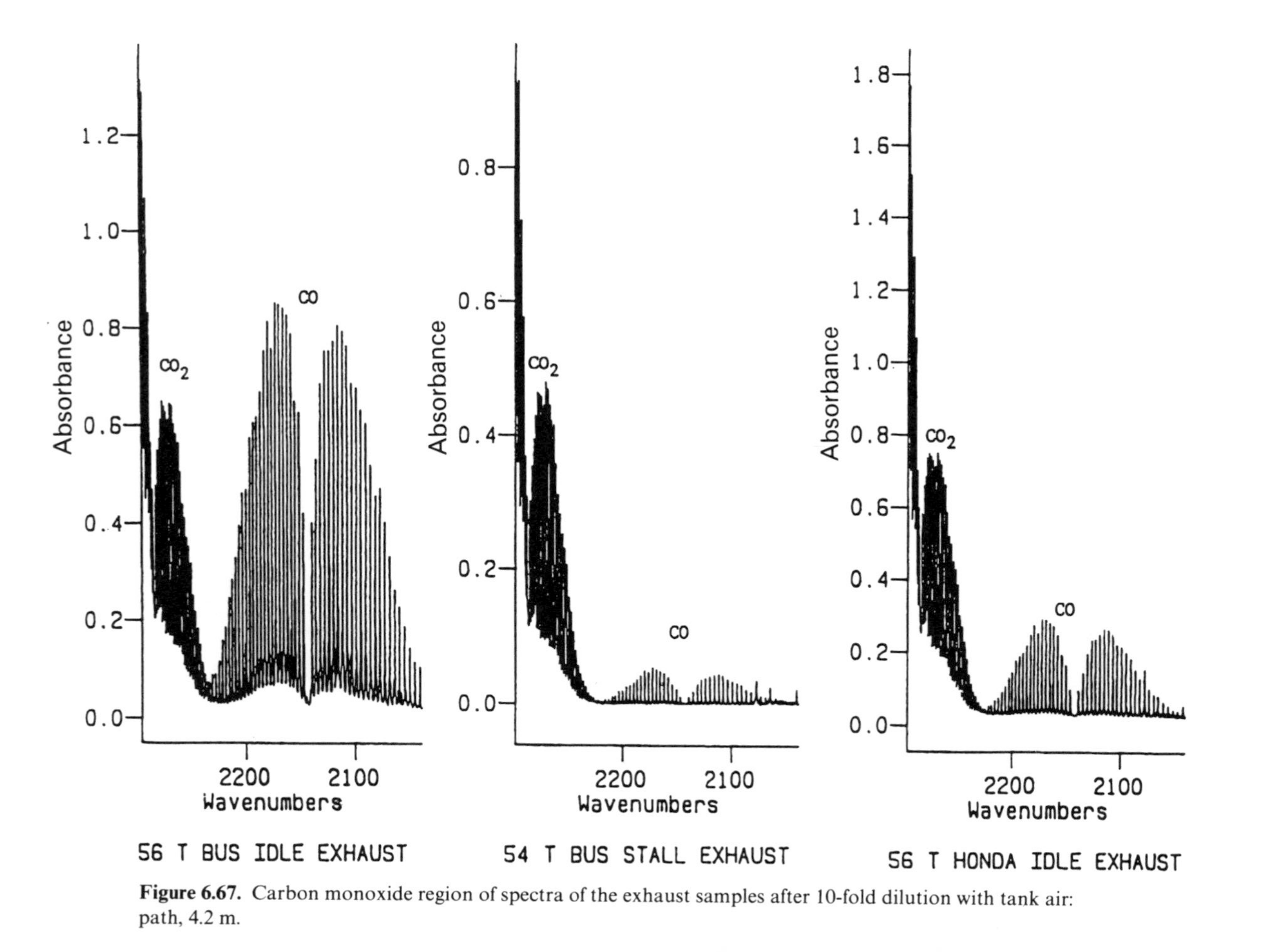

Figure 6.67. Carbon monoxide region of spectra of the exhaust samples after 10-fold dilution with tank air: path, 4.2 m.

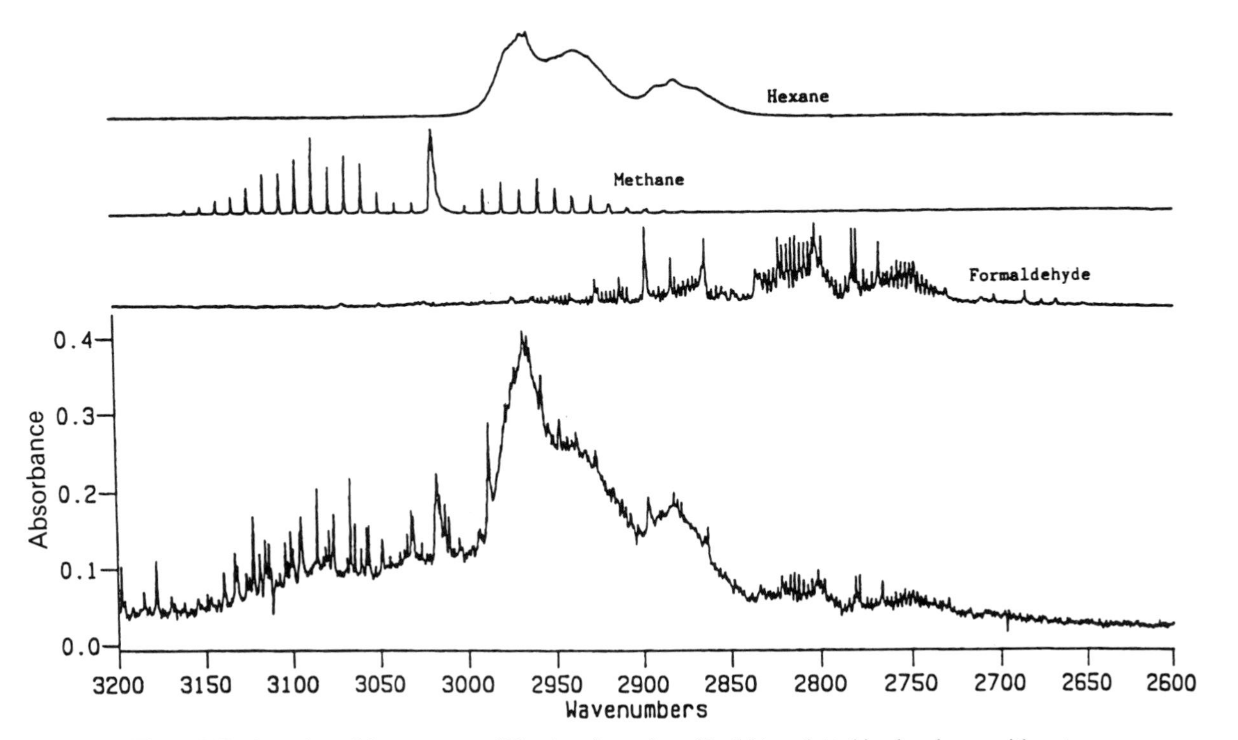

Figure 6.68. A portion of the spectrum of Honda exhaust from Fig.6.64, replotted in absorbance with water lines removed (lower); reference spectra used for quantitative analysis (upper).

the C—H vibration of organic compounds. Before Fig. 6.68 was plotted, water and CO_2 lines were subtracted from the spectrum. The reference spectra for water and CO_2 subtraction were obtained on pure samples of the compounds mixed in tank air in the 4.2-m absorption cell. The spectrum after subtraction shows bands of methane, formaldehyde, and gasoline-type hydrocarbons. The bands are indicated by the reference spectra plotted in the upper part of Fig. 6.68. These are the reference spectra that were used in the quantitative subtraction process.

Figure 6.69 shows the high-frequency region of the spectrum for the methanol-powered bus when its motor is idling (replotted from Fig. 6.66). This spectrum has had water and CO_2 lines removed. The reference spectra for methane, formaldehyde, and methanol are shown. Evidence for formaldehyde appears in the two very small spectral lines indicated by the arrow. This exhaust had a much lower concentration of formaldehyde than did the Honda automobile exhaust. Detailed examination of the data in Fig. 6.66 shows that the methanol-powered bus operating under heavy load did not produce any detectable amount of formaldehyde and not much methane either.

Figure 6.70 shows measurement of nitric oxide and nitrogen dioxide in the automobile exhaust. This spectrum is replotted from Fig. 6.64, with water and CO_2 lines removed. Also identified are lines of formaldehyde (its carbonyl band), although these were not used for measurement.

Figure 6.71 shows three plots of the NO_2 band that were extracted from the data of Figs. 6.64–6.66. Also shown is the reference spectrum of the NO_2 that corresponds to 20 ppm in the 4.2-m cell. In all three exhaust spectra the water lines were subtracted as well as possibile. A methyl nitrite band was also subtracted from the two bus exhaust spectra, and a formaldehyde band was subtracted from automobile exhaust spectrum. In all three cases the NO_2 concentration is measurable with an uncertainty of about 2 ppm.

In Fig. 6.72 the lower frequency end of the spectrum of the Honda exhaust is shown, replotted from Fig. 6.64, with water and CO_2 lines removed. Evident here are bands of methane, ethylene, acetylene, and two small features due to propylene and isobutylene.

Figure 6.73 shows the low-frequency end of the spectrum of the exhaust from the bus when operating under heavy engine load. This is a portion of the spectrum in Fig. 6.66, replotted in absorbance with water and CO_2 lines removed. The large band centered at 1032 wavenumbers is due to unburned methanol. Also in Fig. 6.73 we see that methyl nitrite has been produced. This compound is formed when the alcohol reacts with NO_2, which happens both in the exhaust system and in the collection vessel (Roby and Harrington, 1981).

Although the main purpose of the study just described was to demonstrate the efficacy of the infrared method for exhaust analysis, the results on exhaust

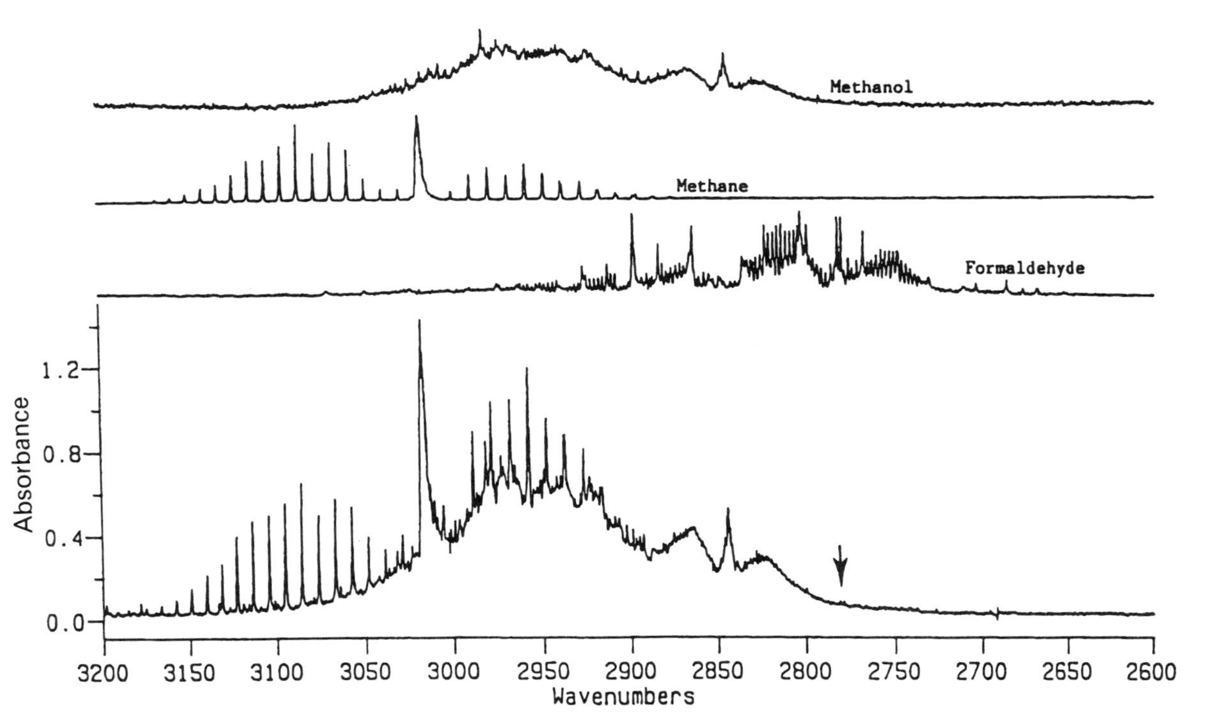

Figure 6.69. A portion of the spectrum of bus exhaust (idle condition) from Fig. 6.65, replotted in absorbance with water lines removed (lower); reference spectra used for quantitative analysis (upper).

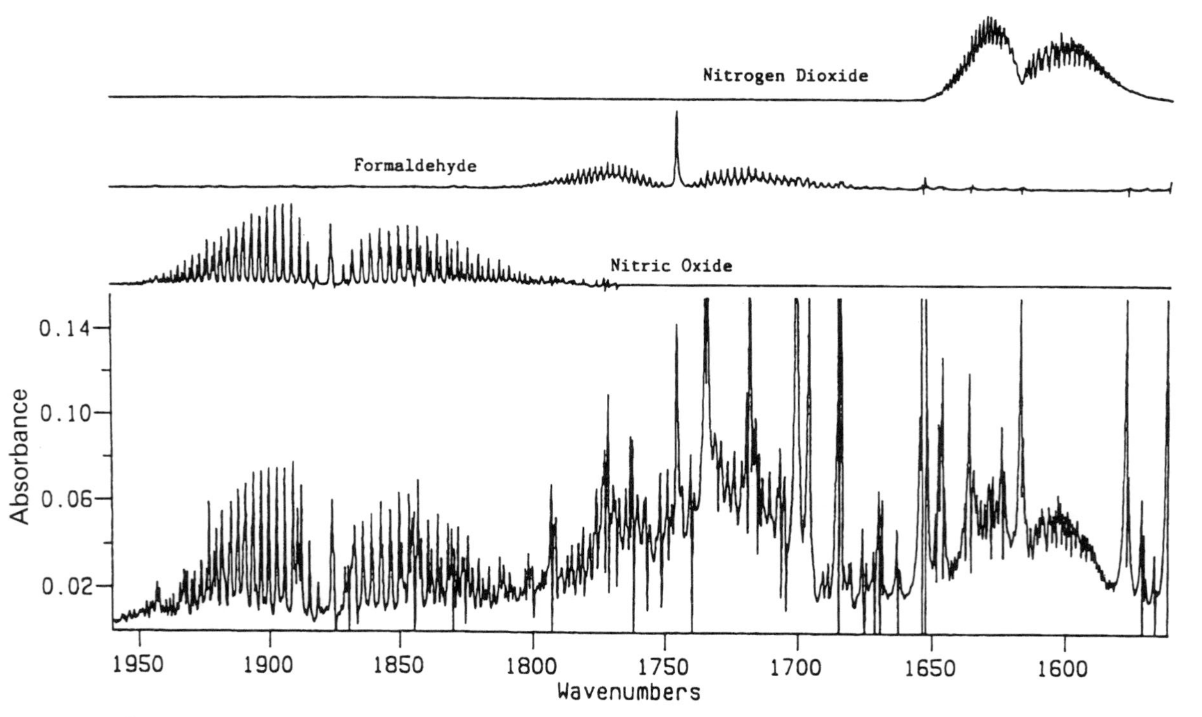

Figure 6.70. A portion of the spectrum of Honda exhaust from Fig. 6.64 replotted in absorbance with water lines removed (lower); reference spectra (upper).

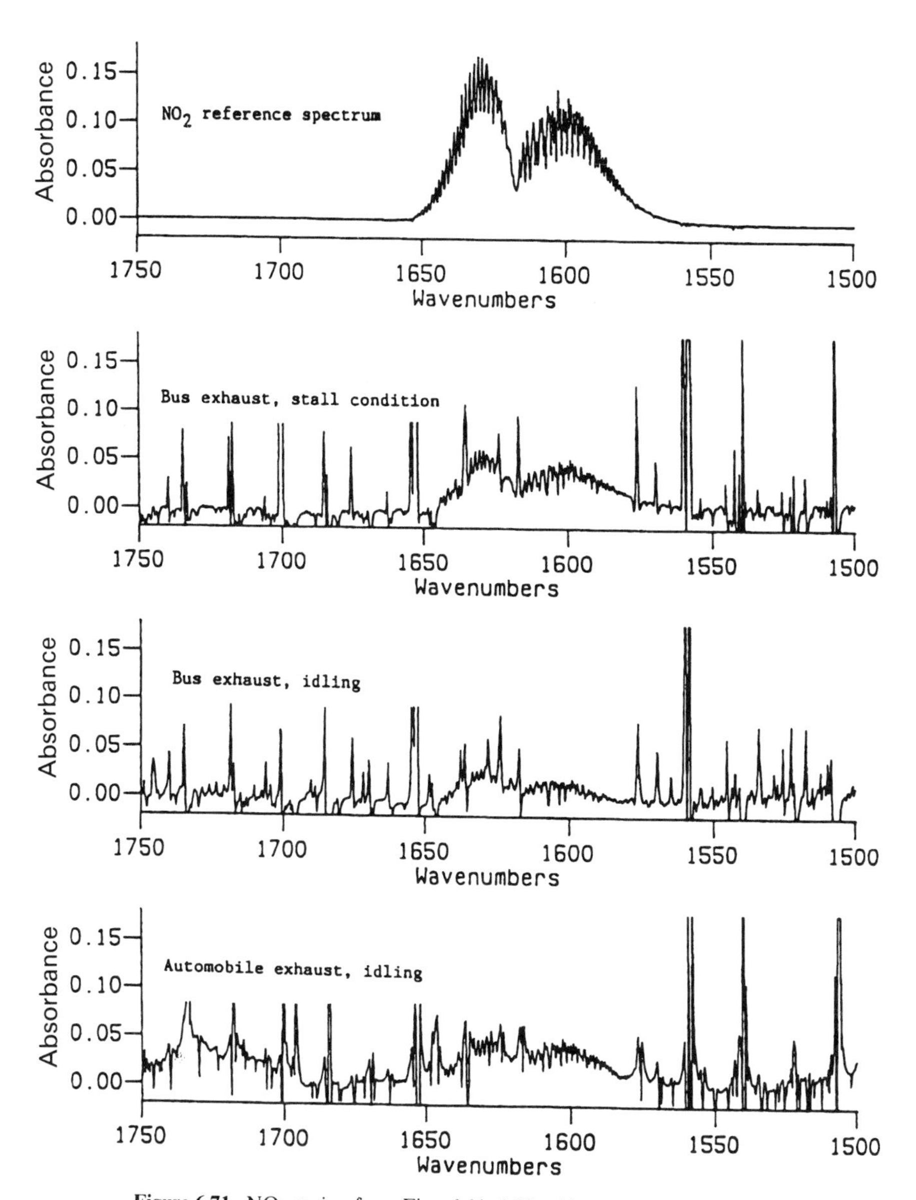

Figure 6.71. NO_2 region from Figs. 6.64–6.66, with water lines removed.

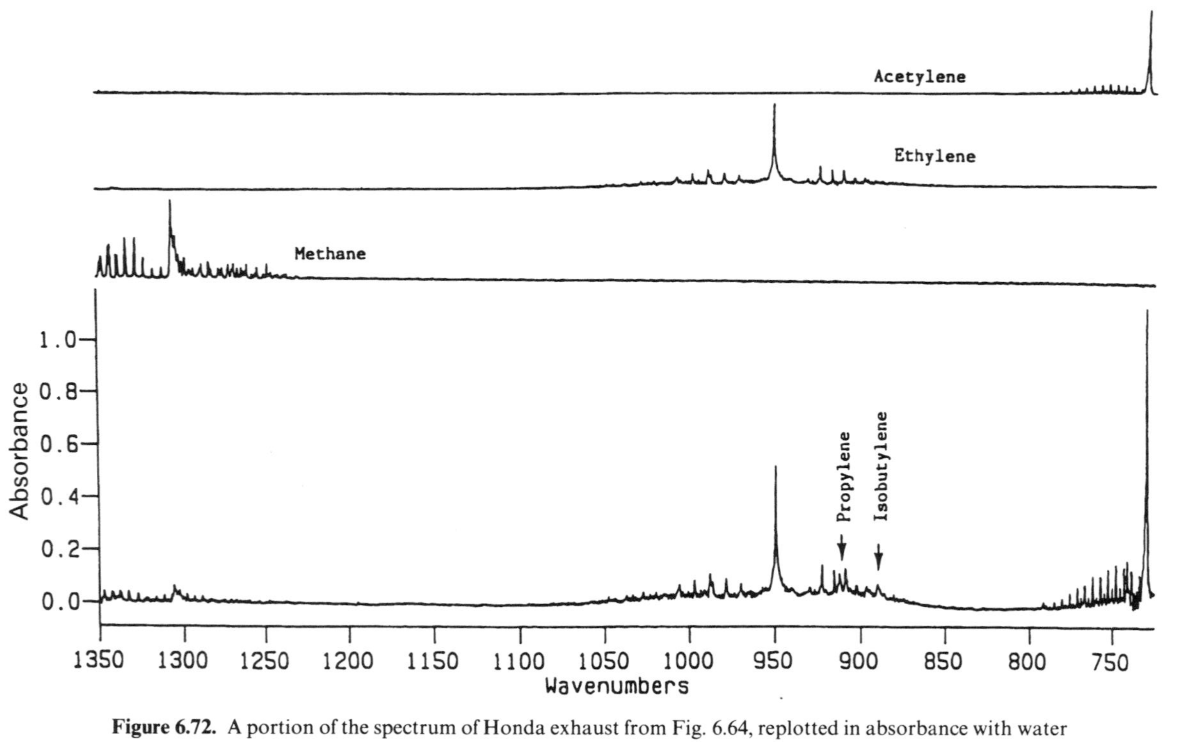

Figure 6.72. A portion of the spectrum of Honda exhaust from Fig. 6.64, replotted in absorbance with water and CO_2 lines removed (lower); reference spectra used for quantitative analysis (upper).

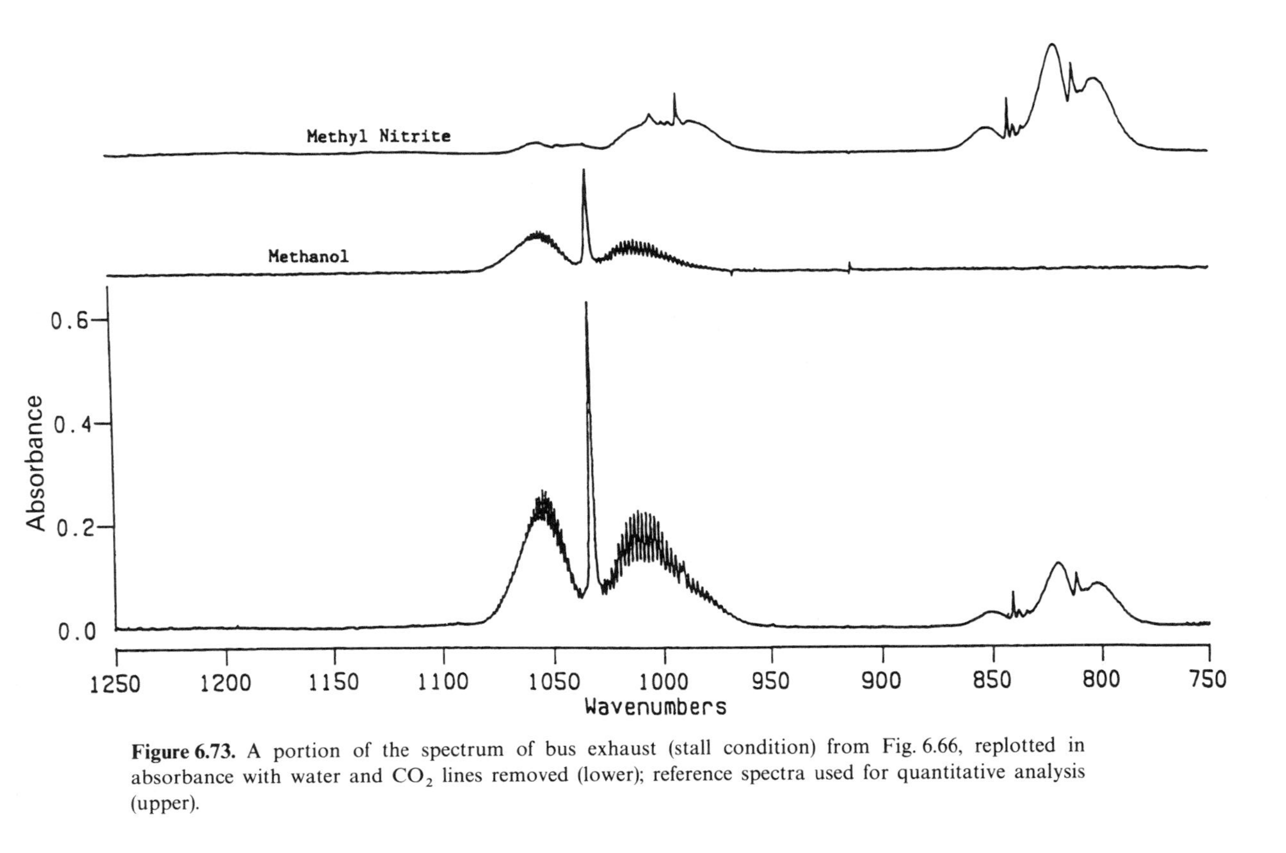

Figure 6.73. A portion of the spectrum of bus exhaust (stall condition) from Fig. 6.66, replotted in absorbance with water and CO_2 lines removed (lower); reference spectra used for quantitative analysis (upper).

compostion have intrinsic value. These results are listed in Table 6.5, which reveals the following:

1. The idling bus, without a catalytic muffler, produced about six times more carbon monoxide per unit exhaust volume than did the idling automobile.
2. The idling automobile, operating on gasoline, produced a higher concentration of nitrogen oxides than did the idling bus, operating on methanol.
3. The gasoline-powered automobile produced a higher concentration of formaldehyde than did the methanol-powered bus.
4. The methanol-burning engine produced methyl nitrite, a compound that would be extremely detrimental to air quality.

6.5.6.4. Conclusion

Exhaust analysis by the method of infrared absorption appears to be a simple and reliable process. The samples must be handled and possibly diluted in such a way that the analytical bands have low enough absorbance for the use of quantitative reference spectra. The analyst is not required to record either

Table 6.5. Concentration of Exhaust Components (ppb)[a]

Compound	Methanol-Powered Bus		Gasoline-Powered 1983 Honda Auto
	Idle	Stall	Idle
Carbon monoxide	11400	340	1830
Methanol	1470	280	< 2
Non-methane hydrocarbons (carbons)	< 10	< 10	440
Formaldehyde	13	< 5	35
Nitric oxide	18	14	80
Nitrogen dioxide	7	12	8
Methane	450	< 5	44
Ethylene	25	1	130
Acetylene	8	0.1	40
Propylene	< 2	< 2	49
Isobutylene	< 2	< 2	49
Methyl nitrite	46	69	< 5

[a] The "less than" symbol (<) means the compound was not seen, with the numbers indicating the lower limit of detection.

her/his own reference spectra or the spectra for subtraction of water and CO_2. These can be obtained in digitized form from the instrument manufacturer. All that is required for the quantitative subtraction process to succeed is that the sample spectra be recorded under the same instrumental conditions of resolution, apodization, and zero-filling as were used in recording the reference spectra. The amount of CO_2 detected is an index of the extent to which the sample was diluted by air. Since all other exhaust constituents can be normalized aganist the CO_2, there is no need for quantitative sample collection. The absorption bands of the various exhaust constituents do not interfere with each other. All compounds are measured simultaneously on a single exhaust sample. The spectrum gives a complete analysis, including the detection of unexpected compounds such as methyl nitrite.

6.6. NONDISPERSIVE ANALYSIS

6.6.1. Introductory Remarks

Nondispersive analyzers have been used for measurement of gases for more than 40 years, and they, like dispersive instruments, have undergone many improvements. A discussion of such instrumentation was presented by one of us some 20 years ago (Hanst, 1971).

Nondispersive methods do not involve a spatial separation of the infrared frequencies such as happens with interferometers and grating spectrometers. Instead, the radiation sources and detectors are matched to the spectra of the absorbing compounds either by their inherent response characteristics or by the use of filters. Laser sources may be tuned to the absorption lines of the compound being measured, or narrow-band interference filters may be used. However, it is spectrum matching through the use of gas filters that has found the widest application.

When compared with instruments that use gratings and slits, a nondispersive instrument can show advantages of high-energy throughput, a high degree of spectral multiplexing, and an unlimited degree of spectral resolution. These advantages result from the use of optical components that are inherently responsive to spectral characteristics of compounds being measured, while passing broad bands of frequencies. The performance of the nondispersive optical system depends on correlations between the infrared spectrum of the gas being measured and the spectral response characteristics of the system elements. Nondispersive analyzers have been discussed by Luft (1943), Fastie and Pfund (1947), Wright and Herscher (1946), Bartle et al. (1972), Fowler (1949), Hill and Powell (1968), and Sebacher (1977).

All nondispersive analyzers have as basic components a radiation source, filter, a sample cell, a radiation detector, and electronics that process and display the detector signal. Nearly all systems have at least one rotating light chopper. The systems currently in use do not differ basically from those studied by A. H. Pfund and others in the 1940s. Although the principles of operation of the systems have not changed in the last 50 years or so, capabilities of system components have been greatly expanded. Filters, detectors, and electronics have been especially improved. When these improvements are coupled with gas-filled filter cells for interference removal, it is found that the sensitivity of the nondispersive technique becomes great enough for direct measurement of pollutants in the ambient air.

6.6.2. System Operational Principles

The type of nondispersive instrument that is most widely used at present is diagrammed in Fig. 6.74. Radiation sources in instruments described in the literature have included hot wires, globars, Nernst glowers, heated gas, and the sun. The source spectrum may be altered by spectral filters that can be positioned nearly anywhere in the optical train. Typically, source filtering might be done with a narrow-band interference filter in combination with a gas-filter cell. The emitted radiation is passed through a sample cell which may be a single-pass type or the multiple-pass type (White cell). In field studies, the sample has been the open atmosphere.

The gas correlation filter cells have usually been just a few centimeters in length, containing the filter gas at a partial pressure of several torr. The optical trimmer is a neutral density filter that can be adjusted to balance the two beams. Wedges or screens can be used for this. Beam-combining optics and a photoelectric detector complete the train of optical components. The beam alternator is the only moving part in many systems. Instead of beam alternation, one could choose to modulate each beam at a different frequency and to monitor the intensity ratio of the two frequency components.

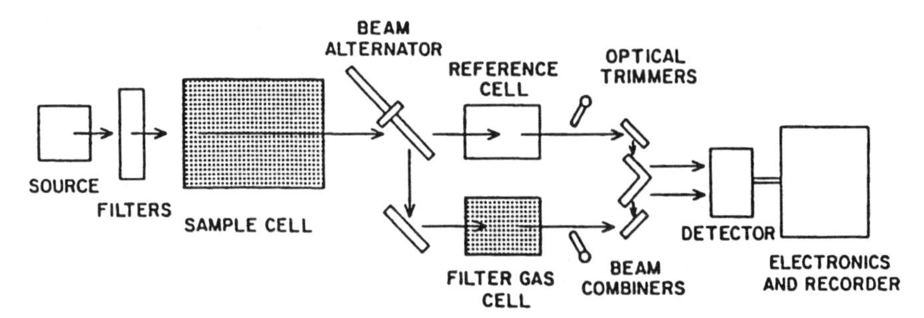

Figure 6.74. Nondispersive analyzer in negative filter configuration.

System operation can be understood by considering the spectra and detector signal in the absence and presence of the gas being measured (object gas). Let it be the case that the source and its filters produce a range of frequencies that encompasses two spectral lines of the object gas. Figure 6.75 (upper half) shows the spectrum in the two channels of the instrument when there is no object gas in the sample cell. The absorption lines appear only in the filter gas channel, and the optical trimmers are adjusted to give equal total intensities with a consequent zero detector signal at the beam-alternation frequency. Then, when the object gas does appear in the sample cell (lower half of Fig. 6.75), its absorption lines appear in both channels, the intensities at the detector become unequal, and a signal with the beam-alternation frequency appears. The system is called *negative filter* because the intensity reduction is greater in the reference channel than in the gas-filter channel.

Interfering gases will increase the signal when their absorption lines overlap with the lines of the object gas and will decrease the signal when the lines do

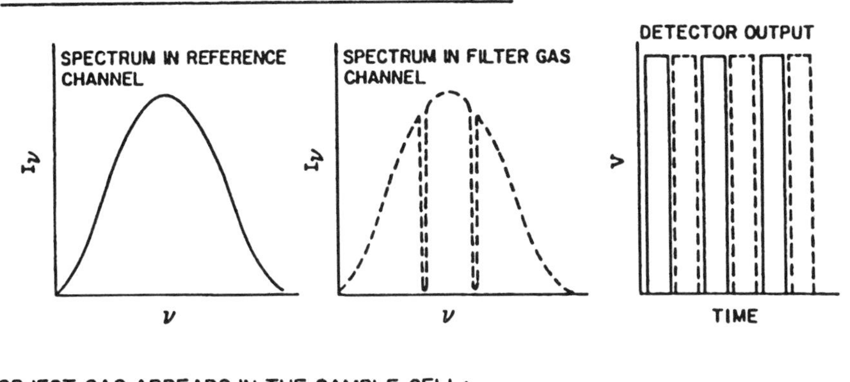

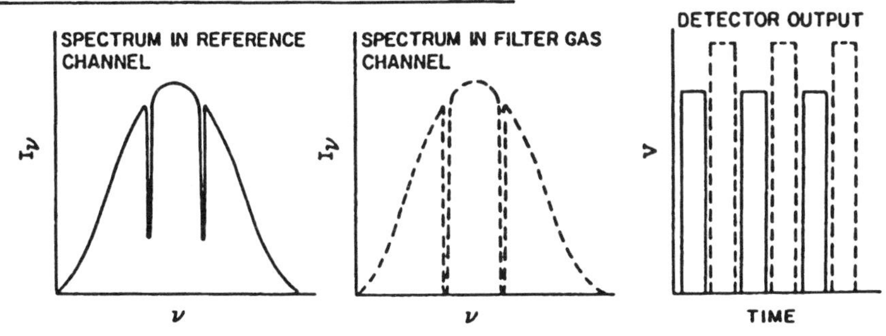

Figure 6.75. Spectra and detector signals in nondispersive analyzer.

not overlap. If the interfering gas has many lines in the spectral region being used, the positive and negative interferences might nearly cancel each other.

Proper choice of source, detector, and filter will center the band pass around a chosen band of the object gas, thus maximizing the integrated absorption coefficient. If the object gas has a broad absorption band in a spectral region relatively free from interference, then the best system design would include a filter with a band pass slightly wider than the spectral band of the object gas. This would be the case for NO_2 or SO_2, or when using the C—H band to measure hydrocarbons.

If the object gas has a band with only a small number of lines widely spaced, then most of the energy will fall between the spectral lines and the use of a continuum source with a band pass filter will not yield a high absorption coefficient. In this case a heated sample of object gas could be used as source. The emitted lines would match the absorption lines quite well, with little energy falling between the lines, and therefore the integrated absorption coefficient would be high. This technique can be used for detecting HCl, CO, NH_3, NO, and other thermally stable small molecules.

6.6.3. Removal of Interferences

When another gas has no lines within the band pass selected for detection of the object gas, there will of course be no interference. If there are such lines, they can be blanked out of both beams by placing enough of said gas in a filter cell located in the combined beam portion of the optical train. The system will then be "blinded" to the interfering gas. Interference between pollutants is only an occasional problem because the main absorption bands of the major gaseous pollutants do not generally overlap each other and, besides, the absorption by each pollutant is very slight. The absorption by atmospheric water vapor, however, is very great, and the main bands of several important pollutants such as NO_2, NO, and SO_2 do fall within strong regions of absorption by water. For the detection of these pollutants the system must be blinded to water vapor.

A small cell cannot be used to blind the system to water vapor because at ordinary temperatures the vapor pressure of water is no higher than about 0.03 atm. Instead, a long-path filter cell is required. Preferably, the product of water vapor concentration times path length should be greater in the filter cell than in the sample cell. Thus the system requires two long-path cells in tandem.

6.6.4. Detection Sensitivity

Detection sensitivity may be examined in terms of the Beer–Lambert law given in Eq. (6.3). The detection limit is assumed to be reached when the signal reduction due to sample absorption equals the noise level (N) of the total signal (S). Since one is always working in the small absorption range, log I_0/I at the detection limit is approximately equal to the noise-to-signal ratio, N/S. There is no apparent reason why the N/S of a properly designed system cannot be 10^{-4} or smaller.

The system should be designed to develop the largest possible absorption coefficient, α, for the object gas. The higher the value of α, the shorter the required path length. For a strongly absorbing species, when the pass band is properly centered on the absorption band, the value of α can be as high as $50\ \text{cm}^{-1} \cdot \text{atm}^{-1}$. If the absorption is weak or if there is difficulty in controlling the pass band, α might be as low as $1\ \text{cm}^{-1} \cdot \text{atm}^{-1}$.

Because the system always operates in the small absorption range the output signal is directly proportional to the concentration of the object gas.

The detection limit, C, is listed in Table 6.6 as a function of the noise-to-signal level, absorption coefficient, and path length. Tests have shown that for many important air contaminants, the bottom-line detection limit of 10^{-9} atm is achievable. If the system designers are only seeking ppm sensitivity rather than ppb, then they can allow themselves the indicated shorter path lengths and higher values of the noise-to-signal ratio.

6.7. AVAILABLE LONG-PATH CELLS AND GAS MEASUREMENT SYSTEMS

Infrared Analysis, Inc. manufactures and sells a wide variety of long-path absorption cells. These cells may be used on Fourier transform, laser, or dispersive spectrometers. The company also sells complete gas measurement systems and the digitized reference spectra that are necessary for the quantitative analysis of gas mixtures.

The cells have transfer optics that allow them to be used in the sample compartments of commercial spectrometers without modifications. Installation or removal is usually just a matter of lifting the cell in or out of the sample compartment.

The optical beam is reflected within the long-path cells by the three-mirror White system, which gives multiple passing in increments of four passes. The path lengths are variable from 0.6 to > 100 m. Operating pressure is from vacuum to 4 atm. The beam is folded by three spherical mirrors, each with the same radius of curvature, which is the cell base path length, as was depicted in Fig. 6.14.

The radiation from the source is focused in the entrance aperture of the cell. From there the diverging beam passes to the first objective mirror, which focuses an image onto the left mirror (Fig. 6.14), which then directs the beam

Table 6.6. Detection Limit as a Function of System Parameters

Noise-to-Signal Ratio (N/S)	Absorption Coefficient α $(cm^{-1} \cdot atm^{-1})$	Path Length L (cm)	Detection Limit C (atm)
10^{-3}	1	10^{+3}	10^{-6}
		10^{+4}	10^{-7}
	10	10^{+3}	10^{-7}
		10^{+4}	10^{-8}
10^{-4}	1	10^{+3}	10^{-7}
		10^{+4}	10^{-8}
	10	10^{+3}	10^{-8}
		10^{+4}	10^{-9}

either out of the cell (four passes) or back to the field mirror for additional mutliple passing. The images line up on the field mirror in two rows, as was shown in Fig. 6.15.

Stainless steel vacuum valves and pressure relief valves are used. The cell bodies are of borosilicate glass for chemical inertness and to allow in situ photochemistry. All cells but the smallest also have a vaccum/pressure gauge. The cells are usable in the ultraviolet, visible, and infrared spectral regions. Available mirror coatings include aluminum, silver, and gold. Windows are KBr flats, 25 mm in diameter. Each cell has transfer optics matched to the sample compartment in which it is to be used.

Special cell designs and cell sizes are available for gas pressures up to 100 atm and path lengths up to 2 km. Also offered are super cells with doubled or tripled paths, permanently aligned cells, photochemical research facilities, and complete air pollution measurement systems.[1]

Available cell accessories include a laser illuminator for verifying path length and cell alignment, photochemical lamp banks, gas-sampling systems, purge covers, short-path cells, calibration samples, gas-handling tubes, crystal windows and spare mirrors.

For ultraviolet use, Infrared Analysis, Inc. offers mirrors coated with aluminum and magnesium fluoride. These mirrors have a reflection coefficient of about 0.90 in the near-ultraviolet region, giving a maximum signal at 12

[1] Details of the cells and their prices are contained in the current version of the company catalog, which is available on request from Infrared Analysis, Inc., 11629 Deborah Drive, Potomac, Maryland 20854.

passes. For infrared use, Infrared Analysis, Inc. offers ceramic-protected silver mirrors that reflect more than 99% that allow the use of 100 reflections or more. These mirrors are superior to gold in reflection coefficient and abrasion resistance. They are also corrosion resistant. If they become soiled they may be washed with cotton, detergent, and water.

6.8. AN ATLAS OF QUANTITATIVE REFERENCE SPECTRA

Infrared Analysis, Inc. has published an atlas of the infrared absorption spectra of gases. This atlas is available in analog form (on paper) and in digitized form (on computer diskettes). All the gases previously listed in Section 6.1.4 (on detection limits) are included in the collection. This atlas of infrared spectra presents the basic physical data that may be used to identify and quantify gases of all kinds, as they are seen by all types of infrared spectrometers. Computer smoothing of these spectra to lower resolution closely reproduces the smoothing that results from instrumental band pass. The computer can in fact convert a single master spectrum to a family of master spectra that applies to a wide range of concentration–path length products. When this family of spectra has its lines broadened by smoothing, the new family of spectra shows the absorption law failure just as it would be encountered with a lower resolution spectrometer. These digitized spectra therefore relieve laboratory workers of the need to handle calibration gases. Calibration and quantification are derived from the software. For air analysis, subtraction spectra are supplied for removal of the interfering lines of water and other infrared absorbing constituents of the atmosphere.

Spectral Region. The infrared spectra of this atlas extend across the spectral region 3700–500 cm^{-1} (except for HF). This is the fundamental infrared region, where the rotations and vibrations of the molecules give rise to the infrared absorption. For gas measurement there is nothing of any value on either side of the fundamental region. HF is the only molecule in the atlas that has a strong band at frequencies higher than 3700 cm^{-1}. At higher frequencies there are only overtones, which have extremely small absorption coefficients. At lower frequencies there may be good strong rotation lines, but because of extremely strong absorption by water vapor they are quite inaccessible.

Resolution Considerations. The spectra were recorded at high resolution to avoid the inaccuracies of absorption coefficient that arise when there is unresolved fine structure in the spectrum. When a spectral line is narrower in wavenumbers than the resolution bandwidth of the spectrometer, the line will appear wider and shorter than it actually is. For lines that have small maximum absorbance, the area under the unresolved line will be the same as when the line is fully resolved. For lines that have large maximum absorbance, however, line shape distortion makes the area under the unresolved line smaller than it should be. This is a result of the logarithmic nature of the absorption law [see Eq. (6.3)]. Measurements with varying concentrations of an absorber that has distorted spectral lines will show a deviation from this absorption law. In such cases the absorption law is not obeyed at high absorbance values, but it is nearly always obeyed at low absorbance values. One way to avoid error in quantitative analysis of gases is to work only in the low absorbance range. If the lines being used for measurement absorb strongly, the sample should be diluted, or the path should be shortened.

When it is not possible to work only in the low absorbance range, a calibration curve must be constructed showing absorbance against concentration. This curve may apply either to the area under a spectral feature or to the peak absorbance of the feature.

When the fine structure in a spectrum is fully resolved, the absorption law is always obeyed, even at high absorbance values. A fully resolved spectrum can be de-resolved to match any lesser degree of resolution used in the study of gases.

Spectrometer. Most of the spectra were recorded on an Analect RFX-65 Fourier transform spectrometer (KVB/Analect, Irvine, CA) equipped with a HgCdTe detector. This instrument has a resolution bandwidth of 0.125 cm^{-1}. Since the width of a single absorption line at normal temperature and pressure is about 0.2 cm^{-1} (fwhm), the spectral details are well reproduced. Higher resolution would not significantly change the detailed spectral structure. These master spectra may be smoothed by a computer program to any chosen degree of lower resolution. The computer smoothing closely reproduces the smoothing that results from instrumental band pass. The smoothed family of spectra can be created as a calibration set that is the same as the set of spectra that would be created on a spectrometer with a set of calibration gases. Some of the spectra—usually of heavy molecules that do not have spectral fine structure—were recorded on a Digilab FTS-40 spectrometer working at 0.5 cm^{-1}.

Pressure and Temperature Considerations. Spectral line widths vary with pressure. If the sample pressure is lowered, the absorption lines get taller and narrower. Deviations from the logarithmic absorption law then become greater. A set of quantitative calibration spectra created at 1 atm total pressure will not apply to a sample at 0.5 atm.

The effect of temperature changes on a sample is to change the distribution of intensities among the rotational lines. This causes changes in the size and shapes of spectral features. Calibration spectra recorded at normal temperature may be invalid when the sample temperature is lower or higher. This atlas is intended to apply to samples at ordinary temperature and pressure; that is, a temperature near 70 °F (21 °C) and pressure near 1 atm. Temperature changes within the range in which people live will not invalidate the quantitative nature of the spectra. Likewise, pressure changes within the normal living range will not invalidate the spectra. Some corrections to the absorption coefficients might be considered for samples taken at locations at especially high altitudes, such as Denver or Mexico City.

Gas-Handling Technique. The gaseous samples were handled in a glass vacuum line. Some gases were obtained from lecture bottles, but in most cases vapors were obtained from pure liquids or solids. Some of the gases were generated in chemical reactions conducted in an attachment to the vacuum line. In these cases, a purification step was introduced by trapping the gas at liquid nitrogen temperature and then revaporizing it.

Pressure measurements were made on a mercury manometer and a silicone-oil manometer. A single-pass absorption cell of 11-cm length was used in most cases, allowing a manometer reading of several centimeters. A measurement accuracy of 5% or better was sought. Nitrogen or oxygen was added to the sample to raise the total pressure to 1 atm. Gases that dimerize, such as the aliphatic acids, were measured at ppm mixing ratios in a long-path cell, usually set for an 11-m path. When a gas could only be seen in a mixture with other infrared absorbing gases (such as HNO_2 in equilibrium with NO, NO_2, and H_2O), the unwanted bands of the other gases were subtracted from the spectrum before it was plotted. Reference spectra of water vapor for long-path measurements in air were recorded in a large multiple-pass cell of 3-m base path, used at paths up to 288 m.

Software. The spectra were processed in and plotted from LAB CALC, the data processing program of Galactic Industries, Inc.

Identification of Gaseous Spectra. In gas analysis, identification of the absorbing species is not usually the main problem. Most of the spectra are easily recognized. If an unknown spectrum is encountered, the number of possible absorbing species is not great. This atlas will in most cases contain the unknown. Some consideration of functional group frequencies will often assist in finding the molecule. For example, a strong band in the region 1800–1700 cm^{-1} will indicate a compound with a carbonyl group, and a band near 3000 cm^{-1} will say that there are C—H groups in the molecule. Also, for example, nitrate groups and halogen atoms have characteristic strong bands. The amount of fine structure in a spectrum is an indication of molecular size and symmetry. If there are resolvable lines, the molecule is small. If there are regular arrays of lines, the molecule is linear. If the lines are bunched in a regular way, the molecule is probably a symmetric top. If the bands are smooth and structureless, the molecule is large and nonsymmetrical.

Finally, if the various clues from the spectrum do not identify the molecule, the analyst can look for the bands while leafing through the atlas. In just a few minutes time, every page can be examined, with a high probability of encountering the spectrum being sought. An automatic computer search program is also available from Infrared Analysis, Inc.

The Problem of Absorbance Nonlinearities. Across a single absorption line the variation of the absorption coefficient α with frequency ν gives the line shape. This shape changes with total pressure P as implied by Eq. (6.1), but when α is integrated over a whole line the integral is constant, independent of total pressure. If an instrument can reveal the true shape of spectral features, the true absorbance αCL is revealed, and it always obeys Eq. (6.3). Usually, however, the instrumental resolving power is not high enough to reveal the true spectrum and one sees only an apparent absorbance. Apparent absorbance does not obey Eq. (6.3). The equation actually only applies to a single frequency or to a band of frequencies that all have the same absorption coefficient. The equation is not valid when a number of frequencies with various absorption coefficients are seen simultaneously. In these cases the relationship between absorbance ($\log I_0/I$) and the concentration–path length product ($C \times L$) must be determined empirically. In other words, a calibration curve must be constructed.

Availability of Spectra in Digitized Calibration Sets. The spectra are available in digitized form at any resolution from 0.125 to 4 cm^{-1}. For each compound, the sets of spectra are calculated from the fully resolved master spectrum that appears in this atlas. All absorbance nonlinearities are reproduced just as they would appear if the sets had been recorded at the chosen resolution. A set may include as many as five spectra, each calculated for a different level of absorbance. The calibration sets may be used to cover wide ranges of absorbance either for the full spectrum or parts of the spectrum. They may be used for analysis by interactive subtraction or by automated calculations.

These calibration sets eliminate any need for reference gas samples. In using these sets, the only chemical that is handled by the analyst is the sample itself.

The sets of calibration spectra and subtraction spectra can be matched to the resolution and frequency peculiarities of a particular instrument. For this purpose a sample spectrum recorded in air must be supplied to Infrared Analysis, Inc. An example is given in Fig. 6.76. The bottom spectrum is a portion of a spectrum of water in room air, recorded on an Analect Transept spectrometer at a resolution of 2 cm^{-1}. The center spectrum is a portion of our

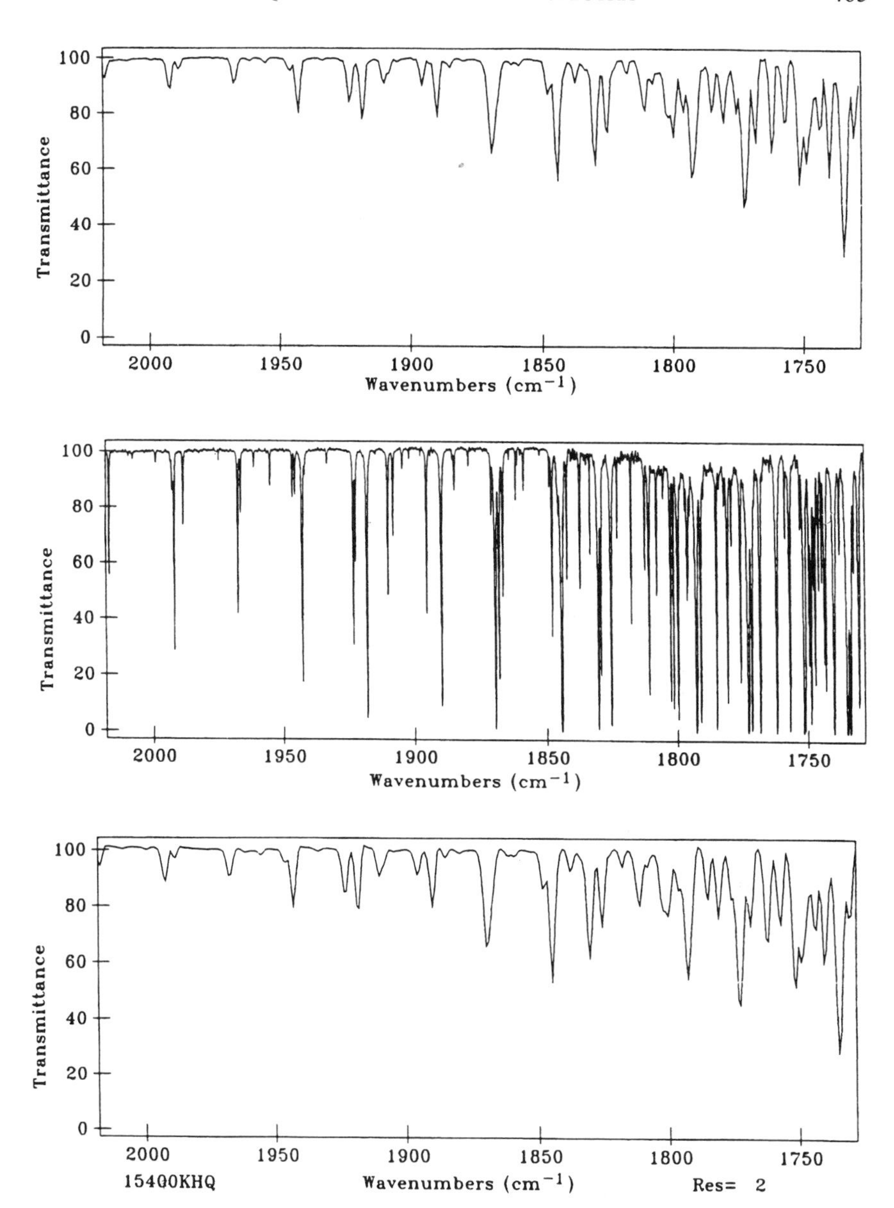

Figure 6.76. Sample spectra of water vapor (bottom to top): Analect Transept; master; derived reference. (see the text.)

master spectrum of water vapor. The top spectrum is the matching reference spectrum that was created from the master spectrum.

Besides giving the correct answers in the quantitative analysis, the calibration sets allow subtraction of the spectra of individual compounds from the spectrum of a complex mixture. When a detailed spectrum like water is being subtracted, its pattern of absorbance nonlinearities must be the same as in the sample spectrum. Since the pattern changes as the concentration–path length product changes, to cover a wide range of humidities and optical pathlengths in air will require a large number of water reference spectra.

For air analysis, Infrared Analysis, Inc. supplies families of water reference spectra. Also offered are dry-air reference spectra that are made by combining reference spectra for carbon dioxide, methane, nitrous oxide, and carbon monoxide in the clean-air proportions of 350, 1.7, 0.3, and 0.2 ppm, respectively. When one of these dry-air reference spectra is subtracted from a real-air spectrum, any CO_2, CH_4, N_2O, or CO absorption that is due to pollution will show as a residual, along with the bands of other pollutants.

REFERENCES

Ackerman, M., Frimout, D., Girard, A., Gottignies, M., and Muller, C. (1976). Stratospheric HCl from infrared spectra. *Geophys. Res. Lett.* **3**, 81–83.

Bartle, E. R., Kays, S., and Mackstroth, E. A. (1972). An in-situ monitor for HCl and HF. *J. Spacecr. Rockets* **9** (11), 836–841.

Blatherwick, R. D., Goldman, A., Murcray, F. J., Cook, G. R., and Van Allen, J. W. (1980). Simultaneous mixing ratio profiles of stratospheric NO and NO_2 as derived from balloon-borne infrared solar spectra. *Geophys. Res. Lett.* **7**, 471–473.

Camy-Peyret, C., Flaud, J. M., Laurent, J., and Stokes, G. M. (1983). First infrared measurement of atmospheric NO_2 from the ground. *Geophys. Res. Lett.* **10**, 35–38.

Chan, W. H., Nordstrom, R. J., Calvert, J. G., and Shaw, J. H. (1976). Kinetic study of HONO formation and decay reactions in gaseous mixtures of HONO, NO, NO_2, H_2O and N_2. *Environ. Sci. Technol.* **10**, 674–682.

Coffey, M. T., Mankin, W. G., and Goldman, A. (1981). Simultaneous spectroscopic determination of latitudinal, seasonal, and diurnal variability of stratospheric N_2O, NO, NO_2, and HNO_3. *J. Geophys. Res.* **86C**, 7331–7341.

Darley, E. F., Stephens, E. R., Middleton, J. T., and Hanst, P. L. (1959). Oxidant plant damage from ozone-olefin reactions. *Int. J. Air Pollut.* **1**, 155–162.

Farmer, C. B. (1985). High resolution infrared spectroscopy from space: Preliminary results from the ATMOS experiment on Spacelab-3. In *Fourier and Computerized Infrared Spectroscopy* (J. G. Grasselli and D. G. Cameron, Eds.), SPIE **553**, 87.

Farmer, C. B., Raper, O. F., and Norton, R. H. (1976). Spectroscopic detection and vertical distribution of HCl in the troposphere and stratosphere. *Geophys. Res. Lett.* **3**, 13–16.

Farmer, C. B., Raper, O. F., Robbins, B. D., Toth, R. A., and Muller, C. (1980). Simultaneous spectroscopic measurement of stratospheric species: O_3, CH_4, CO, CO_2, N_2O, H_2O, HCl, and HF at northern and southern mid-latitudes. *J. Geophys. Res.* **85C**, 1621–1632.

Fastie, W. G., and Pfund, A. H. (1947). Selective infrared gas analyzers. *J. Opt. Soc. Am.* **37**, 762–768.

Fontanella, J. C., Girard, A., Gramont, L., and Louisnard, N. (1975). Vertical distribution of NO, NO_2, and HNO_3 as derived from stratospheric absorption infrared spectra. *Appl. Opt.* **14**, 825–839.

Fowler, R. C. (1949). A rapid infrared gas analyzer. *Rev. Sci. Instrum.* **20**, 175–178.

Gay, B. W., Hanst, P. L., Bufalini, J., and Noonan, R. (1976a). Atmospheric oxidation of chlorinated ethylenes. *Environ. Sci. Technol.* **10**, 58–67.

Gay, B. W., Noonan, R. C., Bufalini, J., and Hanst, P. L. (1976b). Photochemical synthesis of peroxyacyl nitrates in the gas phase via chlorine-aldehyde reaction. *Environ. Sci. Technol.* **10**, 82–85.

Goldman, A., Fernald, F. G., Williams, W. J., and Murcray, D. G. (1978). Vertical distribution of NO_2 in the stratosphere as determined from balloon measurements of solar spectra in the 4500 A region. *Geophys. Res. Lett.* **5**, 257–260.

Goldman, A., Murcray, F. H., Murcray, D. G., and Rinsland, C. P. (1984). A search for formic acid in the upper troposphere: A tentative identification of the 1105 cm^{-1} ν_6 band Q-branch in high-resolution balloon-borne solar absorption spectra. *Geophys. Res. Lett.* **11**, 307–310.

Grant, W. B., Kagann, R. H., and McClenny, W. A. (1992). Optical remote measurement of toxic gases. *J. Air Waste Manage. Assoc.* **42**, 18–30.

Hanst, P. L. (1971). Spectroscopic methods for air pollution measurement. *Adv. Environ. Sci. Technol.* **2**, 91–213.

Hanst, P. L (1978). Air pollution measurement by Fourier transform spectroscopy. *Appl. Opt.* **17**, 1360–1366.

Hanst, P. L., and Gay, B. W. (1977). Photochemical reactions among formaldehyde, chlorine and nitrogen dioxide in air. *Environ. Sci. Technol.* **11**, 1105–1109.

Hanst, P. L., and Hanst, S. T. (1992). *Infrared Spectra for Quantitative Analysis of Gases.* Infrared Analysis, Inc., 11629 Deborah Drive, Potomac, MD 20854.

Hanst, P. L., Lefohn, A. S., and Gay, B. W. (1973). Detection of atmospheric pollutants at parts-per-billion levels by infrared spectroscopy. *Appl. Spectros.* **27**, 188–198.

Hanst, P. L., Spiller, L. L., Watts, D. M., Spence, J. W., and Miller, M. F. (1975a). Infrared measurement of fluorocarbons, carbon tetrachloride, carbonyl sulfide and other atmospheric trace gases. *J. Air Pollut. Control. Assoc.* **25**, 1220–1226.

Hanst, P. L., Wilson, W. E., Patterson, R. K., Gay, B. W., Chaney, L. W., and Burton, C. S. (1975b). *A Spectroscopic Study of California Smog*, Publ. EPA 650/4-75-006. U. S. Environmental Protection Agency, Research Triangle Park,

Hanst, P. L., Wong, N. W., and Bragin, J, (1982). A long-path infrared study of Los Angeles smog. *Atmos. Environ.* **16**, 969–981.

Herget, W. F. (1979). Air pollution: Ground-based sensing of source emissions. In

Fourier Transform Infrared Spectroscopy (J. Ferraro and L. Basile, Eds.), Vol. 2. Chapter 3. Academic Press, New York.

Hill, D. W., and Powell, T. (1968). *Non-Dispersive Infrared Gas Analysis in Science, Medicine, and Industry*. Plenum, New York.

Horn, D., and Pimentel, G. C. (1971). 2.5 km low temperature multiple-reflection cell, *Appl. Opt.* **10**, 1892–1898.

Heuss, J. M., and Glasson, W. A. (1968). Hydrocarbon reactivity and eye irritation, *Environ. Sci. Technol.* **2**, 1109–1116.

Infrared Analysis, Inc. (1992). Catalog 92. 1424 North Central Park Avenue, Anaheim, CA 92802.

Jouve, P. (1980). Telemesure de HCl, HF, ozone et formaldehyde dans l'atmosphère moyenne par infrarouge haute résolution à partir du sol. *Proc. 1st Eur. Symp. Physico-Chemical Behavior of Atmospheric Pollutants*, Ispra, Italy, 1979 (B. Versino and H. Ott, Eds.). Brussels, Luxembourg.

Leighton, P. A. (1961). *Photochemistry of Air Pollution.* Academic Press, New York.

Luft, K. F. (1943). Über eine neue Methode der registrierenden Gasanalyse mit Hilfe der Absorption ultraroter Strahlung ohne spektrale Zerlegung. *Z. Tech. Phys.* **24**(5), 97–104.

Mankin, W. G., Coffey, M. T., and Griffith, D. W. T. (1979). Spectroscopic measurement of carbonyl sulfide (COS) in the stratosphere. *Geophys. Res. Lett.* **6**, 853–856.

Migeotte, M., Neven, L., and Swensonn, J. (1956). The solar spectrum from 2.8 to 23.7 microns, Part 1. Photometric atlas. *Mem. Soc. R. Sci. Liege, Spec. Vol.* **1**.

Molina, M. J., and Rowland, F. S. (1974). Stratospheric sink for chlorofluoromethanes: Chlorine atom catalyzed destruction of ozone. *Nature (London)* **249**, 810–812.

Murcray, D. G., Murcray, F. H., Williams, W. J., Kyle, T. G., and Goldman, A. (1969). Variation of the infrared solar spectrum between 700 cm^{-1} and 2240 cm^{-1} with altitude. *Appl. Opt.* **8**, 2519–2536.

Murcray, D. G., Goldman, A., Williams, W. J. Murcray, F. H., Bonomo, F. S., Bradford, C. M., Cook, G. R., Hanst, P. L., and Molina, M. J. (1977). Upper limits for stratospheric $ClONO_2$ from balloon-borne infrared measurements. *Geophys. Res Lett.* **4**, 227–230.

Murcray, D. G., Goldman, A., Murcray, F. H., Murcray, F. J., and Williams, W. J. (1979). Stratospheric distribution of $ClONO_2$, *Geophys. Res. Lett.* **6**, 857–859.

Niki, H., Maker, P. D., Savage, C. M., and Breitenbach, L. P. (1977a). Fourier transform IR spectroscopic observation of pernitric acid formed via $HO_2 + NO_2 \rightarrow HOONO_2$. *Chem. Phys. Lett.* **45**, 564–566.

Niki, H., Maker, P. D., Savage, C. M., and Breitenbach, L. P. (1977b). Fourier transform IR spectroscopic observation of propylene ozonide in the gas phase reaction of ozone–cis-2-butene–formaldehyde. *Chem. Phys. Lett.* **46**, 327–330.

Park, J. H., Zander, R., Farmer, C. B., Rinsland, C. P., Russell, J. M., III, Norton, R. H., and Raper, O. F. (1986). Spectroscopic detection of CH_3Cl in the upper troposphere and lower stratosphere. *Geophys. Res. Lett.* **13**, 765–768.

Perner, D., and Platt, H. (1979). Detection of nitrous acid in the atmosphere by differential optical absorption. *Geophys. Res. Lett.* **6**, 917–920.

Pitts, J. N., McAfee, J. M., Long, W. D., and Winer, A. M. (1976). Long-path infrared spectroscopic investigation of ambient concentrations of the 2% neutral buffered potassium iodide method for the determination of ozone. *Environ. Sci. Technol.* **10**, 787–793.

Rinsland, C. P., and Levine, J. S. (1985). Free tropospheric carbon monoxide concentrations in 1950 and 1951 deduced from infrared total column amount measurements. *Nature (London)* **318**, 250–253.

Rinsland, C. P., Levine, J. S. and Miles, T. (1985). Concentration of methane in the troposphere deduced from 1951 infrared solar spectral. *Nature (London)* **318**, 245–249.

Rinsland, C. P., Zander, R., Farmer, C. B., Norton, R. H., Brown, L. R., Russell, J. M., III, and Park, J. H. (1986a) Evidence for the presence of the 802.7 cm^{-1} band Q branch of HO_2NO_2 in high resolution solar absorption spectra of the stratosphere. *Geophys. Res. Lett.* **13**, 761–764.

Rinsland, C. P., Zander, R., Brown, L. R., Farmer, C. B., Park, J. H., Norton, R. H., Russell, J. M., III, and Raper, O. F. (1986b). Detection of carbonyl fluoride in the stratosphere. *Geophys. Res. Lett.* **13**, 769–772.

Ritz, S., Hausmann, M., and Platt, U. (1993). An improved open-bath multi-reflection cell for the measurement of NO_2 and NO_3. In *Optical Methods in Atmospheric Chemistry* (H. I. Schiff and U. Platt, Eds.), SPIE **1715**, 200–211.

Roby, R. J., and Harrington, J. A. (1981). Organic nitrites in aged exhaust from alcohol-fueled vehicles. *J. Air Pollut. Control Assoc.* **31**, 995–996.

Scott, W. E., Stephens, E. R., Hanst, P. L., and Doerr, R. C. (1957). Further developments in the chemistry of the atmosphere. *Proc. Am. Pet. Inst., Sect. III* **37**, 171.

Sebacher, D. I. (1977). *A Gas Filter Correlation Monitor for CO, CH_4 and HCl*, NASA Tech. Pap. No. 1113. NASA Sci. Tech. Off., Washington, DC.

Smith, M. A. H., and Rinsland, C. P. (1985). Spectroscopic measurement of atmospheric HCN at northern and southern latitudes. *Geophys. Res. Lett.* **12**, 5–8.

Stephens, E. R., Hanst, P. L., Doerr, R. C., and Scott, W. E. (1956). Reactions of nitrogen dioxide and organic compounds in air. *Ind. Eng. Chem.* **48**, 1498–1504.

Stephens, E. R., Hanst, P. L., Doerr, R. C., and Scott, W. E. (1959). Auto exhaust: Composition and photolysis products. *J. Air Pollut. Control Assoc.* **8**, 333–335.

Tuazon, E. C., Graham, R. A., Winer, A. M., Easton, R. R., Pitts, J. N., and Hanst, P. L. (1978). A kilometer pathlength Fourier transform infrared system for the study of trace pollutants in ambient and synthetic atmospheres. *Atmos. Environ.* **12**, 865–875.

White, J. U. (1942). Long optical paths of large aperture. *J. Opt. Soc. Am.* **32**, 285–288.

White, J. U. (1976). Very long optical paths in air. *J. Opt. Soc. Am.* **66**, 411–416.

Williams, W. J., Kosters, J. J., Goldman, A., and Murcray, D. G. (1976a). Measurements of stratospheric halocarbon distributions using infrared techniques. *Geophys. Res. Lett.* **3**, 379–382.

Williams, W. J., Kosters, J. J., Goldman, A., and Murcray, D. G. (1976b). Measurements of stratospheric mixing ratio of HCl using infrared absorption techniques. *Geophys. Res. Lett.* **3**, 383–385.

Wright, N., and Herscher, L. W. (1946). Recording infrared analyzers for butadiene and styrene plant streams. *J. Opt. Soc. Am.* **36**, 195–202.

Zander, R., Rinsland, C. P., Farmer, C. B., Brown, L. R., and Norton, R. H. (1986). Observation of several chlorine nitrate ($ClONO_2$) bands in stratospheric infrared spectra. *Geophys. Res. Lett.* **13**, 757–760.

Zimmerman, P. R., Chatfield, R. B., Fishman, J., Crutzen, P. J., and Hanst, P. L. (1978). Estimates on the production of CO and H_2 from the oxidation of hydrocarbon emissions from vegetation. *Geophys. Res. Lett.* **5**, 679–681.

CHAPTER

7

MATRIX ISOLATION SPECTROSCOPY IN ATMOSPHERIC CHEMISTRY

DAVID W. T. GRIFFITH

Department of Chemistry,
University of Wollongong,
Wollongong NSW 2522, Australia

7.1. INTRODUCTION

In 1954, Whittle, Dows, and Pimentel at the University of California, Berkeley, introduced the technique of matrix isolation (MI) to the laboratory spectroscopist. The technique was based on trapping the target chemical species in a cold, inert, rigid, and transparent matrix, and provided a means by which unstable or reactive chemical species could be isolated and held for minutes to hours in the laboratory for a detailed and relatively leisurely spectroscopic analysis. Matrix isolation made the spectroscopic analysis of many free radicals and other normally short-lived species accessible for the first time and has grown to become an important spectroscopic technique. It has been combined with most types of spectroscopy, such as infrared (IR), Raman, ultraviolet (UV)/visible, electron spin resonance (ESR), and Mossbauer, of which the IR type has been the most widely used. The great majority of matrix isolation studies have been carried out on radicals, high-temperature species, weakly bonded complexes, intermediates, and other labile species in matrices of argon or nitrogen below 20 K. There is now a very large literature of matrix isolation spectroscopy, including a number of comprehensive reviews (Milligan and Jacox, 1972; Downs and Peake, 1973; Chadwick, 1975, 1979), several books (Meyer, 1971; Hallam, 1973; Barnes et al., 1981; Almond and Downs, 1989), and a continuing series of biannual international conferences.

The use of matrix isolation spectroscopy as a quantitative method for the analysis of mixtures of stable gases was first reported by Rochkind (1967,

Air Monitoring by Spectroscopic Techniques, Edited by Markus W. Sigrist. Chemical Analysis Series, Vol. 127.
ISBN 0-471-55875-3

1968a,b, 1971), who used matrix isolation in N_2 at 10 K as a sampling method for the infrared analysis of multicomponent mixtures of small hydrocarbons. The lack of spectral congestion in the MI compared to gas–phase spectra allowed the simultaneous analysis of a wide range of small hydrocarbons and their isotopes with good sensitivity. An independent study also showed similar potential for the analysis of airborne amines (Ball and Purnell, 1976). At a time when Fourier transform infrared (FTIR) spectrometers with high resolution and sensitivity were not yet readily available, these studies showed that matrix isolation offered significant advantages over gas–phase IR spectroscopy for multicomponent gas analysis.

The ability of matrix isolation spectroscopy to trap labile chemical species and to be a quantitative analytical technique suggested its potential suitability for atmospheric trace gas analysis. In the 1970s, the threat of nitrogen oxides and chlorine containing compounds to the stratospheric ozone layer were recognized (Crutzen, 1970; Johnson, 1971; Molina and Rowland, 1974) and a major international effort was made to understand the chemistry of ozone and the stratosphere. Many of the key species in stratospheric chemistry remained either undetected or at least only poorly quantified, yet detailed measurements were required to validate the chemical models of the stratosphere. This was particularly true for free radicals, such as OH, HO_2, NO, and NO_2, and the so-called reservoir species, such as $ClONO_2$, HO_2NO_2, and H_2O_2. Matrix isolation promised to be a very valuable technique in this area.

The first attempts to apply matrix isolation sampling and spectroscopy directly to atmospheric trace gas analysis were made by Snelson (1974, 1975). A modified commercial closed-cycle refrigerator was installed in a high-altitude military aircraft to collect whole air samples at low temperature (13 K) at altitudes of 12–18 km. After the sampling flight, the cryostat was transported to the laboratory under power to maintain the low temperature and various analyses were carried out on the frozen sample. In the earlier work, whole-air samples of 0.5–4 L were frozen onto a sapphire cold-finger in a cavity that could be mated to an ESR spectrometer in the laboratory for the analysis of free radical species (NO_2, OH, and HO_2). In laboratory tests, however, ESR spectra of NO_2 were found to be so broadened by molecular oxygen, itself paramagnetic and constituting 21% of the matrix, that signals due to NO_2 could not be observed. Although MI–ESR measurements of the cryogenically collected air samples were actually not made, the samples were nevertheless warmed up and analyzed for stable gases by gas chromatographic (GC) techniques.

In the second incarnation of this sampler, the sapphire cold-finger of the same cryogenic sampler was replaced with a CsBr window suitable for optical spectroscopy (Snelson, 1975). Laboratory tests of infrared spectra of H_2O, CO_2, O_3, and CH_4 in air matrices at 13 K indicated sub-ppm sensitivity for all four gases, which was sufficient for H_2O, CO_2, and CH_4 at atmospheric levels, but not for O_3. Unfortunately, owing to the lack of aircraft availability, the cryostat was never flown in this configuration and thus never tested. The principle, nevertheless, was established.

Since that time, both ESR- and FTIR-based matrix isolation spectroscopy have been successfully developed and used in atmospheric trace gas analysis by Mihelcic, Helten, and co-workers (ESR), and Griffith and co-workers (FTIR). In both cases the key to success has been the adoption of a nonconventional matrix gas: H_2O or D_2O in the case of ESR studies, and CO_2 for FTIR. These studies will form the main part of this chapter and are described in detail in Sections 7.3 and 7.4 after a general description of matrix isolation spectroscopy (Section 7.2). Section 7.5 briefly describes further applications of MI–FTIR as a GC detector for polluted air analyses, and in studies of the kinetics and product distributions of reactions of atmospheric importance.

7.2. MATRIX ISOLATION SPECTROSCOPY

Matrix isolation spectroscopy has been described and reviewed in detail in several books and review articles (see Section 7.1). This chapter therefore describes only the salient features, especially as they relate to applications in atmospheric chemistry, and the interested reader is referred to the existing works for more detailed information.

In matrix isolation spectroscopy, the chemical species to be investigated are isolated in a solid diluent matrix at low temperature. The matrix should be inert so that it does not react with the sample, rigid so that it does not allow diffusion of sample molecules through the matrix, and transparent to the radiation used. For the common case of optical spectroscopy, molecular nitrogen and the rare gases have been the matrices of choice because they show no absorption throughout the IR, visible, or UV regions and are chemically inert. Their main disadvantage is that temperatures below 20 K, and preferably below 10 K, are required to ensure rigidity and to prevent diffusion. A liquid-helium-cooled cryostat is therefore required.

For effective matrix isolation, the ratio of sample to matrix molecules should normally be at least 1:100, and preferably 1:1000, which as a rule of thumb separates sample molecules by an average of 10 matrix molecules in each direction. Matrices may be grown by slowly spraying a mixture of sample and matrix onto a cooled window, mirror, or other substrate over a period of minutes to hours or in a series of short pulses (Perutz and Turner, 1973; Chadwick, 1975). There are advantages to both methods, but the deposition techniques remain something of an art form. For stable molecules, sample and matrix gases can be premixed before deposition. Radicals, high-temperature molecules, and other labile species must be produced in a furnace, flow tube, or other dynamic system, dynamically diluted, and flowed onto the cooled matrix substrate within the chemical lifetime of the species. Alternatively, matrix-isolated species may be produced by in situ photochemical reactions in the matrix.

The advantages of matrix isolation lie both in the sampling method and in the spectroscopy.

Sampling Advantages

- Each sample molecule is in a chemically stable and inert environment, even when it may itself be very reactive. Once grown, the matrix is stable for many hours and amenable to detailed spectroscopic analysis. For short-lived species, this may offer the only practicable means to study their spectra.
- Several potentially reactive species can be studied in a single sample matrix because they are isolated and do not chemically interact.
- There are only weak interactions between sample molecules and their nearest neighbors. Intermolecular interactions that complicate the spectra, especially in pure crystals, are minimized or completely absent. In optical spectra there is normally a small matrix or solvent shift in the frequencies of absorption bands relative to their gas phase values.
- If an air sample is collected by condensation into a matrix near 77 K, the most abundant atmospheric components, N_2, O_2, and Ar, are not collected, resulting in a large preconcentration of the trace species under investigation.

Spectroscopic Advantages

- Molecular rotation, which occurs freely in the gas phase, is usually constrained by the rigid matrix. In gas-phase vibrational spectra a single molecular vibration normally appears as a structured band of rotational lines tens to hundreds of wavenumbers wide. In MI spectra this structure is absent and a single vibration normally appears as a single absorption line. Depending on the sample and matrix molecules, infrared absorption lines of matrix-isolated species vary in width from less than 0.1 up to several wavenumbers, but are typically less than one wavenumber wide. In ESR spectra, free rotation in the gas phase smears out dipole–dipole interactions and thus the hyperfine structure of the spectra which allows differentiation of radicals present in the sample. The lack of rotational structure significantly reduces congestion in multicomponent spectra. Thus MI spectra may yield information about many species in a single sample.

- Spectra are further simplified by the absence of temperature-dependent absorption bands, which are precluded by the low temperature of the matrix.

Figure 7.1 illustrates the features of matrix isolation with a simple example from infrared spectroscopy. Trace (a) in Fig. 7.1 shows the absorbance spectrum of N_2O in the gas phase near the ν_1 symmetric stretching band centered at 1284.9 cm^{-1}. This band displays the classic shape of a parallel band of a linear molecule and has a total width on the order of 100 cm^{-1}. Trace (b) shows the spectrum of N_2O isolated in a CO_2 matrix at 77 K at a matrix ratio of 1000:1, approximately the same ratio as occurs in clean air. The rotational structure of the gas phase spectrum is clearly absent in the MI spectrum. The ν_1 vibration of the parent isotope of N_2O appears as the very strong absorption line at 1296.9 cm^{-1}, a matrix shift of 12 cm^{-1} relative to the gas phase. In addition there are several other features in the spectrum, mostly due to naturally occurring isotopes. Lines at 1257.8, 1281.5, and 1292.4 cm^{-1} are the ν_1 vibrations of the isotopomers of N_2O, $^{14}N^{14}N^{18}O$, $^{15}N^{14}N^{16}O$, and $^{14}N^{15}N^{16}O$, respectively. The sharp lines at 1253.3 and 1264.6 cm^{-1} are due to the ν_1 vibrations (in Fermi resonance with $2\nu_2$) of the isotopically unsymmetrical $^{16}O^{12}C^{18}O$ and $^{16}O^{12}C^{17}O$, respectively. This vibration is IR

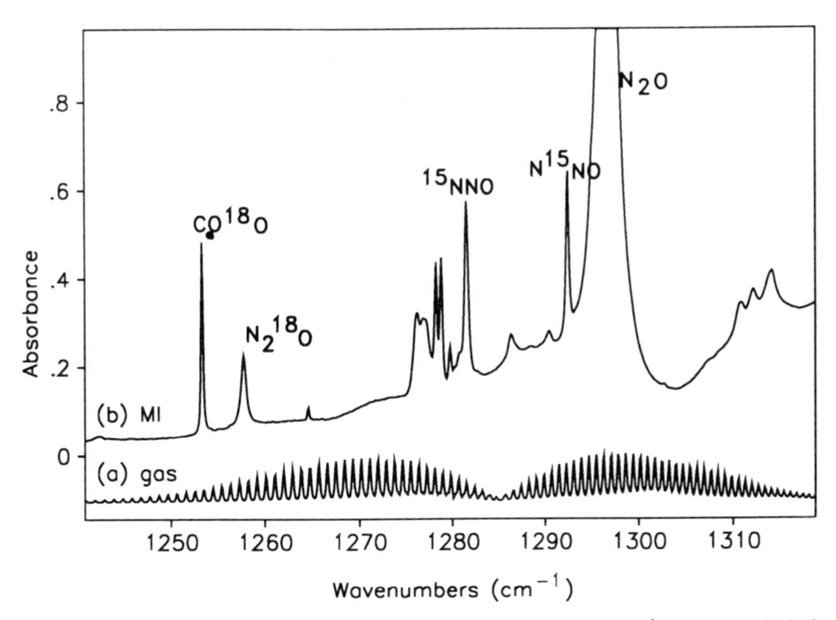

Figure 7.1. (a) Gas phase spectrum of N_2O near the ν_1 band at 1284 cm^{-1}. (b) Matrix isolation spectrum of 1:1000 N_2O in CO_2 in the same region. See text for discussion.

forbidden in the parent CO_2 because of symmetry, but allowed weakly in the unsymmetrically substituted isotopomers. The complex band just below 1280 cm^{-1} is, however, due to the parent CO_2. Although still forbidden by symmetry in a perfect crystal, this band becomes weakly allowed for CO_2 molecules that are in an unsymmetrical environment, such as next to a guest molecule or at a crystal defect. As a result, the intensity of this band is sensitive to the total amount of other trace gases in the matrix. Finally, the broad absorption above 1300 cm^{-1} is the onset of absorption due to lattice vibrations of the CO_2 crystal in combination with the (forbidden) ν_1 intramolecular vibration.

This spectrum illustrates several of the advantages of matrix isolation infrared spectroscopy in trace gas analysis. First, the lack of congestion and the ability to easily distinguish adjacent lines are evident. Secondly, the sensitivity of the method can be easily assessed as follows. The spectrum in Fig. 7.1 can be routinely obtained from the CO_2 and N_2O molecules contained in about 12 L [at STP (standard temperature and pressure)] of clean air, approximately 200 μmol CO_2 and 200 nmol N_2O. The ^{15}N isotope has a natural abundance of 0.4‰, so the amounts of the ^{15}N isotopomers are correspondingly 800 pmol. Alternatively, in terms of mixing ratios, N_2O has a clean-air atmospheric mixing ratio of approximately 300 ppbv [parts per billion (10^9) by volume], and the ^{15}N isotopic species correspondingly 1.2 ppbv (Houghton et al., 1990). The noise level indicates that the minimum detectable amount of ^{15}NNO would be about 100 times less than these figures, that is, about 12 pptv [parts per trillion (10^{12}) by volume], or 8 pmol.

7.3. MI–FTIR SPECTROSCOPY IN ATMOSPHERIC TRACE GAS ANALYSIS

The infrared region is particularly well suited to atmospheric trace gas analysis because most atmospheric gases exhibit infrared spectra, and modern FTIR instruments are sensitive, accurate, and precise in quantitative analysis. Direct measurements of the FTIR absorption or emission spectra of the atmosphere have indeed been a very powerful tool in atmospheric composition measurement and are the subject of Chapter 6 of this volume. Spectra taken over atmospheric paths ranging from a few centimeters to the full depth of the atmosphere, from the ground, aircraft, balloons, and spacecraft, have been studied and, in those spectral regions where the atmosphere is not opaque owing to water, CO_2, or other absorption, are extremely valuable. These remote sensing methods have many advantages, the main one being that the air to be analyzed need not be collected. In one important respect, however, they can be limited: the infrared absorption by many important trace species is weak, even over the full depth of the atmosphere, and sensitivity can be insufficient. Normal detection limits are on the order of tens of ppbv for ground-based long-path FTIR spectroscopy and hundreds of pptv for solar measurements.

The spectra of Fig. 7.1, on the other hand, illustrate the potentially high sensitivity of MI–FTIR spectroscopy for atmospheric trace gas analysis, in the low-pptv range. In essence, the MI–FTIR technique involves three main steps:

1. Cryogenic collection of air samples near 77 K so that all condensable gases, but not N_2, O_2, or Ar, are frozen out in the sampler
2. Recording of the FTIR spectrum of the sampled trace gases in a matrix of the naturally occurring atmospheric CO_2
3. Quantitative analysis of the resultant FTIR spectra based on laboratory calibration spectra

The key to sensitivity is in using the naturally present atmospheric CO_2. Most trace gases are present in much lower concentrations than CO_2 and are therefore effectively diluted and isolated in the CO_2 matrix. The main features of the technique using natural CO_2 as the matrix gas can be summarized as follows:

- *Preconcentration.* CO_2 has a mixing ratio of about 350 ppmv in clean air (Houghton et al., 1990). If atmospheric samples are collected by passing air through a trap at 77 K (or higher), the main atmospheric constituents N_2, O_2, and Ar are not collected, and after water vapor is removed or reduced, CO_2 is the most abundant species. This represents a preconcentration of the trace gases in the sample of a factor of $1/(350 \times 10^{-6})$, or about 2900:1. Thus a species with a mixing ratio of 1 ppbv in clean air is concentrated to 3 ppmv in CO_2, a much more accessible concentration range in which to analyze and calibrate gas mixtures. After CO_2, the most abundant condensable trace gas in clean air is N_2O, with a mixing ratio of about 300 ppbv (Houghton et al., 1990). N_2O is thus diluted in CO_2 by more than 1000:1, sufficient for good matrix isolation. Other trace gases are normally more diluted, and good matrix isolation is achieved despite the high degree of preconcentration.
- *Temperature.* Liquid nitrogen (77.4 K) or liquid argon (87.6 K) temperature is sufficient to completely trap CO_2 and most trace gases, while not trapping the major species, N_2, O_2, and the rare gases. CH_4, H_2, CO, NO, and O_3 are condensed either partially or not at all. CO_2 matrices at 77 K are rigid, stable, and show no evidence of diffusion or reaction of embedded trace gas species, although there may be some diffusion and

guest–guest interactions during the matrix growth from the gas phase. The experimental and cost advantages of working with liquid nitrogen rather than liquid helium as cryogen are considerable, especially in fieldwork situations.

- *Internal Calibration.* Unlike the traditional matrix gases such as argon and nitrogen, CO_2 shows some absorption in the infrared. This absorption can be turned to advantage because suitable weak CO_2 absorption lines can be used as an internal reference to determine the effective thickness of the matrix for quantitative analysis. This dispenses with the need to measure the amount of air collected in a sample precisely.
- *Transparency.* IR absorption due to CO_2 makes some parts of the spectrum unusable for analysis, but fortunately this is not a severe problem. Figure 7.2 shows a survey mid-IR spectrum of a 400-μm-thick matrix of CO_2 at 77 K, typical of the thickness used in atmospheric trace gas analysis. There is strong absorption below 800 cm^{-1} (ν_2 band), from 2200 to 2400 cm^{-1} (ν_3), and above 3600 cm^{-1} ($\nu_1 + \nu_3$, $2\nu_2 + \nu_3$). There are also strong, narrower absorption regions near 2000 cm^{-1} due to the $\nu_1 + \nu_2$ and $3\nu_2$ combination bands. Few other molecules, however, show fundamental absorptions in these regions, except below 800 cm^{-1}, so that CO_2

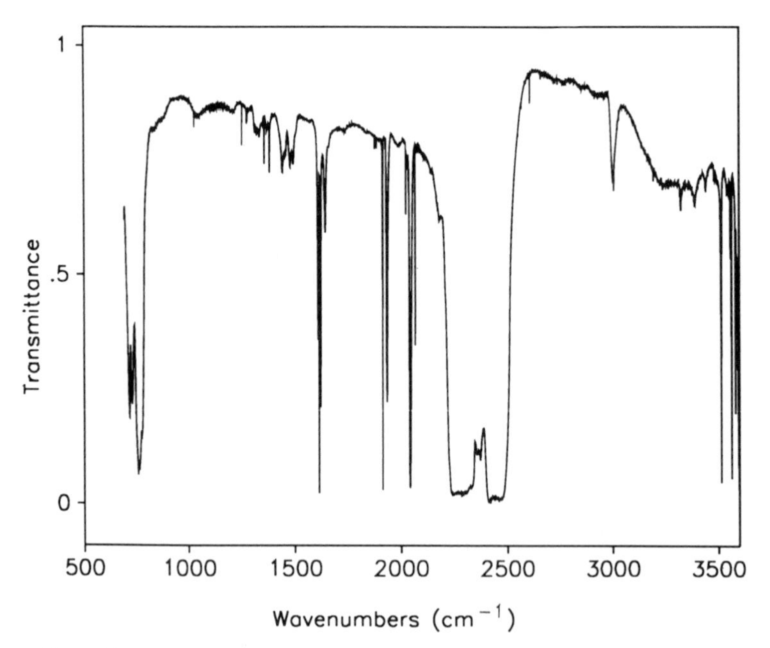

Figure 7.2. Transmittance spectrum of a 400-μm-thick matrix of CO_2 at 77 K.

absorptions do not present a major limitation. Other absorptions due to lattice vibrations from 1300 to 1500 cm^{-1} (Jacox and Milligan, 1961; Mannik and Allin, 1972) are weak and broad. The relatively sharp features of matrix-isolated species on top of this absorption can be readily distinguished, so this part of the spectrum is still usable for trace gas analysis.

- *Sampling.* The principal disadvantages of the method relate to the need to collect and dry samples of air to be analyzed. In collecting and drying sampled air, chemical changes may occur, some species may be lost, and others produced. These problems are common to any sampling-based analytical technique and are described in more detail as necessary below.

We have developed MI-FTIR spectroscopy under the acronym MISST (matrix isolation sampling of the stratosphere and troposphere), with the motto

Wer mit MISST misst, misst keinen Mist.[1]

We have applied MISST in several areas of atmospheric trace gas analysis from ground level to the lower stratosphere. The method has been described in some detail (Ihrig, 1985; Griffith and Schuster, 1987; Berger et al., 1989), and the interested reader is referred where necessary to these earlier works. In what follows, the main features of the technique, advances since the original publications, modifications and variations for specific purposes, and examples of new applications are presented.

7.3.1. Sample Collection

7.3.1.1. Direct Sample Collection and Measurement

The most desirable sampling method would be to trap the condensable gases directly from the air into a dry CO_2 matrix at 77 K in the field and measure the FTIR spectrum of the collected matrix without further sample treatment. This direct method would minimize chemical interactions in the sample and maximize the integrity of the analysis, as proven in the MI–ESR studies that will be described in Section 7.4. This approach has met with difficulties for FTIR work, however, and has not to date proved feasible for trace gas analysis with useful detection limits. The difficulties stem first from the necessity to reduce the water vapor mixing ratio in the sample air to below 1 ppmv, which

[1]He who measures with MISST, measures no garbage!

is necessary to obtain CO_2 matrices of good optical quality; secondly, good-quality CO_2 matrices have not been obtained when the CO_2 is frozen directly from the large excess of N_2, O_2, and Ar in a whole-air stream.

Ihrig (1985) described an in situ air-sampling and -drying system and a portable liquid nitrogen cryostat for direct sampling and MI–FTIR analysis. The sampled air stream was dried by being passed in series through two cylindrical drying tubes that relied on diffusion of water vapor to the walls. These were designed to dry the air stream with minimum wall contact before deposition in the cryostat. The inside surface of the first tube was lined with silica gel packed behind wire mesh. The walls of the second tube were held at approximately $-100\,^{\circ}C$. A heated cylindrical concentric wire mesh inside the second tube ensured that the air did not become supercooled and allow water vapor to condense into aerosol form. At flow rates below $1\ L \cdot min^{-1}$ water mixing ratios below 1 ppmv could be achieved by this combination.

The dried air was then introduced through a glass jet onto a liquid-nitrogen-cooled gold mirror surface in a portable vacuum cryostat, so that a matrix of condensable gases, mainly CO_2, was formed on a small area of the cold mirror. Matrices up to 600 μm thick could be grown over times of minutes to hours, although sticking efficiencies of only about 30% were obtained. The cryostat design allowed 36 such sample matrices to be collected by rotating successive clean facets of the disk-shaped cold mirror to the inlet jet. After such a set of samples had been collected, the cryostat was coupled to an FTIR spectrometer with suitable transfer optics and the absorption spectra of the matrices were recorded by reflection–absorption spectroscopy. The system was used both in ground level and laboratory tests and in a short series of flights in the upper troposphere and lower stratosphere.

Despite varying deposition parameters such as the temperature, deposition rate, and deposition gas pulsing patterns over a very wide range, matrices of good optical quality were never obtained. The matrices always displayed poor transmission, strong optical scattering, and broadened, asymmetric absorption line shapes, in direct contrast to the high-quality matrices obtained by the two-step method described below. The poor optical quality most likely stems from the microcrystalline structure of the CO_2 matrix formed in the presence of a large excess of noncondensable N_2, O_2, and Ar in the whole-air stream. These excess gases bring with them a large thermal load that cannot be dissipated quickly through the growing matrix and cryostat. As a result, the growing matrix surface warms significantly, allowing exchange of CO_2 with the gas phase and migration of trace gases at the growing surface. This accounts for the low sticking efficiency. Secondly, a significant amount of the noncondensable gases may be trapped as impurities by the growing CO_2 matrix, resulting in many imperfections in the solid structure. Regardless of whether the trapped gases eventually escape and are pumped away, the crystal imperfections may remain, and no amount of annealing produces an optically better crystal. It is not clear whether the CO_2 matrices are microcrystalline, glassy, or a mixture of both.

7.3.1.2. Two-Step Sample Collection

The difficulty of direct sampling for MISST analysis has meant that all measurements to date have been made with a two-step sampling procedure. This involves collection of the sample in a cryosampler in the field, followed by gas-phase transfer and recondensation of the sample as a transparent matrix in the measurement cryostat in the laboratory. This procedure allows the growth of CO_2 matrices of good optical quality well suited to FTIR analysis, but allows the possibility of chemical changes in the sample while it is in the gas phase. Removal of water is necessary from all samples, and this may also cause changes in sample composition.

The cryosampler is coil made from 1.5–2.5 m of 8-mm-o.d. glass tubing closed by two greaseless glass/Teflon stopcocks immersed in liquid nitrogen or argon. Air samples of 10–80 L are typically drawn through the coil at 3 $L \cdot min^{-1}$ with an oil-free membrane pump, such that all condensable gases are trapped in the coil. the sample coils are then returned to the laboratory, where the sample is evaporated and transferred to the measurement cryostat as described below.

Water content must be reduced in the sample to a level well below 1% of CO_2; otherwise matrices that both scatter and absorb IR radiation strongly are obtained. This corresponds to an effective water mixing ratio in air of much less than 3.5 ppmv. In the upper troposphere and stratosphere, H_2O mixing ratios are on the order of 3–5 ppmv. At altitudes below about 12 km, water content increases strongly to be on the order of 1% at the earth's surface. The drying technique employed thus depends on the application and region of the atmosphere being studied. For upper tropospheric and stratospheric samples only minimal drying is necessary and is simply achieved by allowing the sampling coil to warm to only $-50\,°C$ during transfer of the sample to the measurement cryostat (Wilson et al., 1988). The residual water remains trapped by the low temperature in the coil. In the lower troposphere, this technique alone is not sufficient because of the large excess of water vapor over CO_2. The sampled air may be dried before collection, then transferred for analysis as described above. Various chemical drying agents such as Drierite™ ($CaSO_4$), $MgClO_4$, or P_2O_5 or a cold trap can be used. Drying increases the risk of chemical change to the sample and losses of some species, and must be tested for each species analyzed. Alternatively, the whole-air sample may be collected without predrying. In this case, the sample is transferred to the measurement cryostat via a low temperature trap below $-50\,°C$. Careful control of the transfer procedure and temperature results in suitably dry samples for FTIR analysis.

Any ozone present in the air is also partially trapped in the cryosampler and, as a strong oxidant, can cause chemical changes in the sample during gas-phase transfer. In the upper troposphere and lower stratosphere, ozone levels are relatively high (>100 ppbv) and the problem is potentially more serious. Ozone trapping is significantly reduced by using liquid argon as cryogen during sample collection. The vapor pressure of ozone is 0.0018 torr at 77.4 K (liquid nitrogen) and 0.044 torr at 87.6 K (liquid argon) (Hanson and Mauersberger, 1986); ozone would not be trapped at all if the sample collection were a thermodynamically controlled process. It is also possible to scrub the ozone from the sampled gas stream with agents such as silver wool, sodium sulfite, or potassium iodide, but at the risk of causing other chemical changes to the sample.

7.3.2. Matrix Growth and FTIR Spectrum Measurement

After collection and return to the laboratory, the sample must be transferred to the measurement cryostat and its FTIR absorption spectrum recorded. We have used two optical configurations for measurement of the spectrum—simple reflection–absorption spectroscopy and, for lower detection limits, an integrating sphere.

7.3.2.1. Reflection–Absorption Spectroscopy

In reflection–absorption spectroscopy (Griffths and de Haseth, 1986) the sample is deposited on a mirror surface and the FTIR beam is reflected from the mirror, thus double-passing the sample and producing an absorption spectrum. The reference spectrum is the spectrum of the mirror without sample. We have improved on our original design for a reflection–absorption cryostat suitable for MI spectroscopy (Griffith and Schuster, 1987) with a more compact cryostat and transfer optics shown schematically in Fig. 7.3. In this cryostat, the IR beam from the interferometer is both focused onto the sample cold mirror and collected from it by a single ellipsoidal mirror. The sample location on the cold mirror is coincident with one focus (F2) of the ellipsoidal mirror, and the other focus (F1) is located in the spectrometer sample compartment. Transfer optics divert the normal sample compartment beam, refocus it onto an aperture located at F1 of the ellipsoid, and collect it after transmission through the cryostat. The alignment is very stable because of the property of an ellipsoid that all light passing through one focus must pass through the other and vice versa. Thus all input rays passing through F1 must ultimately return through F1 after reflection at the mirror at F2. The

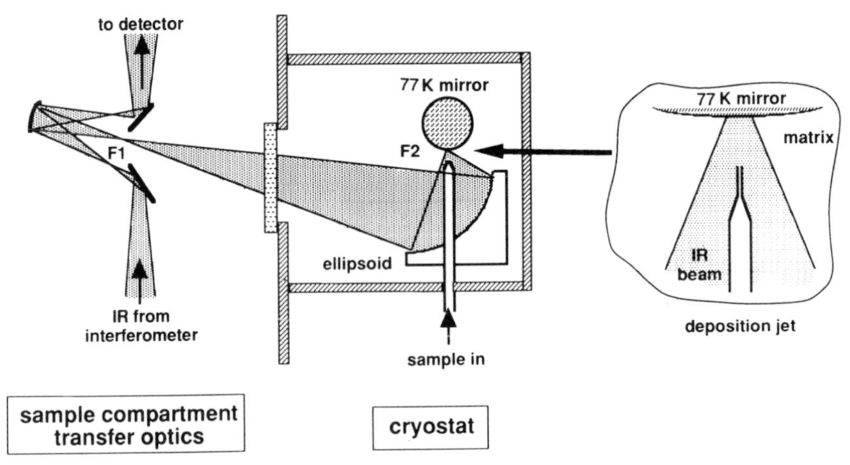

Figure 7.3. Schematic diagram of a liquid-nitrogen-cooled cryostat and beam-reducing optics suited to matrix isolation studies.

input and return beams are separated in the vertical plane, and the transfer optics are symmetrical with respect to input and output beams. There is an effective beam reduction of approximately 6 × at the cold mirror focus F2. The cryostat is permanently mounted to the side plate of a Bomem DA3 FTIR spectrometer; the transfer optics are kinematically mounted in the sample compartment and can be easily removed and replaced. The transmission of the whole system is about 50%.

The cryogenically collected sample is evaporated to room temperature, transferred quickly through a greaseless glass vacuum manifold via a cold trap to trap water vapor if necessary, into a 1-L glass deposition bulb. A typical clean atmospheric sample of 15 L at STP contains about 4 torr·L of CO_2. The sample is deposited from the deposition bulb in a single pulse through a 6-mm-o.d. glass tube with a 1-mm-i.d. end constriction onto the cold mirror in the cryostat precisely at the focus F2 (Fig. 7.3). Typically 3 torr·L of CO_2 are deposited in 90 s to form a transparent matrix 300 μm thick at the center and 1–2 mm in diameter. The FTIR beam is focused to less than 1 mm diameter at the center of the matrix. Samples can thus be deposited and spectra recorded without any need for movement of the cryostat, deposition jet, cold-finger, or optics, leading to stable spectral baselines and reproducible spectra. A helium–neon laser directed at the center of the growing matrix can be used to measure growth rate and matrix thickness directly by counting interference fringes as the matrix grows (Groner et al., 1973; Griffith and Schuster, 1987).

Spectra are normally recorded with a globar source, Ge–KBr beamsplitter and narrow-band HgCdTe(mercury–cadmium–telluride) detector (cutoff:

$>700\,cm^{-1}$) from 800 to $3600\,cm^{-1}$ at $0.2\text{-}cm^{-1}$ resolution. One hundred scans require approximately 10 min and result in a noise level of less than 0.0001 absorbance units (base *e*), as illustrated in Fig. 7.1.

7.3.2.2. An Integrating Sphere

We have also used an integrating sphere for the measurement of absorption spectra, a technique which has two major advantages over simple reflection–absorption spectroscopy (Berger, 1986; Berger et al., 1989): first, the IR beam passes through the sample matrix many times, resulting in enhanced absorption; secondly, the matrix can be grown much more rapidly, minimizing chemical changes in the sample during the time it is in the gas phase.

The integrating sphere cryostat is depicted in Fig. 7.4. The cold end of the cryostat is a copper block containing a gold-coated, polished spherical cavity 12 mm in diameter with three orthogonal apertures. One aperture is the entrance aperture for the IR beam, one is for the exit beam, and one is for introduction of the sample gas. An HgCdTe detector chip, cooled by thermal contact with the main copper block, is mounted immediately in front of the exit aperture and intercepts the strongly divergent exit beam. The cryostat thus contains its own detector and is mounted at the detector position of the FTIR spectrometer with the beam focused into the entrance aperture. The sample to be analyzed is transferred in a few seconds from the gas phase into the sphere, resulting in a thin matrix film over the entire surface of the sphere. Any optical scattering by the fast-grown matrix is less important than in the reflection–absorption technique because scattered light remains in the sphere

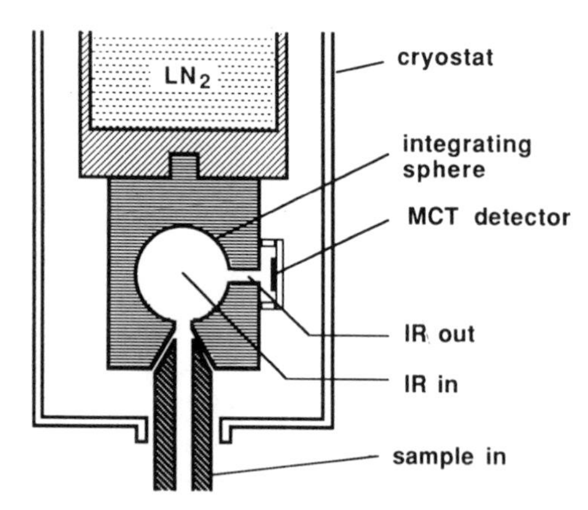

Figure 7.4. Schematic diagram of an integrating sphere cryostat and detector for matrix isolation studies. (The IR entrance aperture is at right angles to the page; MCT, mercury–cadmium–telluride; LN_2, liquid nitrogen.)

and is not lost from the optical system. The spectra obtained do not show evidence of detrimental light scattering; line shapes are symmetrical, as they are for reflection–absorption spectra, and there is little falloff in the 100% line at high frequency.

The cryostat can be cooled to below 70 K by pumping on the liquid nitrogen cryogen. Deposition at below 70 K reduces migration and aggregation of trace gases during matrix growth, in particular with water molecules. Increased scattering in matrices grown below 70 K precludes using this technique in the reflection–absorption cryostat.

A simple optical description of the integrating sphere shows that the effective absorbance due to sample absorption is given by

$$A_{\mathrm{eff}} = \log\left(\frac{1-(1-h)\rho t}{(1-(1-h)\rho)t}\right) \tag{7.1}$$

where h is the fractional area of the sphere's surface occupied by exit and entrance apertures; ρ is the internal surface reflectivity; and t is the transmittance of the sample matrix for a single pass. The effective absorbance is enhanced over that for a single reflection, $-\log(t)$, by a factor equal to the average number of reflections of the beam before leaving the sphere. This number is smaller for wavelengths that are strongly absorbed, because there is little intensity remaining after only a few reflections. Conversely, the average number of reflections and the enhancement of absorbance is larger at wavelengths that are weakly absorbed, because the intensity is diminished less per reflection and on average a larger number of reflections can occur. Thus weak absorption lines are more strongly enhanced than strong lines, and the sensitivity to trace absorptions is correspondingly increased. Enhancement of a factor of 10–20 has been demonstrated for weak absorption lines.

The effective absorbance is thus not proportional to concentration of absorber; that is, the Beer–Lambert law is not obeyed. However Eq. (7.1) provides a basis for linearising the absorbance with respect to concentration. Inversion of Eq. (7.1) yields

$$A_1 = \log(1 + \beta \cdot 10^{A_{\mathrm{eff}}}) - \log(1 + \beta) \tag{7.2}$$

where A_1 is the absorption for a single pass of the matrix ($= 10^{-t}$) and $\beta = [(1/(1-h)\rho)] - 1$; A_1 varies linearly with concentration of absorber according to the Beer–Lambert law.

The integrating sphere is most useful for measuring trace species in clean air when the highest sensitivity and minimum sample perturbation are required. It has thus been used mainly in studies of the upper troposphere/lower stratosphere (Wilson et al., 1988, 1989; Helas et al., 1989), and of clean air from ground-based clean-air stations (S. R. Wilson and D. W. T. Griffith, unpublished). To improve sensitivity, larger samples can be used than for reflection–absorption (90-L air samples are typical), leading to a matrix thickness of ca. 70 μm in the sphere without significant detrimental effects of scattering or matrix quality on the spectrum.

7.3.3. Calibration and Quantitative Analysis

In reflection–absorption measurements, the Beer–Lambert law is well obeyed for weaker CO_2 and trace gas absorption up to matrix thicknesses of at least 500 μm (Griffith and Schuster, 1987). The Beer–Lambert law is the basis for calibration and is applied in the form

$$A = \alpha \cdot \mu \cdot d \tag{7.3}$$

where A is the integrated absorbance of a trace gas absorption line in cm^{-1}; μ is the mixing ratio of absorber in CO_2 in the matrix in ppm (parts per million, by mole); d is the effective thickness of the matrix to the IR beam in micrometers; and α is the integrated absorption coefficient in $cm^{-1} \cdot \mu m^{-1} \cdot ppm^{-1}$. Trace gas concentration measurements are thus determined as mixing ratios relative to CO_2; to calculate the mixing ratios in air, the mixing ratio of CO_2 in the sampled air must be known.

The thickness of the matrix is determined from weak absorption lines of CO_2. This dispenses with the need to measure the volume of air collected accurately and is an important advantage of using CO_2 as the matrix. The mixing ratio of CO_2 in the matrix is effectively unity; so from Eg. (7.3) we obtain

$$A_{CO_2} = \alpha_{CO_2} \cdot d \tag{7.4}$$

with α_{CO_2} in units of $cm^{-1} \cdot \mu m^{-1}$. Table 7.1 lists absorption coefficients for seven absorption lines of solid CO_2 suitable for thickness measurement from 50 to 500 μm. (All absorbances are calculated using base e logarithms.) The absorption coefficients of the lines at 1253.3, 1358.6, and 2065.7 cm^{-1} were determined initially from integrated areas of absorption lines and direct measurement of matrix thickness by He–Ne laser interference (Griffith and Schuster, 1987). We have subsequently made a large number of measurements of line areas in the cryostat shown in Fig. 7.3 without making direct laser thickness measurements. Values from a sample set of 34 spectra are reported in Table 7.1. For each line, the mean ratio and 95% confidence limit of the line area to that of the line at 1358.6 cm^{-1} are given in columns 2 and 3. The line at 1358.6 cm^{-1} is arbitrarily selected as the reference and its absolute value of α, 1.65×10^{-4} $cm^{-1} \cdot \mu m^{-1}$, is taken from Griffith and Schuster (1987). The

Table 7.1. Integrated Absorption Coefficients of Seven Absorption Lines of Solid CO_2

Line (cm^{-1})	Mean Ratio Relative to 1358.6 cm^{-1}	95% Confidence Limit	Absorption Coefficient α ($cm^{-1} \cdot \mu m^{-1}$)
1253.3	0.557	0.0050	9.52×10^{-5}
1358.6	—	—	1.65×10^{-4}
1876.0	0.328	0.0051	5.40×10^{-5}
1885.6	0.200	0.0038	3.31×10^{-5}
1913.7	35.6	0.51	5.87×10^{-3}
1934.3	143.1	1.39	2.36×10^{-2}
2065.7	7.71	0.056	1.27×10^{-3}

absorption coefficients α in column 4 are then calculated from the ratios and this absorption coefficient of the 1358.6 cm^{-1} line. Measurements of the relative strengths of lines are quite precise, with 95% confidence limits on the order of 1–2%.

Calibration for a trace gas is essentially the determination of the absorption coefficients of suitable absorption lines in the spectrum. Mixtures of trace gas in CO_2 are made up manometrically on a greaseless vacuum line, typically in the 0.1–10 ppm range, depending on atmospheric levels to be analyzed. Spectra are recorded for a range of mixing ratios, and the areas of CO_2 lines and suitable trace gas lines are measured from the local baselines. The nominal matrix thickness d is determined as an average of CO_2 line areas divided by the absorption coefficients of Table 7.1. A "best set" of CO_2 lines is chosen from the lines in Table 7.1 based mainly on the lack of interferences from other lines. In a clean spectrum with no interferences, only the three most precisely determined lines (1253.3, 1358.6, and 2065.7 cm^{-1}) are used. For each trace gas absorption line, a Beer–Lambert law plot of integrated absorbance per micrometer (A/d) vs. mixing ratio (μ) yields the absorption coefficients (α) in Eq. (7.3) as the slope. The standard error of the determination of absorption coefficient is typically 2–5% for sharp absorption lines but can be greater for broad features. The calibration procedure is very stable and reproducible, because it is essentially a measurement of an intrinsic molecular property, the integrated absorption coefficient. Variations in experimental parameters such as matrix shape and thickness on IR beam alignment are effectively compensated by the use of CO_2 absorption lines to determine the effective thickness of the matrix as seen by the analyzing IR beam.

In the integrating sphere, calibration follows essentially the same procedure, except that the measured integrated absorbances are first linearized using Eq. (7.2) before determining effective absorption coefficients.

Analysis of the spectra of air samples follows Eq. (7.1) with the values of α known from calibration. The thickness of the matrix is determined from CO_2 lines in the spectrum as described above, and the mixing ratio of a trace gas in CO_2 is then determined from

$$\mu = \frac{A}{\alpha \cdot d} \tag{7.5}$$

for each suitable trace gas absorption line in the spectrum. The mixing ratio in air can be determined from the mixing ratio of CO_2 in air. For clean air, this can usually be taken as the current background value (currently ca. 350 ppmv) with sufficient precision. For polluted air, the CO_2 mixing ratio must be independently measured. We do this routinely by gas phase FTIR absorption spectroscopy of 0.6-L samples collected concurrently with cryogenic samples (Hurst et al., 1993).

We have developed an extensive library of calibrated spectra of molecules in CO_2 matrices, summarized in Table 7.2. We have recorded uncalibrated spectra of approximately 50 further species including hydrocarbons, aldehydes, chlorofluorocarbons (CFCs), and nitrogen oxides. Calibration of these spectra and extension of the library is progressing. It should be noted that spectra of atmospheric samples, particularly from polluted environments, contain many absorption lines that we are at present unable to identify. This provides strong impetus for expanding the range of species covered by the spectral library.

7.3.4. Applications

The MISST method was originally conceived with the primary purpose of stratospheric measurements of labile species and "reservoir" gases such as $ClONO_2$, H_2O_2, and HNO_4. These measurements would only be possible using direct sampling into the matrix from a high-altitude platform such as a balloon or aircraft, since such species would not be expected to survive the two-step sampling procedure. The impracticality of direct sampling to date has limited the application of MISST to more stable species, although many species otherwise difficult to sample and analyze are still amenable to MISST analysis. The ability to measure many species simultaneously remains a major advantage of the method.

7.3.4.1. Preliminary Studies

Griffith and Schuster (1987) described measurements at ground level as an initial demonstration of the technique using reflection–absorption measurement. They measured N_2O, $CFCl_3$ [Freon-11 (F-11)], CF_2Cl_2 [Freon-12

Table 7.2. Positions and Integrated Absorption Coefficients for Selected Absorption Lines in CO_2 Matrices[a]

Species	Formula	Line Center (cm^{-1})	Absorption Coefficient ($cm^{-1} \cdot ppm^{-1} \cdot \mu m^{-1}$)
Acetaldehyde	CH_3CHO	1127.4	1.5×10^{-5}
		1349.9	8.9×10^{-6}
		1739.3	6.6×10^{-5}
Acetone	$(CH_3)_2CO$	1718.6	5.7×10^{-5}
Acetonitrile	CH_3CN	1054.7	2.1×10^{-5}
		1457.2	7.5×10^{-6}
Acetylene	C_2H_2	3259.6	6.1×10^{-6}
Benzene	C_6H_6	1488.5	4.5×10^{-6}
Carbon disulfide	CS_2	1524.6	6.0×10^{-4}[b]
Carbonyl sulfide	OCS	2056.5	3.2×10^{-4}[b]
Ethane	C_2H_6	2886.7	7.9×10^{-6}
Formaldehyde	CH_2O	1734.6	4.3×10^{-5}
Freon-11	$CFCl_3$	835.1	4.1×10^{-4}
		1092.0	7.0×10^{-5}
Freon-12	CF_2Cl_2	889.1	1.8×10^{-5}
		918.9	1.6×10^{-4}
		1081.1	1.6×10^{-4}
		1130.4	1.7×10^{-4}
Freon-21	$CHCl_2F$	1049.7	4.0×10^{-5}
		1069.3	8.0×10^{-5}
Hydrogen cyanide	HCN	3313.8	5.9×10^{-6}
Nitrous oxide	N_2O	1166.5	9.8×10^{-7}
		1281.5	1.6×10^{-7}
		1890.5	9.2×10^{-8}
		2586.3	6.0×10^{-6}
		2822.7	4.6×10^{-7}
PAN	$CH_3(CO)OONO_2$	1733.9	1.4×10^{-4}
		1828.1	5.1×10^{-5}
Sulfur dioxide	SO_2	1347.9	1.7×10^{-4}[b]
Toluene	C_7H_8	1498.9	2.5×10^{-6}

[a] The absorption coefficients depend slightly on the integration range used and in some cases on the residual water concentration in the matrix. They should be treated only as semiquantitative, with an accuracy of ca. 20%.

[b] Includes satellite line due to ^{34}S isotope.

(F-12)], OCS, CS_2, SO_2, and peroxyacetyl nitrate (PAN) from the ground over a 24-h period at Schauinsland in the Black Forest, Germany, in February 1984. Samples were collected undried and passed through a cold trap at −50 °C during transfer to the measurement cryostat to remove water. N_2O was measured with a mean mixing ratio of 305 ± 13 ppbv in all samples, in good agreement with the accepted atmospheric value at that time (Weiss, 1981). This measurement provides validation of the accuracy of the technique. Other measurements were also consistent with accepted or expected values; F-12 was measured at the accepted baseline value of 300 pptv (Cunnold et al., 1983b) early in the day. It increased strongly in the afternoon as the inversion level rose bringing polluted urban air to the measurement site. F-11 was measured with lower than accepted background values, attributed to clathrate formation in ice as water was removed in the −50 °C cold trap (Cunnold et al., 1983a; Davison, 1983). A feature of this set of measurements was the many unidentified absorption lines, indicating the potentially large number of species that could be measured simultaneously from a single sample.

7.3.4.2. Upper Atmospheric Studies

Following these validation measurements, a program of upper atmospheric measurements was undertaken over several winters from 1985 to 1988, with samples collected from a Lear jet (Wilson et al., 1988, 1989; Helas et al., 1989). Samples were collected in the upper troposphere and the lower stratosphere to 14-km altitude between 50 °N and 80 °N. A flow-through inlet system, depicted in Fig. 7.5, was used in the aircraft. The inlet system was sealed after being evacuated and heated before flight to remove water from wall surfaces, then opened to the air in the dry upper atmosphere. Once opened, air flowed under ram pressure through the loop and a small fraction was withdrawn into sampling coils as described previously. Handled with care, the collected samples were sufficiently dry that only minimal water removal required, as described above. Samples were analyzed in the integrating sphere cryostat with rapid sample transfer to minimize chemical change.

A number of species were measured during these flights, including N_2O, F-11, F-12, F-22 (Freon-22), NO_2, and SO_2. The highlight was however the detection of $COCl_2$, COFCl, and COF_2 in the atmosphere at mixing ratios of approximately 20 pptv in the upper troposphere increasing with height in the lower stratosphere (Wilson et al., 1988). Figure 7.6 shows part of an MI–FTIR spectrum of an air sample with the spectrum of $COCl_2$ for reference. These carbonyl halides, formed from the oxidative degradation of chlorocarbons and CFCs in the atmosphere, have sufficiently long atmospheric lifetimes to transport chlorine from the troposphere to the stratosphere. They are photolyzed in the stratosphere, releasing free chlorine atoms that are active in ozone

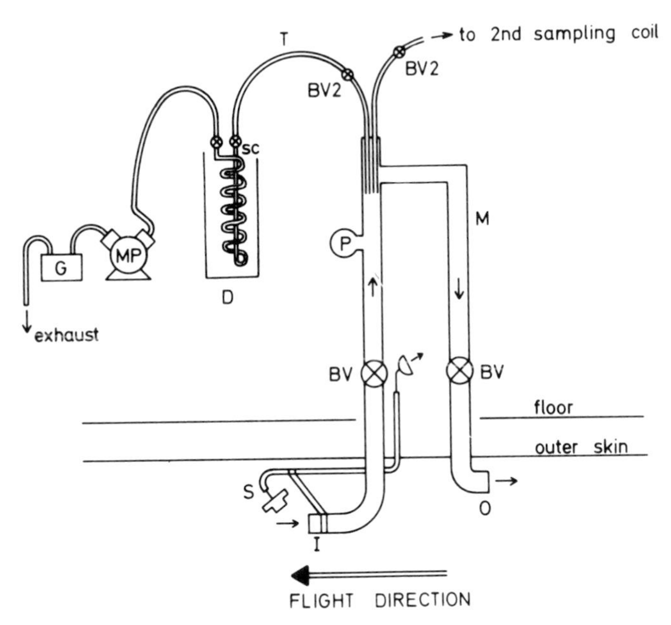

Figure 7.5. Inlet system for cryogenic air sampling from a pressurized aircraft. *Key*: I, inlet; O, outlet; S, retractable stopper; BV, 25-mm ball valve; BV2, $\frac{1}{4}$-in. ball valve; P, pressure gauge; M, 25-mm manifold; T, $\frac{1}{4}$-in. Teflon tubing; SC, sampling coil; D, LN_2 Dewar; MP, membrane pump; G, gas meter.

depletion mechanisms (Warneck, 1988; Helas and Wilson, 1992). This has an impact that is presently not included in calculations of the ozone depletion potentials of chlorocarbons and CFCs; molecules oxidized in the troposphere are normally assumed to have no impact on stratospheric chlorine loadings.

7.3.4.3. Biomass Burning Studies

Most recently, we have used MISST extensively in studies of the emissions of trace gases to the atmosphere from biomass burning. MISST is well suited to these studies because of its ability to make simultaneous measurements of many species in a complex sample. It is also normal practice in biomass burning studies to quantify trace gas emissions as ratios relative to emitted CO_2, the major combustion product, as a way of normalizing the emission data (e.g., Crutzen and Andreae, 1990). After subtraction of backgrounds, MISST provides these emission ratios directly as a natural consequence of the quantitative procedure.

For biomass burning studies, smoke and air samples are normally collected from a light aircraft but may also be collected from the ground. We have

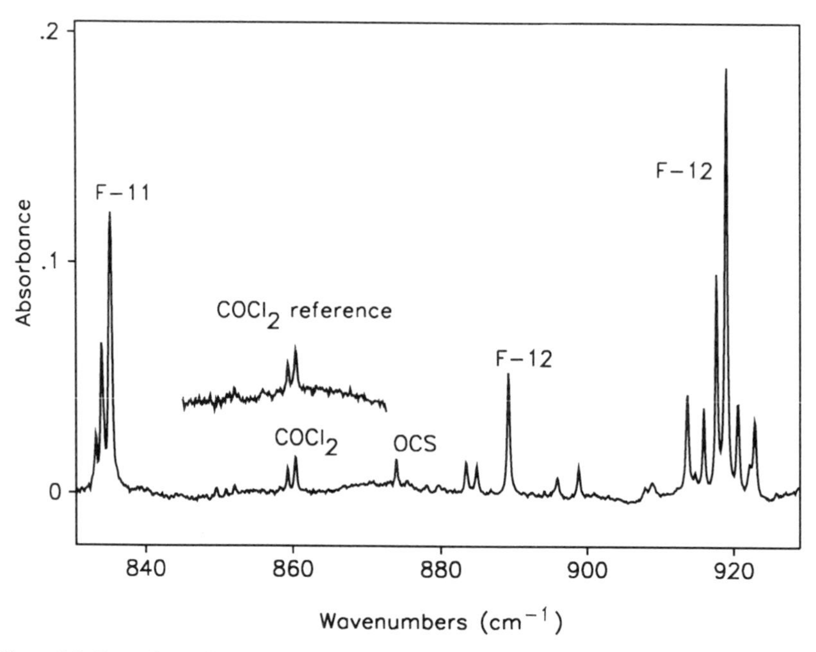

Figure 7.6. Part of an MI–FTIR spectrum of an air sample collected in the upper troposphere from a Lear jet. A reference spectrum equivalent to ca. 20 pptv $COCl_2$ is shown for comparison.

analyzed samples from wild and prescribed fires in tropical savannas and temperate forests and from agricultural waste burning. Aircraft samples are collected in a 40-L PTFE [poly(tetrafluoroethylene)] bag through a 50 mm-o.d. inlet tube under ram pressure in about 1 s during a low altitude traversal of the plume. A 10-L sample for MISST analysis is immediately withdrawn from the bag into a liquid-nitrogen-cooled sample coil. In addition, a 0.6-L gas phase sample is collected from the bag in an evacuated glass bulb. Using a small aircraft, one can normally collect six to eight samples from each fire with all sampling equipment being portable, battery powered, and requiring only a few minutes installation time on the aircraft. On return to the laboratory, the MISST samples are transferred to the measurement cryostat through a $-60\,^{\circ}C$ cold trap to remove water. The gas-phase sample is analyzed by quantitative FTIR absorption spectroscopy in a 5.6-m White cell. This analysis provides absolute values for the CO_2 mixing ratio in the sample, as well as for the noncondensable gases CO and CH_4, which are both important products of biomass burning (Hurst et al., 1993).

As may be expected, the MISST spectra contain many absorption features, and an example from a savanna grass fire is shown in Fig. 7.7. We are able to

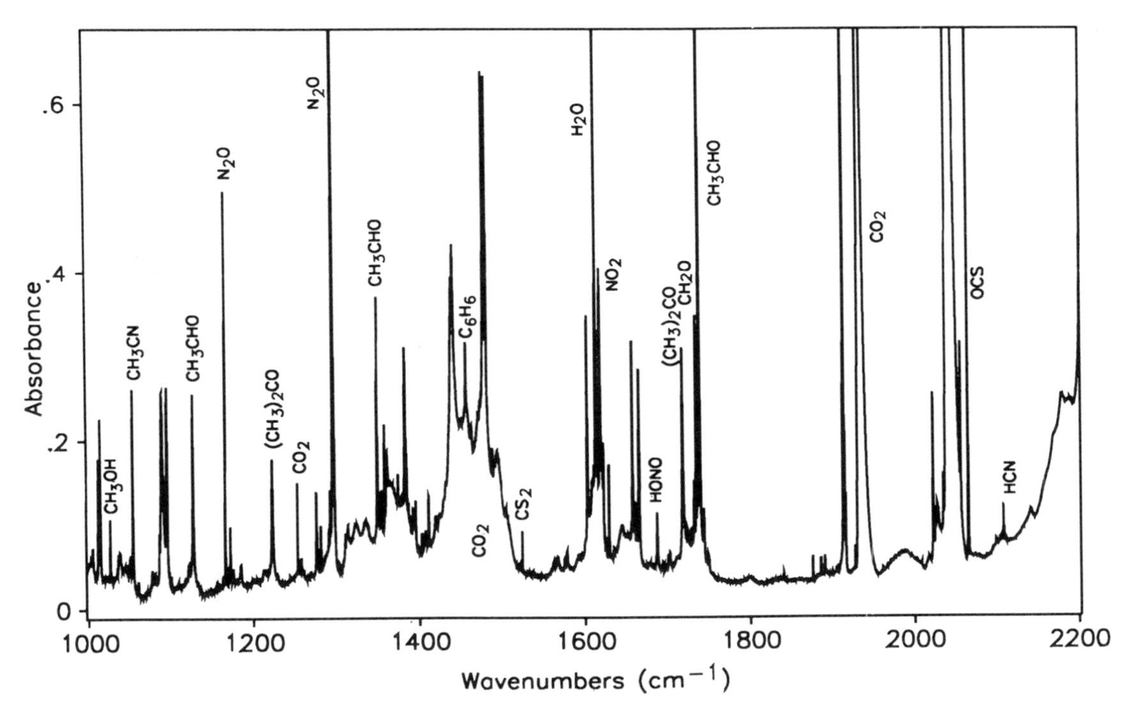

Figure 7.7. Portion of a typical MI–FTIR spectrum of a sample of smoke from biomass burning.

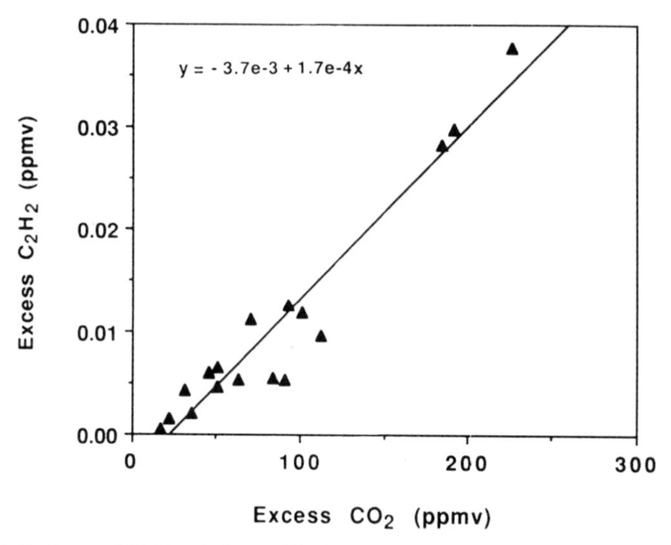

Figure 7.8. Emission of C_2H_2 relative to CO_2 from biomass burning derived from simultaneous MI–FTIR analysis. The slope of the plot gives the C_2H_2/CO_2 emission ratio directly.

identify and quantify many of the absorption lines, as indicated in the figure, but at least as many absorption lines remain unidentified. Figure 7.8 illustrates the measurement of emission ratios directly from MISST results for one particular example, C_2H_2. The measured mixing ratio (as excess over the background value) of C_2H_2 is plotted against the CO_2 excess mixing ratio determined from gas-phase FTIR analysis. The background values are determined from analysis of an air sample collected upwind of the smoke plume. The gradient of the graph in Fig. 7.8 provides the emission ratio of C_2H_2 to CO_2 directly. The data show that C_2H_2 emission is well correlated with that of CO_2 and equal to approximately 0.017% of CO_2 emission. Such emission ratios are used extensively in quantifying the emissions of trace gases to the atmosphere and in determining the fate of individual elements such as C, N, and S in the biomass fuel.

7.4. MI–ESR MEASUREMENTS OF ATMOSPHERIC FREE RADICALS

Free radicals play a very important role in atmospheric chemistry by initiating and maintaining the reaction cycles and pathways of the more stable molecules. With a few exceptions, notably NO and NO_2, their high reactivities mean that they exist only at very low concentrations. Mixing ratios of pptv and lower are typical. Low concentrations and high reactivities present a difficult measurement problem. Normal sampling methods cannot be used, and remote sensing or in situ methods are for the most part too insensitive to be useful. One of the great successes of matrix isolation, on the other hand, has been in the trapping and spectroscopic study of free radicals. The application of MI sampling combined with electron spin resonance (ESR) spectroscopy for detection and quantification has provided some of the few direct measurements of atmospheric free radical concentrations.

In the first attempts, described in the introduction (Snelson, 1974; 1975), matrix isolation of the target radicals was carried out by simply freezing out whole air below 20 K. However O_2, which is paramagnetic, so broadened the spectra of any free radicals present that the measurements were not successful. Mihelcic and co-workers (1978) obtained the first free radical measurements by MI–ESR by using liquid nitrogen as the cryogen to obtain matrices of CO_2 and H_2O, thus avoiding the collection of O_2 and gaining the same benefits of preconcentration as described for FTIR spectroscopy. In these measurements, NO_2 and HO_2 radicals were identified in a cryogenic sample collected from a balloon in the stratosphere at 32-km altitude, and subsequent development of the technique has yielded further measurements of NO_2, NO_3, HO_2, and organic peroxy radicals in both the stratosphere and the troposphere. The technique is described in outline below; for more detail the reader is referred to the original technical papers (Helten et al., 1984; Mihelcic et al., 1985; 1990).

7.4.1. Sampling and Matrix Growth

The reactivity of free radicals puts severe constraints on sampling and matrix growth procedures: first, the sampled air stream must enter the sampling system and be condensed into an inert matrix within the chemical lifetime of the radicals under study, which may be less than 1 s; secondly, the sampled air should not come into contact with any surfaces before condensation in the matrix; and, thirdly, the matrix must remain cold and inert at all times between sampling and measurement to prevent diffusion and reaction of the radicals. The early work (Mihelcic et al., 1978) described a stratospheric sampler for this purpose that sampled air directly onto a copper cold-finger at 68–77 K, forming a matrix of CO_2 containing 1–2% water. Although all constraints were apparently satisfied with this sampler, the collection efficiency of the radicals was low, and varied with the ambient water concentration. This problem was subsequently avoided by adding H_2O or D_2O vapor to the sampled air, increasing the collection efficiency to over 90%. The resultant ice matrix at 77 K provides an inert and rigid matrix for trapped radicals that is well suited to the subsequent ESR spectroscopic analysis. The use of D_2O

rather than H_2O leads to sharper ESR spectra, better detection limits, and better discrimination between different radical species.

Mihelcic, Helten, and their co-workers have developed sampling systems for collection of tropospheric samples from the ground or from aircraft (Mihelcic et al., 1985, 1990) and for balloon-borne stratospheric sampling (Chatzipetros and Helten, 1984; Helten et al., 1984). The sampler of Mihelcic et al. is illustrated in Fig. 7.9. It consists of a liquid-nitrogen-cooled, gold-plated copper cold-finger, 20 mm in diameter, in a vacuum shroud cryo-

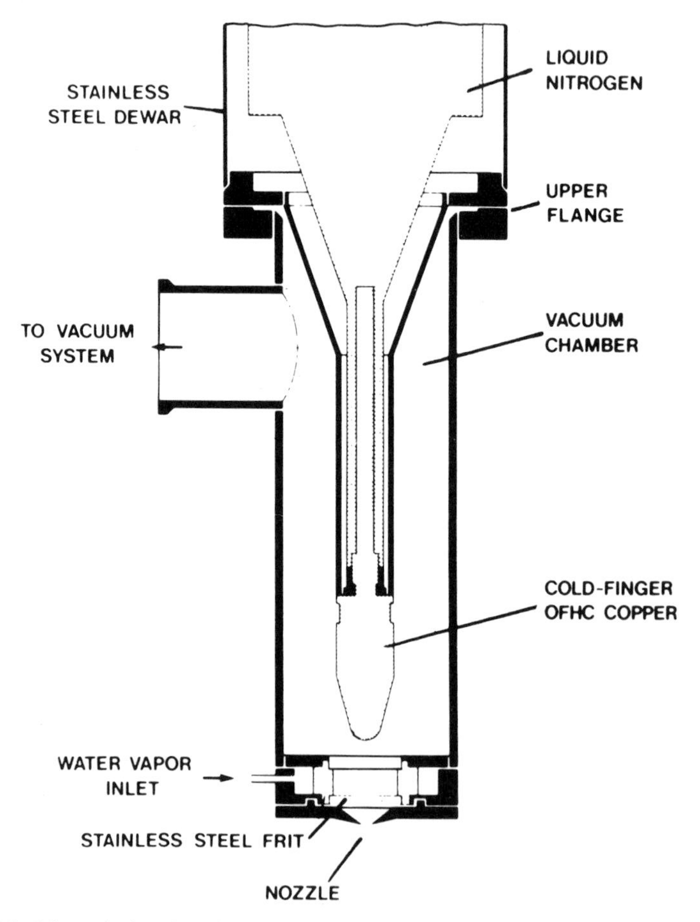

Figure 7.9. Schematic drawing of a cryosampler for MI–ESR studies. *Key*: OFHC, oxygen free, high conductivity. From Mihelcic et al. (1985). Reprinted by permission of Kluwer Academic Publishers.

pumped by a liquid-nitrogen-cooled molecular sieve. Sampled air is admitted to the vacuum chamber through a critical orifice and forms a well-defined jet directed onto the conical end of the cold-finger. Inside the vacuum chamber and surrounding the inlet orifice, an annular frit adds a controlled amount of H_2O (or D_2O) vapor from a thermostatted reservoir to the stream of sampled air. The added H_2O or D_2O forms an ice matrix on the cold-finger, diluting the incoming air such that radical mixing ratios relative to the ice matrix are 10^{-7} or less. The conical design of the orifice minimizes contact of the sampled air with walls and surfaces.

The entire apparatus consists of six such samplers with a common vacuum manifold. In aircraft sampling, the samplers project 70 cm through the fuselage into the free airstream and the inlet orifices can be opened and closed by remote control. The orifice size is selected to provide a sampling rate of about 15 L of air at STP per hour, and samples of 8 L at STP are collected over 30 min into an ice matrix of about 0.6 g. After collection of six samples, the apparatus is returned to the laboratory and the individual cold-fingers are transferred under vacuum or dry nitrogen into a storage cryostat and held at 77 K until ESR measurement. Samples can be stored at 77 K for several weeks without loss of radicals.

The balloon-borne sampler is similar in concept to that described by Mihelcic (1985) with all operations being carried out by telemetry. Normally some 8–14 samples, overlapping in time, are collected during a slow balloon descent, with the inlet orifices projecting from the bottom of the gondola to avoid contamination. On return to the earth's surface, the sampler design is such that the liquid nitrogen cryogen is not lost from the sampler, regardless of the orientation in which the sampler lands.

7.4.2. ESR Spectra and Calibration

ESR measurements of the collected samples are made on a commercial X-band ESR spectrometer (Varian E-line, 9.5 GHz, 12-in. magnet, Varian V4535 large sample cavity with cryostat). The sample cold-finger is transferred through a vacuum lock into an evacuated quartz measurement cryostat, and the spectrum is recorded over a 200-G scan width, digitally averaging up to 250 spectra each of 2-min duration. Spectra are recorded at 4–5 K (Helten et al., 1984) or 77 K (Mihelcic et al., 1985, 1990). Under these conditions, sensitivity is approximately 10^{12} spins, corresponding to detection limits of a few pptv. Spectra are recorded as first derivatives of the absorption spectrum; double integration of the first derivative spectra is thus the equivalent of the measurement of the area under the absorption peaks and provides the basis of the quantitative analysis.

Calibration involves determination of a number of factors:

1. The response of the ESR spectrometer to a given number of spins in the matrix
2. The collection efficiency of radicals, i.e., the fraction of radicals passing through the inlet orifice that are trapped and detected in the matrix
3. The volume of air sampled
4. The individual spectra of the radicals of interest at known concentrations

The response of the spectrometer is in principle absolute, that is, the response as measured by the double integral of the first derivative spectrum depends only on the total number of spins in the sample cavity. In practice, the geometry of the cold-finger and cavity mean that the response may vary slightly over different regions of the cavity. For this reason, sampling of calibration gases and air samples into matrices on the cold-finger must be carried out as identically as possible to avoid possible variations. With this precaution, the response factor in units of "double integral per spin" determined for any single radical species should be applicable to all others and fixed for constant spectrometer settings. The response factor may be determined by measuring the ESR spectrum of a stable radical such as ATMPO (4-amino-2,2,6,6-tetramethylpiperidine-1-oxyl) and independently determining the total amount of ATMPO in the cavity. This approach has been used by collecting the ATMPO in water solution from the cold-finger after the ESR measurement and measuring the total amount by UV absorption spectroscopy (Mihelcic et al., 1985; Helten et al., 1989).

The collection efficiency of radicals from the sampled air onto the cold-finger depends on the following: (1) losses by gas phase or wall reactions in and after passing through the inlet orifice; (2) reactions on the surface of the growing matrix; (3) losses in the stored matrix prior to measurement; and (4) losses of radicals that do not condense on the cold-finger and are pumped away. Losses in the gas phase can be neglected as the flight time between orifice and cold-finger is less than 10^{-4} s. Extensive tests have also shown that losses during storage at 77 K are negligible over several weeks. The remaining factors may be taken together as "collection efficiency," which may be different for each radical species. The collection efficiency can be measured by independently measuring the concentration of radicals in the gas phase outside the inlet orifice and in the matrix from the ESR spectrum (once the spectrometer response factor is known). The collection efficiencies of the samplers described above have been determined in this way for NO_2 and ATMPO as $90 \pm 10\%$ (Mihelcic et al., 1985) and 62% (Helten et al., 1989), respectively. For other radicals the collection efficiency may be lower, but provided that sample collection is carried out in exactly the same way during both calibration and air sampling, the actual value need not be known explicitly.

The volume of air collected must be known precisely to determine the mixing ratio of radicals in the sampled air. This can be calculated from the sampling time and the known flow rate of air through the critical orifice as a function of temperature and pressure as measured in the laboratory. The precision of the volume measurement is 3–8% for stratospheric samples and 3% for calibration samples (Helten et al., 1984).

Individual calibrated spectra of many atmospheric free radicals have been described in detail (Helten et al., 1984; Mihelcic et al., 1990). The radicals are produced at known concentrations in gas flow systems by a variety of methods. The gas flow is sampled and matrix isolated in exactly the same way as for real-air samples. NO_2 in air can be produced simply by dilution from a calibrated standard gas mixture and checked by long-path UV absorption, but the more reactive radicals must be produced by flow-tube methods. The calibrated concentrations may have uncertainties up to 30% owing to the assumptions and uncertainties in kinetic parameters used to calculate them. The ESR spectra of a number of radicals of interest are reproduced in Fig. 7.10 and 7.11. NO_2 (Fig. 7.10a) is usually the dominant species in air spectra, and its spectrum is a broad triplet due to the hyperfine interaction of the electron spin with the $I = 1$ spin of the ^{14}N nucleus. The spectra of NO_3, HO_2 (hydroperoxy), and CH_3COO_2 (acetylperoxy) radicals shown in Fig. 7.10 also display characteristic splitting patterns. A number of alkylperoxy radicals exhibit very similar ESR spectra (Fig. 7.11), and it is difficult to distinguish between the individual alkylperoxy species in atmospheric spectra. The principal values of the g-tensor and the splitting constants for all species are given in detail by Mihelcic et al. (1990).

In the case of NO_2, the MI–ESR method has also been independently fieldtested by comparison with chemiluminescence measurements (Drummond et al., 1985) and long-path differential UV absorption spectroscopy (e.g., see U. Platt, Chapter 2, this volume). The overall agreement between the methods is on average better than 10%.

The ESR spectra of atmospheric samples are dominated by the absorptions of NO_2, with smaller contributions from other radicals and the intrinsic spectrum of the cold-finger. The intrinsic cold-finger spectrum varies from finger to finger and from sample to sample. It contains contributions from the cold-finger itself and from traces of O_2 trapped in the ice matrix, and so must be explicitly measured for each sample. After measurement of the sample matrix spectrum, the cold-finger is warmed to 150 K to allow all radicals to diffuse and react, then recooled to 77 K and the spectrum remeasured. The second spectrum contains only the background contributions, and it is

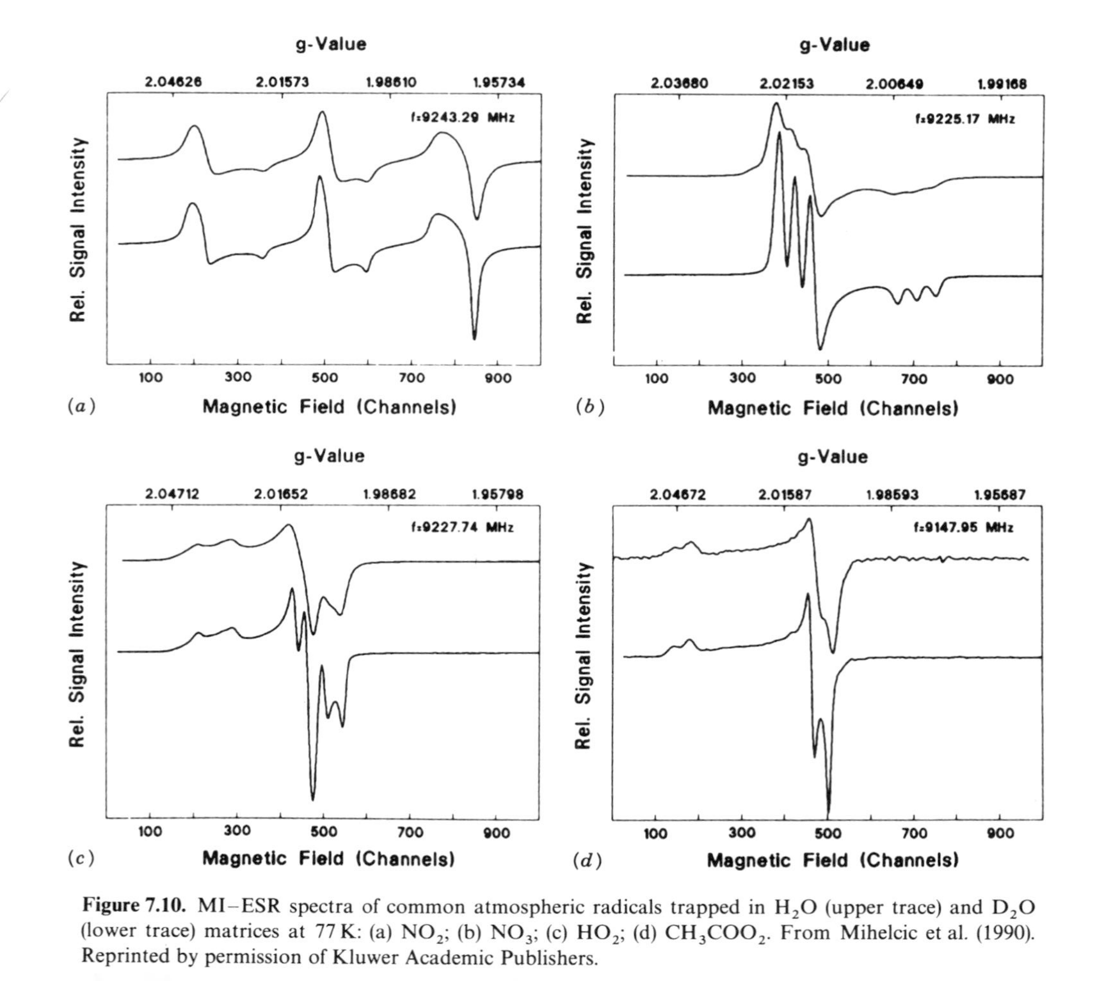

Figure 7.10. MI–ESR spectra of common atmospheric radicals trapped in H_2O (upper trace) and D_2O (lower trace) matrices at 77 K: (a) NO_2; (b) NO_3; (c) HO_2; (d) CH_3COO_2. From Mihelcic et al. (1990). Reprinted by permission of Kluwer Academic Publishers.

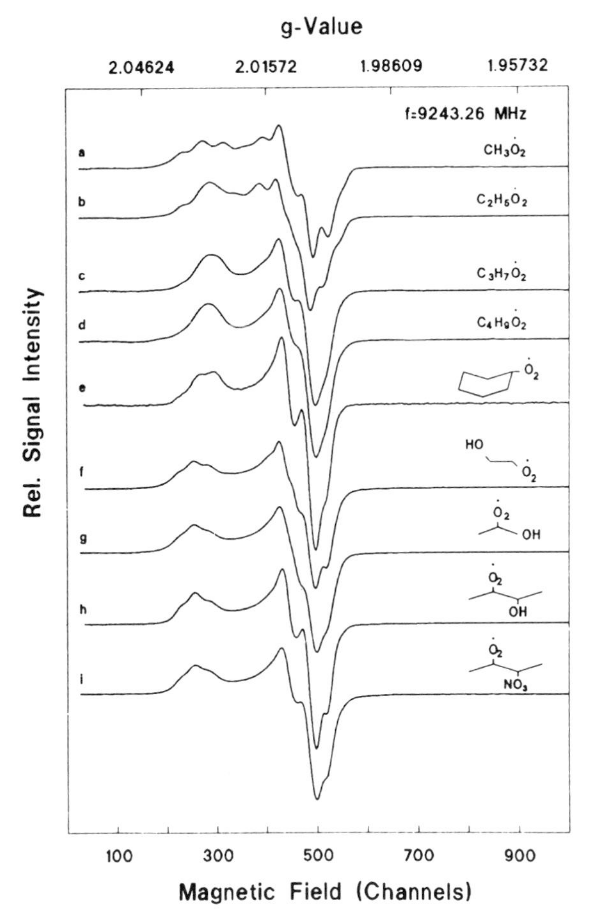

Figure 7.11. MI–ESR spectra of atmospheric peroxy radicals trapped in D_2O matrices. From Mihelcic et al. (1990). Reprinted by permission of Kluwer Academic Publishers.

subtracted from the initial spectrum before further analysis. The spectrum is then analyzed by a numerical procedure of sequentially subtracting fitted single component spectra from the measured spectrum until only a residual of random noise remains (Mihelcic et al., 1990). The analysis procedure is illustrated in Fig. 7.12 for a spectrum of moderately polluted air collected at ground level at Schauinsland, Germany.

Trace A in Figure 7.12 shows the measured spectrum of the air sample, and trace C the spectrum after subtraction of the measured background B. Of the

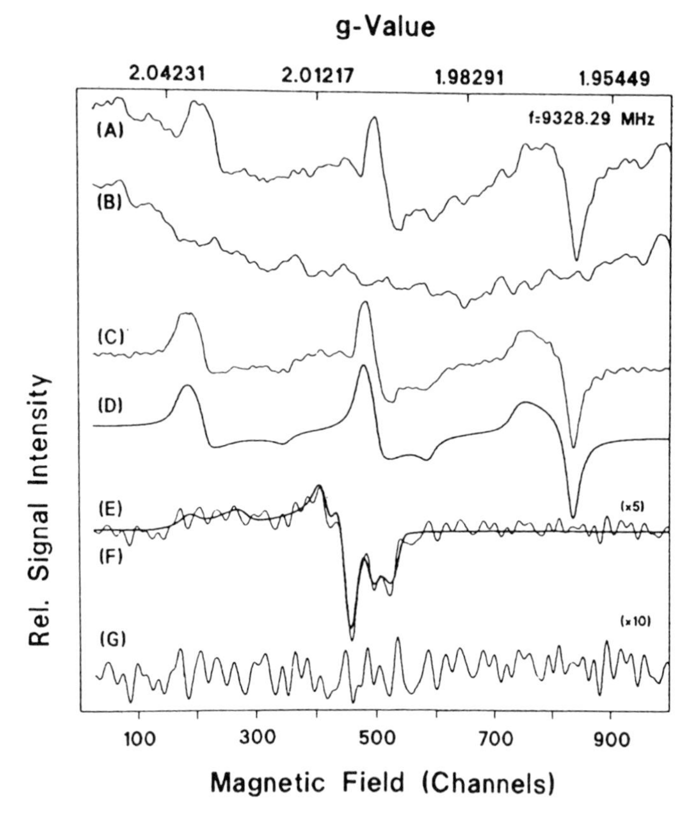

Figure 7.12. MI–ESR spectrum of an air sample in a D_2O matrix at 77 K (A) and its sequential analysis by least fitting (B–G). See text for details. From Mihelcic et al. (1990). Reprinted by permission of Kluwer Academic Publishers.

triplet of NO_2 absorption lines, the line at highest field is not overlapped by any other observed radical absorption, as can be seen from Figs. 7.10 and 7.11. A least squares fit of a calibrated reference spectrum of NO_2 is therefore made to this region of the air sample spectrum, fitting the position, width, and intensity of the absorption line. The position and width of the lines must be fitted to account for the scan-to-scan variability in the spectrometer, which has the effect of both shifting and broadening the absorption lines. The width of the lines is also affected by the H_2O/D_2O ratio in the matrix. The intensity fit yields the amount of NO_2 present in the sample. The fitted NO_2 spectrum D is then subtracted from the sample spectrum over the whole spectral range to yield the residual spectrum E. Spectrum E shows absorption characteristic of the HO_2 radical, F, which dominates the remaining absorption in this sample, while trace G represents the residual random noise. A simultaneous fit of up to six peroxy radical spectra is then carried out over the whole spectral region. Only the intensities are fitted in this step; it is satisfactory to use the shift and width deduced from the earlier NO_2 fit. For the spectrum in Fig. 7.12, the retrieved concentrations in the original air sample were 0.73 ppbv NO_2, 38 pptv HO_2, and 1.4 pptv alkylperoxy radicals. It is not realistic to discriminate among the individual alkylperoxy radicals. In this sample, HO_2 accounted for more than 95% of the total peroxy radical composition of the sample and represented the first spectroscopic measurement of HO_2 in tropospheric air.

The numerical full spectrum least squares fitting procedure allows measurements of radical absorptions even when peak absorption-to-noise ratios are as low as 1:1, resulting in a detection limit and precision of about 6×10^{11} spins, corresponding to 5 pptv in the sampled air. This represents an improvement of an order of magnitude over simple peak-matching spectral subtraction. The added performance is a result of the fact that the fitting procedure uses the quantitative information contained in all points in the spectrum rather than just a few points near the spectrum peaks. This gain is typical of many of the multivariate chemometric methods that have recently been developed for quantitative spectral analysis (e.g., see Martens and Naes, 1989; Haaland, 1990).

7.4.3. Applications

7.4.3.1. Stratospheric Measurements

The first atmospheric measurements with MI–ESR spectroscopy were made in 1976 at 31.8-km altitude with a prototype sampler using a CO_2 matrix without added water (Mihelcic et al., 1978). The spectra of NO_2 and HO_2 were identified and quantified with an error estimate of about a factor of 3 owing to the crude calibrations available at the time. The NO_2 measurement, $5 \times 10^8\ cm^{-3}$, or 2 ppbv, was generally in agreement with the few existing data from other methods, and the HO_2 measurement, $3 \times 10^7\ cm^{-3}$, or 120 pptv, was in general agreement with model-calculated values.

Following these initial flights, a new sampler (Chatzipetros and Helten, 1984) has been used in a series of balloon flights since 1980. All flights have been made from Aire sur l'Adour in southern France (44 °N) at altitudes of 15–35 km. In a typical flight, the balloon is allowed to outgas for 30 min at maximum altitude before beginning a controlled descent at $1\ m \cdot s^{-1}$. Samples are collected for up to 1 h over a 3.6-km altitude range, with some overlap between successive samples. The sampler allows collection of 8, 10, or 14 samples per flight and is mounted such that the inlet orifices are located below the gondola, pointing downward, so that they sample only oncoming air that is not contaminated by the gondola.

The flights are described in a number of papers (Helten et al., 1984, 1985, 1989) and provide profiles of stratospheric NO_2, HO_2, and NO_3. In general, NO_2 profiles agree with those measured by independent chemiluminescence or infrared techniques. There is, however, significant disagreement between measurements of HO_2 from MI–ESR, resonance fluorescence measurements (Anderson et al., 1981), microwave remote sensing (DeZafra et al., 1984), and model calculations. The reasons for the differences are not clear, although much of the variation may be due simply to natural variability, which will be high for such a short-lived species.

On two flights in 1983, samples were collected around sunrise and provided some of the very few available measurements of stratospheric NO_3 during the night. NO_3 forms at night by oxidation of NO_2 by ozone:

$$NO_2 + O_3 \longrightarrow NO_3 + O_2$$

It is photolyzed rapidly by visible light and disappears quickly at sunrise. NO_2 is photolyzed only by UV radiation below 410 nm. The MI–ESR measurements were able to monitor the disappearance of both NO_3 and NO_2 and ascertain their dependences on solar zenith angle. They demonstrated that the loss of NO_3 began significantly before the geometric sunrise, whereas NO_2 photolysis began somewhat later. The observation can be understood on the basis that as the sun rises the solar UV flux is slower to increase than the solar visible flux owing to the higher absorption and scattering of shorter wavelength UV radiation.

7.4.3.2. Tropospheric Measurements

Using H_2O or D_2O as the matrix gas, MI–ESR may also be used for tropospheric measurements, since the high concentrations of tropospheric water vapor simply add to the matrix gas and do not disturb the measurement except for a slight broadening of the absorption lines. Early measurements were made in the boundary layer at ground level in Bonn and Deuselbach, Germany (Mihelcic et al., 1982, 1985), yielding diurnal measurements of NO_2 and total peroxy radical (RO_2) concentrations. The NO_2 results agreed well with independent measurements by chemiluminescence and long-path UV absorption, providing satisfactory validation of the method. The spectral analysis procedure in these measurements did not include the detailed least squares fitting described above, and only total, unspeciated peroxy radical measurements were obtained. Total RO_2 mixing ratios of up to 800 pptv were observed, and in general total RO_2 was anticorrelated with NO_2. Surprisingly, RO_2 concentrations at ground level remained high throughout the night.

NO_2 and RO_2 were also measured in clean air sampled from an aircraft, yielding vertical profiles from altitudes of 1–6 km. In most cases, NO_2 concentrations were high in the boundary layer (> 1 ppbv) but dropped off with an effective scale height of 3 km above it. Again, simultaneous measurements of NO_2 by chemiluminescence were in good agreement. RO_2 profiles normally decreased slowly with height and, in contrast to the boundary layer measurements, were positively correlated with NO_2. RO_2 mixing ratios decreased typically from 200 pptv at the top of the boundary layer to about 50 pptv at 6 km.

The dual technical improvements of using D_2O matrices and least squares spectral fitting for spectral analysis allowed further speciation of RO_2 measurements to distinguish HO_2 and CH_3COO_2 from other alkylperoxy radicals. These improvements were used in analysis of samples collected at Schauinsland, a site a 1200 m above sea level in the Black Forest of Germany (Mihelcic et al., 1990). Measurements of HO_2 are quite variable, from below the detection limit of about 5 pptv to a maximum of 38 pptv. The ratio of HO_2 to other peroxy radicals also varied widely. In the sample illustrated in Fig. 7.12, HO_2 constituted more than 95% of all peroxy radicals, whereas in other samples it was undetectable in the presence of > 100 pptv of other peroxy radicals. The reasons behind this odd chemical behavior remain unclear. Model calculations suggest that HO_2 should normally account for about 50% of the total peroxy radical concentration.

In summary, MI–ESR provides measurements of NO_2 and a range of peroxy radicals with useful sensitivity, accuracy, and precision. NO_2 measurements are in good agreement with simultaneous determinations using the "standard" method of chemiluminescence, a comparison that lends credence to the overall method and calibration procedures. There are very few other methods capable of making peroxy radical measurements, and there have been no side-by-side intercomparisons. The chemical amplifier technique (Stedman and Cantrell, 1982; Cantrell et al., 1984) and laser-induced fluorescence measurements of HO_2 following conversion to OH (Chan et al., 1990; Hard et al., 1992) are the only other currently feasible methods for tropospheric measurements. In the stratosphere, HO_2 has also been measured as OH by laser-induced fluorescence (Anderson et al., 1981; Stimpfle et al., 1990), from the ground by microwave emission (DeZafra et al., 1984), and in IR limb emission spectra measured from balloons (Traub et al., 1990). MI–ESR provides a valuable addition to the range of measurement techniques to these all-important atmospheric species.

7.5. OTHER APPLICATIONS

7.5.1. Gas Chromatography–MI–FTIR

In the wider field of analytical chemistry, matrix isolation has also found a number of useful applications, which have been reviewed in the 1980s (Mamantov et al., 1982; Wehry and Mamantov, 1987; Wilson and Childers, 1989). These include monitoring of atmospheric toxic chemicals such as polyaromatic hydrocarbons (PAHs), dioxins, and polychlorinated biphenyls (PCBs). While these species are generally stable and amenable to gas chromatographic (GC) and GC/mass spectrometric (GC–MS) analyses, in many cases these two techniques cannot distinguish similar structures and isomers within families of species. Similar structures and isomers may however have very dissimilar biological effects; for example, 1,2,3,4-tetrachlorodibenzo-*p*-dioxin (1,2,3,4-TCDD) is a million times less toxic to animals than its isomer 2,3,7,8-TCDD (Wilson and Childers, 1989). An analytical method that can distinguish the individual species in small amounts is therefore required.

Infrared spectroscopy excels at distinguishing between molecules of similar chemical structure, and matrix isolation provides good sensitivity and selectivity. Thus MI–FTIR spectroscopy has found an important niche in this area of environmental analysis. MI–FTIR analysis has in most cases been combined with GC for these studies—the gas chromatograph makes an initial partial separation of the analyte mixture, and MI–FTIR spectroscopy completes the analysis. These applications have been recently and comprehensively reviewed (Wilson and Childers, 1989), so that in the following discussion only an outline of the methods will be given.

PAHs, PCBs, and TCDDs have only low volatility and are usually bound to particulates and aerosols in the atmosphere. Particulates and aerosols can be collected on filters in a high-volume sampler, and the target compounds solvent extracted from the filters. The concentrated extract is then separated by chromatography coupled with a MI–FTIR detection system to analyze the eluting GC peaks.

GC–MI–FTIR spectroscopy was first developed at the Argonne National Laboratory (Bourne et al., 1979; Reedy et al., 1979, 1985), and subsequently as the Cryolect commercial instrument (Bourne et al., 1984). The effluent Ar or N_2 carrier gas from the GC is directed through a heated transfer tube and small deposition nozzle to the mirror-surfaced edge of a slowly rotating metal disk held at ca. 10 K in a cryostat. The effluent gas freezes out in a narrow strip of solid Ar or N_2 around the edge of the disk. The eluting compounds from the GC are thus matrix-isolated in the solid carrier gas. At a displaced location around the disk, an IR beam from an FTIR spectrometer is focused through the matrix strip onto the underlying cold mirror surface, where it is reflected and subsequently refocused onto the IR detector, resulting in a reflection–absorption spectrum of the matrix strip. By exact indexing of the disk rotation with time and an optional concurrent GC trace of gas split from the GC effluent to a conventional detector, any peak of the chromatogram can be rotated to the FTIR beam and analyzed by FTIR spectroscopy. By keeping the matrix trace as narrow as possible (0.3 mm) and finely focusing the FTIR beam, detection limits well below 1 ng can be obtained.

Applications of this method for PAHs, PCBs, and dioxins have been comprehensively reviewed up to 1988 (Wilson and Childers, 1989). There have been further studies of dioxins (Grainger et al., 1989; Wurrey et al., 1989a,b) and of PAHs in diesel exhaust, wood smoke, and urban air (Childers et al., 1989a,b; Blum et al., 1991). Schneider and co-workers (1991) have more recently evaluated the quantitative aspects of GC–MI–FTIR for semivolatile pollutants such as dioxins and PAHs; their results suggest that the technique is suited only to semiquantitative work, with precision on the order of 50%. They did not identify the causes of the lack of precision but suggested losses in the long transfer line from GC to cryostat and the irreproducible alignment of the FTIR beam with the matrix.

7.5.2. Shpol'skii Spectroscopy

Many fluorescent PAHs display remarkably narrow emission spectra when dissolved in *n*-alkanes at low temperature if the chain length of the alkane equals the length of one axis of the PAH molecule. Anthracene in *n*-heptane is the archetypal example of this behavior, where it is assumed that the matching size of alkane and PAH molecules allows simple substitution of the PAH molecule into the polycrystalline alkane lattice. Spectra of PAHs in "non-matching" alkanes display much broader emission features. The effect has been called the Shpol'skii effect after its original discoverers (Shpol'skii et al., 1952). Applications of the Shpol'skii effect to the analysis of PAHs have been reviewed by Wehry and Mamantov (1987). Suitable choice of *n*-alkane solvent combined with the very high sensitivity of fluorescence spectroscopy can make this a very sensitive method for selected PAHs while discriminating against those for which the chosen Shpol'skii matrix is not optimal. The Shpol'skii matrix may be grown by freezing liquid solutions or from the gas phase by spray-on matrix isolation methods. Wehry and Mamantov (1987) describe the application of this specialized form of matrix isolation spectroscopy as a GC detector for PAHs.

7.5.3. Kinetic Studies

One final application of matrix isolation spectroscopy, while somewhat outside the scope of this volume, is deserving of brief mention. Matrix isolation has been used in a number of studies of the kinetics and mechanisms of gas-phase reactions of atmospheric interest. Trapping of a reacting gas mixture into a low-temperature matrix quenches the chemical reactions, and subsequent MI–FTIR analysis of the matrix allows investigations of the structure and distribution of intermediate and final products of the reaction. Günthard and co-workers first used this technique in studies of the products of olefin-ozone reactions (see Kühne et al., 1978, 1980). The reaction took place in a conventional flow-tube reactor, the product distribution being investigated as a function of the time of reaction. Ironically, this is one of the most complex of reaction systems on which to begin with a novel technique! The products of the $ClO + NO_2$ reaction were investigated by trapping the products in N_2 matrices (Burrows et al., 1985). At the time of this work, there was uncertainty about the possible formation of a photolytically unstable isomer of the product chlorine nitrate, $ClONO_2$, which could have had a potentially large impact on stratospheric chlorine chemistry. No evidence of an isomer other than $ClONO_2$ was found. The first step of the oxidation of SO_2 by OH radicals was shown to be the formation of the $HOSO_2$ radical by matrix isolation of the reaction products in an argon matrix (Hashimoto et al., 1984). Finally, a number of studies have employed MI–FTIR spectroscopy in elucidating the mechanisms of the atmospheric oxidation of hydrocarbon molecules (Horie and Moortgat, 1989a,b; 1991a, b; Horie et al., 1990). The technique has thus proved a useful adjunct to the usual methods available for studies of gas-phase reaction kinetics and mechanisms.

7.6. CONCLUSIONS

This chapter has been mainly concerned with the technique of matrix isolation for air sampling combined with FTIR or ESR spectroscopy for analysis of the composition of the sampled air. MI–FTIR is applicable at present to a very wide range of stable and moderately labile atmospheric trace gases. Measurements of many species may be made simultaneously in a single sample with detection limits generally on the order of 1–100 pptv from a 10- to 100-L sample. Quantitative accuracy and precision are good, on the order of 5%, and interferences are negligible (or at worst easily recognisable) because of the resolved nature of the MI–FTIR spectra. Equipment requirements consist of sampling glassware, a general-purpose vacuum line for gas manipulations, an FTIR spectrometer, and a cryostat suited to liquid nitrogen temperatures and reduced IR beam size. The work to date has all been carried out at a spectral resolution of 0.2 cm^{-1}, requiring a research-grade FTIR spectrometer. However, there is no compelling reason that a lower resolution (1 cm^{-1}) could not be used, allowing the use of a much cheaper and more portable spectrometer.

Some additional overlapping of adjacent lines, especially in polluted samples, may result from the reduced resolution, but in most cases this should not be a major limitation because only the narrowest of absorption lines have widths of less than 0.5 cm^{-1}. Once sampling and matrix growth procedures are established, the method is relatively simple to operate. Analysis of a single sample typically requires 1–2 h, depending largely on the signal-to-noise level (and hence detection limits) required.

The spectroscopic principles and practice of the MI–FTIR technique are well established. High-quality matrices can be grown reliably and reproducibly, and their IR spectra routinely obtained and largely assigned and interpreted. In clean air samples most absorption features can be identified, but in heavily polluted air such as the smoke from bushfires there are still many unidentified absorption lines. The principal limitation of the technique stems from the need to remove water vapor from the sampled air and the need to transfer the collected sample from the cryogenic sampler to the matrix for measurement. Both processes may lead to the loss or generation of some chemical species in the sample, so that sampling and drying procedures must be validated for each species analyzed. For stable, non-water-soluble gases there are few problems, but water-soluble and polar compounds present more difficulties. Despite these shortcomings, the technique has shown its worth in a number of studies of atmospheric trace gas composition.

MI–ESR spectroscopy is suited to the analysis of free radicals in the atmosphere. It has been mostly applied to measurements of NO_2, HO_2, and other peroxy radicals, and NO_3 in both the stratosphere and the troposphere. Sensitivity is adequate, on the order of 10^{12} spins or about 5 pptv at 1-atm pressure, and precision is better than 10%. Stringent control must be exercised over sample collection and handling procedures because of the reactivity of the radical species, necessitating specialized and mechanically complex cryogenic samplers and sample handling facilities. Spectroscopic analysis can be carried out on a commercially available ESR spectrometer. The ESR spectra of atmospheric radicals all overlap to some extent, requiring full spectrum least squares fitting procedures for best results in quantitative analysis of the spectra. With such procedures, analyses of several radicals in a single air sample can be obtained simultaneously. Because of the reactivity and low concentrations of most atmospheric free radicals, there are few techniques available for their measurement: MI–ESR has provided a significant proportion of the available measurements.

ACKNOWLEDGMENTS

I would like to thank Dr. Dale Hurst and Dr. Stephen Wilson for their detailed and critical reading of the manuscript.

REFERENCES

Almond, M. J., and Downs, A. J. (1989). Spectroscopy of matrix isolated species. *Adv. Spectrosc.* **17**.

Anderson, J. G., Grassl, H. J., Shetter, R. E., and Margitan, J. J. (1981). HO_2 in the stratosphere: Three *in situ* observations. *Geophys. Res. Lett.* **8**, 289–292.

Ball, D. F., and Purnell, C. J. (1976). The use of low temperature matrix isolation infrared spectroscopy for the identification and measurement of airborne amines. *Int. J. Environ. Stud.* **9**, 131–138.

Barnes, A. J., Orville-Thomas, W. J., Müller, A., and Gaufrès, R., Eds. (1981). *Matrix Isolation Spectroscopy.* Reidel, Dordrecht, The Netherlands.

Berger, E., (1986). Einsatz einer Ulbricht'schen Kugel zur Messung von atmosphärischen Spurengasen mit MI–FTIR–Spektroskopie. Diploma Thesis, University of Mainz, Germany.

Berger, E., Griffith, D. W. T., Schuster, G., and Wilson, S. R. (1989). Spectroscopy of matrices and thin films with an integrating sphere. *Appl. Spectrosc.* **43**, 320–324.

Blum, T., Frahne, D., and Herrmann, D. (1991). Analysis of PAH by MI–GC–FTIR spectroscopy. *Fat. Sci. Technol.* **10**, 394–402.

Bourne, S., Reedy, G. T., and Cunningham, P. T. (1979). Gas chromatography/matrix isolation/infrared spectroscopy: An evaluation of the performance potential. *J. Chromatogr. Sci.* **17**, 460–463.

Bourne, S., Reedy, G., Coffey, P., and Mattson, D. (1984). Matrix isolation GC/FTIR. *Am. Lab.* **16**(6), 90–101.

Burrows, J. P., Griffith, D. W. T., Moortgart, G. K., and Tyndall, G. S. (1985). Matrix isolation–FTIR study of the products of the reaction between ClO and NO_2. *J.Phys. Chem.* **89**, 266–271.

Cantrell, C. A., Stedman, D. H., and Wendel, G. J. (1984). Measurements of atmospheric peroxy radicals by chemical amplification. *Anal. Chem.* **56**, 1496–1502.

Chadwick, B. M., (1975). Matrix isolation. In *Molecular Spectroscopy: Chemical Society Specialist Periodical Reports*, pp. 281–382. Chemical Society, London.

Chadwick, B. M., (1979). Matrix isolation. In *Molecular Spectroscopy: Chemical Society Specialist Periodical Reports*, pp. 72–135. Chemical Society, London.

Chan, C. Y., Hard, T. M., Mehrabzadeh, A. A., George, L. A., and O'Brien, R. J. (1990). Third generation FAGE instrument for tropospheric hydroxyl measurement. *J.Geophys. Res.* **95D**, 18569–18576.

Chatzipetros, J., and Helten, M. (1984). Cryogenic pump and air sampler. U.S. Pat. 4,425,811.

Childers, J. W., Wilson, N. K., and Barbour, R. K. (1989a). GC–MI–IR spectrometry for the identification of PAH's in urban particulate matter. *Appl. Spectrosc.* **43**, 1344–1349.

Childers, J. W., Wilson, N. K., and Barbour, R. K. (1989a). Analysis of environmental air sample extracts by GC–MI–IR spectroscopy. In *Proc. 7th Int. Conf. on Fourier Transform Spectroscopy* (D. G. Cameron, Ed.), SPIE **1145,** 611–612.

Crutzen, P. J. (1970). The influence of nitrogen oxides on the atmospheric ozone content. *Q. J. R. Meteorol. Soc.* **96**, 320–325.

Crutzen, P. J., and Andreae, M. O. (1990). Biomass burning in the tropics: Impact on atmospheric chemistry and biogeochemical cycles. *Science* **250**, 1669–1678.

Cunnold, D. M., Prinn, R. G., Rasmussen, R. A., Simmons, P. G., Alyea, F. N., Cardelino, C. A., Crawford, A. J., Fraser, P. J., and Rosen, R. D. (1983a). The atmospheric lifetime experiment. 3. Lifetime, methodology and application to 3 years of $CFCl_3$ data. *J. Geophys. Res.* **88D**, 8379–8400.

Cunnold, D. M., Prinn, R. G., Rasmussen, R. A., Simmons, P. G., Alyea, F. N., Cardelino, C. A., and Crawford, A. J. (1983b). The atmospheric lifetime experiment. 4. Results for CF_2Cl_2 based on three years of data. *J. Geophys. Res.* **88D**, 8401–8414.

Davison, D. W. (1983). Clathrate hydrates. In *Water: A Comprehensive Treatise* (F. Franks, Ed.), pp. 115–234. Plenum, New York.

DeZafra, R. L., Parrish, A., Salomon, P. M., and Barrett, J. W. (1984). A measurement of stratospheric HO_2 by ground-based millimeter-wave spectroscopy. *J. Geophys. Res.* **89D**, 1321–1326.

Downs, A. J., and Peake, S. C. (1973). Matrix isolation. In *Molecular Spectroscopy: Chemical Society Specialist Periodical Reports*, pp. 523–607. Chemical Society, London.

Drummond, J., Volz, A., and Ehhalt, D. H. (1985). An optimized chemiluminescence detector for tropospheric NO measurements. *J. Atmos. Chem.* **2**, 287–306.

Grainger, J., Patterson, D. G., and Presser, D. (1989). Structure–retention time correlations for the 22 tetrachlorodibenzodioxin isomers by GC–MI–FTIR spectroscopy. *Chemosphere* **19**, 1513–1520.

Griffith, D. W. T., and Schuster, G. (1987). Atmospheric trace gas analysis using matrix isolation Fourier transform infrared spectroscopy. *J. Atmos. Chem.* **5**, 59–81.

Griffiths, P. R., and de Haseth, J. A. (1986). *Fourier Transform Infrared Spectrometry*. Wiley, New York.

Groner, P., Stolkin, I., and Günthard, H. H. (1973). Measurement of deposition rates in matrix spectroscopy with a small laser. *J. Phys. E* **6**, 122–123.

Haaland, D. M. (1990). Multivariate calibration methods applied to quantitative FTIR analyses. In *Practical Fourier Transform Infrared Spectroscopy* (J. R. Ferraro and K. Krishnan, Eds.), Chapter 8. Academic Press, San Diego.

Hallam, H. E., Ed. (1973). *Vibrational Spectroscopy of Trapped Species*. Wiley, New York.

Hanson, D., and Mauersberger, K. (1986). The vapor pressures of solid and liquid ozone. *J. Chem. Phys.* **85**, 4669–4672.

Hard, T. M., Chan, C. Y., Mehrabzadeh, A. A., and Obrien, R. J. (1992). Diurnal HO_2 cycles at clean air and urban sites in the troposphere. *J. Geophys. Res.* **97D**, 9785–9794.

Hashimoto, S., Inoue, G., and Akimoto, H. (1984). Infrared spectroscopic detection of the hydrosulfonyl ($HOSO_2$) radical in argon matrix at 11 K. *Chem. Phys. Lett.* **107**, 198–202.

Helas, G., and Wilson, S. R. (1992). On sources and sinks of phosgene in the troposphere. *Atmos. Environ.* **26A**, 2975–2982.

Helas, G., Schuster, G., and Wilson, S. R. (1989). Measurements of ozone and other trace components near the tropopause over northern Europe. In *Ozone in the Atmosphere* (R. D. Bojkov and P. Fabian, Eds.), pp. 345–350. A. Deepak, Hampton, VA.

Helten, M., Pätz, W., Trainer, M., Farke, H., Klein, E., and Ehhalt, D. H. (1984). Measurements of stratospheric HO_2 and NO_2 by matrix isolation and ESR spectroscopy. *J. Atmos. Chem.* **2**, 191–202.

Helten, M., Pätz, W., Ehhalt, D. H., and Roth, E. P. (1985). Measurements of nighttime NO_3 and NO_2 in the stratosphere by matrix isolation and ESR spectroscopy. In *Atmospheric Ozone* (C. S. Zerefos and A. Ghazi, Eds.), pp. 196–200. Reidel, Dordrecht, The Netherlands.

Helten, M., Fark, H., Pätz, H. W., and Kley, D. (1989). HO_2 measurements in the lower stratosphere using matrix isolation and ESR spectroscopy. In *Ozone in the Atmosphere* (R. D. Bojkov and P. Fabian, Eds.), pp. 306–309. A. Deepak, Hampton, VA.

Horie, O., and Moortgat, G. K. (1989a). A new transitory product in the ozonolysis of *trans*-2-butene at atmospheric pressure. *Chem. Phys. Lett.* **156**, 39–46.

Horie, O., and Moortgat, G. K. (1989b). Ozonolysis of alkenes under atmospheric conditions. In *Ozone in the Atmosphere* (R. D. Bojkov and P. Fabian, eds.), pp. 698–701. A. Deepak, Hampton, VA.

Horie, O., and Moortgat, G. K. (1991a). Analysis of reaction products in the oxidation reactions of hydrocarbons by means of matrix isolation FTIR spectroscopy. *Fresenius' Z. Anal. Chem.* **340**, 641–645.

Horie, O., and Moortgat, G. K. (1991b). Decomposition pathways of the excited Criegee intermediates in the ozonolysis of simple alkenes. *Atmos. Environ.* **25A**, 1881–1896.

Horie, O., Crowley, J. N., and Moortgat, G. K. (1990). Methyl peroxy self-reaction: Products and branching ratio between 223 and 333 K. *J. Phys. Chem.* **94**, 8198–8203.

Houghton, J. T., Jenkins, G. J., and Ephraums, J. J., Eds. (1990). *Climate Change: The IPCC Scientific Assessment.* Cambridge University Press, Cambridge.

Hurst, D. F., Griffith, D. W. T., and Cook, G. D. (1993). *J. Geophys. Res.* (to be published).

Ihrig, D. (1985). Messung Atmosphärischer Spurengase durch Matrix-Isolations-FTIR-Spektroskopie. Ph.D. Thesis, University of Mainz, Germany.

Jacox, M. E., and Milligan, D. E. (1961). The infrared spectra of thick films of CO_2 and $CO_2 + H_2O$ at low temperatures. *Spectrochim. Acta* **17**, 1196–1202.

Johnson, H. (1971). Reduction of stratospheric ozone by nitrogen oxide catalysts from supersonic transport exhausts. *Science* **173**, 517–522.

Kühne, H., Vaccani, S., Bauder, A., and Günthard, H.-H. (1978). Linear reactor-infrared matrix and microwave spectroscopy of the gas phase ethylene ozonolysis. *Chem. Phys.* **28**, 11–29.

Kühne, H., Forster, M., Hulliger, J., Ruprecht, H., Bauder, A., and Günthard, H.-H. (1980). Linear-reactor-infrared-matrix and microwave spectroscopy of the *cis*-2-butene gas-phase ozonolysis. *Helv. Chim. Acta* **63**, 1971–1999.

Mamantov, G., Garrison, A. A., and Wehry, E. L. (1982). Analytical applications of matrix isolation FTIR spectroscopy. *Appl. Spectrosc.* **36**, 339–347.

Mannik, L., and Allin, E. J. (1972). The $(\nu_1, 2\nu_2)$ vibron-phonon infrared absorption band of solid CO_2. *Can. J. Phys.* **50**, 2105–2110.

Martens, H., and Naes, T. (1989). *Multivariate Calibration.* Wiley, New York.

Meyer, B. (1971) *Low Temperature Spectroscopy*, Elsevier, Amsterdam.

Mihelcic, D., Ehhalt, D. H., Klomfass, J. Kulessa, G. F., Schmidt, U., and Trainer, M. (1978). Measurements of free radicals in the atmosphere by matrix isolation and electron paramagnetic resonance. *Ber. Bunsenges. Phys. Chem.* **82**, 16–19.

Mihelcic, D., Helten, M., Fark, H., Müsgen, P., Pätz, H. W., Trainer, M., Kempa, D., and Ehhalt, D. H. (1982). Tropospheric airborne measurements of NO_2 and RO_2 using the technique of matrix isolation and electron spin resonance. In *Second Symposium on the Composition of the Non-Urban Troposphere, Williamsburg, VA.* Am. Meteorol. Soc. pp. 327–329.

Mihelcic, D., Müsgen, P. and Ehhalt, D. H. (1985). An improved method of measuring tropospheric NO_2 and RO_2 by matrix isolation and electron spin resonance. *J. Atmos. Chem.* **3**, 341–361.

Mihelcic, D., Volz-Thomas, A., Pätz, H. W., and Kley, D. (1990). Numerical analysis of ESR spectra from atmospheric samples. *J. Atmos. Chem.* **11**, 271–297.

Milligan, D. E, and Jacox, M. E. (1972). Infrared and ultraviolet spectroscopic studies of free radicals and molecular ions isolated in inert solid matrices. In *Molecular Spectroscopy: Modern Research* (K. N. Rao and C. W. Mathews, Eds.), pp. 259–286. Academic Press, New York.

Molina, M. J., and Rowland, F. S. (1974). Stratospheric sink for chlorofluoromethanes: Chlorine atom catalysed destruction of stratospheric ozone. *Nature (London)* **249**, 810–814.

Perutz, R. N., and Turner, J. J. (1973). Pulsed matrix isolation: A comparative study. *J. Chem. Soc., Faraday Trans. 2* **69**, 452–461.

Reedy, G. T., Bourne, S., and Cunningham, P. T. (1979). Gas chromatography-infrared matrix isolation spectrometry. *Anal. Chem.* **51**, 1535–1540.

Reedy, G. T., Ettinger, D. G., Schneider, J. F., and Bourne, S. (1985). High resolution gas chromatography-matrix isolation infrared spectrometry. *Anal. Chem.* **57**, 1602–1609.

Rochkind, M. M. (1967). Infrared analysis of multicomponent gas mixtures. *Anal. Chem.* **39**, 567–574.

Rochkind, M. M. (1968a). Infrared pseudo matrix isolation spectroscopy. *Science* **160**, 196–197.

Rochkind, M. M. (1968b). Quantitative infrared analysis of mixtures of isotopically labelled gases. *Anal. Chem.* **40**, 762–768.

Rochkind, M. M. (1971). IR chemical analysis using pseudo matrix isolation: A practical application for cryogenic spectroscopy. *Spectrochim. Acta* **27A**, 547–568.

Schneider, J. F., Schneider, K. R., Spiro, S. E., Bierma, D. R., and Sytsma, L. F. (1991). Evaluation of GC-MI-IR spectroscopy for the quantitative analysis of environmental samples. *Appl. Spectrosc.* **45**, 566–571.

Shpol'skii, E. V., Al'ina, A. A., and Klimova, L. A. (1952). *Dokl. Akad. Nauk SSSR* **87**, 935–938.

Snelson, A. (1974). *Chemical Composition of the Atmosphere at* 40,000–65,000 *Feet Using a Modified Matrix Isolation Technique*, Air Force Off. Sci. Res. TR-74–1727. IIT Research Institute, Chicago, IL.

Snelson, A., (1975). *Composition of the Atmosphere at 10,000–40,000 Feet Using a Modified Matrix Isolation Technique*, Air Force Off. Sci. Res. TR-75-1171. IIT Research Institute, Chicago, IL.

Stedman, D. H., and Cantrell, C. A. (1982). Laboratory studies of an HO_2/RO_2 detector. In *Second Symposium on the Composition of the Nonurban Troposphere, Williamsburg, VA*, pp 68–71. Am. Meteorol. Soc., Boston, MA.

Stimpfle, R. M., Wennberg, P. O., Lapson, L. B., and Anderson, J. G. (1990). Simultaneous *in situ* measurements of OH and HO_2 in the stratosphere. *Geophys. Res. Lett.* **17**, 1905–1908.

Traub, W. A., Johnson, D. G., and Chance, K. V. (1990). Stratospheric hydroperoxyl measurements. *Science* **247**, 446–449.

Warneck, P. (1988). *Chemistry of the Natural Atmosphere.* Academic Press, San Diego.

Wehry, E. L., and Mamantov, G. (1987). Matrix isolation molecular spectrometry in chemical analysis. *Prog. Anal. Spectrosc.* **10**, 507–527.

Weiss, R. F. (1981). The temporal and spatial distribution of tropospheric nitrous oxide. *J. Geophys. Res.* **86D**, 7185–7195.

Whittle, E., Dows, D. A., and Pimentel, G. C. (1954). Matrix isolation method for the experimental study of unstable species. *J. Chem. Phys.* **22**, 1943.

Wilson, N. K., and Childers, J. W. (1989). Recent advances in matrix isolation infrared spectrometry of organic compounds. *Appl. Spectrosc. Rev.* **25**, 1–61.

Wilson, S. R., Crutzen, P. J., Schuster, G., Griffith, D. W. T., and Helas, G. (1988). Phosgene measurements in the upper troposphere and the lower stratosphere. *Nature (London)* **334**, 689–691.

Wilson, S. R., Schuster, G., and Helas, G. (1989). Measurements of COClF and $COCl_2$ near the tropopause. In *Ozone in the Atmosphere* (R. D. Bojkov and P. Fabian, Eds.), pp. 302–305. A Deepak. Hampton, VA.

Wurrey, C. J., Fairless, B. J., and Kimball, H. E. (1989a). Analysis of halogenated dibenzo-*p*-dioxins and dibenzofurans using matrix isolation GC/FTIR. In *Proc. 7th Int. Conf. on Fourier Transform Spectroscopy* (D. G. Cameron, Ed.), SPIE **1145**, 250–251.

Wurrey, C. J., Fairless, B. J., and Kimball, H. E. (1989b). GC–MI–FTIR spectra of the laterally chlorinated dibenzo-*p*-dioxins and dibenzofurans. *Appl. Spectrosc.* **43**, 1317–1324.

INDEX